Student Solutions Manual

for

Whitten, Davis, Peck, and Stanley's

Chemistry
Eighth Edition

Wendy L. Keeney-Kennicutt
Texas A&M University

THOMSON
BROOKS/COLE

Australia • Brazil • Canada • Mexico • Singapore • Spain • United Kingdom • United States

Printed in the United States of America
  2 3 4 5 6 7   10 09 08 07 06

Printer: Thomson/West

0-495-01454-0
Cover image credit: Stuart Westmorland/Getty (ocean) and 2064design (molecule)

Thomson Higher Education
10 Davis Drive
Belmont, CA 94002-3098
USA

For more information about our products, contact us at:
Thomson Learning Academic Resource Center
1-800-423-0563

For permission to use material from this text or product, submit a request online at
http://www.thomsonrights.com.
Any additional questions about permissions can be submitted by email to **thomsonrights@thomson.com.**

Table of Contents

Foreword to the Students

This Solutions Manual supplements the textbook, *General Chemistry*, eighth edition, by Kenneth W. Whitten, Raymond E. Davis, M. Larry Peck and George Stanley. The solutions of the 1355 even-numbered problems at the end of the chapters have been worked out in a detailed, step-by-step fashion.

Your learning of chemistry serves two purposes: (1) to accumulate fundamental knowledge in chemistry which you will use to understand the world around you, and (2) to enhance your ability to make logical deductions in science. This ability comes when you know how to reason in a scientific way and how to perform the mathematical manipulations necessary for solving certain problems. The excellent textbook by Whitten, Davis, Peck and Stanley provides you with a wealth of chemical knowledge, accompanied by good solid examples of logical scientific deductive reasoning. The problems at the end of the chapters are a review, a practice and, in some cases, a challenge to your scientific problem-solving abilities. It is the fundamental spirit of this Solutions Manual to help you to understand the scientific deductive process involved in each problem.

In this manual, I provide you with a solution and an answer to the numerical problems, but the emphasis lies on providing the step-by-step reasoning behind the mathematical manipulations. In some cases, I present as many as three different approaches to solve the same problem, since I do understand that each of you has your own unique learning style. In stoichiometry as well as in many other types of calculations, the "unit factor" method is universally emphasized in general chemistry textbooks. I think that the over-emphasis of this method may train you to regard chemistry problems as being simply mathematical manipulations in which the only objective is to cancel units and get the answer. My goal is for you to understand the principles behind the calculations and hopefully to visualize with your mind's eye the chemical processes and the experimental techniques occurring as the problem is being worked out on paper. And so I have dissected the "unit factor" method for you and introduced chemical meaning into each of the steps.

I gratefully acknowledge the tremendous help provided by Frank Kolar in the preparation of this manuscript, as well as the guidance given me by my co-author in past editions, Yi-Noo Tang.

Wendy L. Keeney-Kennicutt

Department of Chemistry
Texas A&M University

1   The Foundations of Chemistry

1-2. *Refer to the Introduction to Chapter 1 and a dictionary.*

(a) Organic chemistry is the study of the chemical compounds of carbon and hydrogen.

(b) Forensic chemistry deals with the chemistry involved in solving crimes, including chemical analyses of crime scene artifacts, such as paint chips, dirt, liquids, blood, and hair.

(c) Physical chemistry is the study of the part of chemistry that applies the mathematical theories and methods of physics to the properties of matter and to the study of chemical processes and the accompanying energy changes.

(d) Medicinal chemistry is the study of the chemistry and biochemistry dealing with all aspects of the medical field.

1-4. *Refer to the Sections 1-1, 1-4, 1-8, 1-13 and the Key Terms for Chapter 1.*

(a) Weight is a measure of the gravitational attraction of the earth for a body. Although the mass of an object remains constant, its weight will vary depending on its distance from the center of the earth. One kilogram of mass at sea level weighs about 2.2 pounds (9.8 newtons), but that same one kilogram of mass weighs less at the top of Mt. Everest. In more general terms, it is a measure of the gravitational attraction of one body for another. The weight of an object on the moon is about 1/7th that of the same object on the earth..

(b) Potential energy is the energy that matter possesses by virtue of its position, condition, or composition. Your chemistry book lying on a table has potential energy due to its position. Energy is released if it falls from the table.

(c) Temperature is a measurement of the intensity of heat, *i.e.* the "hotness" or "coldness" of an object. The temperature at which water freezes is 0°C or 32°F.

(d) An endothermic process is a process that absorbs heat energy. The boiling of water is a physical process that requires heat and therefore is endothermic.

(e) An extensive property is a property that depends upon the amount of material in a sample. Extensive properties include mass and volume.

1-6. *Refer to Section 1-1.*

Einstein's equation, written as $E = mc^2$, tells us that the amount of energy released when matter is transformed into energy is the product of the mass of matter transformed and the speed of light squared. From this equation, we see that energy and matter are equivalent. Known as the Law of Conservation of Matter and Energy, we can use this equation to calculate the amount of energy released in a nuclear reaction because it is equivalent to the difference in mass between the products and the reactants. The energy released (in joules) equals the mass difference (in kilograms) times the square of the speed of light (in m/s).

1-8. *Refer to Section 1-1.*

Electrical motors are less than 100% efficient in the conversion of electrical energy into useful work, since a part of that energy is converted into frictional heat which radiates away.

However, the Law of Conservation of Energy still applies:

$$\text{electrical energy} = \text{useful work} + \text{heat}$$

Solids: are rigid and have definite shapes;

they occupy a fixed volume and are thus very difficult to compress;

the hardness of a solid is related to the strength of the forces holding the particles of a solid together; the stronger the forces, the harder is the solid object.

Liquids: occupy essentially constant volume but have variable shape;

they are difficult to compress;

particles can pass freely over each other;

their boiling points increase with increasing forces of attraction among the particles.

Gases: expand to fill the entire volume of their containers;

they are very compressible with relatively large separations between particles.

The three states are alike in that they all exhibit definite mass and volume under a given set of conditions. All consist of some combination of atoms, molecules or ions. The differences are stated above. Additional differences occur in their relative densities:

$$\text{gases} <<< \text{liquids} < \text{solids}.$$

Molecular representations of these three phases can be seen in Figure 1-8.

(a) A substance is a kind of matter in which all samples have identical chemical composition and physical properties, e.g., iron (Fe) and water (H_2O).

(b) A mixture is a sample of matter composed of two or more substances in variable composition, each substance retaining its identity and properties, e.g., soil (minerals, water, organic matter, living organisms, etc.) and seawater (water, different salts, dissolved gases, organic compounds, living organisms, etc.).

(c) An element is a substance that cannot be decomposed into simpler substances by chemical means, e.g., nickel (Ni) and nitrogen (N).

(d) A compound is a substance composed of two or more elements in fixed proportions. Compounds can be decomposed into their constituent elements by chemical means. Examples include water (H_2O) and sodium chloride (NaCl).

(a) Gasoline is a homogeneous liquid mixture of organic compounds distilled from oil.

(b) Tap water is a homogeneous liquid mixture, called an aqueous solution, containing water, dissolved salts, and gases such as chlorine and oxygen.

(c) Calcium carbonate is a compound, $CaCO_3$, consisting of the elements Ca, C and O in the fixed atomic ratio, 1:1:3.

(d) Ink from a ball-point pen is a homogeneous mixture of solvent, water and dyes.

(e) Toothpaste is a heterogeneous mixture of water, organic and inorganic compounds to aid in cleaning, whitening and preventing cavities, which can include calcium hydrogen phosphate ($CaHPO_4$), glycerin sorbitol, cellulose gum and sodium fluoride.

(f) Aluminum foil is composed of the metallic element, Al.

The coin is a heterogeneous mixture of gold and copper because it consists of two distinguishable compounds that can be recognized on sight.

(a) Striking a match, causing it to burst into flames, is a chemical property, since a change in composition is occurring of the substances in the match head and new substances including carbon dioxide gas and water vapor, are being formed.

(b) The hardness of steel is a physical property. It can be determined without a composition change.

(c) The density of gold is a physical property, since it can be observed without any change in the composition of the gold.

(d) The ability of iron to dissolve in hydrochloric acid with the evolution of hydrogen gas is a chemical property of iron, since during the reaction, its composition is changing and a new substance is being formed.

(e) The ability of fine steel wool to burn in air is a chemical property of steel wool since a compositional change in the steel wool occurs and heat is released.

(f) The ripening of fruit is a chemical property. When the temperature of the fruit decreases when put into a refrigerator, the rate of the chemical reaction slows. So, the lowering of the fruit's temperature is a physical change, but temperature has a definite effect on the chemical properties of the fruit.

(a) A wet towel drying in the sun is a physical change since the water evaporates from the towel.

(b) When lemon juice is added to tea, a color change occurs. This is a chemical change; the lemon (containing citric acid) is reacting with some component of the tea.

(c) Hot air rising over a radiator is a physical change. The heat causes the air to expand and its density to decrease; the air then rises.

(d) When coffee is brewed, the hot water dissolves the soluble components in the ground coffee. Therefore, this is a physical change. (Note: there are some chemists who consider dissolution a chemical change.)

(e) When an antacid tablet decreases the acidity of an over acid stomach, there is a chemical change. The alkaline substances in the tablet react with the excess acid in the stomach, making it less acidic.

(b), (d) and (e) are examples of potential energy. An inflated balloon (b) possesses energy which will be released if it is popped. The stored chemical energy in a flashlight battery (d) will convert to electrical energy, then into kinetic energy once it is put to use. A frozen lake (e) is stored energy. Once spring comes, the water molecules will be free to move, the lake will be circulating and the energy will convert to kinetic energy. However, a lake can also be a source of potential energy that can be converted into kinetic energy if the water is released via a dam.

(a), (c) and (f) are all examples of kinetic energy due to their motion.

(a) Combustion is an exothermic process in which a chemical reaction releases heat.

(b) The freezing of water is an exothermic process. Heat must be removed from the molecules in the liquid state to cause solidification.

(c) The melting of ice is an endothermic process. The system requires heat to break the attractive forces that hold solid water together.

(d) The boiling of water is an endothermic process. Molecules of liquid water must absorb energy to break away from the attractive forces that hold liquid water together in order to form gaseous molecules.

(e) The condensing of steam is an exothermic process. The heat stored in water vapor must be removed for the vapor to liquefy. The condensation process is the opposite of boiling which requires heat.

(f) The burning of paper is an exothermic process. The heat generated can be used to light the wood in a fireplace.

1-26.   *Refer to Section 1-5.*

When the sulfur is heated, some of it obviously became a gas. However, there is not enough information to tell whether or not this was the result of a physical or a chemical change.

Hypothesis 1:   Solid sulfur could be changing directly into gaseous sulfur. This is a physical change called sublimation.

Hypothesis 2:   Solid sulfur could be reacting with oxygen in the air to form a gaseous compound consisting of sulfur and oxygen. This would be a chemical change. The sharp odor may indicate the presence of SO_2, but the smell test is not conclusive.

To verify which hypothesis is correct, we need to identify the gas that is produced.

1-28.   *Refer to Section 1-9.*

(a) $423.\underline{0}6$ mL $= 4.2306 \times 10^2$ mL   (5 significant figures)

(b) $0.001\underline{073040}$ g $= 1.073040 \times 10^{-3}$ g   (7 significant figures)

(c) $1\underline{0}81.2$ pounds $= 1.0812 \times 10^3$ pounds   (5 significant figures)

1-30.   *Refer to Section 1-9.*

(a) 50600   (c) 0.1610   (e) 90000.

(b) 0.00040010   (d) 0.000206   (f) 0.0009000

1-32.   *Refer to Section 1-9.*

? volume $(cm^3) = 252.56$ cm $\times$ 18.23 cm $\times$ 6.50 cm $= 29927 = \mathbf{2.99 \times 10^4\ cm^3}$ (3 significant figures based on 6.50 cm)

1-34.   *Refer to Section 1-10, the conversion factors from Tables 1-6 and 1-8, and Example 1-5.*

(a) ? km $= 25.5$ m $\times \dfrac{1 \text{ km}}{1000 \text{ m}} = \mathbf{2.55 \times 10^{-2}\ km}$

(b) ? m $= 36.3$ km $\times \dfrac{1000 \text{ m}}{1 \text{ km}} = \mathbf{3.63 \times 10^4\ m}$

(c) ? g $= 487$ kg $\times \dfrac{1000 \text{ g}}{1 \text{ kg}} = \mathbf{4.87 \times 10^5\ g}$

(d) ? mL $= 1.32$ L $\times \dfrac{1000 \text{ mL}}{1 \text{ L}} = \mathbf{1.32 \times 10^3\ mL}$

(e) $? L = 55.9 \text{ dL} \times \dfrac{1 \text{ L}}{10 \text{ dL}} = \textbf{5.59 L}$

(f) $? \text{ cm}^3 = 6251 \text{ L} \times \dfrac{1000 \text{ cm}^3}{1 \text{ L}} = \textbf{6.251} \times \textbf{10}^6 \textbf{ cm}^3$ (Note: $1 \text{ cm}^3 = 1 \text{ mL}$)

1-36.   *Refer to Section 1-10, the conversion factors from Table 1-8, and Examples 1-9 and 1-11.*

(a) $? L = 0.500 \text{ ft}^3 \times \dfrac{(12 \text{ in})^3}{(1 \text{ ft})^3} \times \dfrac{(2.54 \text{ cm})^3}{(1 \text{ in})^3} \times \dfrac{1 \text{ L}}{1000 \text{ cm}^3} = \textbf{14.2 L}$

(b) $? \text{ pints} = 1.00 \text{ L} \, \dfrac{1.057 \text{ qt}}{1 \text{ L}} \times \dfrac{2 \text{ pt}}{1 \text{ qt}} = \textbf{2.11 pt}$

(c) $? \dfrac{\text{km}}{\text{L}} = \dfrac{1 \text{ mile}}{1 \text{ gal}} \times \dfrac{1.609 \text{ km}}{1 \text{ mile}} \times \dfrac{1 \text{ gal}}{4 \text{ qt}} \times \dfrac{1.057 \text{ qt}}{1 \text{ L}} = \textbf{0.4252} \, \dfrac{\textbf{km}}{\textbf{L}}$

    Therefore, to convert miles per gallon to kilometers per liter, one multiplies the miles per gallon by the factor, 0.4252.

1-38.   *Refer to Section 1-10, the conversion factors listed in Table 1-8, and Example 1-11.*

$? \text{ cents/L} = \dfrac{\$2.119}{1 \text{ gal}} \times \dfrac{1 \text{ gal}}{4 \text{ qt}} \times \dfrac{1.057 \text{ qt}}{1 \text{ L}} \times \dfrac{100 \text{ cents}}{\$1} = \textbf{55.99 cents/L}$

1-40.   *Refer to Section 1-2, Example 1-2, and Figure 1-8.*

(a) Box (i) represents the very ordered, dense solid state.
(b) Box (iii) represents the less ordered, slightly less dense liquid state.
(c) Box (ii) represents the disordered, much less dense gaseous state.
(d) The physical states rank from most dense to least dense: solid state > liquid state >> gaseous state

---

1-42.   *Refer to Section 1-12 and Example 1-13.*

$\text{Density (mg/mm}^3) = \dfrac{m}{V} = \dfrac{6.080 \text{ mg}}{(2.20 \text{ mm} \times 1.36 \text{ mm} \times 1.23 \text{ mm})} = 1.65 \text{ mg/mm}^3$

$\text{Density (g/cm}^3) = \dfrac{1.65 \text{ mg}}{1 \text{ mm}^3} \times \dfrac{1 \text{ g}}{1000 \text{ mg}} \times \dfrac{(10 \text{ mm})^3}{(1 \text{ cm})^3} = \textbf{1.65 g/cm}^3$

1-44.   *Refer to Section 1-12 and Example 1-14.*

(a) Method 1: $D = \dfrac{m}{V}$; $V (\text{cm}^3) = \dfrac{m \text{ (g)}}{D \text{ (g/cm}^3)} = \dfrac{743 \text{ g}}{10.5 \text{ g/cm}^3} = \textbf{70.8 cm}^3$ since $0.743 \text{ kg} \equiv 743 \text{ g}$

     Method 2:   Dimensional Analysis

         $? \text{ cm}^3 \text{ silver} = 0.743 \text{ kg} \times \dfrac{1000 \text{ g}}{1 \text{ kg}} \times \dfrac{1 \text{ cm}^3}{10.5 \text{ g}} = \textbf{70.8 cm}^3$

(b) length of each edge (cm) $= \sqrt[3]{V} = \sqrt[3]{70.8 \text{ cm}^3} = \textbf{4.14 cm}$

(c) length of each edge (in) $= 4.14 \text{ cm} \times \dfrac{1 \text{ in}}{2.54 \text{ cm}} = \textbf{1.63 in}$

1-46. *Refer to Section 1-12.*

(a) We know that Density $\left(\dfrac{g}{cm^3}\right) = \dfrac{m}{V} = \dfrac{\text{mass of substance (g)}}{\text{volume of substance (cm}^3)}$

    Therefore, volume of container = volume of water $= \dfrac{\text{mass of water (g)}}{\text{density of water (g/cm}^3)}$

$$= \dfrac{\text{total mass(container + water) - mass of container}}{1.0000 \text{ g/cm}^3}$$

$$= \dfrac{(99.646 \text{ g - 77.664 g})}{1.0000 \text{ g/cm}^3}$$

$$= \mathbf{21.982 \ cm^3}$$

(b) mass of metal = total mass (container + metal) - mass of container = 85.308 g - 77.664 g = **7.644 g**

(c) mass of water added = total mass (container + metal + water) - mass (container + metal)
$$= 106.442 \text{ g - 85.308 g}$$
$$= \mathbf{21.134 \ g}$$

(d) Since $D = \dfrac{m}{V}$, then $V\,(cm^3) = \dfrac{m \text{ (g)}}{D \text{ (g/cm}^3)} = \dfrac{21.134 \text{ g}}{1.0000 \text{ g/cm}^3} = \mathbf{21.134 \ cm^3}$

(e) volume of metal = total volume of container [from (a)] - volume of water [from (d)]
$$= 21.982 \text{ cm}^3 - 21.134 \text{ cm}^3$$
$$= \mathbf{0.848 \ cm^3}$$

(f) Therefore, the density of the metal $= \dfrac{\text{mass (g)}}{\text{volume (cm}^3)} = \dfrac{7.644 \text{ g}}{0.848 \text{ cm}^3} = \mathbf{9.01 \ g/cm^3}$

1-48. *Refer to Sections 1-12 and 3-6.*

                  (1)   (2)   (3)
Plan: L solution $\Rightarrow$ mL solution $\Rightarrow$ g solution $\Rightarrow$ g iron(III) chloride

Using 3 unit factors,

(1) Convert liters to milliliters using 1000 mL = 1 liter,
(2) Convert mL of solution to mass of solution using density, then
(3) Convert mass of solution to mass of iron(III) chloride using the definition of % by mass.

? g iron(III) chloride = 2.50 L soln $\times \dfrac{1000 \text{ mL soln}}{1 \text{ L soln}} \times \dfrac{1.149 \text{ g soln}}{1 \text{ mL soln}} \times \dfrac{40.0 \text{ g Fe(III) chloride}}{100 \text{ g soln}} = \mathbf{1.15 \times 10^3 \ g}$

1-50. *Refer to Section 1-13, and Examples 1-18 and 1-19.*

In determining the correct number of significant figures, note that the following values are exact: $32°F$, $1°C/1.8°F$, and $1°C/1 \text{ K}$.

(a) $? \ °C = \dfrac{1°C}{1.8°F} \times (15°F - 32°F) = \mathbf{-9.4°C}$

(b) $? \ °C = \dfrac{1°C}{1.8°F} \times (32.6°F - 32.0°F) = 0.3°C$ (1 sig. fig. since 32°F has an infinite number of significant figures)

    $? \ K = \dfrac{1 \text{ K}}{1°C} \times (0.3°C + 273.2°C) = \mathbf{273.5 \ K}$ since $0°C = 273.15 \text{ K}$

(c) $? \ °C = \dfrac{1°C}{1 \text{ K}} \times (328 \text{ K} - 273 \text{ K}) = 55°C$

    $? \ °F = \left(55°C \times \dfrac{1.8°F}{1°C}\right) + 32°F = \mathbf{130°F}$ (2 sig. figs.)

(d) $? \ °F = \left(11.3°C \times \dfrac{1.8°F}{1°C}\right) + 32°F = \mathbf{52.3°F}$

	Freezing Point of Water (FP)	Boiling Point of Water (BP)
Celsius Scale	0°C	100°C
Fahrenheit Scale	32°F	212°F
Rėamur Scale	0°R	80°R

(a) $\dfrac{BP_{water} - FP_{water} \text{ on Celsius Scale}}{BP_{water} - FP_{water} \text{ on Rėamur Scale}} = \dfrac{100°C - 0°C}{80°R - 0°R} = \dfrac{100°C}{80°R} = \dfrac{1.0°C}{0.8°R} = \dfrac{5°C}{4°R}$

Therefore, since both scales set the freezing point of water = 0°, then $? \ °C = \left(x°R \times \dfrac{5°C}{4°R} \right)$

(b) $\dfrac{BP_{water} - FP_{water} \text{ on Fahrenheit Scale}}{BP_{water} - FP_{water} \text{ on Rėamur Scale}} = \dfrac{212°F - 32°F}{80°R - 0°R} = \dfrac{180°F}{80°R} = \dfrac{9°F}{4°R}$

Therefore, $? \ °F = \left(x°R \times \dfrac{9°F}{4°R} \right) + 32°F$

Note that we must add 32°F to account for the fact that 0°R is equivalent to 32°F.

(c) From (a), $? \ °C = \left(x°R \times \dfrac{5°C}{4°R} \right)$

Rearranging, we have $? \ °R = \left(x°C \times \dfrac{4°R}{5°C} \right)$

$BP_{mercury} (°R) = 356.6°C \times \dfrac{4°R}{5°C} = \mathbf{285.3°R}$

For Al: $? \ °C = \dfrac{1°C}{1 \ K} \times (933.6 \ K - 273.2 \ K) = \mathbf{660.4°C}$

$? \ °F = \left(660.4°C \times \dfrac{1.8°F}{1°C} \right) + 32°F = \mathbf{1221°F}$

For Ag: $? \ °C = \dfrac{1°C}{1 \ K} \times (1235.1 \ K - 273.2 \ K) = \mathbf{961.9°C}$

$? \ °F = \left(961.9°C \times \dfrac{1.8°F}{1°C} \right) + 32°F = \mathbf{1763°F}$

$? \ °C = \dfrac{1°C}{1.8°F} \times (102.0°F - 32.0°F) = \mathbf{38.9°C}$

$? \ K = 38.9°C + 273.2° = \mathbf{312.1 \ K}$

amount of heat *gained* (J) = (mass of substance)(specific heat)(temp. change)

$= 35.1 \ g \times 0.895 \ J/g·°C \times (62.5°C - 27.0°C)$

$= \mathbf{1120 \ J}$ (3 sig. figs.)

(a) amount of heat *gained* (J) = (mass of substance)(specific heat)(temp. change)

$$= (69,700 \text{ g})(0.818 \text{ J/g·°C})(41.0°C - 25.0°C)$$

$$= \mathbf{9.12 \times 10^5 \text{ J}}$$

(b) Note that we will follow the convention of representing temperature (°C) as *t* and temperature (K) as *T*.

In any insulated system, the Law of Conservation of Energy states:

the amount of heat lost by Substance 1 = amount of heat gained by Substance 2

As will be discussed in later chapters, "heat lost" is a negative quantity and "heat gained" is a positive quantity. However, the "*amount* of heat lost" and the "*amount* of heat gained" quoted here call for *absolute* quantities without a sign associated with them:

$$\left| \text{the amount of heat lost by Substance 1} \right| = \left| \text{amount of heat gained by Substance 2} \right|$$

$$\left| (\text{mass})(\text{Sp. Ht.})(\text{temp. change}) \right|_1 = \left| (\text{mass})(\text{Sp. Ht.})(\text{temp. change}) \right|_2$$

In this exercise,

$$\left| (\text{mass})(\text{Sp. Ht.})(\text{temp. change}) \right|_{\text{limestone}} = \left| (\text{mass})(\text{Sp. Ht.})(\text{temp. change}) \right|_{\text{air}}$$

Since any "change" is always defined as the final value minus the initial value, we have

$$(\text{temp. change})_{\text{limestone}} = (30.0°C - 41.0°C) \text{ and } (\text{temp. change})_{\text{air}} = (t_{\text{final}} - 10.0°C)$$

for the limestone, $\left| 30.0°C - 41.0°C \right| = \left| \text{negative value} \right| = (41.0°C - 30.0°C) = 11.0°C$

for the interior air, $\left| t_{\text{final}} - 10.0°C \right| = \left| \text{positive value} \right| = (t_{\text{final}} - 10.0°C)$

Before we start, we must first calculate the mass of air inside the house:

$$? \text{ g air} = 2.83 \times 10^5 \text{ liters} \times \frac{1000 \text{ mL}}{1 \text{ L}} \times \frac{1.20 \times 10^{-3} \text{ g}}{1 \text{ mL}} = 3.40 \times 10^5 \text{ g}$$

$$69,700 \text{ g limestone} \times 0.818 \text{ J/g·°C} \times (41.0°C - 30.0°C) = 3.40 \times 10^5 \text{ g air} \times 1.004 \text{ J/g·°C} \times (t_{\text{final}} - 10.0°C)$$

$$6.27 \times 10^5 \text{ J} = (3.41 \times 10^5 \times t_{\text{final}}) \text{ J} - 3.41 \times 10^6 \text{ J}$$

$$4.04 \times 10^6 \text{ J} = (3.41 \times 10^5 \text{ J/°C}) \times t_{\text{final}}$$

$$t_{\text{final}} = \mathbf{11.8°C}$$

$$\left| \text{the amount of heat lost by Substance 1} \right| = \left| \text{amount of heat gained by Substance 2} \right|$$

$$\left| (\text{mass})(\text{Sp. Ht.})(\text{temp. change}) \right|_{\text{metal}} = \left| (\text{mass})(\text{Sp. Ht.})(\text{temp. change}) \right|_{\text{water}}$$

$$50.0 \text{ g} \times (\text{Sp. Ht.}) \times (75.0°C - 18.3°C) = 100. \text{ g} \times 4.18 \text{ J/g°C} \times (18.3°C - 15.0°C)$$

$$(\text{Sp. Ht.}) \times 2835 \text{ (remember: it has only 3 sig. figs.*)} = 1379 \text{ (only 2 sig. figs.)}$$

Sp. Ht. of the metal = **0.49 J/g°C** (2 significant figures set by the temperature change of the water)

* Note: it is better to carry all the numbers in your calculator and do your rounding to the correct number of significant figures at the end.

(a) ? tons ore = 6.40 tons hematite $\times \dfrac{100 \text{ tons ore}}{9.24 \text{ tons hematite}}$ = **69.3 tons ore**

(b) ? kg ore = 6.40 kg hematite $\times \dfrac{100 \text{ kg ore}}{9.24 \text{ kg hematite}}$ = **69.3 kg ore**

1-66. *Refer to Section 1-10, the conversion factors from Table 1-8, and Example 1-5.*

? m = 23.5 ft $\times \dfrac{12 \text{ in}}{1 \text{ ft}} \times \dfrac{2.54 \text{ cm}}{1 \text{ in}} \times \dfrac{1 \text{ m}}{100 \text{ cm}}$ = **7.16 m**

1-68. *Refer to Section 1-10 and Table 1-8.*

? lethal dose = 165 lb body wt $\times \dfrac{453.6 \text{ g body wt}}{1 \text{ lb body wt}} \times \dfrac{1 \text{ kg body wt}}{1000 \text{ g body wt}} \times \dfrac{1.5 \text{ mg drug}}{1 \text{ kg body wt}}$ = **110 mg drug** (2 sig. figs.)

1-70. *Refer to Sections 1-12 and 3-6.*

Plan: g ammonia $\overset{(1)}{\Rightarrow}$ g solution $\overset{(2)}{\Rightarrow}$ mL solution

Using 2 unit factors, (1) Convert mass of ammonia to mass of solution using the definition of % by mass, then
(2) Convert mass of solution to volume (in mL) of solution using density

? L solution = 25.8 g ammonia $\times \dfrac{100 \text{ g soln}}{5 \text{ g ammonia}} \times \dfrac{1 \text{ mL soln}}{1.006 \text{ g soln}}$ = **500 mL** (1 significant figure)

1-72. *Refer to Section 1-6.*

(a) Oil and vinegar can be physically separated by simply pouring off the less dense oil which is sitting on the vinegar layer. A scientific piece of equipment made especially for this kind of separation of two immiscible liquids is called a separatory funnel.

(b) A mixture of salt and pepper can separated by adding water which dissolves only the table salt, sodium chloride, filtering the mixture to collect the pepper, then evaporating the water to recover the salt.

1-74. *Refer to Sections 1-4 and 1-5.*

Physical properties: zinc metal is a gray and shiny solid
zinc metal piece can be cut with scissors
copper chloride solution is blue in color
the new product is brown and granular

Physical changes: the zinc pieces reduced in size when cut with scissors
the zinc pieces reduced in size during the reaction
the solution became colorless and became warmer

Chemical changes: some of the zinc disappeared. It must have reacted, because zinc metal is not soluble in water
a new brown granular product formed
the reaction is exothermic and heat was released, making the flask warm to the touch

1-76. *Refer to Sections 1-4 and 1-5, and Exercise 1-75.*

Water is more dense than ice at 0°C because a cube of ice (less dense) will float in a glass of water (more dense). The first drawing shows the water molecules in a very rigid, ordered structure, whereas the second drawing depicts liquid water molecules that are disorganized and slightly closer together. When a sample has more mass per unit volume, it is more dense.

1-78. *Refer to your life story.*

Chemical vocabulary and understanding can come from many experiences, besides the classroom. Perhaps you visited a science museum, or had a chemistry "magic show" come to your school. You may have been given a chemistry set as a present. There are many science-related shows on television and the World Wide Web has many, many links to science pages. Use your own life experiences to answer this question.

1-80. *Refer to Section 1-10, Table 1-8 for conversion factors, and Example 1-6.*

Each cesium atom has a diameter = 2 x 2.65 Å = 5.30 Å

$$? \text{ Cs atoms} = 1.00 \text{ inch} \times \frac{2.54 \text{ cm}}{1 \text{ in}} \times \frac{1 \text{ m}}{100 \text{ cm}} \times \frac{1 \text{ Å}}{10^{-10} \text{ m}} \times \frac{1 \text{ atom}}{5.30 \text{ Å}} = \textbf{4.79 x } 10^7 \textbf{ atoms}$$

1-82. *Refer to Section 1-5 and your common sense.*

As a student writes out an End-of-Chapter Exercise, the direct chemical changes that occur include (1) reactions (including irreversible adsorption) of the ink in the pen with the paper, (2) the body's biochemical reactions, (3) the creation of new neural pathways in the student's brain due to the new information she/he is learning. More indirect chemical changes include the burning of coal or natural gas to provide the power for electricity, heat and light. If the student is doing a problem outside on a beautiful day, chemical changes might involve photosynthesis occurring in the plants around her/him providing oxygen for the student to breathe and the fusion reactions in the sun which provide heat and light, etc. The complete answer is limited only by the student's imagination and understanding of the meaning of chemical changes. So, definitely yes, the answer involves knowledge not covered in Chapter 1.

1-84. *Refer to Section 1-13 and Example 1-18.*

$$? \text{ °C of iron} = \frac{1 \text{°C}}{1.8 \text{°F}} \times (65 \text{°F} - 32 \text{°F}) = \textbf{18°C}$$

Therefore, the **water sample** at 65°C has a higher temperature than the iron sample at only 18°C.

2 Chemical Formulas and Composition Stoichiometry

2-2. *Refer to Section 2-1 and the Key Terms for Chapter 2.*

Allotropes are defined as different forms of the same element in the same physical state. Two examples of allotropes are:

    (1) oxygen, O_2 (a diatomic molecule) and ozone, O_3, and

    (2) carbon as graphite, $C_{graphite}$, and carbon as diamond, $C_{diamond}$.

2-4. *Refer to Section 2-1 and Figure 2-1.*

The structural formulas and ball-and-stick models of water and ethanol are given in Figure 2-1. You can see that the general shape and bond angles are similar around the oxygen atom.

2-6. *Refer to Section 2-1 and Figure 2-1.*

Organic compounds can be distinguished from inorganic compounds because organic compounds contain C-C or C-H bonds or both. Refer to Figure 2-1. According to this definition, water, H_2O, hydrogen peroxide, H_2O_2, and carbon tetrachloride, CCl_4, are considered inorganic molecules, whereas ethanol, C_2H_5OH, is an organic molecule.

2-8. *Refer to Section 2-1, Table 2-1, and Figure 1-5.*

Ball-and-stick model of ethane, CH_3CH_3:

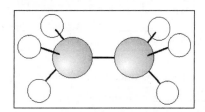

2-10. *Refer to Section 2-1 and Table 2-1.*

(a) O_3, HNO_3, SO_3 (b) H_2, H_2O, H_2O_2, H_2SO_4 (c) H_2O_2, NH_3, SO_3

(d) CH_3COOH, C_2H_6 (e) $CH_3CH_2CH_3$, $CH_3CH_2CH_2OH$

2-12. *Refer to Sections 2-1 and 2-2, and Tables 2-1 and 2-2.*

(a) HNO_3 nitric acid   (b) C_5H_{12} pentane   (c) NH_4^+ ammonium ion   (d) CH_3OH methanol

---

2-14. *Refer to Section 2-2 and Table 2-2.*

(a) Mg^{2+} monatomic cation   (b) SO_3^{2-} polyatomic anion   (c) Cu^+ monatomic cation

(d) NH_4^+ polyatomic cation   (e) O^{2-} monatomic anion

$CaBr_2$	calcium bromide	$Ca(CH_3COO)_2$	calcium acetate
NaBr	sodium bromide	$NaCH_3COO$	sodium acetate

(a) $CuCO_3$ copper(II) carbonate   (b) $SrBr_2$ strontium bromide

(c) $(NH_4)_2CO_3$ ammonium carbonate   (d) ZnO   zinc oxide

(e) $Fe_2(SO_4)_3$ iron(III) sulfate

(a) Na_2CO_3 (b) $MgCl_2$ (c) $Zn(OH)_2$ (d) $(NH_4)_2S$ (e) NaI

(a) NaCl   sodium chloride   (b) $MgBr_2$ magnesium bromide

(c) SO_2 sulfur dioxide or SO_3 sulfur trioxide   (d) CaO   calcium oxide

(e) K_2S potassium sulfide   (f) $AlBr_3$ aluminum bromide

The mass ratio of a rubidium atom (85.4678 amu) to a bromine atom (79.904 amu) is 85.4678/79.904 = 1.0696 (to 5 significant figures) or **1.070** (to 4 significant figures).

(a) The atomic weight of an element is the weighted average of the masses of all the element's constituent isotopes.

(b) Atomic weights can be referred to as relative numbers, because all atomic weights are determined relative to the mass of a particular carbon isotope, called carbon-12. The atomic mass unit (amu) is defined as exactly 1/12 of the mass of carbon-12.

(a) bromine, Br_2 2 x Br = 2 x 79.904 amu = **159.808 amu**

(b) hydrogen peroxide, H_2O_2

$$2 \times H = 2 \times 1.0079 \text{ amu} = 2.0158 \text{ amu}$$
$$2 \times O = 2 \times 15.9994 \text{ amu} = 31.9988 \text{ amu}$$
$$\text{formula weight} = \textbf{34.0146 amu}$$

(c) saccharin, $C_7H_5NSO_3$

$$7 \times C = 7 \times 12.011 \text{ amu} = 84.077 \text{ amu}$$
$$5 \times H = 5 \times 1.0079 \text{ amu} = 5.0395 \text{ amu}$$
$$1 \times N = 1 \times 14.0067 \text{ amu} = 14.0067 \text{ amu}$$
$$1 \times S = 1 \times 32.066 \text{ amu} = 32.066 \text{ amu}$$
$$3 \times O = 3 \times 15.9994 \text{ amu} = 47.9982 \text{ amu}$$
$$\text{formula weight} = \textbf{183.187 amu}$$

(d) potassium chromate, K_2CrO_4

$$2 \times K = 2 \times 39.0983\,\text{amu} = 78.1966 \quad \text{amu}$$
$$1 \times Cr = 1 \times 51.9961\,\text{amu} = 51.9961 \quad \text{amu}$$
$$\underline{4 \times O = 4 \times 15.9994\,\text{amu} = 63.9976 \quad \text{amu}}$$
$$\text{formula weight} = \textbf{194.1903} \quad \textbf{amu}$$

2-30. *Refer to Section 2-6 and Example 2-8.*

All atomic weights are rounded to 2 decimal places.

(a) hydrogen sulfide, H_2S

$$2 \times H = 2 \times 1.01 \quad \text{amu} = 2.02 \quad \text{amu}$$
$$\underline{1 \times S = 1 \times 32.07 \quad \text{amu} = 32.07 \quad \text{amu}}$$
$$\text{formula weight} = \textbf{34.09} \quad \textbf{amu}$$

(b) phosphorus trichloride, PCl_3

$$1 \times P = 1 \times 30.97 \quad \text{amu} = 30.97 \quad \text{amu}$$
$$\underline{3 \times Cl = 3 \times 35.45 \quad \text{amu} = 106.3_5 \quad \text{amu*}}$$
$$\text{formula weight} = \textbf{137.3} \quad \textbf{amu}$$

(c) hypochlorous acid, $HClO$

$$1 \times H = 1 \times 1.01 \quad \text{amu} = 1.01 \quad \text{amu}$$
$$1 \times Cl = 1 \times 35.45 \quad \text{amu} = 35.45 \quad \text{amu}$$
$$\underline{1 \times O = 1 \times 16.00 \quad \text{amu} = 16.00 \quad \text{amu}}$$
$$\text{formula weight} = \textbf{52.46} \quad \textbf{amu}$$

(d) hydrogen iodide, HI

$$1 \times H = 1 \times 1.01 \quad \text{amu} = 1.01 \quad \text{amu}$$
$$\underline{1 \times I = 1 \times 126.90 \quad \text{amu} = 126.90 \quad \text{amu}}$$
$$\text{formula weight} = \textbf{127.91} \quad \textbf{amu}$$

* The number was not rounded to the correct number of significant figures until after addition.

2-32. *Refer to Section 2-6.*

Method 1: Use the units of formula weight to derive a formula relating grams, moles and formula weight:

$$\text{formula weight, FW} \left(\frac{\text{g}}{\text{mol}}\right) = \frac{\text{grams of compound}}{\text{moles of compound}}$$

Therefore, grams of compound = moles of compound $\times$ FW

(1) $? \text{ g } CCl_4 = 1.922 \text{ mol } CCl_4 \times 153.8 \text{ g/mol} = \textbf{295.6 g } \mathbf{CCl_4}$

(2) $? \text{ kg } CCl_4 = 295.6 \text{ g } CCl_4 \times \dfrac{1 \text{ kg}}{1000 \text{ g}} = \textbf{0.2956 kg } \mathbf{CCl_4}$

Method 2: Dimensional Analysis

(2) $? \text{ kg } CCl_4 = 1.922 \text{ mol } CCl_4 \times \dfrac{153.8 \text{ g } CCl_4}{1 \text{ mol } CCl_4} \times \dfrac{1 \text{ kg}}{1000 \text{ g}} = \textbf{0.2956 kg } \mathbf{CCl_4}$

2-34. *Refer to Section 2-6, and Examples 2-10 and 2-11.*

The molecular mass of C_3H_8 is 44.1 g/mol. Each C_3H_8 molecule contains 8 hydrogen atoms.

Plan: $\text{g } C_3H_8 \Rightarrow \text{mol } C_3H_8 \Rightarrow \text{molecules } C_3H_8 \Rightarrow \text{atoms H}$

$$? \text{ H atoms} = 135 \text{ g } C_3H_8 \times \frac{1 \text{ mol } C_3H_8}{44.1 \text{ g } C_3H_8} \times \frac{6.02 \times 10^{23} \text{ } C_3H_8 \text{ molecules}}{1 \text{ mol } C_3H_8} \times \frac{8 \text{ H atoms}}{1 \text{ } C_3H_8 \text{ molecule}} = \textbf{1.47 x } \mathbf{10^{25}} \textbf{ H atoms}$$

Method 1: Use the units of formula weight to derive a formula relating grams, moles and formula weight:

$$\text{formula weight, FW}\left(\frac{g}{mol}\right) = \frac{\text{grams of substance}}{\text{moles of substance}}$$

Therefore, moles of substance $= \dfrac{\text{grams of substance}}{\text{formula weight (g/mol)}}$

$$? \text{ mol } NH_3 = \frac{10.45 \text{ g}}{17.03 \text{ g/mol}} = \textbf{0.6136 mol } NH_3$$

(Note: be sure you use <u>at least</u> as many significant figures in the formula weight as you have significant figures in your data.)

Method 2: Dimensional Analysis

$$? \text{ mol } NH_3 = 10.45 \text{ g } NH_3 \times \frac{1 \text{ mol } NH_3}{17.03 \text{ g } NH_3} = \textbf{0.6136 mol } NH_3$$

Plan: g substance $\overset{(1)}{\Rightarrow}$ moles substance $\overset{(2)}{\Rightarrow}$ molecules substance

Method 1: Recall: mol substance $= \dfrac{g \text{ substance}}{\text{formula weight}}$ and Avogadro's Number, $N = 6.02 \times 10^{23}$ molecules/mol

As an example:

(a) (1) $? \text{ mol } CO_2 = \dfrac{g \ CO_2}{FW \ CO_2} = \dfrac{15.5 \text{ g}}{44.0 \text{ g/mol}} = \textbf{0.352 mol } CO_2$

(2) $?$ molecules $CO_2 = 0.352$ mol $CO_2 \times (6.02 \times 10^{23}$ molecules/mol$) = \textbf{2.12} \times \textbf{10}^{\textbf{23}}$ **molecules** CO_2

Method 2: Dimensional Analysis. Each unit factor corresponds to a step in the Plan.

$$\qquad\qquad\qquad\qquad \text{Step 1} \qquad\qquad \text{Step 2}$$

(a) $?$ molecules $CO_2 = 15.5$ g $CO_2 \times \dfrac{1 \text{ mol } CO_2}{44.0 \text{ g } CO_2} \times \dfrac{6.02 \times 10^{23} \text{ molecules } CO_2}{1 \text{ mol } CO_2} = \textbf{2.12} \times \textbf{10}^{\textbf{23}}$ **molecules** CO_2

(b) $?$ molecules $N_2 = 15.5$ g $N_2 \times \dfrac{1 \text{ mol } N_2}{28.0 \text{ g } N_2} \times \dfrac{6.02 \times 10^{23} \text{ molecules } N_2}{1 \text{ mol } N_2} = \textbf{3.33} \times \textbf{10}^{\textbf{23}}$ **molecules** N_2

(c) $?$ molecules $P_4 = 15.5$ g $P_4 \times \dfrac{1 \text{ mol } P_4}{124 \text{ g } P_4} \times \dfrac{6.02 \times 10^{23} \text{ molecules } P_4}{1 \text{ mol } P_4} = \textbf{7.52} \times \textbf{10}^{\textbf{22}}$ **molecules** P_4

(d) $?$ molecules $P_2 = 15.5$ g $P_2 \times \dfrac{1 \text{ mol } P_2}{62.0 \text{ g } P_2} \times \dfrac{6.02 \times 10^{23} \text{ molecules } P_2}{1 \text{ mol } P_2} = \textbf{1.50} \times \textbf{10}^{\textbf{23}}$ **molecules** P_2

(e) $?$ atoms P in (c) $= 7.52 \times 10^{22}$ molecules $P_4 \times \dfrac{4 \text{ atoms P}}{1 \ P_4 \text{ molecule}} = 3.01 \times 10^{23}$ atoms P in (c)

$?$ atoms P in (d) $= 1.50 \times 10^{23}$ molecules $P_2 \times \dfrac{2 \text{ atoms P}}{1 \ P_2 \text{ molecule}} = 3.00 \times 10^{23}$ atoms P in (d)

Yes, there are the same number of P atoms in 15.5 g of pure phosphorus, regardless of whether the phosphorus is in the form of P_4 or P_2. The difference is due to rounding error only.

	Element	Mass of 1 Mole of Atoms (g)	Atomic Weight (g)
(a)	Sn	**118.710**	**118.710**
(b)	**Br**	79.904	**79.904**
(c)	Mg	**24.3050**	**24.3050**
(d)	**Cr**	**51.9961**	51.9961

2-42. **_Refer to Section 2-6 and Table 2-5._**

Moles of compound	Moles of cations	Moles of anions
1 mol $NaClO_4$	**1** **mol Na^+**	**1** **mol ClO_4^-**
2 mol K_2SO_4	**4** **mol K^+**	**2** **mol SO_4^{2-}**
0.2 mol calcium sulfate, $CaSO_4$	**0.2** **mol Ca^{2+}**	**0.2** **mol SO_4^{2-}**
0.25 mol $(NH_4)_2SO_4$	0.50 mol NH_4^+	0.25 mol SO_4^{2-}

2-44. **_Refer to Section 2-6 and Example 2-9._**

Plan: molecules $CH_4 \overset{(1)}{\Rightarrow}$ moles $CH_4 \overset{(2)}{\Rightarrow}$ g CH_4

The molecular mass of CH_4 is 16.0 g/mol.

$$? \text{ g } CH_4 = 8.00 \times 10^6 \text{ molecules } CH_4 \times \frac{1 \text{ mol } CH_4}{6.02 \times 10^{23} \text{ molecules } CH_4} \times \frac{16.0 \text{ g } CH_4}{1 \text{ mol } CH_4} = \mathbf{2.13 \times 10^{-16} \text{ g } CH_4}$$

2-46. **_Refer to Section 2-1 and Figure 2-1._**

$$? \text{ mol atoms in } H_2S = 100.0 \text{ g } H_2S \times \frac{1 \text{ mol } H_2S}{34.09 \text{ g } H_2S} \times \frac{3 \text{ mol atoms in } H_2S}{1 \text{ mol } H_2S} = \mathbf{8.800 \text{ mol atoms}}$$

$$? \text{ mol atoms in } PCl_3 = 100.0 \text{ g } PCl_3 \times \frac{1 \text{ mol } PCl_3}{137.3 \text{ g } PCl_3} \times \frac{4 \text{ mol atoms in } PCl_3}{1 \text{ mol } PCl_3} = \mathbf{2.913 \text{ mol atoms}}$$

$$? \text{ mol atoms in } HClO = 100.0 \text{ g } HClO \times \frac{1 \text{ mol } HClO}{52.46 \text{ g } HClO} \times \frac{3 \text{ mol atoms in } HClO}{1 \text{ mol } HClO} = \mathbf{5.719 \text{ mol atoms}}$$

$$? \text{ mol atoms in } HI = 100.0 \text{ g } HI \times \frac{1 \text{ mol } HI}{127.91 \text{ g } HI} \times \frac{3 \text{ mol atoms in } HI}{1 \text{ mol } HI} = \mathbf{2.345 \text{ mol atoms}}$$

Therefore, 100.0 g **H_2S** contains more moles of atoms that 100.0 g of the other compounds.

2-48. **_Refer to Section 2-7 and Example 2-12._**

mass of 1 mol Ag_2CO_3

2 x Ag = 2 x 107.9 g =	215.8	g
1 x C = 1 x 12.01 g =	12.01	g
3 x O = 3 x 16.00 g =	48.00	g
mass of 1 mol	= 275.8	g

percent Ag by mass

%Ag = (215.8/275.8) x 100%   = **78.25%**

2-50. *Refer to Sections 2-8 and 2-9, and Example 2-17.*

(a) First, we must calculate the % by mass of N in skatole.

$?\ \%\ N = 100.00\% - (\%\ C + \%\ H) = 100.00\% - (82.40\% + 6.92\%) = 10.68\%\ N$

To find the simplest formula, assume 100 g of skatole.

$?\ mol\ C = \dfrac{g\ C}{AW\ C} = \dfrac{82.40\ g}{12.01\ g/mol} = 6.861\ mol\ C$ $Ratio = \dfrac{6.861}{0.7623} = 9$

$?\ mol\ H = \dfrac{g\ H}{AW\ H} = \dfrac{6.92\ g}{1.008\ g/mol} = 6.87\ mol\ H$ $Ratio = \dfrac{6.87}{0.7623} = 9$

$?\ mol\ N = \dfrac{g\ N}{AW\ N} = \dfrac{10.68\ g}{14.01\ g/mol} = 0.7623\ mol\ N$ $Ratio = \dfrac{0.7623}{0.7623} = 1$

The simplest formula is the true formula, C_9H_9N.

(b) The molecular weight of skatole:

$$
\begin{array}{lllll}
9 \times C & = 9 \times & 12.01\ g & = 108.1\ g \\
9 \times H & = 9 \times & 1.008\ g & = 9.07\ g \\
1 \times N & = 1 \times & 14.01\ g & = 14.01\ g \\
\hline
& mass\ of\ 1\ mol\ C_9H_9N & & = \mathbf{131.2\ g}
\end{array}
$$

2-52. *Refer to Sections 2-8 and 2-9, and Examples 2-13 and 2-17.*

(a) Assume 100 g of timolol.

$?\ mol\ C = \dfrac{49.4\ g\ C}{12.0\ g/mol} = 4.12\ mol\ C$ $Ratio = \dfrac{4.12}{0.316} = 13$

$?\ mol\ H = \dfrac{7.64\ g\ H}{1.008\ g/mol} = 7.58\ mol\ H$ $Ratio = \dfrac{7.58}{0.316} = 24$

$?\ mol\ N = \dfrac{17.7\ g\ N}{14.0\ g/mol} = 1.26\ mol\ N$ $Ratio = \dfrac{1.26}{0.316} = 4$

$?\ mol\ O = \dfrac{15.2\ g\ O}{16.0\ g/mol} = 0.950\ mol\ O$ $Ratio = \dfrac{0.950}{0.316} = 3$

$?\ mol\ S = \dfrac{10.1\ g\ S}{32.0\ g/mol} = 0.316\ mol\ S$ $Ratio = \dfrac{0.316}{0.316} = 1$

The simplest formula for timolol is $\mathbf{C_{13}H_{24}N_4O_3S}$ (FW = 316 g/mol)

(b) $MW\ (g/mol) = \dfrac{g\ timolol}{mol\ timolol} = \dfrac{3.16\ g}{0.0100\ mol} = 316\ g/mol$

$n = \dfrac{molecular\ weight}{simplest\ formula\ weight} = \dfrac{316}{316} = 1$

The simplest formula is therefore the true molecular formula, $\mathbf{C_{13}H_{24}N_4O_3S}$.

2-54. *Refer to Section 2-8 and Example 2-13.*

Plan: (1) If percentage composition instead of sample mass is given, assume a 100 g sample.
      (2) Calculate the moles of each element in the 100 g sample.
      (3) Divide each of the mole values by the smallest number obtained as a mole value for the 100 g sample. General Rule: do not round off further than 0.1 from a whole number.
      (4) Determine a whole number ratio.

Let us assume we have a 100.0 g sample of norepinephrine containing 56.8 g C, 6.56 g H, 28.4 g O and 8.28 g N.

$? \text{ mol C} = \dfrac{\text{g C}}{\text{AW C}} = \dfrac{56.8 \text{ g}}{12.01 \text{ g/mol}} = 4.73 \text{ mol}$
$\qquad$
$\text{Ratio} = \dfrac{4.73}{0.591} = 8$

$? \text{ mol H} = \dfrac{\text{g}}{\text{AW H}} = \dfrac{6.56 \text{ g}}{1.008 \text{ g/mol}} = 6.51 \text{ mol}$
$\qquad$
$\text{Ratio} = \dfrac{6.51}{0.591} = 11$

$? \text{ mol O} = \dfrac{\text{g O}}{\text{AW O}} = \dfrac{28.4 \text{ g}}{16.00 \text{ g/mol}} = 1.78 \text{ mol}$
$\qquad$
$\text{Ratio} = \dfrac{1.78}{0.591} = 3.01 = 3$

$? \text{ mol N} = \dfrac{\text{g N}}{\text{AW N}} = \dfrac{8.28 \text{ g}}{14.01 \text{ g/mol}} = 0.591 \text{ mol}$
$\qquad$
$\text{Ratio} = \dfrac{0.591}{0.591} = 1$

Therefore, the simplest formula is **$C_8H_{11}O_3N$**.

2-56. *Refer to Section 2-8 and Example 2-13.*

Let us assume we have a 100.00 g sample of the kitchen product.

It contains 27.37 g Na, 1.20 g H, 14.30 g C and 57.14 g O.

$? \text{ mol Na} = \dfrac{\text{g Na}}{\text{AW Na}} = \dfrac{27.37 \text{ g}}{22.99 \text{ g/mol}} = 1.19 \text{ mol}$
$\qquad$
$\text{Ratio} = \dfrac{1.19}{1.19} = 1$

$? \text{ mol H} = \dfrac{\text{g}}{\text{AW H}} = \dfrac{1.20 \text{ g}}{1.008 \text{ g/mol}} = 1.19 \text{ mol}$
$\qquad$
$\text{Ratio} = \dfrac{1.19}{1.19} = 1$

$? \text{ mol C} = \dfrac{\text{g C}}{\text{AW C}} = \dfrac{14.30 \text{ g}}{12.01 \text{ g/mol}} = 1.191 \text{ mol}$
$\qquad$
$\text{Ratio} = \dfrac{1.191}{1.19} = 1$

$? \text{ mol O} = \dfrac{\text{g O}}{\text{AW O}} = \dfrac{57.14 \text{ g}}{16.00 \text{ g/mol}} = 3.571 \text{ mol}$
$\qquad$
$\text{Ratio} = \dfrac{3.571}{1.19} = 3$

Therefore, the simplest formula is **$NaHCO_3$** or **sodium bicarbonate** (also called sodium hydrogen carbonate.) Its common name is **baking soda**.

2-58. *Refer to Sections 2-8 and 2-9, and Examples 2-13 and 2-17.*

Let us assume we have a 100.0 g sample of lysine.

So, we have 19.2 g N, 9.64 g H, 49.3 g C and 21.9 g O.

$? \text{ mol N} = \dfrac{\text{g N}}{\text{AW N}} = \dfrac{19.2 \text{ g}}{14.01 \text{ g/mol}} = 1.37 \text{ mol}$
$\qquad$
$\text{Ratio} = \dfrac{1.37}{1.37} = 1$

$? \text{ mol H} = \dfrac{\text{g}}{\text{AW H}} = \dfrac{9.64 \text{ g}}{1.008 \text{ g/mol}} = 9.56 \text{ mol}$
$\qquad$
$\text{Ratio} = \dfrac{9.56}{1.37} = 7$

$? \text{ mol C} = \dfrac{\text{g C}}{\text{AW C}} = \dfrac{49.3 \text{ g}}{12.01 \text{ g/mol}} = 4.10 \text{ mol}$
$\qquad$
$\text{Ratio} = \dfrac{4.10}{1.37} = 3$

$? \text{ mol O} = \dfrac{\text{g O}}{\text{AW O}} = \dfrac{21.9 \text{ g}}{16.00 \text{ g/mol}} = 1.37 \text{ mol}$
$\qquad$
$\text{Ratio} = \dfrac{1.37}{1.37} = 1$

Therefore, the simplest formula of lysine is C_3H_7NO (arranging the atoms in alphabetical order).

Since each molecule of lysine has 2 nitrogen atoms, the molecular formula of lysine must be **$C_6H_{14}N_2O_2$**.

Assume 100 g of the compound found in photochemical smog.

$$? \text{ mol C} = \frac{42.9 \text{ g C}}{12.0 \text{ g/mol}} = 3.58 \text{ mol C} \qquad\qquad \text{Ratio} = \frac{3.58}{3.57} = 1$$

$$? \text{ mol O} = \frac{57.1 \text{ g O}}{16.0 \text{ g/mol}} = 3.57 \text{ mol O} \qquad\qquad \text{Ratio} = \frac{3.57}{3.57} = 1$$

The simplest formula for this compound is **CO** (FW = 28 g/mol)

$$n = \frac{\text{molecular weight}}{\text{simplest formula weight}} = \frac{56}{28} = 2$$

The true molecular formula is **C_2O_2**.

2-62. *Refer to Section 2-7 and Example 2-12.*

(a)   mass of 1 mole of dopa, $C_9H_{11}NO_4$

9 x C = 9 x 12.01 g	=	108.1 g	? % C = (108.1 g/197.2 g) x 100% = **54.82 % C**
11 x H = 11 x 1.008 g	=	11.09 g	? % H = (11.09 g/197.2 g) x 100% = **5.624 % H**
1 x N = 1 x 14.01 g	=	14.01 g	? % N = (14.01 g/197.2 g) x 100% = **7.104 % N**
4 x O = 4 x 16.00 g	=	64.00 g	? % O = (64.00 g/197.2 g) x 100% = **32.45 % O**
mass of 1 mol	=	197.2 g	

(b)   mass of 1 mole of Vitamin E, $C_{29}H_{50}O_2$

29 x C = 29 x 12.01 g	=	348.3 g	? % C = (348.3 g/430.7 g) x 100% = **80.87 % C**
50 x H = 50 x 1.008 g	=	50.40 g	? % H = (50.40 g/430.7 g) x 100% = **11.70 % H**
2 x O = 2 x 16.00 g	=	32.00 g	? % O = (32.00 g/430.7 g) x 100% = **7.430 % O**
mass of 1 mol	=	430.7 g	

(c)   mass of 1 mole of vanillin, $C_8H_8O_3$

8 x C = 8 x 12.01 g	=	96.08 g	? % C = (96.08 g/152.14 g) x 100% = **63.15 % C**
8 x H = 8 x 1.008 g	=	8.064 g	? % H = (8.064 g/152.14 g) x 100% = **5.300 % H**
3 x O = 3 x 16.00 g	=	48.00 g	? % O = (48.00 g/152.14 g) x 100% = **31.55 % O**
mass of 1 mol	=	152.14 g	

2-64. *Refer to Section 2-7 and Example 2-12.*

mass of 1 mol $Cu_3(CO_3)_2(OH)_2$ **percent Cu by mass**

3 x Cu = 3 x 63.55 g	=	190.6 g	%Cu = (190.6/344.6) x 100% = 55.31%
2 x C = 2 x 12.01 g	=	24.02 g	
8 x O = 8 x 16.00 g	=	128.0 g	
2 x H = 2 x 1.01 g	=	2.02 g	
mass of 1 mol	=	344.6 g	

mass of 1 mol Cu_2S   **percent Cu by mass**

2 x Cu = 2 x 63.55 g	=	127.1 g	%Cu = (127.1/159.2) x 100% = 79.84%
1 x S = 1 x 32.07g	=	32.07 g	
mass of 1 mol	=	159.2 g	

mass of 1 mol CuFeS$_2$

1 x Cu = 1 x 63.55 g	=	63.55 g
1 x Fe = 1 x 55.85 g	=	55.85 g
2 x S = 2 x 32.07 g	=	64.14 g
mass of 1 mol	=	183.54 g

percent Cu by mass

% Cu = (63.55/183.5) x 100% = 34.63%

mass of 1 mol CuS

1x Cu = 1 x 63.55 g	=	63.55 g
1 x S = 1 x 32.07g	=	32.07 g
mass of 1 mol	=	95.62 g

percent Cu by mass

% Cu = (63.55/95.62) x 100% = 66.46%

mass of 1 mol Cu$_2$O

2 x Cu = 2 x 63.55 g	=	127.1 g
1 x 0 = 1 x 16.00 g	=	16.00 g
mass of 1 mol	=	143.1 g

percent Cu by mass

% Cu = (127.1/143.1) x 100% = 88.82%

mass of 1 mol Cu$_2$CO$_3$(OH)$_2$

2 x Cu = 2 x 63.55 g	=	127.1 g
1 x C = 1 x 12.01 g	=	12.01 g
5 x O = 5 x 16.00 g	=	80.00 g
2 x H = 2 x 1.01 g	=	2.02 g
mass of 1 mol	=	221.1 g

percent Cu by mass

% Cu = (127.1/221.1) x 100% = 57.49%

Therefore, **chalcopyrite, CuFeS$_2$**, has the lowest copper content on a percent by mass basis.

2-66. *Refer to Section 2-8, and Examples 2-15 and 2-16.*

Plan: (1) Use the masses of CO_2 and H_2O to calculate the masses of C and H respectively.
       (2) Calculate the percentages of C and H in the sample.

(1) $? \text{ g C} = 0.3986 \text{ g CO}_2 \times \dfrac{12.01 \text{ g C}}{44.01 \text{ g CO}_2} = \textbf{0.1088 g C}$

    $? \text{ g H} = 0.0578 \text{ g H}_2\text{O} \times \dfrac{2.016 \text{ g H}}{18.02 \text{ g H}_2\text{O}} = \textbf{0.00647 g H}$

(2) $? \text{ g sample} = \text{mass of C} + \text{mass of H} = 0.1088 \text{ g C} + 0.00647 \text{ g H} = 0.1153 \text{ g sample}$

    $? \% \text{ C} = \dfrac{0.1088 \text{ g C}}{0.1153 \text{ g sample}} \times 100\% = \textbf{94.36 \% C}$

    $? \% \text{ H} = \dfrac{0.00647 \text{ g H}}{0.1153 \text{ g sample}} \times 100\% = \textbf{5.61 \% H}$

2-68. *Refer to Section 2-9 and Example 2-16.*

Plan: $\overset{(1)}{\text{g C}_2\text{H}_5\text{OH}} \Rightarrow \overset{(2)}{\text{mol C}_2\text{H}_5\text{OH}} \Rightarrow \overset{(3)}{\text{mol CO}_2} \Rightarrow \text{g CO}_2$

$$? \text{ g CO}_2 = 0.377 \text{ g C}_2\text{H}_5\text{OH} \times \overset{\text{Step 1}}{\dfrac{1 \text{ mol C}_2\text{H}_5\text{OH}}{46.1 \text{ g C}_2\text{H}_5\text{OH}}} \times \overset{\text{Step 2}}{\dfrac{2 \text{ mol CO}_2}{1 \text{ mol C}_2\text{H}_5\text{OH}}} \times \overset{\text{Step 3}}{\dfrac{44.0 \text{ g CO}_2}{1 \text{ mol CO}_2}} = \textbf{0.720 g CO}_2$$

2-70. *Refer to Sections 2-8 and 2-9, and Examples 2-13 and 2-15.*

Plan: (1) Use the masses of CO_2 and H_2O to calculate the masses of C and H respectively.

(2) Calculate the mass of O in the sample by difference: g O = g sample - g C - g H since the compound contains only C, H and O.

(3) Determine the simplest formula.

(1) $? \text{ g C} = 1.913 \text{ g CO}_2 \times \dfrac{12.01 \text{ g C}}{44.01 \text{ g CO}_2} = 0.5220 \text{ g C}$ $? \text{ g H} = 1.174 \text{ g H}_2\text{O} \times \dfrac{2.016 \text{ g H}}{18.02 \text{ g H}_2\text{O}} = 0.1313 \text{ g H}$

(2) $? \text{ g O} = 1.000 \text{ g compound} - 0.5220 \text{ g C} - 0.1313 \text{ g H} = 0.347 \text{ g O}$

(3) $? \text{ mol C} = \dfrac{0.5220 \text{ g C}}{12.01 \text{ g/mol}} = 0.04346 \text{ mol C}$ $\text{Ratio} = \dfrac{0.04346}{0.0217} = 2$

$? \text{ mol H} = \dfrac{0.1313 \text{ g H}}{1.008 \text{ g/mol}} = 0.1303 \text{ mol H}$ $\text{Ratio} = \dfrac{0.1303}{0.0217} = 6$

$? \text{ mol O} = \dfrac{0.347 \text{ g O}}{16.00 \text{ g/mol}} = 0.0217 \text{ mol O}$ $\text{Ratio} = \dfrac{0.0217}{0.0217} = 1$

The simplest formula for this alcohol is **C_2H_6O**.

2-72. *Refer to Section 2-9 and Example 2-18.*

(a) in NO: $? \text{ g O} = 3.00 \text{ g N} \times \dfrac{16.0 \text{ g O}}{14.0 \text{ g N}} = \textbf{3.43 g O}$

(b) in NO_2: $? \text{ g O} = 3.00 \text{ g N} \times \dfrac{32.0 \text{ g O}}{14.0 \text{ g N}} = \textbf{6.86 g O}$

One can easily see that the ratio: $\dfrac{\text{g O in NO}}{\text{g O in NO}_2} = \dfrac{3.43}{6.86} = \dfrac{1}{2}$

This result illustrates the **Law of Multiple Proportions** which states that when elements form more than one compound, the ratio of the masses of one element that combine with a given mass of another element in each of the compounds can be expressed by small whole numbers.

2-74. *Refer to Section 2-9 and Example 2-18.*

(a) in SO_2: $? \text{ g O} = 9.04 \text{ g S} \times \dfrac{32.0 \text{ g O}}{32.066 \text{ g S}} = \textbf{9.02 g O}$

(b) in SO_3: $? \text{ g O} = 9.04 \text{ g S} \times \dfrac{48.0 \text{ g O}}{32.066 \text{ g S}} = \textbf{13.5 g O}$

2-76. *Refer to Section 2-10 and Example 2-19.*

$\qquad\qquad$ (1) $\qquad\quad$ (2) $\qquad\quad$ (3)
Plan: g HgS $\Rightarrow$ mol HgS $\Rightarrow$ mol Hg $\Rightarrow$ g Hg

$\qquad\qquad\qquad$ **Step 1** $\qquad$ **Step 2** $\qquad$ **Step 3**

$? \text{ g Hg} = 725 \text{ g HgS} \times \dfrac{1 \text{ mol HgS}}{232.7 \text{ g HgS}} \times \dfrac{1 \text{ mol Hg}}{1 \text{ mol HgS}} \times \dfrac{200.6 \text{ g Hg}}{1 \text{ mol Hg}} = \textbf{625 g Hg}$

2-78. *Refer to Section 2-10 and Example 2-20.*

$$\text{Plan: } \overset{(1)}{\text{g Mn}} \Rightarrow \overset{(2)}{\text{mol Mn}} \Rightarrow \overset{(3)}{\text{mol KMnO}_4} \Rightarrow \text{g KMnO}_4$$

$$\overset{\textbf{Step 1}}{} \quad \overset{\textbf{Step 2}}{} \quad \overset{\textbf{Step 3}}{}$$

$$? \text{ g KMnO}_4 = 35.0 \text{ g Mn} \times \frac{1 \text{ mol Mn}}{54.9 \text{ g Mn}} \times \frac{1 \text{ mol KMnO}_4}{1 \text{ mol Mn}} \times \frac{158 \text{ g KMnO}_4}{1 \text{ mol KMnO}_4} = \textbf{101 g KMnO}_4$$

2-80. *Refer to Section 2-10 and Example 2-21.*

$$\text{Plan: } \overset{(1)}{\text{lb CuFeS}_2} \Rightarrow \overset{(2)}{\text{lb Cu in CuFeS}_2} = \overset{(3)}{\text{lb Cu in Cu}_2\text{S}} \Rightarrow \text{lb Cu}_2\text{S}$$

Note: Because there is a constant conversion factor between grams and pounds, we can work totally in pounds. Since the formula weights are: $CuFeS_2$ (183.5 g/mol), Cu_2S (159.2 g/mol) and Cu (63.55 g/mol), we have

$$\overset{\textbf{Step 1}}{} \quad \overset{\textbf{Step 2}}{} \quad \overset{\textbf{Step 3}}{}$$

$$? \text{ lb Cu}_2\text{S} = 235 \text{ lb CuFeS}_2 \times \frac{63.55 \text{ lb Cu in CuFeS}_2}{183.5 \text{ lb CuFeS}_2} \times \frac{1 \text{ lb Cu in Cu}_2\text{S}}{1 \text{ lb Cu in CuFeS}_2} \times \frac{159.2 \text{ lb Cu}_2\text{S}}{2 \times 63.55 \text{ lb Cu in Cu}_2\text{S}}$$

$$= \textbf{102 lb Cu}_2\textbf{S}$$

2-82. *Refer to Section 2-10 and Example 2-22.*

(a) Plan: $\text{g CuSO}_4 \cdot 5\text{H}_2\text{O} \Rightarrow \text{mol CuSO}_4 \cdot 5\text{H}_2\text{O} \Rightarrow \text{mol CuSO}_4 \cdot \text{H}_2\text{O} \Rightarrow \text{g CuSO}_4 \cdot \text{H}_2\text{O}$

$$? \text{ g CuSO}_4 \cdot \text{H}_2\text{O} = 675 \text{ g CuSO}_4 \cdot 5\text{H}_2\text{O} \times \frac{1 \text{ mol CuSO}_4 \cdot 5\text{H}_2\text{O}}{249.7 \text{ g CuSO}_4 \cdot 5\text{H}_2\text{O}} \times \frac{1 \text{ mol CuSO}_4 \cdot \text{H}_2\text{O}}{1 \text{ mol CuSO}_4 \cdot 5\text{H}_2\text{O}} \times \frac{177.6 \text{ g CuSO}_4 \cdot \text{H}_2\text{O}}{1 \text{ mol CuSO}_4 \cdot \text{H}_2\text{O}}$$

$$= \textbf{480. g CuSO}_4 \cdot \textbf{H}_2\textbf{O}$$

(b) Plan: $\text{g CuSO}_4 \cdot 5\text{H}_2\text{O} \Rightarrow \text{mol CuSO}_4 \cdot 5\text{H}_2\text{O} \Rightarrow \text{mol CuSO}_4 \Rightarrow \text{g CuSO}_4$

$$? \text{ g CuSO}_4 = 584 \text{ g CuSO}_4 \cdot 5\text{H}_2\text{O} \times \frac{1 \text{ mol CuSO}_4 \cdot 5\text{H}_2\text{O}}{249.7 \text{ g CuSO}_4 \cdot 5\text{H}_2\text{O}} \times \frac{1 \text{ mol CuSO}_4}{1 \text{ mol CuSO}_4 \cdot 5\text{H}_2\text{O}} \times \frac{159.6 \text{ g CuSO}_4}{1 \text{ mol CuSO}_4}$$

$$= \textbf{373 g CuSO}_4$$

2-84. *Refer to Section 2-11 and Example 2-23.*

Plan: $\text{g ore} \Rightarrow \text{g FeCr}_2\text{O}_7 \Rightarrow \text{g Cr present} \Rightarrow \text{g Cr recovered}$ (FW of $FeCr_2O_7$ is 271.85 g/mol)

(1) $? \text{ g Cr present} = 345.0 \text{ g ore} \times \dfrac{55.0 \text{ g FeCr}_2\text{O}_7}{100 \text{ g ore}} \times \dfrac{2 \times 52.0 \text{ g Cr}}{271.85 \text{ g FeCr}_2\text{O}_7} = \textbf{72.6 g Cr}$

(2) $? \text{ g Cr recovered} = 500.0 \text{ g ore} \times \dfrac{72.6 \text{ g Cr present}}{345 \text{ g ore}} \times \dfrac{90.0 \text{ g Cr recovered}}{100.0 \text{ g Cr present}} = \textbf{94.7 g Cr recovered}$

2-86. *Refer to Section 2-11 and Example 2-23.*

(a) $? \text{ lb MgCO}_3 = 562 \text{ lb ore} \times \dfrac{26.7 \text{ lb MgCO}_3}{100 \text{ lb ore}} = \textbf{150. lb MgCO}_3$

(b) $? \text{ lb impurities} = 562 \text{ lb ore} - 150. \text{ lb MgCO}_3 = \textbf{412 lb impurity}$

(c) $? \text{ lb Mg} = 562 \text{ lb ore} \times \dfrac{26.7 \text{ lb MgCO}_3}{100 \text{ lb ore}} \times \dfrac{24.3 \text{ lb Mg}}{84.3 \text{ lb MgCO}_3} = \textbf{43.3 lb Mg}$

2-88. *Refer to Section 2-11 and Example 2-23.*

(a) Let us assume that we have 1 mole of $CuSO_4 \cdot 5H_2O$

$$\% \ CuSO_4 \text{ by mass} = \frac{FW \ CuSO_4}{FW \ CuSO_4 \cdot 5H_2O} \times 100\% = \frac{159.6 \text{ g } CuSO_4}{249.7 \text{ g } CuSO_4 \cdot 5H_2O} \times 100\% = \textbf{63.92\%}$$

(b) $\% \ CuSO_4 \text{ by mass} = \dfrac{74.4 \text{ g } CuSO_4 \cdot 5H_2O}{100.0 \text{ g sample}} \times \dfrac{63.92 \text{ g } CuSO_4}{100.0 \text{ g } CuSO_4 \cdot 5H_2O} \times 100\% = \textbf{47.6\%}$

2-90. *Refer to Sections 2-6 and 2-10, and Example 2-21.*

(a) Formula Weight, $FW \left(\dfrac{g}{mol}\right) = \dfrac{g \text{ substance}}{mol \text{ substance}}$

$? \text{ mol } O_3 = \dfrac{g \ O_3}{FW} = \dfrac{96.0 \text{ g } O_3}{48.0 \text{ g/mol}} = \textbf{2.00 mol } O_3$

(b) Plan: $g \ O_3 \Rightarrow mol \ O_3 \Rightarrow mol \ O$

$? \text{ mol } O = 96.0 \text{ g } O_3 \times \dfrac{1 \text{ mol } O_3}{48.0 \text{ g } O_3} \times \dfrac{3 \text{ mol } O}{1 \text{ mol } O_3} = \textbf{6.00 mol } O$

(c) Plan: $g \ O_3 \Rightarrow mol \ O_3 \Rightarrow mol \ O \Rightarrow mol \ O_2 \Rightarrow g \ O_2$

$? \text{ g } O_2 = 96.0 \text{ g } O_3 \times \dfrac{1 \text{ mol } O_3}{48.0 \text{ g } O_3} \times \dfrac{3 \text{ mol } O}{1 \text{ mol } O_3} \times \dfrac{1 \text{ mol } O_2}{2 \text{ mol } O} \times \dfrac{32.0 \text{ g } O_2}{1 \text{ mol } O_2} = \textbf{96.0 g } O_2$

Note: Samples with the same number of atoms or moles of an element have the same mass.

(d) Plan: $g \ O_3 \Rightarrow mol \ O_3 \Rightarrow molecules \ O_3 = molecules \ O_2 \Rightarrow mol \ O_2 \Rightarrow g \ O_2$

$? \text{ g } O_2 = 96.0 \text{ g } O_3 \times \dfrac{1 \text{ mol } O_3}{48.0 \text{ g } O_3} \times \dfrac{6.02 \times 10^{23} \text{ molecules } O_3}{1 \text{ mol } O_3} \times \dfrac{1 \text{ molecule } O_2}{1 \text{ molecule } O_3} \times \dfrac{1 \text{ mole } O_2}{6.02 \times 10^{23} \text{ molecules } O_2}$

$\times \dfrac{32.0 \text{ g } O_2}{1 \text{ mol } O_2}$

$= \textbf{64.0 g } O_2$

2-92. *Refer to Section 2-8, and Examples 2-15 and 2-16.*

Plan: (1) Use the masses of CO_2 and H_2O to calculate the masses of C and H respectively.

(2) The masses of C and H do not add up to the mass of the sample, therefore there must be O in the sample as well. Determine the mass of O by subtracting the masses of C and H from the mass of the sample.

(3) Determine the empirical (simplest) formula of Vitamin E.

(1) $? \text{ g C} = 1.47 \text{ g } CO_2 \times \dfrac{12.01 \text{ g C}}{44.01 \text{ g } CO_2} = 0.401 \text{ g C}$

$? \text{ g H} = 0.518 \text{ g } H_2O \times \dfrac{2.016 \text{ g H}}{18.02 \text{ g } H_2O} = 0.0580 \text{ g H}$

(2) $? \text{ g O} = \text{mass of sample} - (\text{mass of C} + \text{mass of H}) = 0.497 \text{ g} - (0.401 \text{ g C} + 0.0580 \text{ g H}) = 0.038 \text{ g O}$

(3) $? \text{ mol C} = \dfrac{0.401 \text{ g C}}{12.01 \text{ g/mol}} = 0.0334 \text{ mol C}$ $\text{Ratio} = \dfrac{0.0334}{0.0024} = 14$

$? \text{ mol H} = \dfrac{0.0580 \text{ g H}}{1.008 \text{ g/mol}} = 0.0575 \text{ mol H}$ $\text{Ratio} = \dfrac{0.0575}{0.0024} = 24$

$? \text{ mol O} = \dfrac{0.038 \text{ g O}}{16.00 \text{ g/mol}} = 0.0024 \text{ mol O}$ $\text{Ratio} = \dfrac{0.0024}{0.0024} = 1$

The calculated simplest formula for Vitamin E is $\textbf{C}_{\textbf{14}}\textbf{H}_{\textbf{24}}\textbf{O}$. The actual simplest formula for Vitamin E is $\textbf{C}_{\textbf{29}}\textbf{H}_{\textbf{50}}\textbf{O}_{\textbf{2}}$. If the original data had been measured to 4 significant figures, we could have determined the formula correctly.

Sample 1: $\dfrac{1.60 \text{ g O}}{2.43 \text{ g Mg}} = 0.658$ Sample 2: $\dfrac{0.658 \text{ g O}}{1.00 \text{ g Mg}} = 0.658$ Sample 3: $\dfrac{2.29 \text{ g O}}{3.48 \text{ g Mg}} = 0.658$

All three samples of magnesium oxide had the same O/Mg mass ratio. This is an example of the **Law of Constant Composition.**

2-96.   *Refer to Section 2-9 and Examples 2-13, 2-16 and 2-17.*

(a) Plan: (1) Use the masses of CO_2 and H_2O to calculate the masses of C and H respectively.
       (2) The masses of C and H do not add up to the mass of the sample, therefore there must be O in the sample as well. Determine the mass of O by subtracting the masses of C and H from the mass of the sample.
       (3) Determine the empirical (simplest) formula of adipic acid.

(1) $? \text{ g C} = 2.960 \text{ g } CO_2 \times \dfrac{12.01 \text{ g C}}{44.01 \text{ g } CO_2} = 0.8078 \text{ g C}$ $? \text{ g H} = 1.010 \text{ g } H_2O \times \dfrac{2.016 \text{ g H}}{18.02 \text{ g } H_2O} = 0.1130 \text{ g H}$

(2) $? \text{ g O} = 1.6380 \text{ g adipic acid} - 0.8078 \text{ g C} - 0.1130 \text{ g H} = 0.7172 \text{ g O}$

(3) $? \text{ mol C} = \dfrac{0.8078 \text{ g C}}{12.01 \text{ g/mol}} = 0.06726 \text{ mol C}$ $\text{Ratio} = \dfrac{0.06726}{0.04483} = 1.5$

$? \text{ mol H} = \dfrac{0.1130 \text{ g H}}{1.008 \text{ g/mol}} = 0.1121 \text{ mol H}$ $\text{Ratio} = \dfrac{0.1121}{0.04483} = 2.5$

$? \text{ mol O} = \dfrac{0.7172 \text{ g O}}{16.00 \text{ g/mol}} = 0.04483 \text{ mol O}$ $\text{Ratio} = \dfrac{0.04483}{0.04483} = 1$

A 1.5:2.5:1 ratio converts to 3:5:2 by multiplying by 2. Therefore, the simplest formula for adipic acid is $C_3H_5O_2$ (FW = 73.07 g/mol).

(b) $n = \dfrac{\text{molecular weight}}{\text{simplest formula weight}} = \dfrac{146.1 \text{ g/mol}}{73.07 \text{ g/mol}} = 2$

The true molecular formula for adipic acid is $(C_3H_5O_2)_2 = C_6H_{10}O_4$.

2-98.   *Refer to Section 2-6.*

When organic compounds are combusted, all the hydrogen present is converted to water. So, the moles of water produced are equal to 1/2 the moles of hydrogen in the compound. In other words,

for 1 mole of compound: $? \text{ mol } H_2O = \text{moles H in compound} \times \dfrac{1 \text{ mol } H_2O}{2 \text{ mol H}}$

(a) for CH_3CH_2OH: $? \text{ mol } H_2O = 2.5 \text{ mol compound} \times \dfrac{6 \text{ moles H in compound}}{1 \text{ mol compound}} \times \dfrac{1 \text{ mol } H_2O}{2 \text{ mol H}} = 7.5 \text{ moles } H_2O$

(b) for CH_3OH: $? \text{ mol } H_2O = 2.5 \text{ mol compound} \times \dfrac{4 \text{ moles H in compound}}{1 \text{ mol compound}} \times \dfrac{1 \text{ mol } H_2O}{2 \text{ mol H}} = 5.0 \text{ moles } H_2O$

(c) for CH_3OCH_3: $? \text{ mol } H_2O = 2.5 \text{ mol compound} \times \dfrac{6 \text{ moles H in compound}}{1 \text{ mol compound}} \times \dfrac{1 \text{ mol } H_2O}{2 \text{ mol H}} = 7.5 \text{ moles } H_2O$

CH_3CH_2OH and CH_3OCH_3 will produce the most water (7.5 moles) and CH_3OH will make the fewest (5.0 moles).

2-100. *Refer to Section 2-10.*

Plan:　　g $MgCl_2$ $\Rightarrow$ mol $MgCl_2$ $\Rightarrow$ mol ions $\Rightarrow$ mol NaCl $\Rightarrow$ g NaCl　(FW of $MgCl_2$ is 95.2 g/mol)

? g NaCl $= 365$ g $MgCl_2$ $\times \dfrac{1 \text{ mol } MgCl_2}{95.2 \text{ g } MgCl_2} \times \dfrac{3 \text{ mol ions}}{1 \text{ mol } MgCl_2} \times \dfrac{1 \text{ mol NaCl}}{2 \text{ mol ions}} \times \dfrac{58.4 \text{ g NaCl}}{1 \text{ mol NaCl}} = $ **336 g NaCl**

2-102. *Refer to Section 2-7.*

Plan: Determine the % Zn by mass in each compound. The compound with the greater % Zn for the same price will be the cheaper source of Zn.

$ZnSO_4$:　　　　　　　% Zn $= \dfrac{\text{AW Zn}}{\text{FW } ZnSO_4} \times 100\% = \dfrac{65.39 \text{ g}}{161.46 \text{ g}} \times 100\% = 40.50\%$ Zn

$Zn(CH_3CO_2)_2 \cdot 2H_2O$:　　% Zn $= \dfrac{\text{AW Zn}}{\text{FW } Zn(CH_3CO_2)_2 \cdot 2H_2O} \times 100\% = \dfrac{65.39 \text{ g}}{219.51 \text{ g}} \times 100\% = 29.79\%$ Zn

Therefore, **$ZnSO_4$** is the cheaper source of Zn.

You would get $\dfrac{(40.50 - 29.79)}{29.79} \times 100\% = $ **35.95%** more Zn for your money buying $ZnSO_4$, rather than $Zn(CH_3CO_2)_2 \cdot 2H_2O$.

2-104. *Refer to Sections 2-6 and 2-10.*

(a)　? mol Ag $= 0.8520$ mol $Ag_2S \times \dfrac{2 \text{ mol Ag}}{1 \text{ mol } Ag_2S} = $ **1.704 mol Ag**

(b)　? mol Ag $= 0.8520$ mol $Ag_2O \times \dfrac{2 \text{ mol Ag}}{1 \text{ mol } Ag_2O} = $ **1.704 mol Ag**

(c)　? mol Ag $= 0.8520$ g $Ag_2S \times \dfrac{1 \text{ mol } Ag_2S}{247.8 \text{ g } Ag_2S} \times \dfrac{2 \text{ mol Ag}}{1 \text{ mol } Ag_2S} = $ **6.877 x 10^{-3} mol Ag**

(d)　? mol Ag $= 8.52 \times 10^{22}$ formula units $Ag_2S \times \dfrac{1 \text{ mol } Ag_2S}{6.02 \times 10^{23} \text{ formula units } Ag_2S} \times \dfrac{2 \text{ mol Ag}}{1 \text{ mol } Ag_2S} = $ **0.283 mol Ag**

2-106. *Refer to Sections 2-8 and 2-9, and Examples 2-13, 2-15 and 2-17.*

Plan:　(1)　Use the masses of CO_2 and H_2O to calculate the masses of C and H respectively.

　　　(2)　The masses of C and H do not add up to the mass of the sample, therefore there must be O in the sample as well. Determine the mass of O by subtracting the masses of C and H from the mass of the sample.

　　　(3)　Determine the empirical (simplest) formula

(1)　? g C $= 1.114$ g $CO_2 \times \dfrac{12.01 \text{ g C}}{44.01 \text{ g } CO_2} = 0.3040$ g C

　　? g H $= 0.455$ g $H_2O \times \dfrac{2.016 \text{ g H}}{18.02 \text{ g } H_2O} = 0.0509$ g H

(2)　? g O $= 0.625$ g unknown compound $- 0.3040$ g C $- 0.0509$ g H $= 0.270$ g O

(3)　? mol C $= \dfrac{0.3040 \text{ g C}}{12.01 \text{ g/mol}} = 0.0253$ mol C　　　　　　　Ratio $= \dfrac{0.0253}{0.0169} = 1.50$

　　? mol H $= \dfrac{0.0509 \text{ g H}}{1.008 \text{ g/mol}} = 0.0505$ mol H　　　　　　　Ratio $= \dfrac{0.0505}{0.0169} = 2.99$

　　? mol O $= \dfrac{0.0.270 \text{ g O}}{16.00 \text{ g/mol}} = 0.0169$ mol O　　　　　　　Ratio $= \dfrac{0.0169}{0.0169} = 1.00$

A 1.50:2.99:1.00 ratio converts to 3:6:2 by multiplying by 2. Therefore, the simplest formula for this compound is $C_3H_6O_2$ (FW = 74.1 g/mol). The true molecular formula for the compound is the same, $C_3H_6O_2$, because the true molecular formula is the same as the empirical formula.

2-108. *Refer to Sections 2-5, 2-6 and 1-12.*

Plan: (1) Determine the number of molecules in 225 mL of H_2O.
       (2) Determine the volume of ethanol that contains the same number of molecules.

(1) ? H_2O molecules $= 225 \text{ mL } H_2O \times \dfrac{1.00 \text{ g } H_2O}{1.00 \text{ mL } H_2O} \times \dfrac{1 \text{ mol } H_2O}{18.0 \text{ g } H_2O} \times \dfrac{6.02 \times 10^{23} \text{ } H_2O \text{ molecules}}{1 \text{ mol } H_2O}$

$\qquad\qquad\qquad\quad = 7.52 \times 10^{24} \text{ molecules}$

(2) ? mL ethanol $= 7.52 \times 10^{24} \text{ molecules} \times \dfrac{1 \text{ mol}}{6.02 \times 10^{23} \text{ molecules}} \times \dfrac{46.1 \text{ g}}{1 \text{ mol}} \times \dfrac{1.00 \text{ mL}}{0.789 \text{ g}}$

$\qquad\qquad\qquad\quad = \textbf{730. mL ethanol}$

2-110. *Refer to Sections 2-5, 2-6 and 1-12.*

(a) ? density $NaHCO_3$ (g/mL) $= \dfrac{1 \text{ mol } NaHCO_3}{0.0389 \text{ L}} \times \dfrac{84.0 \text{ g } NaHCO_3}{1 \text{ mol } NaHCO_3} \times \dfrac{1 \text{ L}}{1000 \text{ mL}} = \textbf{2.16 g/mL}$

(b) ? density I_2 (g/mL) $= \dfrac{1 \text{ mol } I_2}{0.05148 \text{ L}} \times \dfrac{253.8 \text{ g } I_2}{1 \text{ mol } I_2} \times \dfrac{1 \text{ L}}{1000 \text{ mL}} = \textbf{4.930 g/mL}$

(c) ? density Hg (g/mL) $= \dfrac{1 \text{ mol Hg}}{0.01476 \text{ L}} \times \dfrac{200.59 \text{ g Hg}}{1 \text{ mol Hg}} \times \dfrac{1 \text{ L}}{1000 \text{ mL}} = \textbf{13.59 g/mL}$

(d) ? density NaCl (g/mL) $= \dfrac{1 \text{ mol NaCl}}{0.02699 \text{ L}} \times \dfrac{58.44 \text{ g NaCl}}{1 \text{ mol NaCl}} \times \dfrac{1 \text{ L}}{1000 \text{ mL}} = \textbf{2.165 g/mL}$

3 Chemical Equations and Reaction Stoichiometry

3-2.   *Refer to Section 3-1.*

The Law of Conservation of Matter provides the basis for balancing a chemical equation. It states that matter is neither created nor destroyed during an ordinary chemical reaction. Therefore, a balanced chemical equation must always contain the same number of each kind of atom on both sides of the equation.

3-4.   *Refer to Section 3-1.*

(a) balanced equation: $2H_2(g) + O_2(g) \rightarrow 2H_2O(g)$

(b)

3-6.   *Refer to Section 3-1.*

When 1 atom of solid sulfur reacts with 1 molecule of oxygen gas, 1 molecule of sulfur dioxide gas is produced.

3-8.   *Refer to Section 3-1 and Example 3-1.*

Hints for balancing equations:
   (1) Use smallest whole number coefficients. However, it may be useful to temporarily use a fractional coefficient, then for the last step, multiply all the terms by a factor to change the fractions to whole numbers.
   (2) Look for special groups of elements that appear unchanged on both sides of the equation, e.g., NO_3, PO_4, SO_4. Treat them as units when balancing.
   (3) Begin by balancing both the special groups and the elements that appear only once on both sides of the equation.
   (4) Any element that appears more than once on one side of the equation is normally the last element to be balanced.
   (5) If free, uncombined elements appear on either side, balance them last. They are always the easiest to balance.
   (6) When an element has an "odd" number of atoms on one side of the equation and an "even" number on the other side, it is often advisable to multiply the "odd" side by 2, then finish balancing. For example, if you have 3 carbon atoms on one side and 2 carbon atoms on the other, multiply the coefficients of the first side by 2 and the other side by 3. This way you'll have 6 carbons on both sides of the equation.

(a)   unbalanced: $K + O_2 \rightarrow K_2O$

           Step 1: $K + O_2 \rightarrow \boxed{2}\,K_2O$ balance O

           Step 2: $\boxed{4}\,K + O_2 \rightarrow 2K_2O$ balance K

(b)   unbalanced: $Mg_3N_2 + H_2O \rightarrow NH_3 + Mg(OH)_2$

      Step 1: $Mg_3N_2 + H_2O \rightarrow NH_3 + \boxed{3}Mg(OH)_2$ balance Mg

      Step 2: $Mg_3N_2 + H_2O \rightarrow \boxed{2}NH_3 + 3Mg(OH)_2$ balance N

      Step 3: $Mg_3N_2 + \boxed{6}H_2O \rightarrow 2NH_3 + 3Mg(OH)_2$ balance H, O

(c)   unbalanced: $LiCl + Pb(NO_3)_2 \rightarrow PbCl_2 + LiNO_3$

      Step 1: $LiCl + Pb(NO_3)_2 \rightarrow PbCl_2 + \boxed{2}LiNO_3$ balance NO_3

      Step 2: $\boxed{2}LiCl + Pb(NO_3)_2 \rightarrow PbCl_2 + 2LiNO_3$ balance Li, Cl

(d)   unbalanced: $H_2O + KO_2 \rightarrow KOH + O_2$

      Step 1: $H_2O + KO_2 \rightarrow \boxed{2}KOH + O_2$ balance H

      Step 2: $H_2O + \boxed{2}KO_2 \rightarrow 2KOH + O_2$ balance K

      Step 3: $H_2O + 2KO_2 \rightarrow 2KOH + \boxed{3/2}O_2$ balance O

      Step 4: $\boxed{2}H_2O + \boxed{4}KO_2 \rightarrow \boxed{4}KOH + \boxed{3}O_2$ multiply by 2
                                                                whole number coefficients

(e)   unbalanced: $H_2SO_4 + NH_3 \rightarrow (NH_4)_2SO_4$

      Step 1: $H_2SO_4 + \boxed{2}NH_3 \rightarrow (NH_4)_2SO_4$ balance N, H

3-10.   *Refer to Section 3-1, Example 3-1 and Exercise 3-8 Solution.*

(a)   unbalanced: $Fe_2O_3 + CO \rightarrow Fe + CO_2$

      Step 1: $Fe_2O_3 + \boxed{3}CO \rightarrow Fe + \boxed{3}CO_2$ balance C, O

      Step 2: $Fe_2O_3 + 3CO \rightarrow \boxed{2}Fe + 3CO_2$ balance Fe

(b)   unbalanced: $Rb + H_2O \rightarrow RbOH + H_2$

      Step 1: $Rb + H_2O \rightarrow RbOH + \boxed{1/2}H_2$ balance H

      Step 2: $\boxed{2}Rb + \boxed{2}H_2O \rightarrow \boxed{2}RbOH + \boxed{1}H_2$ multiply by 2
                                                                whole number coefficients

(c)   unbalanced: $K + KNO_3 \rightarrow K_2O + N_2$

      Step 1: $K + \boxed{2}KNO_3 \rightarrow K_2O + N_2$ balance N

      Step 2: $K + 2KNO_3 \rightarrow \boxed{6}K_2O + N_2$ balance O

      Step 3: $\boxed{10}K + 2KNO_3 \rightarrow 6K_2O + N_2$ balance K

(d)   unbalanced: $(NH_4)_2Cr_2O_7 \rightarrow N_2 + H_2O + Cr_2O_3$

      Step 1: $(NH_4)_2Cr_2O_7 \rightarrow N_2 + \boxed{4}H_2O + Cr_2O_3$ balance H, O

(e)   unbalanced: $Al + Cr_2O_3 \rightarrow Al_2O_3 + Cr$

   Step 1: $\boxed{2}\,Al + Cr_2O_3 \rightarrow Al_2O_3 + Cr$ balance Al

   Step 2: $2Al + Cr_2O_3 \rightarrow Al_2O_3 + \boxed{2}\,Cr$ balance Cr

---

3-12.   *Refer to Section 3-2 and Example 3-2.*

(a) $N_2 + 3H_2 \rightarrow 2NH_3$

(b) ? molecules H_2 = 200 molecules $N_2 \times \dfrac{3 \text{ molecules } H_2}{1 \text{ molecule } N_2}$ = **600 molecules H_2**

(c) ? molecules NH_3 = 200 molecules $N_2 \times \dfrac{2 \text{ molecules } NH_3}{1 \text{ molecule } N_2}$ = **400 molecules NH_3**

3-14.   *Refer to Section 3-2 and Example 3-3.*

(a) $CaCO_3 + 2HCl \rightarrow CaCl_2 + CO_2 + H_2O$

(b) ? mol HCl = 5.4 mol $CaCO_3 \times \dfrac{2 \text{ mol HCl}}{1 \text{ mol } CaCO_3}$ = **11 mol HCl** (2 significant figures)

(c) ? mol H_2O = 5.4 mol $CaCO_3 \times \dfrac{1 \text{ mol } H_2O}{1 \text{ mol } CaCO_3}$ = **5.4 mol H_2O**

3-16.   *Refer to Section 2-10 and Examples 2-19 and 2-20.*

Plan: mol C $\overset{(1)}{\Rightarrow}$ mol $NaHCO_3$ $\overset{(2)}{\Rightarrow}$ g $NaHCO_3$

$\qquad\qquad\qquad\qquad$ **Step 1** $\qquad$ **Step 2**

? g $NaHCO_3$ = 12.2 mol C $\times \dfrac{1 \text{ mol } NaHCO_3}{1 \text{ mol C}} \times \dfrac{84.0 \text{ g } NaHCO_3}{1 \text{ mol } NaHCO_3}$ = **1020 g $NaHCO_3$**   (to 3 significant figures)

3-18.   *Refer to Section 3-2 and Example 3-3.*

(a) balanced equation: $2KClO_3 \rightarrow 2KCl + 3O_2$

   ? mol O_2 = 7.5 mol $KClO_3 \times \dfrac{3 \text{ mol } O_2}{2 \text{ mol } KClO_3}$ = **11 mol O_2**

(b) balanced equation: $2H_2O_2 \rightarrow 2H_2O + O_2$

   ? mol O_2 = 7.5 mol $H_2O_2 \times \dfrac{1 \text{ mol } O_2}{2 \text{ mol } H_2O_2}$ = **3.8 mol O_2**

(c) balanced equation: $2HgO \rightarrow 2Hg + O_2$

   ? mol O_2 = 7.5 mol HgO $\times \dfrac{1 \text{ mol } O_2}{2 \text{ mol HgO}}$ = **3.8 mol O_2**

(d) balanced equation: $2NaNO_3 \rightarrow 2NaNO_2 + O_2$

   ? mol O_2 = 7.5 mol $NaNO_3 \times \dfrac{1 \text{ mol } O_2}{2 \text{ mol } NaNO_3}$ = **3.8 mol O_2**

(e) balanced equation: $KClO_4 \rightarrow KCl + 2O_2$

$$? \text{ mol } O_2 = 7.5 \text{ mol } KClO_4 \times \frac{2 \text{ mol } O_2}{1 \text{ mol } KClO_4} = \textbf{15 mol } \textbf{O}_2$$

3-20.   *Refer to Section 3-2 and Example 3-3.*

unbalanced:	$NH_3 + O_2 \rightarrow NO + H_2O$	
Step 1:	$\boxed{2}NH_3 + O_2 \rightarrow NO + \boxed{3}H_2O$	balance H
Step 2:	$2NH_3 + O_2 \rightarrow \boxed{2}NO + 3H_2O$	balance N
Step 3:	$2NH_3 + \boxed{5/2}O_2 \rightarrow 2NO + 3H_2O$	balance O
Step 4:	$\boxed{4}NH_3 + \boxed{5}O_2 \rightarrow \boxed{4}NO + \boxed{6}H_2O$	whole number coefficients

(a) $? \text{ mol } O_2 = 7.50 \text{ mol } NH_3 \times \dfrac{5 \text{ mol } O_2}{4 \text{ mol } NH_3} = \textbf{9.38 mol } \textbf{O}_2$

(b) $? \text{ mol } NO = 7.50 \text{ mol } NH_3 \times \dfrac{4 \text{ mol } NO}{4 \text{ mol } NH_3} = \textbf{7.50 mol NO}$

(c) $? \text{ mol } H_2O = 7.50 \text{ mol } NH_3 \times \dfrac{6 \text{ mol } H_2O}{4 \text{ mol } NH_3} = \textbf{11.2 mol } \textbf{H}_2\textbf{O}$

3-22.   *Refer to Section 3-2 and Examples 3-5 and 3-6.*

Balanced equation: $CH_4 + 2O_2 \rightarrow CO_2 + 2H_2O$

Method 1:   Plan: $\text{g CH}_4 \overset{(1)}{\Rightarrow} \text{mol CH}_4 \overset{(2)}{\Rightarrow} \text{mol O}_2 \overset{(3)}{\Rightarrow} \text{g O}_2$

(1) $? \text{ mol } CH_4 = \dfrac{\text{g CH}_4}{\text{FW CH}_4} = \dfrac{48.0 \text{ g}}{16.0 \text{ g/mol}} = 3.00 \text{ mol CH}_4$

(2) $? \text{ mol } O_2 = \text{mol CH}_4 \times 2 \text{ mol O}_2 / 1 \text{ mol CH}_4 = 3.00 \text{ mol} \times 2 = 6.00 \text{ mol O}_2$

(3) $? \text{ g } O_2 = \text{mol O}_2 \times \text{FW O}_2 = 6.00 \text{ mol} \times 32.0 \text{ g/mol} = \textbf{192 g } \textbf{O}_2$

Note: To minimize rounding errors, keep all your numbers in your calculator until the end, then round to the appropriate number of significant figures.

Method 2:   Dimensional Analysis (Each unit factor corresponds to a step in Method 1.)

$$? \text{ g } O_2 = 48.0 \text{ g } CH_4 \times \overset{\textbf{Step 1}}{\frac{1 \text{ mol CH}_4}{16.0 \text{ g CH}_4}} \times \overset{\textbf{Step 2}}{\frac{2 \text{ mol O}_2}{1 \text{ mol CH}_4}} \times \overset{\textbf{Step 3}}{\frac{32.0 \text{ g O}_2}{1 \text{ mol O}_2}} = \textbf{192 g } \textbf{O}_2$$

Method 3:   Proportion or Ratio Method

$$\frac{? \text{ g O}_2}{\text{g CH}_4} = \frac{2 \times \text{FW O}_2}{1 \times \text{FW CH}_4}$$

Solving, $? \text{ g } O_2 = \text{g CH}_4 \times \dfrac{2 \times \text{FW O}_2}{1 \times \text{FW CH}_4}$

$$= 48.0 \text{ g} \times \frac{2 \times 32.0 \text{ g}}{1 \times 16.0 \text{ g}}$$

$$= \textbf{192 g } \textbf{O}_2$$

Balanced equation: $Fe_3O_4 + 4H_2 \rightarrow 3Fe + 4H_2O$

Method 1:   Plan: $g\ H_2O \overset{(1)}{\Rightarrow} mol\ H_2O \overset{(2)}{\Rightarrow} mol\ Fe_3O_4 \overset{(3)}{\Rightarrow} g\ Fe_3O_4$

(1) $?\ mol\ H_2O = \dfrac{g\ H_2O}{FW\ H_2O} = \dfrac{36.15\ g}{18.02\ g/mol} = 2.006\ mol\ H_2O$

(2) $?\ mol\ Fe_3O_4 = mol\ H_2O \times 1\ mol\ Fe_3O_4/4\ mol\ H_2O = 2.006\ mol \times 1/4 = 0.5015\ mol\ Fe_3O_4$

(3) $?\ g\ Fe_3O_4 = mol\ Fe_3O_4 \times FW\ Fe_3O_4 = 0.5015\ mol \times 231.55\ g/mol =$ **116.1 g Fe$_3$O$_4$**

Note: Remember to use at least as many significant figures in your formula weights as your data. To minimize rounding errors, keep all your numbers in your calculator until the end, then round to the appropriate number of significant figures.

Method 2:   Dimensional Analysis (Each unit factor corresponds to a step in Method 1.)

$$?\ g\ Fe_3O_4 = 36.15\ g\ H_2O \times \overset{\textbf{Step 1}}{\dfrac{1\ mol\ H_2O}{18.02\ g\ H_2O}} \times \overset{\textbf{Step 2}}{\dfrac{1\ mol\ Fe_3O_4}{4\ mol\ H_2O}} \times \overset{\textbf{Step 3}}{\dfrac{231.55\ g\ Fe_3O_4}{1\ mol\ Fe_3O_4}} = \textbf{116.1 g Fe}_3\textbf{O}_4$$

Method 3:   Proportion or Ratio Method

$$\dfrac{?\ g\ Fe_3O_4}{g\ H_2O} = \dfrac{1 \times FW\ Fe_3O_4}{4 \times FW\ H_2O}$$

Solving, $?\ g\ Fe_3O_4 = g\ H_2O \times \dfrac{FW\ Fe_3O_4}{4 \times FW\ H_2O} = 36.15\ g \times \dfrac{1 \times 231.55\ g}{4 \times 18.02\ g} =$ **116.1 g Fe$_3$O$_4$**

(a) Balanced equation: $2Na + I_2 \rightarrow 2NaI$

(b) *Method 1:*   Plan: $g\ I_2 \overset{(1)}{\Rightarrow} mol\ I_2 \overset{(2)}{\Rightarrow} mol\ NaI \overset{(3)}{\Rightarrow} g\ NaI$

(1) $?\ mol\ I_2 = \dfrac{g\ I_2}{FW\ I_2} = \dfrac{57.73\ g}{253.8\ g/mol} = 0.2275\ mol\ I_2$

(2) $?\ mol\ NaI = mol\ I_2 \times 2\ mol\ NaI/1\ mol\ I_2 = 0.2275\ mol \times 2 = 0.4549\ mol\ NaI$

(3) $?\ g\ NaI = mol\ NaI \times FW\ NaI = 0.4549\ mol \times 149.9\ g/mol =$ **68.19 g NaI**

Note 1: To minimize rounding errors, keep all your numbers in your calculator until the end, then round to the appropriate number of significant figures.

Note 2: The number of significant figures of your calculated formula weight should have at least as many significant figures as the number of significant figures in your data.

Method 2:   Dimensional Analysis (Each unit factor corresponds to a step in Method 1.)

$$?\ g\ NaI = 57.73\ g\ I_2 \times \overset{\textbf{Step 1}}{\dfrac{1\ mol\ I_2}{253.8\ g\ I_2}} \times \overset{\textbf{Step 2}}{\dfrac{2\ mol\ NaI}{1\ mol\ I_2}} \times \overset{\textbf{Step 3}}{\dfrac{149.9\ g\ NaI}{1\ mol\ NaI}} = \textbf{68.19 g NaI}$$

Balanced equation: $C_3H_8 + 5O_2 \rightarrow 3CO_2 + 4H_2O$

Method 1: Plan: mol H_2O $\overset{(1)}{\Rightarrow}$ mol C_3H_8 $\overset{(2)}{\Rightarrow}$ g C_3H_8

(1) ? mol C_3H_8 = mol H_2O x 1 mol C_3H_8/4 mol H_2O = 2.75 mol x 1/4 = 0.688 mol C_3H_8

(2) ? g C_3H_8 = mol C_3H_8 x FW C_3H_8 = 0.688 mol x 44.09 g/mol = **30.3 g C_3H_8**

Note: To minimize rounding errors, keep all your numbers in your calculator until the end, then round to the appropriate number of significant figures.

Method 2: Dimensional Analysis (Each unit factor corresponds to a step in Method 1.)

$$? \text{ g } C_3H_8 = 2.75 \text{ mol } H_2O \times \overset{\textbf{Step 1}}{\frac{1 \text{ mol } C_3H_8}{4 \text{ mol } H_2O}} \times \overset{\textbf{Step 2}}{\frac{44.09 \text{ g } C_3H_8}{1 \text{ mol } C_3H_8}} = \textbf{30.3 g } C_3H_8$$

Balanced equation: $2CO + O_2 \rightarrow 2CO_2$

This is a limiting reactant problem.

(a) Reactants:

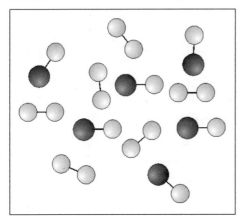

(b) Products:

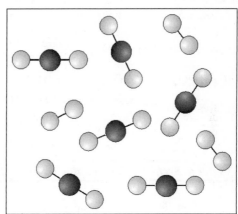

(c) Plan: (1) Find the limiting reactant.
  (2) Do the stoichiometric problem based on amount of limiting reactant.

(1) Convert the mass of reactants to moles and compare the required ratio to the available ratio.

$$? \text{ mol CO} = \frac{\text{g CO}}{\text{FW CO}} = \frac{134.67 \text{ g}}{28.010 \text{ g/mol}} = 4.8079 \text{ mol CO}$$

$$? \text{ mol } O_2 = \frac{\text{g } O_2}{\text{FW } O_2} = \frac{77.25 \text{ g}}{32.00 \text{ g/mol}} = 2.414 \text{ mol } O_2$$

$$\text{Required ratio} = \frac{2 \text{ mol CO}}{1 \text{ mol } O_2} = 2 \qquad \text{Available ratio} = \frac{4.8079 \text{ mol CO}}{2.414 \text{ mol } O_2} = 1.992$$

The available ratio < required ratio. Therefore, we do not have enough CO to react with all the O_2, and so CO is the limiting reactant.

(2) The amount of CO_2 is determined by the amount of limiting reactant, 134.67 g of CO.

$$? \text{ g } CO_2 = 134.67 \text{ g CO} \times \frac{1 \text{ mol CO}}{28.010 \text{ g CO}} \times \frac{2 \text{ mol } CO_2}{2 \text{ mol CO}} \times \frac{44.010 \text{ g } CO_2}{1 \text{ mol } CO_2} = \mathbf{211.60 \text{ g } CO_2}$$

3-32. *Refer to Section 3-3, Examples 3-9 and 3-10, and Exercise 3-30 Solution.*

Balanced equation: $Ca_3(PO_4)_2 + 2H_2SO_4 \rightarrow Ca(H_2PO_4)_2 + 2CaSO_4$

This is a limiting reactant problem.

(1) Convert to moles and compare required ratio to available ratio to find the limiting reactant.

$$? \text{ mol } Ca_3(PO_4)_2 = \frac{\text{g } Ca_3(PO_4)_2}{\text{FW } Ca_3(PO_4)_2} = \frac{200.0 \text{ g}}{310.2 \text{ g/mol}} = 0.6447 \text{ mol } Ca_3(PO_4)_2$$

$$? \text{ mol } H_2SO_4 = \frac{\text{g } H_2SO_4}{\text{FW } H_2SO_4} = \frac{133.5 \text{ g}}{98.08 \text{ g/mol}} = 1.361 \text{ mol } H_2SO_4$$

$$\text{Required ratio} = \frac{2 \text{ mol } H_2SO_4}{1 \text{ mol } Ca_3(PO_4)_2} = 2 \qquad \text{Available ratio} = \frac{1.361 \text{ mol } H_2SO_4}{0.6447 \text{ mol } Ca_3(PO_4)_2} = 2.11$$

Available ratio > required ratio; $Ca_3(PO_4)_2$ is the limiting reactant.

(2) First, find the mass of H_2SO_4 that reacted, then determine the mass of superphosphate (triple phosphate).

The Law of Conservation of Mass states that the mass of reactants that react equal the mass of products formed. Therefore, we can calculate the mass of superphosphate.

$$? \text{ g } H_2SO_4 = 200.0 \text{ g } Ca_3(PO_4)_2 \times \frac{1 \text{ mol } Ca_3(PO_4)_2}{310.2 \text{ g } Ca_3(PO_4)_2} \times \frac{2 \text{ mol } H_2SO_4}{1 \text{ mol } Ca_3(PO_4)_2} \times \frac{98.08 \text{ g } H_2SO_4}{1 \text{ mol } H_2SO_4} = 126.5 \text{ g } H_2SO_4$$

$$? \text{ g } [Ca(H_2PO_4)_2 + 2CaSO_4]_{\text{formed}} = \text{g } [H_2SO_4 + Ca_3(PO_4)_2]_{\text{reacted}}$$
$$= 126.5 \text{ g } H_2SO_4 + 200.0 \text{ g } Ca_3(PO_4)_2$$
$$= \mathbf{326.5 \text{ g superphosphate } (Ca(H_2PO_4)_2 \cdot 2CaSO_4)}$$

3-34. *Refer to Section 3-3, Examples 3-9 and 3-10, and Exercise 3-30 Solution.*

Balanced equation: $Na + KCl \rightarrow NaCl + K$

This is a limiting reactant problem.

(1) Convert to moles and compare the required ratio to the available ratio to find the limiting reactant.

$$? \text{ mol Na} = \frac{\text{g Na}}{\text{AW Na}} = \frac{87.50 \text{ g}}{22.99 \text{ g/mol}} = 3.806 \text{ mol Na}$$

$$? \text{ mol KCl} = \frac{\text{g KCl}}{\text{FW KCl}} = \frac{87.50 \text{ g}}{74.55 \text{ g/mol}} = 1.174 \text{ mol KCl}$$

$$\text{Required ratio} = \frac{1 \text{ mol Na}}{1 \text{ mol KCl}} = 1 \qquad \text{Available ratio} = \frac{3.806 \text{ mol Na}}{1.174 \text{ mol KCl}} = 3.242$$

Available ratio > required ratio; KCl is the limiting reactant.

(2) The mass of K produced is determined by the mass of KCl.

$$? \text{ g K} = 87.50 \text{ g KCl} \times \frac{1 \text{ mol KCl}}{74.55 \text{ g KCl}} \times \frac{1 \text{ mol K}}{1 \text{ mol KCl}} \times \frac{39.10 \text{ g K}}{1 \text{ mol K}} = \mathbf{45.89 \text{ g K}}$$

3-36. *Refer to Section 3-3, Examples 3-9 and 3-10, and Exercise 3-30 Solution.*

Balanced equation: $3Ca(OH)_2 + 2H_3PO_4 \rightarrow Ca_3(PO_4)_2 + 6H_2O$

This is a limiting reactant problem.

(1) Convert to moles and compare the required ratio to the available ratio to find the limiting reactant.

$$? \text{ mol Ca(OH)}_2 = \frac{\text{g Ca(OH)}_2}{\text{FW Ca(OH)}_2} = \frac{12.9 \text{ g}}{74.10 \text{ g/mol}} = 0.174 \text{ mol Ca(OH)}_2$$

$$? \text{ mol H}_3PO_4 = \frac{\text{g H}_3PO_4}{\text{FW H}_3PO_4} = \frac{18.37 \text{ g}}{97.99 \text{ g/mol}} = 0.1875 \text{ mol H}_3PO_4$$

$$\text{Required ratio} = \frac{3 \text{ mol Ca(OH)}_2}{2 \text{ mol H}_3PO_4} = 1.5 \qquad \text{Available ratio} = \frac{0.174 \text{ mol Ca(OH)}_2}{0.1875 \text{ mol H}_3PO_4} = 0.928$$

Available ratio < required ratio; $Ca(OH)_2$ is the limiting reactant.

(2) The mass of $Ca_3(PO_4)_2$ produced is determined by the mass of $Ca(OH)_2$.

$$? \text{ g Ca}_3(PO_4)_2 = 12.9 \text{ g Ca(OH)}_2 \times \frac{1 \text{ mol Ca(OH)}_2}{74.10 \text{ g Ca(OH)}_2} \times \frac{1 \text{ mol Ca}_3(PO_4)_2}{3 \text{ mol Ca(OH)}_2} \times \frac{310.2 \text{ g Ca}_3(PO_4)_2}{1 \text{ mol Ca}_3(PO_4)_2} = \mathbf{18.0 \text{ g Ca}_3(PO_4)_2}$$

3-38. *Refer to Section 3-3, Examples 3-9 and 3-10, and Exercise 3-30 Solution.*

Balanced equation: $2AgNO_3 + BaCl_2 \rightarrow 2AgCl + Ba(NO_3)_2$

(a) Convert to moles and compare the required ratio to the available ratio to find the limiting reactant.

$$? \text{ mol AgNO}_3 = \frac{\text{g AgNO}_3}{\text{FW AgNO}_3} = \frac{62.4 \text{ g}}{169.9 \text{ g/mol}} = 0.367 \text{ mol AgNO}_3$$

$$? \text{ mol BaCl}_2 = \frac{\text{g BaCl}_2}{\text{FW BaCl}_2} = \frac{53.1 \text{ g}}{208.2 \text{ g/mol}} = 0.255 \text{ mol BaCl}_2$$

$$\text{Required ratio} = \frac{2 \text{ mol AgNO}_3}{1 \text{ mol BaCl}_2} = 2 \qquad \text{Available ratio} = \frac{0.367 \text{ mol AgNO}_3}{0.255 \text{ mol BaCl}_2} = 1.44$$

Available ratio < required ratio; **$AgNO_3$** is the limiting reactant.

(b) The reactant in excess will be $BaCl_2$ and its mass depends on the limiting reactant, $AgNO_3$.

$$? \text{ g BaCl}_2 \text{ that reacted} = 62.4 \text{ g AgNO}_3 \times \frac{1 \text{ mol AgNO}_3}{169.9 \text{ g AgNO}_3} \times \frac{1 \text{ mol BaCl}_2}{2 \text{ mol AgNO}_3} \times \frac{208.2 \text{ g BaCl}_2}{1 \text{ mol BaCl}_2} = \mathbf{38.2 \text{ g BaCl}_2}$$

$$? \text{ g BaCl}_2 \text{ that remain} = \text{initial mass - mass that reacted}$$
$$= 53.1 \text{ g} - 38.2 \text{ g}$$
$$= \mathbf{14.9 \text{ g BaCl}_2}$$

(c) The mass of AgCl; produced is determined by the mass of $AgNO_3$.

$$? \text{ g AgCl} = 62.4 \text{ g AgNO}_3 \times \frac{1 \text{ mol AgNO}_3}{169.9 \text{ g AgNO}_3} \times \frac{2 \text{ mol AgCl}}{2 \text{ mol AgNO}_3} \times \frac{143.4 \text{ g AgCl}}{1 \text{ mol AgCl}} = \mathbf{52.7 \text{ g AgCl}}$$

3-40. *Refer to Section 3-4 and Example 3-11.*

Balanced equation: $PCl_3 + Cl_2 \rightarrow PCl_5$

Step 1. Calculate the theoretical yield of PCl_5.

$$\qquad\qquad (1) \qquad\qquad (2) \qquad\qquad (3)$$
Plan: $\text{g PCl}_3 \Rightarrow \text{mol PCl}_3 \Rightarrow \text{mol PCl}_5 \Rightarrow \text{g PCl}_5 \text{ (theoretical)}$

(1) $? \text{ mol PCl}_3 = \dfrac{\text{g PCl}_3}{\text{FW PCl}_3} = \dfrac{96.7 \text{ g}}{137 \text{ g/mol}} = 0.706 \text{ mol PCl}_3$

(2) $? \text{ mol PCl}_5 = \text{mol PCl}_3 = 0.706 \text{ mol PCl}_5$

(3) $? \text{ g PCl}_5 = \text{mol PCl}_5 \times \text{FW PCl}_5 = 0.706 \text{ mol} \times 208 \text{ g/mol} = 147 \text{ g PCl}_5$

Alternatively by dimensional analysis:

$? \text{ g PCl}_5 = 96.7 \text{ g PCl}_3 \times \dfrac{1 \text{ mol PCl}_3}{137 \text{ g PCl}_3} \times \dfrac{1 \text{ mol PCl}_5}{1 \text{ mol PCl}_3} \times \dfrac{208 \text{ g PCl}_5}{1 \text{ mol PCl}_5} = \mathbf{147 \text{ g PCl}_5}$

Step 2.   Solve for the actual yield of PCl$_5$.

$? \text{ \% yield} = \dfrac{\text{actual yield}}{\text{theoretical yield}} \times 100\%$ Substituting, $86.5\% = \dfrac{? \text{ actual yield}}{147 \text{ g}} \times 100\%$

Therefore, $? \text{ actual yield} = \dfrac{86.5\% \times 147 \text{ g}}{100\%} = \mathbf{127 \text{ g PCl}_5}$

3-42.   *Refer to Section 3-4 and Example 3-11.*

Balanced equation: $2\text{KClO}_3 \rightarrow 2\text{KCl} + 3\text{O}_2$

(a) Calculate the theoretical yield of O$_2$.

$\qquad\qquad\qquad\quad$ (1) $\qquad\qquad$ (2) $\qquad\quad$ (3)

Plan: g KClO$_3$ $\Rightarrow$ mol KClO$_3$ $\Rightarrow$ mol O$_2$ $\Rightarrow$ g O$_2$ (theoretical)

(1) $? \text{ mol KClO}_3 = \dfrac{\text{g KClO}_3}{\text{FW KClO}_3} = \dfrac{3.75 \text{ g}}{122.55 \text{ g/mol}} = 0.0306 \text{ mol KClO}_3$

(2) $? \text{ mol O}_2 = 0.0306 \text{ mol KClO}_3 \times \dfrac{3 \text{ mol O}_2}{2 \text{ mol KClO}_3} = 0.0459 \text{ mol O}_2$

(3) $? \text{ g O}_2 = \text{mol O}_2 \times \text{AW O}_2 = 0.0459 \text{ mol} \times 32.0 \text{ g/mol} = \mathbf{1.47 \text{ g O}_2}$ (theoretical yield)

Alternatively by dimensional analysis:

$? \text{ g O}_2 = 3.75 \text{ g KClO}_3 \times \dfrac{1 \text{ mol KClO}_3}{122.55 \text{ g KClO}_3} \times \dfrac{3 \text{ mol O}_2}{2 \text{ mol KClO}_3} \times \dfrac{32.0 \text{ g O}_2}{1 \text{ mol O}_2} = \mathbf{1.47 \text{ g O}_2}$

(b) Calculate the percent yield of O$_2$.

$\text{\% yield} = \dfrac{\text{actual yield}}{\text{theoretical yield}} \times 100\% = \dfrac{1.05 \text{ g}}{1.47 \text{ g}} \times 100\% = \mathbf{71.4\%}$

3-44.   *Refer to Section 3-4 and Example 3-11.*

Balanced equation: $2\text{AgNO}_3 \rightarrow 2\text{Ag} + 2\text{NO}_2 + \text{O}_2$

Step 1.   Calculate the theoretical yield of Ag.

$\qquad\qquad\qquad\quad$ (1) $\qquad\qquad$ (2) $\qquad\quad$ (3)

Plan: g AgNO$_3$ $\Rightarrow$ mol AgNO$_3$ $\Rightarrow$ mol Ag $\Rightarrow$ g Ag (theoretical)

(1) $? \text{ mol AgNO}_3 = \dfrac{\text{g AgNO}_3}{\text{FW AgNO}_3} = \dfrac{1.099 \text{ g}}{169.9 \text{ g/mol}} = 6.469 \times 10^{-3} \text{ mol AgNO}_3$

(2) $? \text{ mol Ag} = \text{mol AgNO}_3 = 6.469 \times 10^{-3} \text{ mol Ag}$

(3) $? \text{ g Ag} = \text{mol Ag} \times \text{AW Ag} = (6.469 \times 10^{-3} \text{ mol}) \times 107.9 \text{ g/mol} = 0.6980 \text{ g Ag}$ (theoretical yield)

Alternatively by dimensional analysis:

$? \text{ g Ag} = 1.099 \text{ g AgNO}_3 \times \dfrac{1 \text{ mol AgNO}_3}{169.9 \text{ g AgNO}_3} \times \dfrac{2 \text{ mol Ag}}{2 \text{ mol AgNO}_3} \times \dfrac{107.9 \text{ g Ag}}{1 \text{ mol Ag}} = 0.6980 \text{ g Ag}$

Step 2. Calculate the percent yield of Ag.

$$\% \text{ yield} = \frac{\text{actual yield}}{\text{theoretical yield}} \times 100\% = \frac{0.665 \text{ g}}{0.6980 \text{ g}} \times 100\% = \textbf{95.3\%}$$

3-46.   *Refer to Section 3-4 and Example 3-11.*

Balanced equation: $CS_2 + 3O_2 \rightarrow CO_2 + 2SO_2$

Step 1. Calculate the theoretical yield of SO_2. To minimize the number of steps, you could work in mg and mmol.

$$\text{Plan: } \underset{(1)}{\text{mg } CS_2} \Rightarrow \underset{(2)}{\text{g } CS_2} \Rightarrow \underset{(3)}{\text{mol } CS_2} \Rightarrow \underset{(4)}{\text{mol } SO_2} \Rightarrow \text{g } SO_2 \text{ (theoretical)} \underset{(5)}{\Rightarrow} \text{mg } SO_2 \text{ (theoretical)}$$

$$? \text{ mg } SO_2 = 82.7 \text{ mg } CS_2 \times \overset{\textbf{Step 1}}{\frac{1 \text{ g } CS_2}{1000 \text{ mg } CS_2}} \times \overset{\textbf{Step 2}}{\frac{1 \text{ mol } CS_2}{76.14 \text{ g } CS_2}} \times \overset{\textbf{Step 3}}{\frac{2 \text{ mol } SO_2}{1 \text{ mol } CS_2}} \times \overset{\textbf{Step 4}}{\frac{64.1 \text{ g } SO_2}{1 \text{ mol } SO_2}} \times \overset{\textbf{Step 5}}{\frac{1000 \text{ mg } SO_2}{1 \text{ g } SO_2}} = \textbf{139 mg SO}_2$$

Step 2. Calculate the percent yield of SO_2.

$$\% \text{ yield} = \frac{\text{actual yield}}{\text{theoretical yield}} \times 100\% = \frac{108 \text{ mg}}{139 \text{ mg}} \times 100\% = \textbf{77.7\%}$$

3-48.   *Refer to Sections 3-3 and 3-4.*

Balanced equation: $Fe_2O_3 + 2Al \rightarrow 2Fe + Al_2O_3$

(a) Plan: (1) Find the limiting reactant.
            (2) Calculate the theoretical yield of Fe, based on the limiting reactant.

(1) $? \text{ mol Al} = \dfrac{\text{g Al}}{\text{AW Al}} = \dfrac{500.0 \text{ g}}{26.98 \text{ g/mol}} = 18.53 \text{ mol Al}$

$? \text{ mol } Fe_2O_3 = \dfrac{\text{g } Fe_2O_3}{\text{FW } Fe_2O_3} = \dfrac{500.0 \text{ g}}{159.7 \text{ g/mol}} = 3.131 \text{ mol } Fe_2O_3$

$\text{Required ratio} = \dfrac{2 \text{ mol Al}}{1 \text{ mol } Fe_2O_3} = 2$ $\text{Available ratio} = \dfrac{18.53 \text{ mol Al}}{3.131 \text{ mol } Fe_2O_3} = 5.918$

Available ratio > required ratio; Fe_2O_3 is the limiting reactant.

(2) The mass of Fe that should be produced from the limiting reactant, Fe_2O_3, is the theoretical yield.

$$? \text{ g Fe (theoretical yield)} = 500.0 \text{ g } Fe_2O_3 \times \frac{1 \text{ mol } Fe_2O_3}{159.7 \text{ g } Fe_2O_3} \times \frac{2 \text{ mol Fe}}{1 \text{ mol } Fe_2O_3} \times \frac{55.85 \text{ g Fe}}{1 \text{ mol Fe}} = \textbf{349.7 g Fe}$$

(b) Calculate the percent yield of Fe.

$$\% \text{ yield} = \frac{\text{actual yield}}{\text{theoretical yield}} \times 100\% = \frac{166.5 \text{ g}}{349.7 \text{ g}} \times 100\% = \textbf{47.61\%}$$

3-50.   *Refer to Section 3-5 and Example 3-12.*

Balanced equations: $TeO_2 + 2OH^- \rightarrow TeO_3^{2-} + H_2O$
$TeO_3^{2-} + 2H^+ \rightarrow H_2TeO_3$

$$\text{Plan: } \underset{(1)}{\text{g } TeO_2} \Rightarrow \underset{(2)}{\text{mol } TeO_2} \Rightarrow \underset{(3)}{\text{mol } TeO_3^{2-}} \Rightarrow \underset{(4)}{\text{mol } H_2TeO_3} \Rightarrow \text{g } H_2TeO_3$$

$$? \text{ g } H_2TeO_3 = 64.2 \text{ g } TeO_2 \times \overset{\textbf{Step 1}}{\frac{1 \text{ mol } TeO_2}{159.6 \text{ g } TeO_2}} \times \overset{\textbf{Step 2}}{\frac{1 \text{ mol } TeO_3^{2-}}{1 \text{ mol } TeO_2}} \times \overset{\textbf{Step 3}}{\frac{1 \text{ mol } H_2TeO_3}{1 \text{ mol } TeO_3^{2-}}} \times \overset{\textbf{Step 4}}{\frac{177.6 \text{ g } H_2TeO_3}{1 \text{ mol } H_2TeO_3}} = \textbf{71.4 g H}_2\textbf{TeO}_3$$

3-52. *Refer to Section 3-5 and Example 3-12.*

Balanced equations:
$$2KClO_3 \rightarrow 2KCl + 3O_2$$
$$CH_4 + 2O_2 \rightarrow CO_2 + 2H_2O$$

Plan: $\underset{(1)}{g\ CH_4 \Rightarrow}\ \underset{(2)}{mol\ CH_4 \Rightarrow}\ \underset{(3)}{mol\ O_2 \Rightarrow}\ \underset{(4)}{mol\ KClO_3 \Rightarrow}\ g\ KClO_3$

$$? \text{ g KClO}_3 = 71.06 \text{ g CH}_4 \times \underset{\text{Step 1}}{\frac{1 \text{ mol CH}_4}{16.04 \text{ g CH}_4}} \times \underset{\text{Step 2}}{\frac{2 \text{ mol O}_2}{1 \text{ mol CH}_4}} \times \underset{\text{Step 3}}{\frac{2 \text{ mol KClO}_3}{3 \text{ mol O}_2}} \times \underset{\text{Step 4}}{\frac{122.6 \text{ g KClO}_3}{1 \text{ mol KClO}_3}} = \textbf{724.2 g KClO}_3$$

3-54. *Refer to Section 3-5 and Example 3-12.*

(a) Balanced equations:
$$4NH_3 + 5O_2 \rightarrow 4NO + 6H_2O$$
$$2NO + O_2 \rightarrow 2NO_2$$
$$3NO_2 + H_2O \rightarrow 2HNO_3 + NO$$
$$HNO_3 + NH_3 \rightarrow NH_4NO_3$$

(b) $? \text{ mol N atoms} = 1 \text{ mol NH}_4\text{NO}_3 \times \dfrac{2 \text{ mol N atoms}}{1 \text{mol NH}_4\text{NO}_3} = \textbf{2 mol N atoms}$

(c) Step 1: Determine how many <u>total</u> moles of NH_3 are necessary to produce 1 mole of NH_4NO_3.
In the last reaction, we know we need 1 mole of NH_3 and 1 mole of HNO_3 to make 1 mole of NH_4NO_3.
To determine the number of moles of NH_3 required to make 1 mole of HNO_3:

$$? \text{ mol NH}_3 \text{ (in first reaction)} = 1 \text{ mol NH}_4\text{NO}_3 \times \frac{1 \text{ mol HNO}_3}{1 \text{ mol NH}_4\text{NO}_3} \times \frac{3 \text{ mol NO}_2}{2 \text{ mol HNO}_3} \times \frac{2 \text{ mol NO}}{2 \text{ mol NO}_2} \times \frac{4 \text{ mol NH}_3}{4 \text{ mol NO}}$$

$$= 1.5 \text{ moles NH}_3$$

Therefore, the total moles of NH_3 required to make 1 mole of NH_4NO_3 is $(1 + 1.5)$ mol = 2.5 mol NH_3
This assumes that the NO formed in the third reaction is lost and is not recycled.

Step 2: Do the normal stoichiometric problem.

$$? \text{ g NH}_3 = 250.0 \text{ g NH}_4\text{NO}_3 \times \underset{\text{Step 1}}{\frac{1 \text{ mol NH}_4\text{NO}_3}{80.05 \text{ g NH}_4\text{NO}_3}} \times \underset{\text{Step 2}}{\frac{2.5 \text{ mol NH}_3}{1 \text{ mol NH}_4\text{NO}_3}} \times \underset{\text{Step 3}}{\frac{17.03 \text{ g NH}_3}{1 \text{ mol NH}_3}} = \textbf{133.0 g NH}_3$$

3-56. *Refer to Section 3-5 and Example 3-13.*

Balanced equations:

ZnS in ore $\rightarrow$ ZnS	flotation	89.6% efficient
$2ZnS + 3O_2 \rightarrow 2ZnO + 2SO_2$	heat in air	100% efficient
$ZnO + H_2SO_4 \rightarrow ZnSO_4 + H_2O$	acid treatment	100% efficient
$2ZnSO_4 + 2H_2O \rightarrow 2Zn + 2H_2SO_4 + O_2$	electrolysis	92.2% efficient

Plan: kg ZnS in ore $\Rightarrow$ kg ZnS $\Rightarrow$ g ZnS $\Rightarrow$ mol ZnS $\Rightarrow$ mol ZnO $\Rightarrow$ mol ZnSO$_4$
$\Rightarrow$ mol Zn (theoretical) $\Rightarrow$ mol Zn (actual) $\Rightarrow$ g Zn $\Rightarrow$ kg Zn

$$? \text{ kg Zn} = 454 \text{ kg ZnS in ore} \times \frac{89.6 \text{ kg ZnS}}{100 \text{ kg ZnS in ore}} \times \frac{1000 \text{ g ZnS}}{1 \text{ kg ZnS}} \times \frac{1 \text{ mol ZnS}}{97.5 \text{ g ZnS}} \times \frac{2 \text{ mol ZnO}}{2 \text{ mol ZnS}}$$

$$\times \frac{1 \text{ mol ZnSO}_4}{1 \text{ mol ZnO}} \times \frac{2 \text{ mol Zn}}{2 \text{ mol ZnSO}_4} \times \frac{0.922 \text{ mol Zn (actual)}}{1 \text{ mol Zn (theoretical)}} \times \frac{65.39 \text{ g Zn}}{1 \text{ mol Zn}} \times \frac{1 \text{ kg Zn}}{1000 \text{ g Zn}} = \textbf{252 kg Zn}$$

3-58. ***Refer to Section 3-6 and Example 3-16.***

We know $$D\left(\frac{g}{mL}\right) = \frac{g\ soln}{mL\ soln} \quad \text{and} \quad \%\ by\ mass = \frac{g\ (NH_4)_2SO_4}{g\ soln} \times 100\%$$

$$?\ g\ (NH_4)_2SO_4 = 775.0\ mL\ soln \times \frac{1.10\ g\ soln}{1.00\ mL\ soln} \times \frac{18.0\ g\ (NH_4)_2SO_4}{100\ g\ soln} = \mathbf{153\ g\ (NH_4)_2SO_4}$$

3-60. ***Refer to Section 3-6, Exercise 3-58, and Example 3-17.***

$$?\ mL\ soln = 135\ g\ (NH_4)_2SO_4 \times \frac{100\ g\ soln}{18.0\ g\ (NH_4)_2SO_4} \times \frac{1.00\ mL\ soln}{1.10\ g\ soln} = \mathbf{682\ mL\ soln}$$

3-62. ***Refer to Section 3-6 and Example 3-18.***

Method 1: Use the units of molarity as an equation: $M\left(\frac{mol}{L}\right) = \frac{mol\ substance}{L\ soln}$

Plan: $g\ Na_3PO_4 \overset{(1)}{\Rightarrow} mol\ Na_3PO_4 \overset{(2)}{\Rightarrow} M\ Na_3PO_4$

(1) $?\ mol\ Na_3PO_4 = \frac{g\ Na_3PO_4}{FW\ Na_3PO_4} = \frac{355\ g}{163.94\ g/mol} = 2.17\ mol\ Na_3PO_4$

(2) $?\ M\ Na_3PO_4 = \frac{mol\ Na_3PO_4}{L\ soln} = \frac{2.17\ mol}{4.50\ L} = \mathbf{0.481\ \mathit{M}\ Na_3PO_4}$

Method 2: Dimensional Analysis

$$?\ M\ Na_3PO_4 = \frac{355\ g\ Na_3PO_4}{4.50\ L\ soln} \times \frac{1\ mol\ Na_3PO_4}{163.94\ g\ Na_3PO_4} = \mathbf{0.481\ \mathit{M}\ Na_3PO_4}$$

3-64. ***Refer to Section 3-6 and Example 3-19.***

(a) *Method 1*: Plan: $M,\ L\ Na_2HPO_4\ soln \overset{(1)}{\Rightarrow} mol\ Na_2HPO_4 \overset{(2)}{\Rightarrow} g\ Na_2HPO_4$

We know $M\ Na_2HPO_4 = \frac{mol\ Na_2HPO_4}{L\ soln}$

(1) $?\ mol\ Na_2HPO_4 = M\ Na_2HPO_4 \times L\ soln = 0.40\ M \times 0.25\ L = 0.10\ mol\ Na_2HPO_4$

(2) $?\ g\ Na_2HPO_4 = mol\ Na_2HPO_4 \times FW\ Na_2HPO_4 = 0.10\ mol \times 142.0\ g/mol = \mathbf{14\ g\ Na_2HPO_4}$

Note: To minimize rounding errors, keep all your numbers in your calculator until the end, then round to the appropriate number of significant figures.

Method 2: Dimensional Analysis

$$?\ g\ Na_2HPO_4 = 0.25\ L\ soln \times \frac{0.40\ mol\ Na_2HPO_4}{1\ L} \times \frac{142\ g\ Na_2HPO_4}{1\ mol\ Na_2HPO_4} = \mathbf{14\ g\ Na_2HPO_4}$$

(b) same as (a)

3-66. *Refer to Section 3-6 and Example 3-20.*

(a) % by mass $= \dfrac{\text{g CaCl}_2}{\text{g soln}} \times 100\% = \dfrac{16.0 \text{ g CaCl}_2}{64.0 \text{ g H}_2\text{O} + 16.0 \text{ g CaCl}_2} \times 100\% = \textbf{20.0\% CaCl}_2$

(b) Assume we have 1 liter of solution

$$\text{Plan: } 1 \text{ L soln} \overset{(1)}{\Rightarrow} \text{g soln in 1 L} \overset{(2)}{\Rightarrow} \text{g CaCl}_2 \text{ in 1 L} \overset{(3)}{\Rightarrow} \text{mol CaCl}_2 \text{ in 1 L} = M \text{ CaCl}_2$$

(1) ? g soln in 1 L $= 1000 \text{ mL} \times \dfrac{1.180 \text{ g}}{\text{mL}} = 1180 \text{ g soln}$

(2) ? g CaCl$_2$ in 1 L $= 1180 \text{ g soln} \times 20.0\% \text{ CaCl}_2 = 236 \text{ g CaCl}_2$

(3) ? mol CaCl$_2$ in 1 L $= \dfrac{236 \text{ g CaCl}_2 \text{ in 1 L soln}}{111 \text{ g/mol}} = 2.13 \text{ mol CaCl}_2 \text{ in 1 L} = \textbf{2.13 } \boldsymbol{M} \textbf{ CaCl}_2$

3-68. *Refer to Section 3-6 and Example 3-18.*

$$\textit{Method 1}: \text{ Plan: } \text{g BaCl}_2\cdot 2\text{H}_2\text{O} \overset{(1)}{\Rightarrow} \text{mol BaCl}_2\cdot 2\text{H}_2\text{O} \overset{(2)}{=} \text{mol BaCl}_2 \overset{(3)}{\Rightarrow} M \text{ BaCl}_2$$

(1) ? mol BaCl$_2\cdot$2H$_2$O $= \dfrac{\text{g BaCl}_2\cdot 2\text{H}_2\text{O}}{\text{FW BaCl}_2\cdot 2\text{H}_2\text{O}} = \dfrac{1.72 \text{ g}}{244 \text{ g/mol}} = 0.00705 \text{ mol BaCl}_2\cdot\cdot 2\text{H}_2\text{O}$

(2) ? mol BaCl$_2$ $=$ mol BaCl$_2\cdot$2H$_2$O $= 0.00705$ mol BaCl$_2$

(3) ? M BaCl$_2$ $= \dfrac{\text{mol BaCl}_2}{\text{L soln}} = \dfrac{0.00705 \text{ mol}}{0.650 \text{ L}} = \textbf{0.0108 } \boldsymbol{M} \textbf{ BaCl}_2$

Method 2:   Dimensional Analysis

$$? M \text{ BaCl}_2 = \dfrac{1.72 \text{ g BaCl}_2\cdot 2\text{H}_2\text{O}}{650. \text{ mL soln}} \times \dfrac{1000 \text{ mL}}{1 \text{ L}} \times \dfrac{1 \text{ mol BaCl}_2\cdot 2\text{H}_2\text{O}}{244 \text{ g BaCl}_2\cdot 2\text{H}_2\text{O}} \times \dfrac{1 \text{ mol BaCl}_2}{1 \text{ mol BaCl}_2\cdot 2\text{H}_2\text{O}} = \textbf{0.0108 } \boldsymbol{M} \textbf{ BaCl}_2$$

3-70. *Refer to Section 3-6, Example 3-20, and Exercise 3-66 Solution.*

$$? M \text{ HF} = \dfrac{1000 \text{ mL soln}}{1 \text{ L}} \times \dfrac{1.17 \text{ g soln}}{1 \text{ mL soln}} \times \dfrac{49.0 \text{ g HF}}{100 \text{ g soln}} \times \dfrac{1 \text{ mol HF}}{20.0 \text{ g HF}} = \textbf{28.7 } \boldsymbol{M} \textbf{ HF}$$

3-72. *Refer to Section 3-7 and Example 3-21.*

For a dilution problem, $M_1 \times V_1 = M_2 \times V_2$

Therefore, $V_1 = \dfrac{M_2 \times V_2}{M_1} = \dfrac{1.50 \, M \times 2.50 \text{ L}}{12.0 \, M} = \textbf{0.312 L conc. HCl soln}$

3-74. *Refer to Section 3-7 and Example 3-21.*

For a dilution problem, $M_1 \times V_1 = M_2 \times V_2$, therefore, $M_2 = M_1 \times \dfrac{V_1}{V_2}$

(a) $M_2 = M_1 \times \dfrac{V_1}{V_2} = 0.500 \times \dfrac{1.00 \text{ mL}}{1.00 \times 10^3 \text{ mL}} = 0.000500 \, M$ solution remaining in flask

(b) $M_2 = 0.500 \, M \times \dfrac{1.00 \text{ mL}}{10.00 \text{ mL}} \times \dfrac{1.00 \text{ mL}}{10.00 \text{ mL}} \times \dfrac{1.00 \text{ mL}}{10.00 \text{ mL}} = 0.000500 \, M$ solution remaining in flask

Remember that V_2 is the total volume = 1.00 mL in flask + 9.00 mL of solvent added = 10.00 mL

(c) Both methods resulted in the same solute concentration for the 1.00 mL remaining in the volumetric flask. However, the single rinse used 1.00 L of solvent, whereas the three rinses used only 27.00 mL of solvent.

(d) The triple rinse technique is much more economical than the single rinse and still gives the same result.

3-76. *Refer to Section 3-7 and Example 3-21.*

For a dilution problem, $M_1 \times V_1 = M_2 \times V_2$

Therefore, $V_2 = \dfrac{M_1 \times V_1}{M_2} = \dfrac{12.0\ M \times 100.\ mL}{5.20\ M} =$ **231 mL NaOH soln**

3-78. *Refer to Section 3-8, and Example 3-23.*

Balanced equation: $KOH + CH_3COOH \rightarrow KCH_3COO + H_2O$

Plan: g $CH_3COOH \overset{(1)}{\Rightarrow}$ mol $CH_3COOH \overset{(2)}{\Rightarrow}$ mol KOH $\overset{(3)}{\Rightarrow}$ L KOH soln

Method 1:

(1) ? mol $CH_3COOH = \dfrac{g\ CH_3COOH}{FW\ CH_3COOH} = \dfrac{0.385\ g}{60.05\ g/mol} = 6.41 \times 10^{-3}$ mol CH_3COOH

(2) ? mol $KOH = 6.41 \times 10^{-3}$ mol $CH_3COOH \times (1\ mol\ KOH/1\ mol\ CH_3COOH) = 6.41 \times 10^{-3}$ mol KOH

(3) ? L KOH soln $= \dfrac{mol\ KOH}{M\ KOH} = \dfrac{6.41 \times 10^{-3}\ mol\ KOH}{0.247\ M} =$ **0.0260 L KOH soln**

Method 2: Dimensional Analysis

? L KOH $= 0.385\ g\ CH_3COOH \times \dfrac{1\ mol\ CH_3COOH}{60.05\ g\ CH_3COOH} \times \dfrac{1\ mol\ KOH}{1\ mol\ CH_3COOH} \times \dfrac{1\ L\ soln}{0.247\ mol\ KOH}$

= **0.0260 L KOH soln**

3-80. *Refer to Section 3-8 and Example 3-24.*

Balanced equation: $Ba(OH)_2 + 2HNO_3 \rightarrow Ba(NO_3)_2 + 2H_2O$

Plan: M, L $Ba(OH)_2$ soln $\overset{(1)}{\Rightarrow}$ mol $Ba(OH)_2 \overset{(2)}{\Rightarrow}$ mol $HNO_3 \overset{(3)}{\Rightarrow}$ L HNO_3 soln

Method 1:

(1) ? mol $Ba(OH)_2 = 0.0515\ M \times 0.04555\ L = 2.35 \times 10^{-3}$ mol $Ba(OH)_2$

(2) ? mol $HNO_3 = 2.35 \times 10^{-3}$ mol $Ba(OH)_2 \times (2\ mol\ HNO_3/1\ mol\ Ba(OH)_2) = 4.69 \times 10^{-3}$ mol HNO_3

(3) ? L $HNO_3 = \dfrac{mol\ HNO_3}{M\ HNO_3} = \dfrac{4.69 \times 10^{-3}\ mol}{0.446\ M} =$ **0.0105 L or 10.5 mL HNO_3 soln**

Method 2: Dimensional Analysis

? L $HNO_3 = 0.04555\ L\ Ba(OH)_2 \times \dfrac{0.0515\ mol\ Ba(OH)_2}{1\ L\ soln} \times \dfrac{2\ mol\ HNO_3}{1\ mol\ Ba(OH)_2} \times \dfrac{1\ L\ soln}{0.446\ mol\ HNO_3}$

= **0.0105 L or 10.5 mL HNO_3 soln**

Balanced equation: $AlCl_3 + 3AgNO_3 \rightarrow 3AgCl + Al(NO_3)_3$

Plan: $g \ AgCl \overset{(1)}{\Rightarrow} mol \ AgCl \overset{(2)}{\Rightarrow} mol \ AlCl_3 \overset{(3)}{\Rightarrow} M \ AlCl_3$

Recall: $M = \dfrac{\text{mol substance}}{\text{L soln}}$

Method 1:

(1) $? \ mol \ AgCl = \dfrac{g \ AgCl}{FW \ AgCl} = \dfrac{0.215 \ g}{143.3 \ g/mol} = 1.50 \times 10^{-3} \ mol \ AgCl$

(2) $? \ mol \ AlCl_3 = mol \ AgCl \times 1 \ mol \ AlCl_3/3 \ mol \ AgCl = 5.00 \times 10^{-4} \ mol \ AlCl_3$

(3) $? \ M \ AlCl_3 = \dfrac{mol \ AlCl_3}{L \ soln} = \dfrac{5.00 \times 10^{-4} \ mol \ AlCl_3}{0.1105 \ L \ soln} = \mathbf{0.00453 \ \textit{M} \ AlCl_3 \ soln}$

Method 2: Dimensional Analysis

$? \ mol \ AlCl_3 = 0.215 \ g \ AgCl \times \dfrac{1 \ mol \ AgCl}{143.3 \ g \ AgCl} \times \dfrac{1 \ mol \ AlCl_3}{3 \ mol \ AgCl} = 5.00 \times 10^{-4} \ mol \ AlCl_3$

$? \ M \ AlCl_3 = \dfrac{mol \ AlCl_3}{L \ soln} = \dfrac{5.00 \times 10^{-4} \ mol \ AlCl_3}{0.1105 \ L \ soln} = \mathbf{0.00453 \ \textit{M} \ AlCl_3 \ soln}$

Balanced equation: $Fe_3O_4 + 2C \rightarrow 3Fe + 2CO_2$

Plan: $g \ Fe \Rightarrow mol \ Fe \Rightarrow mol \ Fe_3O_4 \Rightarrow g \ Fe_3O_4 \Rightarrow \%Fe_3O_4 \ in \ ore$

$? \ g \ Fe_3O_4 = 3.49 \ g \ Fe \ in \ ore \times \dfrac{1 \ mol \ Fe}{55.85 \ g \ Fe} \times \dfrac{1 \ mol \ Fe_3O_4}{3 \ mol \ Fe} \times \dfrac{231.55 \ g \ Fe_3O_4}{1 \ mol \ Fe_3O_4} = 4.82 \ g \ Fe_3O_4$

$? \ \%Fe_3O_4 \ in \ ore = \dfrac{4.82 \ g \ Fe_3O_4}{75.0 \ g \ ore} \times 100\% = \mathbf{6.43\% \ Fe_3O_4 \ in \ ore}$

Balanced equation: $2KBr + Cl_2 \rightarrow 2KCl + Br_2$

Method 1: Plan: $g \ Cl_2 \overset{(1)}{\Rightarrow} mol \ Cl_2 \overset{(2)}{\Rightarrow} mol \ Br_2 \overset{(3)}{\Rightarrow} g \ Br_2$

(1) $? \ mol \ Cl_2 = \dfrac{g \ Cl_2}{FW \ Cl_2} = \dfrac{0.631 \ g}{70.9 \ g/mol} = 8.90 \times 10^{-3} \ mol \ Cl_2$

(2) $? \ mol \ Br_2 = mol \ Cl_2 = 8.90 \times 10^{-3} \ mol \ Br_2$

(3) $? \ g \ Br_2 = mol \ Br_2 \times FW \ Br_2 = (8.90 \times 10^{-3} \ mol) \times 159.8 \ g/mol = \mathbf{1.42 \ g \ Br_2}$

Method 2: Dimensional Analysis (The unit factors correspond to the steps in Method 1.)

$$\text{Step 1} \qquad \text{Step 2} \qquad \text{Step 3}$$

$? \ g \ Br_2 = 0.631 \ g \ Cl_2 \times \dfrac{1 \ mol \ Cl_2}{70.9 \ g \ Cl_2} \times \dfrac{1 \ mol \ Br_2}{1 \ mol \ Cl_2} \times \dfrac{159.8 \ g \ Br_2}{1 \ mol \ Br_2} = \mathbf{1.42 \ g \ Br_2}$

Method 3: Proportion or Ratio Method

$$\frac{?\ g\ Br_2}{g\ Cl_2} = \frac{1 \times FW\ Br_2}{1 \times FW\ Cl_2}$$

Solving, $?\ g\ Br_2 = g\ Cl_2 \times \dfrac{1 \times FW\ Br_2}{1 \times FW\ Cl_2} = 0.631\ g \times \dfrac{159.8\ g}{70.9\ g} = \mathbf{1.42\ g\ Br_2}$

3-88. **_Refer to Section 1-5._**

A chemical reaction involves a chemical change in which

(1) one or more substances are used up (at least partially),
(2) one or more new substances are formed and
(3) energy is absorbed or released.

Here is an example of the chemical reaction, presented in Exercise 3-4:

$$2H_2(g) + O_2(g) \rightarrow 2H_2O(g)$$

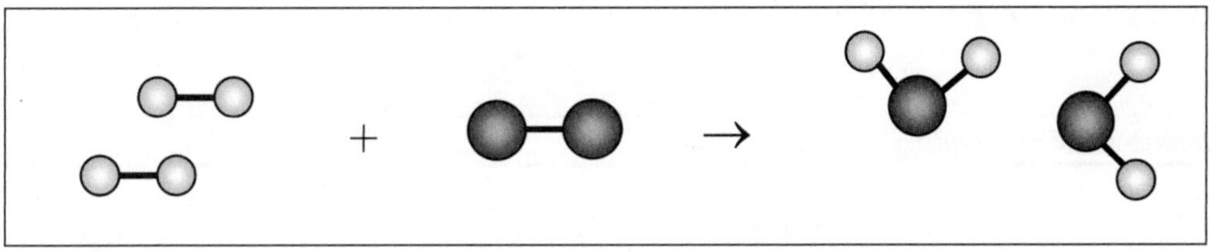

3-90. **_Refer to Sections 3-1 and 3-6._**

A 1.0% solution means there is 1.0 g of solute in 100 g of solution. Since the densities of the solutions are assumed to be nearly identical, as the solute decreases in formula weight, the moles (g/FW) of the solute will increase and the molarity of the solution will increase. In order of increasing molarity:

1.0% $SnCl_2$ (FW=189.6 g/mol) < 1.0% $AlCl_3$ (FW=133.3 g/mol) < 1.0% NaCl (FW=58.4 g/mol)

3-92. **_Refer to Sections 3-2 and 3-6, and Examples 3-6 and 3-16._**

Balanced equation: $Bi + 4HNO_3 + 3H_2O \rightarrow Bi(NO_3)_3 \cdot 5H_2O + NO$

Plan: (1) Calculate the grams of HNO_3 required to react with the Bi,
    (2) Determine the volume of HNO_3 that will contain the desired mass of HNO_3 using the unit factors for % by mass and density.

(1) $?\ g\ HNO_3 = 20.0\ g\ Bi \times \dfrac{1\ mol\ Bi}{209.0\ g\ Bi} \times \dfrac{4\ mol\ HNO_3}{1\ mol\ Bi} \times \dfrac{63.02\ g\ HNO_3}{1\ mol\ HNO_3} = 24.1\ g\ HNO_3$

(2) $?\ mL\ HNO_3 = 24.1\ g\ HNO_3 \dfrac{100\ g\ soln}{30.0\ g\ HNO_3} \times \dfrac{1\ mL\ soln}{1.182\ g\ soln} = \mathbf{68.0\ mL\ HNO_3}$

41

The concentration of particles in each container is determined by counting the number of dissolved particles and dividing by the volume of solution.

Solution	Concentration
A	12 particles/0.5 L = 24 particles/L
B	6 particles/0.5 L = 12 particles/L
C	3 particles/0.5 L = 6 particles/L
D	8 particles/0.5 L = 16 particles/L
E	3 particles/0.25 L = 12 particles/L
F	5 particles/0.25 L = 20 particles/L

(a) **Solution A** is the most concentrated.

(b) **Solution C** is the least concentrated.

(c) **Solutions B and E** have the same concentration.

(d) When solutions E and F are combined, the resulting solution has the concentration:
8 particles/0.5 L = 32 particles/2 L = 16 particles/L, which is the same as **Solution D**.

3-96. *Refer to Sections 3-3 and 3-4, and Exercise 3-48 Solution.*

Balanced equation: $Zn(s) + 2AgNO_3(aq) \rightarrow Zn(NO_3)_2(aq) + 2Ag(s)$

(a) This problem involves both the limiting reactant and the percent yield concepts.

(1) Find the limiting reactant.
(2) Calculate the theoretical yield of Ag, based on the limiting reactant.
(3) Determine the percent yield of Ag.

(1) $? \text{ mol Zn} = \dfrac{\text{g Zn}}{\text{AW Zn}} = \dfrac{100.0 \text{ g}}{65.39 \text{ g/mol}} = 1.529 \text{ mol Zn}$

$? \text{ mol AgNO}_3 = \text{mol Ag}^+ = 1.330 \ M \text{ AgNO}_3 \times 1\text{L} = 1.330 \text{ mol AgNO}_3$

$\text{Required ratio} = \dfrac{1 \text{ mol Zn}}{2 \text{ mol AgNO}_3} = 0.5$ $\qquad\qquad \text{Available ratio} = \dfrac{1.529 \text{ mol Zn}}{1.330 \text{ mol AgNO}_3} = 1.150$

Available ratio > required ratio; $AgNO_3$ is the limiting reactant.

(2) The mass of Ag produced from the limiting reactant, $AgNO_3$, is the theoretical yield.

$? \text{ g Ag (theoretical yield)} = 1.330 \text{ mol AgNO}_3 \times \dfrac{2 \text{ mol Ag}}{2 \text{ mol AgNO}_3} \times \dfrac{107.9 \text{ g Ag}}{1 \text{ mol Ag}} = 143.5 \text{ g Ag}$

(3) % yield of Ag $= \dfrac{\text{actual yield}}{\text{theoretical yield}} \times 100\% = \dfrac{90.0 \text{ g}}{143.5 \text{ g}} \times 100\% = \textbf{62.7\%}$

(b) The yield might be less than 100% because (1) the zinc might not be pure or (2) the technician or student who collected the silver may have been careless. She or he may have not recovered all of it or even may have spilled some. (3) Some of the zinc may not have reacted because it became coated with silver, or (4) some of the zinc may not have been added to the reaction vessel in the first place.

Balanced equation: CH_3COOH + CH_3CH_2OH → $CH_3COOCH_2CH_3$ + H_2O

acetic acid   ethanol   ethyl acetate

(ethyl alcohol)

FW (g/mol)	60.05	46.07	88.10
Density (g/mL)	1.05	0.789	0.902
	(from Table 1-2)	(from Table 1-9)	

(a) Plan: (1) Convert the volume of reactants to the mass of reactants.

         (2) Evaluate the limiting reactant.

(1) ? g CH_3COOH = 20.2 mL CH_3COOH x 1.05 g/mL = 21.2 g CH_3COOH

? g CH_3CH_2OH = 20.1 mL CH_3CH_2OH x 0.789 g/mL = 15.9 g CH_3CH_2OH

(2) Convert the mass of reactants to moles and compare the required ratio to the available ratio.

$$? \text{ mol } CH_3COOH = \frac{g\ CH_3COOH}{FW\ CH_3COOH} = \frac{21.2\ g}{60.05\ g/mol} = 0.353 \text{ mol } CH_3COOH$$

$$? \text{ mol } CH_3CH_2OH = \frac{g\ CH_3CH_2OH}{FW\ CH_3CH_2OH} = \frac{15.9\ g}{46.07\ g/mol} = 0.345 \text{ mol } CH_3CH_2OH$$

$$\text{Required ratio} = \frac{1\ \text{mol } CH_3CH_2OH}{1\ \text{mol } CH_3COOH} = 1 \qquad \text{Available ratio} = \frac{0.345\ \text{mol } CH_3CH_2OH}{0.353\ \text{mol } CH_3COOH} = 0.977$$

Available ratio < required ratio; **CH_3CH_2OH is the limiting reactant**.

(b) Plan: (1) Calculate the actual yield of ethyl acetate, $CH_3COOCH_2CH_3$, in grams.

         (2) Calculate the theoretical yield of ethyl acetate from 0.345 mol CH_3CH_2OH.

         (3) Calculate the percent yield of ethyl acetate.

(1) ? g ethyl acetate = 27.5 mL ethyl acetate x 0.902 g/mL = 24.8 g ethyl acetate (actual yield)

(2) ? mol ethyl acetate = mol CH_3CH_2OH x 1 = 0.345 mol ethyl acetate

? g ethyl acetate = mol ethyl acetate x FW ethyl acetate = 0.345 mol x 88.1 g/mol

         = 30.4 g ethyl acetate (theoretical yield)

$$(3)\ ? \text{ \% yield ethyl acetate} = \frac{\text{actual yield}}{\text{theoretical yield}} \times 100\% = \frac{24.8\ g}{30.4\ g} \times 100\% = \textbf{81.6\%}$$

Plan: This is a complex dilution problem.

    (1) Find the moles of NaCl in each solution

    (2) Add the moles of NaCl together and divide by the total volume to get the new molarity.

(1)   mol NaCl in Solution 1 = 0.375 M x 0.0350 L = 0.0131 mol NaCl

       mol NaCl in Solution 2 = 0.632 M x 0.0475 L = 0.0300 mol NaCl

$$(2)\quad ?\ M\,\text{NaCl} = \frac{\text{total moles of NaCl}}{\text{total volume}} = \frac{(0.0131\ \text{mol} + 0.0300\ \text{mol})}{(0.0350\ L + 0.0475\ L)} = \textbf{0.522}\ \textit{M}\,\textbf{NaCl}$$

3-102. *Refer to Section 3-5 and Example 3-13.*

(a) Balanced equations: $P_4 + 5O_2 \rightarrow P_4O_{10}$ 89.5% efficient

$P_4O_{10} + 6H_2O \rightarrow 4H_3PO_4$ 97.8% efficient

(b) Plan: g P_4 $\Rightarrow$ mol P_4 $\Rightarrow$ mol P_4O_{10} (theoretical) $\Rightarrow$ mol P_4O_{10} (actual) $\Rightarrow$ mol H_3PO_4 (theoretical)

$\Rightarrow$ mol H_3PO_4 (actual) $\Rightarrow$ g H_3PO_4

$? \text{ g } H_3PO_4 = 225 \text{ g } P_4 \times \dfrac{1 \text{ mol } P_4}{124 \text{ g } P_4} \times \dfrac{1 \text{ mol } P_4O_{10} \text{ (theoretical)}}{1 \text{ mol } P_4} \times \dfrac{89.5 \text{ mol } P_4O_{10} \text{ (actual)}}{100. \text{ mol } P_4O_{10} \text{ (theoretical)}}$

$\times \dfrac{4 \text{ mol } H_3PO_4 \text{ (theoretical)}}{1 \text{ mol } P_4O_{10}} \times \dfrac{97.8 \text{ mol } H_3PO_4 \text{ (actual)}}{100. \text{ mol } H_3PO_4 \text{ (theoretical)}} \times \dfrac{98.0 \text{ g } H_3PO_4 \text{ (actual)}}{1 \text{ mol } H_3PO_4}$

$= \textbf{623 g } \textbf{H}_3\textbf{PO}_4$

3-104. *Refer to Sections 3-3 and 3-8.*

Balanced equation: $NaCl(aq) + AgNO_3(aq) \rightarrow AgCl(s) + NaNO_3(aq)$

Plan: (1) Calculate the mass (in grams) of NaCl in the 1.20% solution using density.
(2) Determine the limiting reactant.
(3) Determine the mass of AgCl produced.

(1) $? \text{ g NaCl} = 10.0 \text{ mL NaCl} \times \dfrac{1.02 \text{ g soln}}{1 \text{ mL soln}} \times \dfrac{1.20 \text{ g NaCl}}{100 \text{ g soln}} = 0.122 \text{ g NaCl}$

(2) Convert each reactant to moles and compare required ratio to available ratio to find the limiting reactant.

$? \text{ mol NaCl} = \dfrac{\text{g NaCl}}{\text{FW NaCl}} = \dfrac{0.122 \text{ g}}{58.4 \text{ g/mol}} = 0.00209 \text{ mol NaCl}$

$? \text{ mol AgNO}_3 = M \text{ AgNO}_3 \times L \text{ soln} = (1.21 \times 10^{-2} \ M) \times 0.0500 \text{ L} = 6.05 \times 10^{-4} \text{ mol AgNO}_3$

$\text{Required ratio} = \dfrac{1 \text{ mol NaCl}}{1 \text{ mol AgNO}_3} = 1$ $\text{Available ratio} = \dfrac{0.00209 \text{ mol NaCl}}{6.05 \times 10^{-4} \text{ mol AgNO}_3} = 3.45$

Available ratio > required ratio; $AgNO_3$ is the limiting reactant.

(3) $? \text{ g AgCl} = 0.0500 \text{ L} \times \dfrac{1.21 \times 10^{-2} \text{ mol AgNO}_3}{1 \text{ L soln}} \times \dfrac{1 \text{ mol AgCl}}{1 \text{ mol AgNO}_3} \times \dfrac{143.3 \text{ g AgCl}}{1 \text{ mol AgCl}} = \textbf{0.0867 g AgCl}$

3-106. *Refer to Section 3-2.*

Balanced equations: $Zn + 2HCl \rightarrow ZnCl_2 + H_2$

$2Al + 6HCl \rightarrow 2AlCl_3 + 3H_2$

Assume: (1) 1 mol of H_2 is produced.
(2) Zn costs \$1.00/g Zn; Al costs \$2.00/g Al

Plan: mol H_2 $\Rightarrow$ mol metal $\Rightarrow$ g metal $\Rightarrow$ \$ required/mol H_2

(1) For Zn: $1 \text{ mol } H_2 \times \dfrac{1 \text{ mol Zn}}{1 \text{ mol } H_2} \times \dfrac{65.4 \text{ g Zn}}{1 \text{ mol Zn}} \times \dfrac{\$1.00}{1 \text{ g Zn}} = \textbf{\$65.4/mol } \textbf{H}_2$

(2) For Al: $1 \text{ mol } H_2 \times \dfrac{2 \text{ mol Al}}{3 \text{ mol } H_2} \times \dfrac{27.0 \text{ g Al}}{1 \text{ mol Al}} \times \dfrac{\$2.00}{1 \text{ g Al}} = \textbf{\$36.0/mol } \textbf{H}_2$

Therefore, Al is less expensive for the production of equal amounts of hydrogen gas.

4 Some Types of Chemical Reactions

4-2. *Refer to Section 4-1 and Figure 4-1.*

Mendeleev arranged the known elements in order of increasing atomic weight in sequence so that elements with similar chemical and physical properties fell in the same column or group. To achieve this chemical periodicity, it was necessary for Mendeleev to leave blank spaces for elements undiscovered at that time and to make assumptions concerning atomic weights not known with certainty.

The modern periodic table has elements arranged in order of increasing atomic number so that elements with similar chemical properties fall in the same column.

4-4. *Refer to Section 4-1.*

The atomic weight of an element is a weighted average of the mass of the naturally occurring isotopes of that element. Therefore, the atoms in a naturally occurring sample of argon must be heavier than the atoms in a naturally occurring sample of potassium. Atoms of argon have 18 protons, whereas atoms of potassium have 19 protons. In order for the atomic weight of argon to be greater than that of potassium, argon atoms must have more neutrons.

Consider the isotopes for these elements:

	Isotope	Percent Composition	Number of Protons	Number of Neutrons
argon (AW: 39.948 amu)	$^{40}_{18}Ar$	99.60%	18	22
potassium (AW: 39.0983 amu)	$^{39}_{19}K$	93.1%	19	20
	$^{41}_{19}K$	6.88%	19	22

From these data, we can see that argon would have a higher atomic weight than potassium.

4-6. *Refer to Section 4-1, the Periodic Table and the Handbook of Chemistry and Physics.*

If we look at the densities of the elements in reference to where they are on the periodic table with reference to selenium, we see that the density of Se should be between 2.07 (S) and 6.24 (Te) and between 3.12 (Br) and 5.72 (As). We can estimate it to be the average of the densities of the surrounding elements

	S 2.07 g/cm^3	
As 5.72 g/cm^3	**Se**	**Br** 3.12 g/cm^3
	Te 6.24 g/cm^3	

$$D \text{ of Se} = \frac{2.07 + 6.24 + 5.72 + 3.12}{4} = 4.29 \text{ g/cm}^3$$

The Handbook of Chemistry and Physics gives the density of Se as 4.81 g/cm^3. This averaging method gave us a reasonable estimation of the density with the following relative error:

$$\% \text{ error in the density of Se} = \frac{4.81 - 4.29}{4.81} \times 100$$
$$= 11\%$$

The periodic trends of the element properties also apply to compound containing the elements. Therefore, the melting points of CF_4, CCl_4, CBr_4 and CI_4 should follow a trend.

If we graph the melting points of these compounds versus molecular weight, we can estimate the melting point of CBr_4.

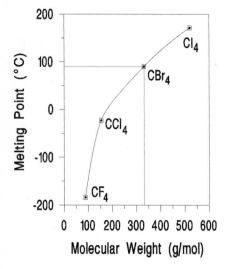

Compound	MW (g/mol)	MP (°C)
CF_4	88.0	−184
CCl_4	153.8	−23
CBr_4	331.6	?
CI_4	519.6	171

The estimated melting point of CBr_4 is about **90°C**. The actual value is 90.1°C according to the *Handbook of Chemistry and Physics*.

Hydride formulas are related to the group number of the central element, e.g.,

2A	3A	4A	5A	6A	7A
BeH_2	BH_3	CH_4	NH_3	H_2O	HF

Arsine, the hydride of arsenic, has the formula AsH_x. Arsenic, the central element, is a 5A element. Therefore, its structure should be similar to NH_3 and is predicted to be AsH_3.

(a) alkaline earth metals: beryllium (Be), magnesium (Mg), calcium (Ca), strontium (Sr), barium (Ba) and radium (Ra)

(b) Group 4A elements: carbon (C), silicon (Si), germanium (Ge), tin (Sn) and lead (Pb)

(c) Group 2B elements: zinc (Zn), cadmium (Cd), mercury (Hg) and ununquadium (Uuq).

(a) Metals are the elements below and to the left of the stepwise division (metalloids) in the upper right corner of the periodic table. They possess metallic bonding. Approximately 80% of the known elements are metals, including potassium (K), calcium (Ca), scandium (Sc) and vanadium (V).

(b) Nonmetals are the elements above and to the right of the metalloids in the periodic table, including carbon (C), nitrogen (N), sulfur (S) and chlorine (Cl).

(c) The halogens, meaning "salt-formers," are the elements of Group 7A. They include fluorine (F), chlorine (Cl), bromine (Br), iodine (I) and astatine (At).

4-16. *Refer to Section 4-2.*

Three major classes of compounds are electrolytes:

	Strong Electrolytes	**Weak Electrolytes**
(1) acids	HCl, $HClO_4$	CH_3COOH, HF
(2) soluble bases	$NaOH$, $Ba(OH)_2$	NH_3, $(CH_3)_3N$
(3) soluble salts	$NaCl$, KNO_3	$Pb(CH_3COO)_2$*

* This is one of the very few soluble salts that is a weak electrolyte.

Therefore, the three classes of compounds which are *strong* electrolytes are strong acids, strong soluble bases and soluble salts.

4-18. *Refer to Sections 4-2 and 4-11.*

A salt is a compound that contains a cation other than H^+ and an anion other than the hydroxide ion, OH^-, or the oxide ion, O^{2-}. A salt is a product of the reaction between a particular acid and base and consists of the cation of the base and the anion of the acid. For example,

$$NaOH + HCl \rightarrow NaCl + H_2O$$
$$\textbf{base} \qquad \textbf{acid} \qquad \textbf{salt}$$

4-20. *Refer to Section 4-2.*

(a) $HCl(aq) \rightarrow H^+(aq) + Cl^-(aq)$

(b) $HNO_3(aq) \rightarrow H^+(aq) + NO_3^-(aq)$

(c) $HClO_2(aq) \rightleftarrows H^+(aq) + ClO_2^-(aq)$

4-22. *Refer to Section 4-2 and Table 4-7.*

Common strong soluble bases include:

lithium hydroxide	LiOH	calcium hydroxide	$Ca(OH)_2$
sodium hydroxide	NaOH	strontium hydroxide	$Sr(OH)_2$
potassium hydroxide	KOH	barium hydroxide	$Ba(OH)_2$
rubidium hydroxide	RbOH		
cesium hydroxide	CsOH		

4-24. *Refer to Section 4-2.*

Household ammonia is the most common weak base.
It ionizes as follows:

$$NH_3(aq) + H_2O(\ell) \rightleftarrows NH_4^+(aq) + OH^-(aq)$$

4-26. *Refer to Section 4-2 and the Key Terms at the end of Chapter 4.*

Ionization refers to the process in which a molecular compound, such as HCl, separates or reacts with water to form ions in solution. Dissociation refers to the process in which a solid ionic compound, such as NaCl, separates into its ions in aqueous solution.

4-28. *Refer to Section 4-2.*

(a) Na_2S is a soluble salt and is a strong electrolyte: $Na_2S(aq) \rightarrow 2Na^+(aq) + S^{2-}(aq)$.

(b) $Ba(OH)_2$ is a strong base and is a strong electrolyte: $Ba(OH)_2(aq) \rightarrow Ba^{2+}(aq) + 2OH^-(aq)$.

(c) CH_3OH, methanol, is a nonelectrolyte.

(d) HCN is a weak acid and is a weak electrolyte.

(e) $Al(NO_3)_3$ is a soluble salt and is a strong electrolyte: $Al(NO_3)_3(aq) \rightarrow Al^{3+}(aq) + 3NO_3^-(aq)$.

4-30. *Refer to Section 4-2, Table 4-8, and the Solubility Guidelines in Section 4-2.*

Ionic Substance	Soluble	Insoluble
chloride	NaCl, KCl	$AgCl$, Hg_2Cl_2
sulfate	Na_2SO_4, K_2SO_4	$BaSO_4$, $PbSO_4$
hydroxide	NaOH, KOH	$Cu(OH)_2$, $Mg(OH)_2$

4-32. *Refer to Sections 4-2 and Tables 4-5, 4-6 and 4-7.*

(a) The acids are: HI, H_2SeO_4, H_2S, H_3PO_3 and HCN.

(b) The soluble bases are: LiOH and NH_3.

4-34. *Refer to Section 4-2 and Table 4-9.*

NaCl	strong electrolyte	CsOH	strong electrolyte	NH_3	weak electrolyte
$MgSO_4$	strong electrolyte	HNO_3	strong electrolyte	KOH	strong electrolyte
HCl	strong electrolyte	HBr	strong electrolyte	$Mg(CH_3COO)_2$	strong electrolyte
$H_2C_2O_4$	weak electrolyte	$Ba(OH)_2$	strong electrolyte	HCN	weak electrolyte
$Ba(NO_3)_2$	strong electrolyte	LiOH	strong electrolyte	$HClO_4$	strong electrolyte
H_3PO_4	weak electrolyte	H_2O	weak electrolyte		

(a) strong acids: HCl, HNO_3, HBr, $HClO_4$

(b) strong bases: CsOH, $Ba(OH)_2$, LiOH, KOH

(c) weak acids: $H_2C_2O_4$ or $(COOH)_2$, H_3PO_4, HCN

(d) weak bases: NH_3

4-36. *Refer to Section 4-2, Example 4-3 and Table 4-8.*

soluble: $Ca(CH_3COO)_2$, NH_4Cl, $AgNO_3$ insoluble: $PbCl_2$ (except in hot water)

4-38. *Refer to Section 4-2.*

Acidic household "chemicals":   vinegar (dilute acetic acid)
                                            vitamin C (ascorbic acid)
                                            lemon juice (citric acid)

Basic household "chemicals":   ammonia
                                            drain cleaner (sodium hydroxide)
                                          Milk of Magnesia (magnesium hydroxide)

4-40. *Refer to Section 4-2.*

Many organic acids, like acetic acid, occur in living systems and are generally weak acids so as not to disrupt living cells. Note that stomach acid, HCl, is a strong acid, but the stomach tissue protects itself with a heavy mucous layer.

4-42. *Refer to Section 4-4 and Example 4-4.*

For a compound, the sum of the oxidation numbers of the component elements must be equal to zero.

(a) Let x = oxidation number of P

PCl_3 $0 = x + 3(\text{ox. no. Cl}) = x + 3(-1) = x - 3$
                  $x = +3$

P_2O_5 $0 = 2x + 5(\text{ox. no. O}) = 2x + 5(-2) = 2x - 10$
                  $x = +5$

P_4O_{10} $0 = 4x + 10(\text{ox. no. O}) = 4x + 10(-2) = 4x - 20$
                  $x = +5$

HPO_3 $0 = 1(\text{ox. no. H}) + x + 3(\text{ox. no. O}) = 1(+1) + x + 3(-2) = x - 5$
                  $x = +5$

H_3PO_3 $0 = 3(\text{ox. no. H}) + x + 3(\text{ox. no. O}) = 3(+1) + x + 3(-2) = x - 3$
                  $x = +3$

$POCl_3$ $0 = x + 1(\text{ox. no. O}) + 3(\text{ox. no. Cl}) = x + 1(-2) + 3(-1) = x - 5$
                  $x = +5$

$H_4P_2O_7$ $0 = 4(\text{ox. no. H}) + 2x + 7(\text{ox. no. O}) = 4(+1) + 2x + 7(-2) = 2x - 10$
                  $x = +5$

$Mg_3(PO_4)_2$ $0 = 3(\text{ox. no. Mg}) + 2x + 8(\text{ox. no. O}) = 3(+2) + 2x + 8(-2) = 2x - 10$
                  $x = +5$

(b) Let x = oxidation number of Br

Br^- $x = -1$

BrO^- $-1 = x + 1(\text{ox. no. O}) = x + 1(-2) = x - 2$
                  $x = +1$

BrO_2^- $-1 = x + 2(\text{ox. no. O}) = x + 2(-2) = x - 4$
                    $x = +3$

BrO_3^- $-1 = x + 3(\text{ox. no. O}) = x + 3(-2) = x - 6$
                    $x = +5$

BrO_4^- $-1 = x + 4(\text{ox. no. O}) = x + 4(-2) = x - 8$
                    $x = +7$

(c) Let x = oxidation number of Mn

MnO $\qquad$ $0 = x + 1(\text{ox. no. O}) = x + 1(-2) = x - 2$
$\qquad$ $x = +2$

MnO_2 $\qquad$ $0 = x + 2(\text{ox. no. O}) = x + 2(-2) = x - 4$
$\qquad$ $x = +4$

$Mn(OH)_2$ $\qquad$ $0 = x + 2(\text{ox. no. O}) + 2(\text{ox. no. H}) = x + 2(-2) + 2(+1) = x - 2$
$\qquad$ $x = +2$

K_2MnO_4 $\qquad$ $0 = 2(\text{ox. no. K}) + x + 4(\text{ox. no. O}) = 2(+1) + x + 4(-2) = x - 6$
$\qquad$ $x = +6$

$KMnO_4$ $\qquad$ $0 = 1(\text{ox. no. K}) + x + 4(\text{ox. no. O}) = 1(+1) + x + 4(-2) = x - 7$
$\qquad$ $x = +7$

Mn_2O_7 $\qquad$ $0 = 2x + 7(\text{ox. no. O}) = 2x + 7(-2) = 2x - 14$
$\qquad$ $x = +7$

(d) Let x = oxidation of O

OF_2 $\qquad$ $0 = x + 2(\text{ox. no. F}) = x + 2(-1) = x - 2$
$\qquad$ $x = +2$

Na_2O $\qquad$ $0 = 2(\text{ox. no. Na}) + x = 2(+1) + x = x + 2$
$\qquad$ $x = -2$

Na_2O_2 $\qquad$ $0 = 2(\text{ox. no. Na}) + 2x = 2(+1) + 2x = 2x + 2$
$\qquad$ $x = -1$

KO_2 $\qquad$ $0 = 1(\text{ox. no. K}) + 2x = 1(+1) + 2x = 2x + 1$
$\qquad$ $x = -1/2$

4-44. *Refer to Section 4-4 and Example 4-4.*

For an ion, the sum of the oxidation numbers of the component elements must equal the charge on the ion.

(a) Let x = oxidation number of N

N^{3-} $\qquad$ $x = -3$

NO_2^- $\qquad$ $-1 = x + 2(\text{ox. no. O}) = x + 2(-2) = x - 4$
$\qquad$ $x = +3$

NO_3^- $\qquad$ $-1 = x + 3(\text{ox. no. O}) = x + 3(-2) = x - 6$
$\qquad$ $x = +5$

N_3^- $\qquad$ $-1 = 3x$
$\qquad$ $x = -1/3$

NH_4^+ $\qquad$ $+1 = x + 4(\text{ox. no. H}) = x + 4(+1) = x + 4$
$\qquad$ $x = -3$

(b) Let x = oxidation number of Cl

Cl_2 $\qquad$ $0 = 2x$
$\qquad$ $x = 0$

HCl $\qquad$ $0 = 1(\text{ox. no. H}) + x = 1(+1) + x = x + 1$
$\qquad$ $x = -1$

HClO $\qquad$ $0 = 1(\text{ox. no. H}) + x + 1(\text{ox. no. O}) = 1(+1) + x + 1(-2) = x - 1$
$\qquad$ $x = +1$

50

$HClO_2$ $0 = 1(\text{ox. no. H}) + x + 2(\text{ox. no. O}) = 1(+1) + x + 2(-2) = x - 3$
$x = +3$

$KClO_3$ $0 = 1(\text{ox. no. K}) + x + 3(\text{ox. no. O}) = 1(+1) + x + 3(-2) = x - 5$
$x = +5$

Cl_2O_7 $0 = 2x + 7(\text{ox. no. O}) = 2x + 7(-2) = 2x - 14$
$x = +7$

$Ca(ClO_4)_2$ $0 = 1(\text{ox. no. Ca}) + 2x + 8(\text{ox. no. O}) = 1(+2) + 2x + 8(-2) = 2x - 14$
$x = +7$

PCl_5 $0 = x + 5(\text{ox. no. Cl}) = x + 5(-1) = x - 5$
$x = +5$

4-46.   *Refer to Section 4-5 and Table 4-11.*

(a) Li^+ lithium ion   (c) Ba^{2+} barium ion   (e) Ag^+ silver ion

(b) Au^{3+} gold(III) ion   (d) Zn^{2+} zinc ion

4-48.   *Refer to Sections 4-5 and 4-6, and Table 4-11.*

(a) bromide ion Br^- (c) telluride ion Te^{2-} (e) nitrite ion NO_2^-

(b) hydrogen sulfide ion HS^- (d) hydroxide ion OH^-

4-50.   *Refer to Sections 4-5 and 4-6, and Table 4-11.*

(a) CuI   copper(I) iodide   (c) Li_3N lithium nitride   (e) $CuCO_3$ copper(II) carbonate

(b) Hg_2Cl_2 mercury(I) chloride   (d) $MnCl_2$ manganese(II) chloride   (f) FeO   iron(II) oxide

---

4-52.   *Refer to Section 4-6.*

(a) copper(II) chlorate $Cu(ClO_3)_2$ (c) barium phosphate $Ba_3(PO_4)_2$ (e) sodium sulfite Na_2SO_3

(b) potassium nitrate KNO_3 (d) copper(I) sulfate Cu_2SO_4

4-54.   *Refer to Section 4-6.*

H_3PO_3 phosphorous acid $H_2PO_3^-$ dihydrogen phosphite ion

HPO_3^{2-} hydrogen phosphite ion PO_3^{3-} phosphite ion

4-56.   *Refer to Section 4-5.*

(a) AlF_3 aluminum fluoride   (c) BrF_5 bromine pentafluoride   (e) Cl_2O_7 dichlorine heptoxide

(b) Br_2O dibromine oxide   (d) CSe_2 carbon diselenide

4-58. *Refer to Section 4-5.*

(a) diboron trioxide B_2O_3 (e) silicon sulfide   SiS

(b) dinitrogen pentasulfide N_2S_5 (f) hydrogen sulfide H_2S

(c) phosphorus triiodide PI_3 (g) tetraphosphorus hexoxide P_4O_6

(d) sulfur hexafluoride SF_6

4-60. *Refer to Sections 4-5 and 4-6.*

(a) CH_4 methane   (b) NH_3 ammonia

4-62. *Refer to Sections 4-5 and 4-6.*

(a) CN^- cyanide ion   (b) NO_3^- nitrate ion

4-64. *Refer to Sections 4-5 and 4-6.*

(a) NH_4Cl ammonium chloride $NiCl_2$ nickel(II) chloride
     NaCl   sodium chloride $FeCl_3$ iron(III) chloride or ferric chloride
     $MgCl_2$ magnesium chloride   AgCl   silver chloride

(b) - $Ni(OH)_2$ nickel(II) hydroxide
     NaOH   sodium hydroxide $Fe(OH)_3$ iron(III) hydroxide or ferric hydroxide
     $Mg(OH)_2$ magnesium hydroxide   AgOH   silver hydroxide

(c) $(NH_4)_2SO_4$ ammonium sulfate $NiSO_4$ nickel(II) sulfate
     Na_2SO_4 sodium sulfate $Fe_2(SO_4)_3$ iron(III) sulfate or ferric sulfate
     $MgSO_4$ magnesium sulfate Ag_2SO_4 silver sulfate

(d) $(NH_4)_3PO_4$ ammonium phosphate $Ni_3(PO_4)_2$ nickel(II) phosphate
     Na_3PO_4 sodium phosphate $FePO_4$ iron(III) phosphate or ferric phosphate
     $Mg_3(PO_4)_2$ magnesium phosphate Ag_3PO_4 silver phosphate

(e) NH_4NO_3 ammonium nitrate $Ni(NO_3)_2$ nickel(II) nitrate
     $NaNO_3$ sodium nitrate $Fe(NO_3)_3$ iron(III) nitrate or ferric nitrate
     $Mg(NO_3)_2$ magnesium nitrate $AgNO_3$ silver nitrate

4-66. *Refer to Section 4-5.*

(a) $N_2(g) + O_2(g) \overset{heat}{\rightarrow} 2NO(g)$ (b) $PbS(s) + PbSO_4(s) \overset{heat}{\rightarrow} 2Pb(s) + 2SO_2(g)$

4-68. *Refer to Section 4-7.*

Due to the Law of Conservation of Matter, electrons cannot be created or destroyed in chemical reactions. The electrons that cause the reduction of one substance must be produced from the oxidation of another substance. Therefore, oxidation and reduction always occur simultaneously in ordinary chemical reactions.

4-70.   *Refer to Section 4-7 and Example 4-5.*

Reaction (b) is the only oxidation-reduction reaction.

In reactions (a), (c) and (d), there are no elements that are changing oxidation number.

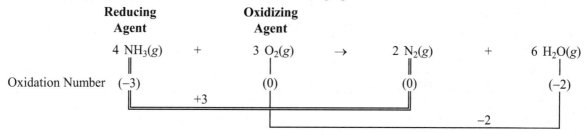

4-72.   *Refer to Section 4-7, Example 4-5 and Exercise 4-69.*

	Oxidizing Agent		Reducing Agent		
(a)	$3H_2SO_4$	+	$2Al$	→	$Al_2(SO_4)_3 + 3H_2$
(b)	N_2	+	$3H_2$	→	$2NH_3$
(c)	$3O_2$	+	$2ZnS$	→	$2ZnO + 2SO_2$
(d)	$4HNO_3$	+	C	→	$4NO_2 + CO_2 + 2H_2O$
(e)	H_2SO_4	+	$2HI$	→	$SO_2 + I_2 + 2H_2O$

4-74.   *Refer to Section 4-7.*

(a) formula unit: $3H_2(g) + Al_2O_3(s) \rightarrow 2Al(s) + 3H_2O(\ell)$

total ionic: $3H_2(g) + Al_2O_3(s) \rightarrow 2Al(s) + 3H_2O(\ell)$

net ionic: $3H_2(g) + Al_2O_3(s) \rightarrow 2Al(s) + 3H_2O(\ell)$

(b) formula unit: $3Zn(s) + 2CoCl_3(aq) \rightarrow 3ZnCl_2(aq) + 2Co(s)$

total ionic: $3Zn(s) + 2Co^{3+}(aq) + 6Cl^-(aq) \rightarrow 3Zn^{2+}(aq) + 6Cl^-(aq) + 2Co(s)$

net ionic: $3Zn(s) + 2Co^{3+}(aq) \rightarrow 3Zn^{2+}(aq) + 2Co(s)$

(c) formula unit: $H_2S(aq) + 8HNO_3(aq) \rightarrow H_2SO_4(aq) + 8NO_2(g) + 4H_2O(\ell)$

total ionic: $H_2S(aq) + 8H^+(aq) + 8NO_3^-(aq) \rightarrow 2H^+(aq) + SO_4^{2-}(aq) + 8NO_2(g) + 4H_2O(\ell)$

net ionic: $H_2S(aq) + 6H^+(aq) + 8NO_3^-(aq) \rightarrow SO_4^{2-}(aq) + 8NO_2(g) + 4H_2O(\ell)$

4-76.   *Refer to Section 4-8.*

(a) $Be + F_2 \rightarrow BeF_2$ (b) $Ca + Br_2 \rightarrow CaBr_2$ (c) $Ba + Cl_2 \rightarrow BaCl_2$

4-78.   *Refer to Sections 4-4 and 4-8.*

(a) $\overset{+6}{3SO_3} + \overset{+3}{Al_2O_3} \rightarrow \overset{+3\ +6}{Al_2(SO_4)_3}$ (b) $\overset{+7}{Cl_2O_7} + H_2O \rightarrow \overset{+7}{2HClO_4}$ (c) $\overset{+2}{CaO} + \overset{+4}{SiO_2} \rightarrow \overset{+2\ +4}{CaSiO_3}$

4-80. *Refer to Sections 4-4 and 4-9.*

$$\overset{-3\ +1\quad +6\ -2}{(\text{NH}_4)_2\text{Cr}_2\text{O}_7(s)} \rightarrow \overset{0}{\text{N}_2(g)} + \overset{+3\ -2}{\text{Cr}_2\text{O}_3(s)} + \overset{+1\ -2}{4\text{H}_2\text{O}(g)}$$

(a) $\overset{-3\ +1\quad +6\ -2}{(\text{NH}_4)_2\text{Cr}_2\text{O}_7(s)} \rightarrow \overset{0}{\text{N}_2(g)} + \overset{+3\ -2}{\text{Cr}_2\text{O}_3(s)} + \overset{+1\ -2}{4\text{H}_2\text{O}(g)}$

(b) $\overset{+1\ +5\ -2}{2\text{NaNO}_3(s)} \rightarrow \overset{+1\ +3\ -2}{2\text{NaNO}_2(s)} + \overset{0}{\text{O}_2(g)}$

4-82. *Refer to Section 4-10, Table 4-14 and Example 4-6.*

Zn, Fe and Ni are more active metals than Cu and will displace Cu from an aqueous solution of CuSO_4.

$\text{Hg}(\ell) + \text{CuSO}_4(aq) \rightarrow$ no reaction

total ionic equation: $\text{Zn}(s) + \text{Cu}^{2+}(aq) + \text{SO}_4{}^{2-}(aq) \rightarrow \text{Zn}^{2+}(aq) + \text{SO}_4{}^{2-}(aq) + \text{Cu}(s)$

net ionic equation: $\text{Zn}(s) + \text{Cu}^{2+}(aq) \rightarrow \text{Zn}^{2+}(aq) + \text{Cu}(s)$

total ionic equation: $\text{Fe}(s) + \text{Cu}^{2+}(aq) + \text{SO}_4{}^{2-}(aq) \rightarrow \text{Fe}^{2+}(aq) + \text{SO}_4{}^{2-}(aq) + \text{Cu}(s)$

net ionic equation: $\text{Fe}(s) + \text{Cu}^{2+}(aq) \rightarrow \text{Fe}^{2+}(aq) + \text{Cu}(s)$

total ionic equation: $\text{Ni}(s) + \text{Cu}^{2+}(aq) + \text{SO}_4{}^{2-}(aq) \rightarrow \text{Ni}^{2+}(aq) + \text{SO}_4{}^{2-}(aq) + \text{Cu}(s)$

net ionic equation: $\text{Ni}(s) + \text{Cu}^{2+}(aq) \rightarrow \text{Ni}^{2+}(aq) + \text{Cu}(s)$

4-84. *Refer to Section 4-10, Table 4-14, Exercise 4-82 and Example 4-7.*

In order of increasing activity: Hg < Cu < Ni < Fe < Zn

4-86. *Refer to Section 4-10, Table 4-14, Exercise 4-85 and Example 4-7.*

In order of increasing activity: Ag < Cr < Na < Ca

4-88. *Refer to Section 4-10, Table 4-14, and Example 4-8.*

Five elements that will react with steam, but not cold water are magnesium (Mg), aluminum (Al), manganese (Mn), zinc (Zn) and chromium (Cr).

4-90. *Refer to Section 4-10 and Example 4-9.*

Each halogen will displace less electronegative (heavier) halogens from their binary salts.
Hence, reactions (b) and (c) will occur and reactions (a) and (d) will not occur.

4-92. *Refer to Section 4-10 and Table 4-14.*

(a) no   (c) no

(b) no   (d) no

(a) formula unit:

$$3CaCl_2(aq) + 2K_3PO_4(aq) \rightarrow Ca_3(PO_4)_2(s) + 6KCl(aq)$$

total ionic:

$$3Ca^{2+}(aq) + 6Cl^-(aq) + 6K^+(aq) + 2PO_4^{3-}(aq) \rightarrow Ca_3(PO_4)_2(s) + 6K^+(aq) + 6Cl^-(aq)$$

net ionic:

$$3Ca^{2+}(aq) + 2PO_4^{3-}(aq) \rightarrow Ca_3(PO_4)_2(s)$$

(b) formula unit:

$$Hg(NO_3)_2(aq) + Na_2S(aq) \rightarrow HgS(s) + 2NaNO_3(aq)$$

total ionic:

$$Hg^{2+}(aq) + 2NO_3^-(aq) + 2Na^+(aq) + S^{2-}(aq) \rightarrow HgS(s) + 2Na^+(aq) + 2NO_3^-(aq)$$

net ionic:

$$Hg^{2+}(aq) + S^{2-}(aq) \rightarrow HgS(s)$$

(c) formula unit:

$$2CrCl_3(aq) + 3Ca(OH)_2(aq) \rightarrow 2Cr(OH)_3(s) + 3CaCl_2(aq)$$

total ionic:

$$2Cr^{3+}(aq) + 6Cl^-(aq) + 3Ca^{2+}(aq) + 6OH^-(aq) \rightarrow 2Cr(OH)_3(s) + 3Ca^{2+}(aq) + 6Cl^-(aq)$$

net ionic:

$$2Cr^{3+}(aq) + 6OH^-(aq) \rightarrow 2Cr(OH)_3(s)$$

therefore,

$$Cr^{3+}(aq) + 3OH^-(aq) \rightarrow Cr(OH)_3(s)$$

(a) formula unit:

$$CH_3COOH(aq) + KOH(aq) \rightarrow KCH_3COO(aq) + H_2O(\ell)$$

total ionic:

$$CH_3COOH(aq) + K^+(aq) + OH^-(aq) \rightarrow K^+(aq) + CH_3COO^-(aq) + H_2O(\ell)$$

net ionic:

$$CH_3COOH(aq) + OH^-(aq) \rightarrow CH_3COO^-(aq) + H_2O(\ell)$$

(b) formula unit:

$$H_2SO_3(aq) + 2NaOH(aq) \rightarrow Na_2SO_3(aq) + 2H_2O(\ell)$$

total ionic:

$$H_2SO_3(aq) + 2Na^+(aq) + 2OH^-(aq) \rightarrow 2Na^+(aq) + SO_3^{2-}(aq) + 2H_2O(\ell)$$

net ionic:

$$H_2SO_3(aq) + 2OH^-(aq) \rightarrow SO_3^{2-}(aq) + 2H_2O(\ell)$$

(c) formula unit:

$$HF(aq) + LiOH(aq) \rightarrow LiF(aq) + H_2O(\ell)$$

total ionic:

$$HF(aq) + Li^+(aq) + OH^-(aq) \rightarrow Li^+(aq) + F^-(aq) + H_2O(\ell)$$

net ionic:

$$HF(aq) + OH^-(aq) \rightarrow F^-(aq) + H_2O(\ell)$$

(a) formula unit:

$$2LiOH(aq) + H_2SO_4(aq) \rightarrow Li_2SO_4(aq) + 2H_2O(\ell)$$

total ionic:

$$2Li^+(aq) + 2OH^-(aq) + 2H^+(aq) + SO_4^{2-}(aq) \rightarrow 2Li^+(aq) + SO_4^{2-}(aq) + 2H_2O(\ell)$$

net ionic:

$$2OH^-(aq) + 2H^+(aq) \rightarrow 2H_2O(\ell)$$

therefore,

$$OH^-(aq) + H^+(aq) \rightarrow H_2O(\ell)$$

(b) formula unit:

$$3Ca(OH)_2(aq) + 2H_3PO_4(aq) \rightarrow Ca_3(PO_4)_2(s) + 6H_2O(\ell)$$

total ionic:

$$3Ca^{2+}(aq) + 6OH^-(aq) + 2H_3PO_4(aq) \rightarrow Ca_3(PO_4)_2(s) + 6H_2O(\ell)$$

net ionic:

$$3Ca^{2+}(aq) + 6OH^-(aq) + 2H_3PO_4(aq) \rightarrow Ca_3(PO_4)_2(s) + 6H_2O(\ell)$$

(c) formula unit:

$$Cu(OH)_2(s) + 2HNO_3(aq) \rightarrow Cu(NO_3)_2(aq) + 2H_2O(\ell)$$

total ionic:

$$Cu(OH)_2(s) + 2H^+(aq) + 2NO_3^-(aq) \rightarrow Cu^{2+}(aq) + 2NO_3^-(aq) + 2H_2O(\ell)$$

net ionic:

$$Cu(OH)_2(s) + 2H^+(aq) \rightarrow Cu^{2+}(aq) + 2H_2O(\ell)$$

4-100. *Refer to Section 4-11 and Example 4-12.*

(a) formula unit: $2HClO_4(aq) + Ca(OH)_2(aq) \rightarrow Ca(ClO_4)_2(aq) + 2H_2O(\ell)$

   total ionic: $2H^+(aq) + 2ClO_4^-(aq) + Ca^{2+}(aq) + 2OH^-(aq) \rightarrow Ca^{2+}(aq) + 2ClO_4^-(aq) + 2H_2O(\ell)$

   net ionic: $2H^+(aq) + 2OH^-(aq) \rightarrow 2H_2O(\ell)$

   therefore, $H^+(aq) + OH^-(aq) \rightarrow H_2O(\ell)$

(b) formula unit: $H_2SO_3(aq) + 2NH_3(aq) \rightarrow (NH_4)_2SO_3(aq)$

   total ionic: $2H_2SO_3(aq) + 2NH_3(aq) \rightarrow 2NH_4^+(aq) + SO_3^{2-}(aq)$

   net ionic: $2H_2SO_3(aq) + 2NH_3(aq) \rightarrow 2NH_4^+(aq) + SO_3^{2-}(aq)$

(c) formula unit: $2CH_3COOH(aq) + Cu(OH)_2(s) \rightarrow Cu(CH_3COO)_2(aq) + 2H_2O(\ell)$

   total ionic: $2CH_3COOH(aq) + Cu(OH)_2(s) \rightarrow Cu^{2+}(aq) + 2CH_3COO^-(aq) + 2H_2O(\ell)$

   net ionic: $2CH_3COOH(aq) + Cu(OH)_2(s) \rightarrow Cu^{2+}(aq) + 2CH_3COO^-(aq) + 2H_2O(\ell)$

4-102. *Refer to Section 4-11 and Example 4-12.*

(a) formula unit: $H_2S(aq) + 2NaOH(aq) \rightarrow Na_2S(aq) + 2H_2O(\ell)$

   total ionic: $H_2S(aq) + 2Na^+(aq) + 2OH^-(aq) \rightarrow 2Na^+(aq) + S^{2-}(aq) + 2H_2O(\ell)$

   net ionic: $H_2S(aq) + 2OH^-(aq) \rightarrow S^{2-}(aq) + 2H_2O(\ell)$

(b) formula unit: $H_3PO_4(aq) + Al(OH)_3(s) \rightarrow AlPO_4(s) + 3H_2O(\ell)$

   total ionic: $H_3PO_4(aq) + Al(OH)_3(s) \rightarrow AlPO_4(s) + 3H_2O(\ell)$

   net ionic: $H_3PO_4(aq) + Al(OH)_3(s) \rightarrow AlPO_4(s) + 3H_2O(\ell)$

(c) formula unit: $H_2CO_3(aq) + Pb(OH)_2(s) \rightarrow PbCO_3(s) + 2H_2O(\ell)$

   total ionic: $H_2CO_3(aq) + Pb(OH)_2(s) \rightarrow PbCO_3(s) + 2H_2O(\ell)$

   net ionic: $H_2CO_3(aq) + Pb(OH)_2(s) \rightarrow PbCO_3(s) + 2H_2O(\ell)$

4-104. *Refer to Section 4-11 and Example 4-12.*

(a) $Mg(OH)_2(s) + 2HNO_3(aq) \rightarrow Mg(NO_3)_2(aq) + 2H_2O(\ell)$

(b) $2Al(OH)_3(s) + 3H_2SO_3(aq) \rightarrow Al_2(SO_3)_3(aq) + 6H_2O(\ell)$

(c) $2KOH(aq) + H_2CO_3(aq) \rightarrow K_2CO_3(aq) + 2H_2O(\ell)$

(d) $Zn(OH)_2(s) + 2HClO_3(aq) \rightarrow Zn(ClO_3)_2(aq) + 2H_2O(\ell)$

(e) $LiOH(aq) + CH_3COOH(aq) \rightarrow LiCH_3COO(aq) + H_2O(\ell)$

4-106. *Refer to Sections 4-2 and 4-11, the Solubility Guidelines, and Table 4-8.*

(a) $2AgNO_3(aq) + CaCl_2(aq) \rightarrow 2AgCl(s) + Ca(NO_3)_2(aq)$

(b) The reaction goes to completion because the ions, Ag^+ and Cl^-, are being removed from solution due to the precipitation of AgCl.

	In Formula Unit Equation	**In Total Ionic Equation**
(a)	$(NH_4)_2SO_4(aq)$	$2NH_4^+(aq) + SO_4^{2-}(aq)$
(b)	$NaBr(aq)$	$Na^+(aq) + Br^-(aq)$
(c)	$SrCl_2(aq)$	$Sr^{2+}(aq) + 2Cl^-(aq)$
(d)	$MgF_2(s)$	$MgF_2(s)$
(e)	$Na_2CO_3(aq)$	$2Na^+(aq) + CO_3^{2-}(aq)$

4-110.   *Refer to Section 4-2 and the Solubility Guidelines summarized in Table 4-8.*

(a) $BaSO_4$ insoluble   (c) CuS insoluble   (e) $Ca(CH_3COO)_2$ soluble

(b) $Al(NO_3)_3$ soluble   (d) Na_3AsO_4 soluble

4-112.   *Refer to Section 4-2 and the Solubility Guidelines summarized in Tables 4-8 and 4-9.*

(a) $KClO_3$ soluble   (c) NH_3 soluble   (e) PbS insoluble

(b) NH_4Cl soluble   (d) HNO_2 soluble

4-114.   *Refer to Section 4-11 and Example 4-14.*

(a) formula unit:
$$Cu(NO_3)_2(aq) + Na_2S(aq) \rightarrow CuS(s) + 2NaNO_3(aq)$$

total ionic:
$$Cu^{2+}(aq) + 2NO_3^-(aq) + 2Na^+(aq) + S^{2-}(aq) \rightarrow CuS(s) + 2Na^+(aq) + 2NO_3^-(aq)$$

net ionic:
$$Cu^{2+}(aq) + S^{2-}(aq) \rightarrow CuS(s)$$

(b) formula unit:
$$CdSO_4(aq) + H_2S(aq) \rightarrow CdS(s) + H_2SO_4(aq)$$

total ionic:
$$Cd^{2+}(aq) + SO_4^{2-}(aq) + H_2S(aq) \rightarrow CdS(s) + 2H^+(aq) + SO_4^{2-}(aq)$$

net ionic:
$$Cd^{2+}(aq) + H_2S(aq) \rightarrow CdS(s) + 2H^+(aq)$$

(c) formula unit:
$$Bi_2(SO_4)_3(aq) + 3(NH_4)_2S(aq) \rightarrow Bi_2S_3(s) + 3(NH_4)_2SO_4(aq)$$

total ionic:
$$2Bi^{3+}(aq) + 3SO_4^{2-}(aq) + 6NH_4^+(aq) + 3S^{2-}(aq) \rightarrow Bi_2S_3(s) + 6NH_4^+(aq) + 3SO_4^{2-}(aq)$$

net ionic:
$$2Bi^{3+}(aq) + 3S^{2-}(aq) \rightarrow Bi_2S_3(s)$$

4-116.   *Refer to Section 4-11, the Solubility Guidelines in Section 4-2, and Examples 4-13 and 4-14.*

Plan: We know that at the time of mixing, the solution contains only ions because the compounds are water soluble. A precipitate will only form if a combination of these ions (in reasonable concentrations) produces an insoluble substance.

(a) ions in solution: $NH_4^+(aq)$, $Br^-(aq)$, $Hg_2^{2+}(aq)$ and $NO_3^-(aq)$
new possible combinations of ions: Hg_2Br_2 and NH_4NO_3

Solubility Guideline 4 says that Hg_2Br_2 is insoluble, while Solubility Guidelines 2 and 3 say that NH_4NO_3 is soluble. Therefore, a **precipitate will form** and it is **Hg_2Br_2**.

(b) ions in solution: $K^+(aq)$, $OH^-(aq)$, $Na^+(aq)$ and $S^{2-}(aq)$

new possible combinations of ions: K_2S and $NaOH$

Solubility Guidelines 2, 6 and 8 say that both K_2S and $NaOH$ are soluble. Therefore a **precipitate will not form**.

(c) ions in solution: $Cs^+(aq)$, $SO_4^{2-}(aq)$, $Mg^{2+}(aq)$ and $Cl^-(aq)$

new possible combinations of ions: $CsCl$ and $MgSO_4$

Solubility Guidelines 2, 4 and 5 say that $CsCl$ and $MgSO_4$ are soluble. Therefore, a **precipitate will not form**.

4-118.   *Refer to Section 4-11.*

The acid-base reactions are (a) and (k) only, in which an acid reacts with a base to give a salt and water. In all acid-base reactions, no oxidation or reduction is involved.

(a) $H_2SO_4(aq) + 2KOH(aq) \rightarrow K_2SO_4(aq) + 2H_2O(\ell)$

(k) $RbOH(aq) + HNO_3(aq) \rightarrow RbNO_3(aq) + H_2O(\ell)$

4-120.   *Refer to Sections 4-7, 4-8, 4-10 and 4-13.*

The oxidation-reduction reactions are:

Net Ionic Equation	Oxidizing Agent	Reducing Agent
(b) $2Rb(s) + Br_2(\ell) \overset{heat}{\rightarrow} 2RbBr(s)$	$Br_2(\ell)$	$Rb(s)$
(c) $2I^-(aq) + F_2(g) \rightarrow 2F^-(aq) + I_2(s)$	$F_2(g)$	$I^-(aq)$
(e) $S(s) + O_2(g) \overset{heat}{\rightarrow} SO_2(g)$	$O_2(g)$	$S(s)$
(g) $HgS(s) + O_2(g) \overset{heat}{\rightarrow} Hg(\ell) + SO_2(g)$	$\underline{Hg}S(s), O_2(g)$	$Hg\underline{S}(s)$
(i) $Pb(s) + 2H^+(aq) + 2Br^-(aq) \rightarrow PbBr_2(s) + H_2(g)$	$H^+(aq)$	$Pb(s)$
(j) $2H^+(aq) + 2I^-(aq) + H_2O_2(aq) \rightarrow I_2(s) + 2H_2O(\ell)$	$H_2\underline{O}_2(aq)$	$I^-(aq)$
(m) $H_2O(g) + CO(g) \overset{heat}{\rightarrow} H_2(g) + CO_2(g)$	$\underline{H}_2O(g)$	$\underline{C}O(g)$
(o) $PbSO_4(s) + PbS(s) \overset{heat}{\rightarrow} 2Pb(s) + 2SO_2(g)$	$\underline{Pb}\,SO_4(s), \underline{Pb}S(s)$	$Pb\underline{S}(s)$

4-122.   *Refer to Section 4-11.*

The metathesis (also called double displacement) reactions are those in which the positive and negative ions in two compounds "change partners," with no change in oxidation numbers, to form two new compounds. There are three:

(a) $H_2SO_4(aq) + 2KOH(aq) \rightarrow K_2SO_4(aq) + 2H_2O(\ell)$

(h) $AgNO_3(aq) + HCl(aq) \rightarrow AgCl(s) + HNO_3(aq)$

(k) $RbOH(aq) + HNO_3(aq) \rightarrow RbNO_3(aq) + H_2O(\ell)$

4-124.   *Refer to Section 4-9.*

The decomposition reactions can be identified easily as one compound breaking down, i.e., decomposing, to other compounds, elements or a combination of element(s) and compound(s). There is only one: (f).

(f) $BaCO_3(s) \overset{heat}{\rightarrow} BaO(s) + CO_2(s)$

(a) The oxidation-reduction reactions that form gaseous products are (e), (g), (i), (m) and (o).

(b) The redox reactions that also fit the definition of gas-formation reactions because they do not have any gaseous reactants are only (i) and (o).

4-128. *Refer to Sections as stated.*

(a) Copper metal is formed by a displacement reaction. (Refer to Section 4-10, Example 4-6 and Table 4-14.)

$$Cu(NO_3)_2(aq) + Mg(s) \rightarrow Mg(NO_3)_2(aq) + Cu(s)$$

(b) Solid barium phosphate is formed in a precipitation reaction.
(Refer to Sections 4-2, 4-11, Example 4-14 and the Solubility Guidelines given in Table 4-8.)

$$3Ba(NO_3)_2(aq) + 2Na_3PO_4(aq) \rightarrow Ba_3(PO_4)_2(s) + 6NaNO_3(aq)$$

(c) There is no reaction because Al is a less active metal than Ca.
(Refer to Section 4-10, Example 4-6 and Table 4-14.)

(d) Solid silver iodide is formed in a precipitation reaction.
(Refer to Sections 4-2, 4-11, Example 4-14 and the Solubility Guidelines given in Table 4-8.)

$$AgNO_3(aq) + NaI(aq) \rightarrow AgI(s) + NaNO_3(aq)$$

4-130. *Refer to Sections as stated.*

(a) Both 1.2 M CH_3COOH and 0.12 M CH_3COOH are equally weak acids, only their concentrations are different (Section 4-2). In fact, the more dilute CH_3COOH solution actually ionizes a little more into its ions (Chapter 18).

(b) The salt produced when nitric acid, HNO_3, reacts with potassium hydroxide, KOH, is KNO_3, not KNO_4 (Section 4-11).

(c) The first two statements are correct - nickel reacts with HCl and not steam, but magnesium is active enough to react with steam. Therefore magnesium is more reactive than nickel. (Section 4-10)

4-132. *Refer to Section 4-4, Example 4-4 and Table 4-10.*

$$\overset{-2.5\ +1}{} \qquad \overset{0}{} \qquad \overset{+4\ -2}{} \qquad \overset{+1\ -2}{}$$
(a) $2 C_4H_{10}(g) + 13 O_2(g) \rightarrow 8 CO_2(g) + 10 H_2O(g)$ \qquad (O_2 is in excess)

$$\overset{-2.5\ +1}{} \qquad \overset{0}{} \qquad \overset{+2\ -2}{} \qquad \overset{+1\ -2}{}$$
(b) $2 C_4H_{10}(g) + 9 O_2(g) \rightarrow 8 CO(g) + 10 H_2O(g)$ \qquad (O_2 is limited)

$$\overset{-2.5\ +1}{} \qquad \overset{0}{} \qquad \overset{-1\ +1}{} \qquad \overset{+2\ -2}{} \qquad \overset{0}{}$$
(c) $C_4H_{10}(g) + O_2(g) \rightarrow C_2H_2(g) + 2 CO(g) + 4 H_2(g)$ \qquad (O_2 is very limited)

4-134. *Refer to Sections 4-2 and 4-9.*

(a) Calcite is primarily $CaCO_3$, while dolomite is a mixture of $CaCO_3$ and $MgCO_3$. This is not surprising because both Ca^{2+} and Mg^{2+} behave similarly. They are neighboring 2A elements than have lost 2 electrons to form their stable ions. In excess CO_3^{2-} ions, both $CaCO_3$ and $MgCO_3$ would precipitate out of solution as insoluble carbonates.

(b) The bubbles resulting when cold HCl(aq) is applied to limestone are composed of carbon dioxide gas formed in the metathesis gas-forming reaction: $CaCO_3(s) + 2HCl(aq) \rightarrow CaCl_2(aq) + H_2O(\ell) + CO_2(g)$

4-136. *Refer to Chapter 3.*

Balanced equation: $P_4(s) + 6Cl_2(g) \rightarrow 4PCl_3(\ell)$

Explanation (particle level): When 1 molecule of solid phosphorus (P_4) reacts with 6 molecules of gaseous chlorine (Cl_2), 4 molecules of liquid phosphorus trichloride (PCl_3) are formed.

Explanation (mole level): When 1 mole of solid phosphorus molecules reacts with 6 moles of gaseous chlorine molecules, 4 moles of liquid phosphorus trichloride molecules are formed.

4-138. *Refer to Section 4-11.*

(a) acid-base reaction:

$$Ba(OH)_2(aq) + H_2SO_4(aq) \rightarrow BaSO_4(s) + 2H_2O(\ell)$$

(b) precipitation reaction:

$$Ba(OH)_2(aq) + Na_2SO_4(aq) \rightarrow BaSO_4(s) + 2NaOH(aq)$$

(c) gas-forming reaction:

$$BaCO_3(s) + H_2SO_4(aq) \rightarrow BaSO_4(s) + CO_2(g) + H_2O(\ell)$$

4-140. *Refer to Chapter 3.*

(a) Balanced equation: $2KClO_3(s) \rightarrow 2KCl(s) + 3O_2(g)$

$$? \text{ mol } O_2 = 15.0 \text{ g KClO}_3 \times \frac{1 \text{ mol KClO}_3}{122.6 \text{ g KClO}_3} \times \frac{3 \text{ mol } O_2}{2 \text{ mol KClO}_3} = \textbf{0.184 mol O}_2$$

(b) Balanced equation: $2H_2O_2(aq) \rightarrow 2H_2O(\ell) + O_2(g)$

$$? \text{ mol } O_2 = 15.0 \text{ g H}_2O_2 \times \frac{1 \text{ mol H}_2O_2}{34.02 \text{ g H}_2O_2} \times \frac{1 \text{ mol } O_2}{2 \text{ mol H}_2O_2} = \textbf{0.220 mol O}_2$$

(c) Balanced equation: $2HgO(s) \rightarrow 2\,Hg(\ell) + O_2(g)$

$$? \text{ mol } O_2 = 15.0 \text{ g HgO} \times \frac{1 \text{ mol HgO}}{216.6 \text{ g HgO}} \times \frac{1 \text{ mol } O_2}{2 \text{ mol HgO}} = \textbf{0.0346 mol O}_2$$

5 The Structure of Atoms

5-2. *Refer to Section 5-2 and Figure 5-2.*

If any oil droplets in Millikan's oil drop experiment had possessed a deficiency of electrons, the droplets would have been positively charged and would have been attracted to, not repelled by, the negatively charged plate. There would have been no voltage setting possible where the electrical and gravitational forces on the drop would have balanced.

5-4. *Refer to Sections 5-2 and 5-3, and Figures 5-1 and 5-3.*

(a) Canal rays, also produced in the cathode ray tube, move toward the cathode (the negative electrode). Therefore, they must be positively charged. Canal rays are positively charged ions created when cathode rays knock electrons from the gaseous atoms in the tube.

(b) Cathode rays are electrons and are independent of source. Canal rays are the positive ions from the specific gas used after a loss of electrons; they are therefore dependent upon the gas used.

5-6. *Refer to Sections 5-2 and 5-3, and Figure 5-2.*

(a) We must modify the Millikan oil drop experiment in order to determine the charge-to-mass ratio of the positively charged whizatron by
    (1) in some way producing an excess of whizatrons on the oil droplets, and
    (2) switching the leads to the plates to make the bottom plate positively charged.

The positively charged whizatrons on the oil droplets will be repulsed by the plate.

(b) Since all of the charges on the droplets will be integral multiples of the charge on the whizatron, we will identify the droplet with the smallest charge and test to see if the other droplets have charges that are multiples of its charge.

From the results shown in the table below, we can deduce that the charge on the whizatron is 1/2 of the smallest observed charge:

$$1/2 \times (2.44 \times 10^{-19}) = 1.22 \times 10^{-19} \text{ coulombs.}$$

All the droplets have charges that are integral multiples of **1.22×10^{-19} coulombs**.

Charge on Droplets (coulombs)	Ratio
4.88×10^{-19}	$\dfrac{4.88 \times 10^{-19}}{1.22 \times 10^{-19}} = 4.0$
6.10×10^{-19}	$\dfrac{6.10 \times 10^{-19}}{1.22 \times 10^{-19}} = 5.0$
2.44×10^{-19}	$\dfrac{2.44 \times 10^{-19}}{1.22 \times 10^{-19}} = 2.0$
8.53×10^{-19}	$\dfrac{8.53 \times 10^{-19}}{1.22 \times 10^{-19}} = 7.0$
7.32×10^{-19}	$\dfrac{7.32 \times 10^{-19}}{1.22 \times 10^{-19}} = 6.0$

Alpha particles, which are the positively-charged nuclei of helium atoms, were thought to be much more dense than gold. Hence, according to Thomson's "plum pudding" model of the atom, it was expected that these particles would pass easily through the gold foil with little deflection. The fact that some particles were greatly deflected astounded Rutherford. This led him to conclude that all the positive charges were found in one central place, instead of scattered throughout the atom.

5-10.   *Refer to Section 5-4.*

Volume of a hydrogen atom $= (4/3)\pi r^3 = (4/3)\pi(5.29 \times 10^{-11} \text{ m})^3 = 6.20 \times 10^{-31} \text{ m}^3$ $(1 \text{ nm} = 1 \times 10^{-9} \text{ m})$

Volume of a hydrogen nucleus = volume of a proton $= (4/3)\pi r^3 = (4/3)\pi(1.5 \times 10^{-15} \text{ m})^3 = 1.4 \times 10^{-44} \text{ m}^3$

Therefore, the fraction of space in a hydrogen atom occupied by the nucleus is:

$$\frac{V_{\text{hydrogen nucleus}}}{V_{\text{hydrogen atom}}} = \frac{1.4 \times 10^{-44} \text{ m}^3}{6.20 \times 10^{-31} \text{ m}^3} = \mathbf{2.3 \times 10^{-14}}$$

As can be seen, an atom is mostly empty space.

5-12.   *Refer to Sections 5-2 and 5-7.*

Calculation of the charge-to-mass ratio:

Species	Charge	Mass Number	Charge-to-Mass Ratio
$^{12}\text{C}^+$	+1	12	$1/12 = 0.0833$
$^{12}\text{C}^{2+}$	+2	12	$2/12 = 0.167$
$^{14}\text{N}^+$	+1	14	$1/14 = 0.0714$
$^{14}\text{N}^{2+}$	+2	14	$2/14 = 0.143$

The order of increasing charge-to-mass ratios is $^{14}\text{N}^+ < {}^{12}\text{C}^+ < {}^{14}\text{N}^{2+} < {}^{12}\text{C}^{2+}$.

5-14.   *Refer to Sections 5-5, 5-6 and 5-7, Table 5-1, and Example 5-1.*

A neutral atom of $^{58}_{28}\text{Ni}$ contains 28 electrons, 28 protons and $(58-28) = 30$ neutrons.

If we assume that the mass of the atom is simply the sum of the masses of its subatomic particles, then

$$\begin{aligned}
\text{mass of } ^{58}\text{Ni} &= (28 \; e^- \times \text{mass } e^-) + (28 \; p \times \text{mass } p) + (30 \; n \times \text{mass } n) \\
&= (28 \times 0.00054858 \text{ amu}) + (28 \times 1.0073 \text{ amu}) + (30 \times 1.0087 \text{ amu}) \\
&= 58.481 \text{ amu/atom}
\end{aligned}$$

(a) % by mass $e^- = \dfrac{\text{mass } e^-}{\text{mass } ^{58}\text{Ni}} \times 100\% = \dfrac{28 \; e^- \times 0.00054858 \text{ amu}/e^-}{58.481 \text{ amu}} \times 100\% = \mathbf{0.026265\%}$

(b) % by mass $p = \dfrac{\text{mass } p}{\text{mass } ^{58}\text{Ni}} \times 100\% = \dfrac{28 \; p \times 1.0073 \text{ amu}/p}{58.481 \text{ amu}} \times 100\% = \mathbf{48.228\%}$

(c) % by mass $n = \dfrac{\text{mass } n}{\text{mass } ^{58}\text{Ni}} \times 100\% = \dfrac{30 \; n \times 1.0087 \text{ amu}/n}{58.481 \text{ amu}} \times 100\% = \mathbf{51.745\%}$

5-16. *Refer to the Key Terms for Chapter 5.*

(a) The atomic number of an element is the integral number of protons in the nucleus. It defines the identity of that element. For example, oxygen has an atomic number of 8 and therefore has 8 protons. All oxygen atoms have exactly 8 protons and there is no other element that has 8 protons in its nucleus (**Section 5-5**).

(b) Isotopes are two or more forms of atoms of the same element with different masses. In other words, they are atoms containing the same number of protons but they have different numbers of neutrons. ^{16}O and ^{17}O are isotopes since both have 8 protons but ^{16}O has (16 - 8) = 8 neutrons while ^{17}O has (17 - 8) = 9 neutrons (**Section 5-7**).

(c) The mass number of an element is the integral sum of the numbers of protons and neutrons in that atom. The mass number of ^{17}O is 17, the sum of protons and neutrons in the nucleus (**Section 5-7**).

(d) Nuclear charge refers to the number of protons or positive charges in the nucleus. The nuclear charge of all oxygen atoms is +8.

5-18. *Refer to Section 5-7, Table 5-2 and the Periodic Table on the inside front cover of the textbook.*

From the Periodic Table, we see that the atomic number of strontium is 38.
Therefore, each strontium atom has 38 protons.
If it is a neutral atom then it also has 38 electrons.
If we assume that these isotopes are neutral, then

Isotope	Number of Protons	Number of Electrons	Number of Neutrons (Mass Number - Atomic Number)
$^{84}_{38}Sr$	38	38	46 (= 84 - 38)
$^{86}_{38}Sr$	38	38	48 (= 86 - 38)
$^{87}_{38}Sr$	38	38	49 (= 87 - 38)
$^{88}_{38}Sr$	38	38	50 (= 88 - 38)

5-20. *Refer to Section 5-7.*

Remember: atomic number = number of protons = number of electrons in a neutral atom
mass number = number of protons + number of neutrons

Kind of Atom	Atomic Number	Mass Number	Isotope	Number of Protons	Number of Electrons	Number of Neutrons
cobalt	<u>27</u>	<u>59</u>	$^{59}_{27}Co$	<u>27</u>	<u>27</u>	32
<u>boron</u>	<u>5</u>	<u>11</u>	$^{11}_{5}B$	<u>5</u>	<u>5</u>	<u>6</u>
<u>manganese</u>	<u>25</u>	<u>55</u>	$^{55}_{25}Mn$	<u>25</u>	25	30
<u>platinum</u>	<u>78</u>	182	$^{182}_{78}Pt$	<u>78</u>	78	<u>104</u>

5-22. *Refer to Section 5-7 and Example 5-1.*

	Symbol of Species	Number of Protons	Number of Neutrons	Number of Electrons
(a)	$^{24}_{12}Mg$	12	12	12
(b)	$^{51}_{23}V$	23	28	23
(c)	$^{91}_{40}Zr$	40	51	40
(d)	$^{27}_{13}Al$	13	14	13
(e)	$^{65}_{30}Zn^{2+}$	30	35	28
(f)	$^{108}_{47}Ag^{+}$	47	61	46

5-24. *Refer to Section 5-7.*

	Number of Protons	Number of Neutrons	Number of Electrons	Z	A	Charge	Symbol
(a)	24	28	24	24	52	0	$^{52}_{24}Cr$
(b)	20	20	20	20	40	0	$^{40}_{20}Ca$
(c)	33	42	33	33	75	0	$^{75}_{33}As$
(d)	53	74	53	53	127	0	$^{127}_{53}I$

5-26. *Refer to Section 5-9 and Example 5-3.*

We know: AW Li = (mass ^{6}Li x fraction of ^{6}Li) + (mass ^{7}Li x fraction of ^{7}Li)

let x = fraction of ^{6}Li

then $(1-x)$ = fraction of ^{7}Li

Substituting,

$$6.941 \text{ amu} = (6.01512 \text{ amu})x + (7.01600 \text{ amu})(1 - x)$$
$$= 6.01512x + 7.01600 - 7.01600x$$
$$0.075 = 1.00088x$$

fraction of ^{6}Li = x = 0.075 % abundance of ^{6}Li = **7.5%**

5-28. *Refer to Section 5-9 and Example 5-2.*

Plan: (1) For each isotope, convert the % abundance to a fraction.
(2) Multiply the fraction of each isotope by its mass and add the terms to obtain the atomic weight of Br.

$$\text{AW (amu)} = \sum \frac{\text{(relative abundance)}}{\text{(total relative abundance)}} \times \text{isotope mass (amu)}$$

? AW Br = (mass ^{79}Br x fraction of ^{79}Br) + (mass ^{81}Br x fraction of ^{81}Br)

$= (78.9183 \text{ amu} \times \frac{50.69}{100}) + (80.9163 \text{ amu} \times \frac{49.31}{100})$

= **79.90 amu** (to 4 significant figures)

Refer to Section 5-9 and Example 5-2.

Plan: (1) For each isotope, convert the % abundance to a fraction.
(2) Multiply the fraction of each isotope by its mass and add the terms to obtain the atomic weight of Fe.

? AW Fe = (mass ^{54}Fe x fraction of ^{54}Fe) + (mass ^{56}Fe x fraction of ^{56}Fe) + (mass ^{57}Fe x fraction of ^{57}Fe)
+ (mass ^{58}Fe x fraction of ^{58}Fe)

$$= (53.9396 \text{ amu} \times \frac{5.82}{100}) + (55.9349 \text{ amu} \times \frac{91.66}{100}) + (56.9354 \text{ amu} \times \frac{2.19}{100})$$

$$+ (57.9333 \text{ amu} \times \frac{0.33}{100})$$

= 3.14 amu + 51.27 amu + 1.25 amu + 0.19 amu

= **55.85 amu** (to 4 significant figures)

5-32. **Refer to Section 5-9 and Example 5-3.**

We know: AW Cu = (mass ^{63}Cu x fraction of ^{63}Cu) + (mass ^{65}Cu x fraction of ^{65}Cu)

let $\quad$ x = fraction of ^{63}Cu
then $\quad$ (1-x) = fraction of ^{65}Cu

Substituting,
$\quad$ 63.546 amu = (62.9298 amu)x + (64.9278 amu)(1 - x)
$\quad\quad\quad\quad\quad$ = 62.9298x + 64.9278 - 64.9278x
$\quad\quad$ 1.998x = 1.382
fraction of ^{63}Cu = x = 0.6917 $\quad\quad\quad$ % abundance of ^{63}Cu = **69.17%**

5-34. **Refer to Section 5-9 and Table 5-3.**

Plan: (1) For each isotope, convert the % abundance to a fraction.
(2) Multiply the fraction of each isotope by its mass. Add terms to obtain the atomic weight of the element.

? AW O = (mass ^{16}O x fraction of ^{16}O) + (mass ^{17}O x fraction of ^{17}O) + (mass ^{18}O x fraction of ^{18}O)

$$= (15.99492 \text{ amu} \times \frac{99.762}{100}) + (16.99913 \text{ amu} \times \frac{0.038}{100}) + (17.99916 \text{ amu} \times \frac{0.200}{100})$$

= **15.999 amu** (to 5 significant figures)

? AW Cl = (mass ^{35}Cl x fraction of ^{35}Cl) + (mass ^{37}Cl x fraction of ^{37}Cl)

$$= (34.96885 \text{ amu} \times \frac{75.770}{100}) + (36.96590 \text{ amu} \times \frac{24.230}{100})$$

= **35.453 amu** (to 5 significant figures)

Yes, the answers agree with the atomic weights given in Table 5-3.

5-36. **Refer to Section 5-9 and Example 5-2.**

If the mass spectrum were complete for germanium, the calculated atomic weight would be the weighted average of the isotopes:

$$AW \text{ (amu)} = \sum \frac{\text{(relative abundance)}}{\text{(total relative abundance)}} \times \text{isotope mass (amu)}$$

= (5.49/15.90)71.9217 + (1.55/15.90)72.9234 + (7.31/15.90)73.9219 + (1.55/15.90)75.9219

= **73.3 amu**

However, the true atomic weight of germanium is 72.61 amu. The observed data gives a value that is too high. Therefore, the spectrum given is incomplete and data must have been lost at the low end of the plot when the recorder jammed.

5-38. *Refer to Section 5-9 and Example 5-2.*

Plan: (1) For each isotope, convert the % abundance to a fraction.
(2) Multiply the fraction of each isotope by its mass and add the terms to obtain the atomic weight of Cr.

? AW Cr = (mass ^{50}Cr x fraction of ^{50}Cr) + (mass ^{52}Cr x fraction of ^{52}Cr) + (mass ^{53}Cr x fraction of ^{53}Cr)
+ (mass ^{54}Cr x fraction of ^{54}Cr)

? AW Cr = (49.9461 amu x $\frac{4.35}{100}$) + (51.9405 amu x $\frac{83.79}{100}$) + (52.9406 amu x $\frac{9.50}{100}$)

+ (53.9389 amu x $\frac{2.36}{100}$)

= **52.0 amu** (to 3 significant figures)

5-40. *Refer to Section 5-10.*

For electromagnetic radiation: frequency x wavelength = speed of light

$$\nu \ (s^{-1}) \ x \ \lambda \ (m) = c \ (m/s)$$

$$\nu \ (s^{-1}) = \frac{c \ (m/s)}{\lambda \ (m)}$$

(a) λ (m) = 8973 Å x $\frac{10^{-10} \ m}{1 \ Å}$ = 8.973 x 10^{-7} m

$\nu \ (s^{-1}) = \frac{3.00 \ x \ 10^8 \ m/s}{8.973 \ x \ 10^{-7} \ m}$ = **3.34 x 10^{14} s^{-1}**

(b) λ (m) = 442 nm x $\frac{10^{-9} \ m}{1 \ nm}$ = 4.42 x 10^{-7} m

$\nu \ (s^{-1}) = \frac{3.00 \ x \ 10^8 \ m/s}{4.42 \ x \ 10^{-7} \ m}$ = **6.79 x 10^{14} s^{-1}**

(c) λ (m) = 4.92 cm x $\frac{1 \ m}{100 \ cm}$ = 0.0492 m

$\nu \ (s^{-1}) = \frac{3.00 \ x \ 10^8 \ m/s}{0.0492 \ m}$ = **6.10 x 10^9 s^{-1}**

(d) λ (m) = 4.55 x 10^{-9} cm x $\frac{1 \ m}{100 \ cm}$ = 4.55 x 10^{-11} m

$\nu \ (s^{-1}) = \frac{3.00 \ x \ 10^8 \ m/s}{4.55 \ x \ 10^{-11} \ m}$ = **6.59 x 10^{18} s^{-1}**

5-42. *Refer to Sections 5-10 and 5-12, Examples 5-5 and 5-6, and Figure 5-13b.*

(a) $\nu \ (s^{-1}) = \frac{c \ (m/s)}{\lambda \ (m)} = \frac{3.00 \ x \ 10^8 \ m/s}{670.8 \ nm \ x \ 10^{-9} \ m/1 \ nm}$ = **4.47 x 10^{14} s^{-1}**

(b) $E = h\nu$ = (6.63 x 10^{-34} J·s)(4.47 x 10^{14} s^{-1}) = **2.96 x 10^{-19} J/photon**

(c) From Figure 5-13b, the color corresponding to λ = 670.8 nm or 6708Å is **red**.

5-44. *Refer to Sections 5-10 and 5-12, and Example 5-6.*

E (J/photon) $= \frac{hc}{\lambda} = \frac{(6.63 \ x \ 10^{-34} \ J·s)(3.00 \ x \ 10^8 \ m/s)}{3400Å \ x \ (10^{-10} \ m/1 \ Å)}$ = **5.85 x 10^{-19} J/photon**

E (J/mol) = 5.85 x 10^{-19} J/photon x 6.02 x 10^{23} photons/mol = **3.52 x 10^5 J/mol or 352 kJ/mol**

5-46. *Refer to Tables 1-6 and 1-8.*

Plan: (1) Use dimensional analysis to determine how far (in miles) light travels in one year, which is a light year.
 (2) Use this as a unit factor to determine the distance in miles between Alpha Centauri and our solar system.

(1) $? \text{ miles/light yr} = \dfrac{3.00 \times 10^8 \text{ m}}{1 \text{ s}} \times \dfrac{100 \text{ cm}}{1 \text{m}} \times \dfrac{1 \text{ in}}{2.54 \text{ cm}} \times \dfrac{1 \text{ ft}}{12 \text{ in}} \times \dfrac{1 \text{ mile}}{5280 \text{ ft}} \times \dfrac{60 \text{ s}}{1 \text{ min}} \times \dfrac{60 \text{ min}}{1 \text{ hr}} \times \dfrac{24 \text{ hr}}{1 \text{ day}} \times \dfrac{365 \text{ d}}{1 \text{ yr}}$

$= 5.88 \times 10^{12} \text{ miles/light yr}$

Therefore in 1 year, light will travel 5.88×10^{12} miles

(2) $? \text{ miles} = 4.3 \text{ light years} \times \dfrac{5.88 \times 10^{12} \text{ miles}}{1 \text{ light year}}$

$= \mathbf{2.5 \times 10^{13} \text{ miles}}$

5-48. *Refer to Section 5-11.*

The photoelectric effect is the emission of an electron from a metal surface caused by impinging electromagnetic radiation. This radiation must have a certain minimum energy, i.e., its frequency must be greater than the threshold frequency, which is characteristic of a particular metal, for current to flow. If the frequency is below the threshold frequency, no current flows. As long as this criterion is met, the current increases with increasing intensity (brightness) of the light.

5-50. *Refer to Sections 5-10, 5-11 and 5-12, Figure 5-13 and Exercise 5-49.*

We know: $E = h\nu = \dfrac{hc}{\lambda}$

Therefore, $\lambda \text{ (m)} = \dfrac{hc}{E} = \dfrac{(6.63 \times 10^{-34} \text{ J·s})(3.00 \times 10^8 \text{ m/s})}{(3.89 \text{ eV})(1.60 \times 10^{-19} \text{ J/eV})} = 3.20 \times 10^{-7} \text{ m}$

$\lambda \text{ (nm)} = 3.20 \times 10^{-7} \text{ m} \times \dfrac{1 \text{ nm}}{10^{-9} \text{ m}} = \mathbf{320 \text{ nm}} \quad \mathbf{(3200 \text{ Å} = ultraviolet)}$

5-52. *Refer to Section 5-12 and Figure 5-17b.*

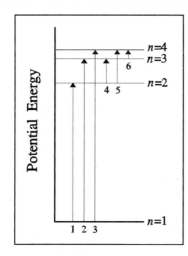

Transitions for Absorption Spectrum

1	$n=1 \rightarrow n=2$
2	$n=1 \rightarrow n=3$
3	$n=1 \rightarrow n=4$
4	$n=2 \rightarrow n=3$
5	$n=2 \rightarrow n=4$
6	$n=3 \rightarrow n=4$

5-54. *Refer to Section 5-12 and Example 5-6.*

The energy loss due to 1 atom emitting a photon is $E = hc/\lambda$.
The energy loss due to 1 mole of atoms each emitting a photon is $E = (hc/\lambda)N$, where N is Avogadro's Number.

Substituting, $E = \dfrac{(6.63 \times 10^{-34} \text{ J·s})(3.00 \times 10^{8} \text{ m/s})}{(5.50 \times 10^{3} \text{ Å})(1 \times 10^{-10} \text{ m/Å})} \times 6.02 \times 10^{23} \text{ mol}^{-1} = 2.18 \times 10^{5} \text{ J/mol} = \textbf{218 kJ/mol}$

5-56. *Refer to Sections 5-10 and 5-12.*

Using the equations, $E = h\nu$ and $c = \lambda\nu$, we can qualitatively deduce the relationships between the energies of photons, E, their wavelengths, λ, and frequency, ν. Consider the diagram at the right, drawn not entirely to scale.

(a) The photon with the smallest energy is produced by an electron in the $n = 7$ major energy level falling to the $n = 2$ major energy level. This photon therefore also has the smallest frequency since $E \propto \nu$ and the longest wavelength since $\nu \propto 1/\lambda$.

(b) The photon with the highest frequency is produced by the transition, $n = 7 \rightarrow n = 1$, since this transition involves the release of the most energy ($\nu \propto E$)

(c) The photon with the shortest wavelength also has the highest frequency ($\lambda \propto 1/\nu$) and releases the most energy. It is produced by an electron in the $n = 7$ major energy level falling to the $n = 1$ major energy level.

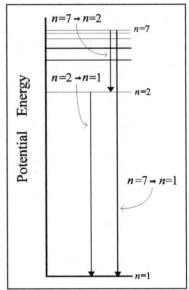

(d) Plan: (1) Use the Rydberg equation, $\dfrac{1}{\lambda} = R\left(\dfrac{1}{n_1{}^2} - \dfrac{1}{n_2{}^2}\right)$ where $R = 1.097 \times 10^7 \text{ m}^{-1}$ to evaluate $\dfrac{1}{\lambda}$.

(2) Solve for λ and ν.

(1) $\dfrac{1}{\lambda} = 1.097 \times 10^7 \text{ m}^{-1} \left(\dfrac{1}{1^2} - \dfrac{1}{6^2}\right) = (1.097 \times 10^7 \text{ m}^{-1})(0.9722) = 1.067 \times 10^7 \text{ m}^{-1}$

(2) $\lambda = 9.376 \times 10^{-8} \text{ m}$ and $\nu = \dfrac{c}{\lambda} = \dfrac{3.00 \times 10^8 \text{ m/s}}{9.376 \times 10^{-8} \text{ m}} = \textbf{3.20} \times \textbf{10}^{\textbf{15}} \textbf{ s}^{-1}$

5-58. *Refer to Sections 5-10 and 5-12, and Figures 5-13b and 5-16.*

(a)	lithium	$\lambda = 4603$ Å	**blue**
(b)	neon	$\lambda = 540.0$ nm or 5400 Å	**greenish-yellow**
(c)	calcium	$\lambda = 6573$ Å	**red**
(d)	potassium	$\nu = 3.90 \times 10^{14} \text{ s}^{-1}$	**red/infrared**

5-60. *Refer to Section 5-12.*

The energy emitted by 1 photon is $E = hc/\lambda$. The energy emitted by *n* photons is $E = (hc/\lambda)n$.

The energy emitted by this laser in 2 seconds is

$E = \text{power} \times \text{time} = 515 \text{ milliwatts} \times \dfrac{1 \text{ watt}}{1000 \text{ milliwatts}} \times \dfrac{1 \text{ J/s}}{1 \text{ watt}} \times 2.00 \text{ s} = 1.03 \text{ J}$

Substituting,

$$1.03 \text{ J} = \frac{(6.63 \times 10^{-34} \text{ J·s})(3.00 \times 10^8 \text{ m/s})}{(488.0 \text{ nm})(10^{-9} \text{ m/nm})} \times n$$

$n = 2.53 \times 10^{18}$ photons

5-62. *Refer to Section 5-13, Example 5-7 and Figure 6-1.*

(a) The de Broglie wavelength is given by $\lambda \text{ (m)} = \dfrac{h(\text{J·s})}{m(\text{kg})v(\text{m/s})}$ where h is Planck's constant, v is velocity

The units are as stated because $1 \text{ J} = 1 \text{ kg·m}^2/\text{s}^2$.

The mass of a proton is $1.67 \times 10^{-24} \text{ g} \times 1 \text{ kg}/1000 \text{ g} = 1.67 \times 10^{-27} \text{ kg}$.

The velocity of the proton, $v = 2.50 \times 10^7 \text{ m/s}$ (1/12 of the speed of light)

Substituting,

$$\lambda = \frac{h}{mv} = \frac{6.63 \times 10^{-34} \text{ J·s}}{(1.67 \times 10^{-27} \text{ kg})(2.50 \times 10^7 \text{ m/s})} = \mathbf{1.59 \times 10^{-14} \text{ m}}$$

(b) For the stone,   mass (kg) $= 30.0 \text{ g} \times 1 \text{ kg}/1000 \text{ g} = 0.0300 \text{ kg}$

$$v \text{ (m/s)} = \frac{2.00 \times 10^3 \text{ m}}{1 \text{ h}} \times \frac{1 \text{ h}}{60 \text{ min}} \times \frac{1 \text{ min}}{60 \text{ s}} = 0.556 \text{ m/s}$$

$$\lambda = \frac{h}{mv} = \frac{6.63 \times 10^{-34} \text{ J·s}}{(0.0300 \text{ kg})(0.556 \text{ m/s})} = \mathbf{3.97 \times 10^{-32} \text{ m}}$$

(c) Radii of atoms range from 0.4 Å (4×10^{-11} m) to 3 Å (3×10^{-10} m).

The wavelength of a proton, calculated in (a), is 4 orders of magnitude smaller than the radius of a typical atom.

The wavelength of a 30 g stone is much smaller by 22 orders of magnitude than a typical atom's radius.

5-64. *Refer to Section 5-13.*

Plan: (1) Calculate the mass (M) of an alpha particle (He nucleus, He^{2+}) in kilograms.

(2) Calculate the velocity (v) from $\lambda = \dfrac{h}{mv}$

(Note: $1 \text{ J} = 1 \text{ kg·m}^2/\text{s}^2$)

(1) M $= 4.003 \text{ g/mol He}^{2+} \times \dfrac{1 \text{ mol He}^{2+}}{6.02 \times 10^{23} \text{ He}^{2+} \text{ ions}} \times \dfrac{1 \text{ kg}}{10^3 \text{ g}}$

$= 6.65 \times 10^{-27} \text{ kg/He}^{2+} \text{ ion}$

(2) $v = \dfrac{h}{m\lambda} = \dfrac{6.63 \times 10^{-34} \text{ J·s}}{(6.65 \times 10^{-27} \text{ kg})(0.529 \text{ Å})(1 \times 10^{-10} \text{ m/Å})}$

$= \dfrac{6.63 \times 10^{-34} \text{ kg·m}^2/\text{s}^2\text{·s}}{(6.65 \times 10^{-27} \text{ kg})(5.29 \times 10^{-11} \text{ m})}$

$= \mathbf{1.88 \times 10^3 \text{ m/s}}$

5-66. *Refer to Section 5-15.*

The angular momentum quantum number, ℓ, for a particular energy level as defined by the principle quantum number, n, depends on the value of n. ℓ can take integral values from 0 up to and including $(n - 1)$.

For example, when $n = 3$, $\ell = 0$, 1, or 2.

5-68. *Refer to Section 5-16, and Figures 5-23 and 5-24.*

An orbital described by $n = 3$, $\ell = 1$, $m_\ell = -1$, and $m_s = -1/2$ is a **3p orbital**. See Figures 5-23 and 5-24 for illustrations.

5-70. *Refer to Section 5-15.*

The maximum number of electrons as designated by sets of quantum numbers:

(a) 6 electrons since quantum numbers $n = 3$ and $\ell = 1$ represent the 3p subshell with 3 orbitals.

(b) 10 electrons since $n = 3$ and $\ell = 2$ represent the 3d subshell with 5 orbitals.

(c) 2 electrons since $n = 3$, $\ell = 0$ and $m_\ell = 0$ represent the 3s orbital.

(d) 2 electrons since $n = 3$, $\ell = 1$ and $m_\ell = -1$ represent one of the 3p orbitals.

(e) 1 electron since $n = 3$, $\ell = 1$, $m_\ell = 0$ and $m_s = -1/2$ represent one of the electrons in a 3p orbital.

5-72. *Refer to Sections 5-14, 5-15 and 5-16, Table 5-4, and Figures 5-21, 5-22, 5-23, 5-24 and 5-25.*

(a) There are **3** energy sublevels in the third major energy level, $n = 3$, corresponding to $\ell = 0$, 1, and 2.

(b) Both $\ell = 1$ and $\ell = 2$ have sets of equivalent orbitals. The $\ell = 1$ (3p) energy sublevel has 3 equivalent orbitals and the $\ell = 2$ (3d) energy sublevel has 5 equivalent orbitals.

(c) Illustrations of these orbitals can be found in Figures 5-21 through 5-25, and 5-27.

5-74. *Refer to Sections 5-14 and 5-15, and Table 5-4.*

There are 9 individual orbitals in the third shell, $n = 3$

n	ℓ	m_ℓ	orbital
3	0	0	3s
3	1	-1	3p
3	1	0	3p
3	1	1	3p

n	ℓ	m_ℓ	orbital
3	2	-2	3d
3	2	-1	3d
3	2	0	3d
3	2	+1	3d
3	2	+2	3d

5-76. *Refer to Sections 5-15 and 5-16, and Table 5-4.*

(a) $n = 3$, $\ell = 0$ 3s subshell   (c) $n = 5$, $\ell = 0$ 5s subshell

(b) $n = 3$, $\ell = 1$ 3p subshell   (d) $n = 3$, $\ell = 2$ 3d subshell

5-78. *Refer to Section 5-16.*

	Designation	Number of Orbitals			Designation	Number of Orbitals
(a)	$4p$	3		(e)	$6d$	5
(b)	$3s$	1		(f)	$5d$	5
(c)	$3p_x$	1		(g)	$5f$	7
(d)	$n = 3$	9		(h)	$6s$	1

5-80. *Refer to Section 5-16 and Figures 5-21, 5-22, 5-23, 5-24, 5-25 and 5-27.*

(a) A $1s$ and a $2s$ orbital, like all s orbitals, can be described as *spherically symmetrical*, i.e., round like a ball. A $2s$ orbital is larger than a $1s$ orbital.

(b) A $3p_x$ orbital resembles an equal-arm dumbbell centered on the nucleus and lying along the x-axis, whereas a $2p_y$ orbital resembles an equal-arm dumbbell as well centered on the nucleus but lying along the y-axis. They are both shaped the same. The $3p$ orbital is larger than the $2p$ orbital and they lie along different axes.

5-82. *Refer to Sections 5-16 and 5-17.*

		$1s$	$2s$	$2p$	$3s$	$3p$	$3d$	$4s$	$4p$
(a)	$_{15}$P	↑↓	↑↓	↑↓ ↑↓ ↑↓	↑↓	↑ ↑ ↑			
(b)	$_{28}$Ni	↑↓	↑↓	↑↓ ↑↓ ↑↓	↑↓	↑↓ ↑↓ ↑↓	↑↓ ↑↓ ↑↓ ↑ ↑	↑↓	
(c)	$_{31}$Ga	↑↓	↑↓	↑↓ ↑↓ ↑↓	↑↓	↑↓ ↑↓ ↑↓	↑↓ ↑↓ ↑↓ ↑↓ ↑↓	↑↓	↑ _ _
(d)	$_{48}$Cd	↑↓	↑↓	↑↓ ↑↓ ↑↓	↑↓	↑↓ ↑↓ ↑↓	↑↓ ↑↓ ↑↓ ↑↓ ↑↓	↑↓	↑↓ ↑↓ ↑↓

	$4d$	$4f$	$5s$
	↑↓ ↑↓ ↑↓ ↑↓ ↑↓	_ _ _ _ _ _ _	↑↓

5-84. *Refer to Sections 5-16 and 5-17.*

(a) The ground state for Si was incorrect because the $3s$ electrons should have been paired up. The corrected electronic configuration is below.

(b) Ni is not an exception to the filling rule, so the 4s electrons should be paired, with two unpaired electrons in the $3d$ energy level. See below for the corrected electronic configuration.

(c) For S, Hund's Rule must be obeyed. The electrons should fill up the $3p$ energy level singly before pairing begins. See below.

		$1s$	$2s$	$2p$	$3s$	$3p$	$3d$	$4s$	$4p$
(a)	$_{14}$Si	↑↓	↑↓	↑↓ ↑↓ ↑↓	↑↓	↑ ↑ _			
(b)	$_{28}$Ni	↑↓	↑↓	↑↓ ↑↓ ↑↓	↑↓	↑↓ ↑↓ ↑↓	↑↓ ↑↓ ↑↓ ↑ ↑	↑↓	
(c)	$_{16}$S	↑↓	↑↓	↑↓ ↑↓ ↑↓	↑↓	↑↓ ↑ ↑			

5-86. *Refer to Sections 5-16 and 5-17, and Appendix B.*

From the information given, we can deduce the following ground state configuration: $1s^2\,2s^2\,2p^6\,3s^2\,3p^6\,4s^2$.

(a) The element has 20 electrons. Because it is neutral, there must be 20 protons. So, the element has atomic number 20 and is calcium, Ca.

(b) The element appears in Period 4.

(c) The element appears in Group 2A.

(d) Calcium has a total of 8 s electrons.

(e) Calcium has a total of 12 p electrons.

(f) Calcium has no d electrons.

5-88. *Refer to Sections 5-16 and 5-17, and Appendix B.*

The first five elements that have an unpaired electron in an s orbital are:

$$_1\text{H},\ _3\text{Li},\ _{11}\text{Na},\ _{19}\text{K and }_{24}\text{Cr.}$$

All but $_{24}$Cr can be found in Group 1A.

5-90. *Refer to Sections 5-16 and 5-17, Table 5-5 and Appendix B.*

(a) 1 outer electron in the first shell: H

(b) 3 outer electrons in the second shell: B

(c) 3 outer electrons in the third shell: Al

(d) 2 outer electrons in the seventh shell: Ra, Ac, Th, Pa, U, Np, Pu, Am, Cm, Bk, Cf, Es, Fm, Md, No, Lr, Rf, Db, Sg, Bh, Hs, Mt

(e) 4 outer electrons in the third shell: Si

(f) 8 outer electrons in the fifth shell: Xe

5-92. *Refer to Sections 5-16 and 5-17, and Exercise 5-82.*

(a) $_{15}$P $1s^2 2s^2 2p^6 3s^2 3p^3$ or $[\text{Ne}]3s^2 3p^3$

(b) $_{28}$Ni $1s^2 2s^2 2p^6 3s^2 3p^6 3d^8 4s^2$ or $[\text{Ar}]3d^8 4s^2$

(c) $_{31}$Ga $1s^2 2s^2 2p^6 3s^2 3p^6 3d^{10} 4s^2 4p^1$ or $[\text{Ar}]3d^{10} 4s^2 4p^1$

(d) $_{48}$Cd $1s^2 2s^2 2p^6 3s^2 3p^6 3d^{10} 4s^2 4p^6 4d^{10} 4f^0 5s^2$ or $[\text{Kr}]4d^{10} 5s^2$

5-94. *Refer to Section 5-17.*

Hund's Rule states that electrons in the ground state must occupy all the orbitals of a given subshell singly before pairing begins. These unpaired electrons have parallel spins.

The electron configurations that violate Hund's Rule are

(b) $1s^2 2s^2 2p_x^2$. The second electron in $2p_x$ should have occupied other $2p$ orbitals before pairing.

(d) $1s^22s^12p_x^12p_z^1$. There should be 2 electrons in the $2s$ orbital before moving on to the $2p$ orbitals.

(e) $1s^22s^12p_x^22p_y^12p_z^1$. There should be 2 electrons in the $2s$ orbital before moving on to the $2p$ orbitals. The $2p$ orbitals were filled properly, according to Hund's Rule.

5-96. *Refer to Sections 5-17 and 5-18.*

The paramagnetic elements with atomic numbers of 11 or less include:

H, Li, B, C, N, O, F and Na.

5-98. *Refer to Section 5-17.*

$_{29}$Cu	$1s^22s^22p^63s^23p^63d^{10}4s^1$	or	$[Ar]3d^{10}4s^1$
$_8$O	$1s^22s^22p^4$	or	$[He]2s^22p^4$
$_{57}$La	$1s^22s^22p^63s^23p^63d^{10}4s^24p^64d^{10}5s^25p^65d^16s^2$	or	$[Xe]5d^16s^2$
$_{39}$Y	$1s^22s^22p^63s^23p^63d^{10}4s^24p^64d^15s^2$	or	$[Kr]4d^15s^2$
$_{56}$Ba	$1s^22s^22p^63s^23p^63d^{10}4s^24p^64d^{10}5s^25p^66s^2$	or	$[Xe]6s^2$
$_{81}$Tl	$1s^22s^22p^63s^23p^63d^{10}4s^24p^64d^{10}4f^{14}5s^25p^65d^{10}6s^26p^1$	or	$[Xe]4f^{14}5d^{10}6s^26p^1$
$_{83}$Bi	$1s^22s^22p^63s^23p^63d^{10}4s^24p^64d^{10}4f^{14}5s^25p^65d^{10}6s^26p^3$	or	$[Xe]4f^{14}5d^{10}6s^26p^3$

5-100. *Refer to Section 5-17 and Appendix B.*

(a) $_{35}$Br

(b) $_{84}$Po

(c) $_{102}$No

(d) $_{43}$Tc

(e) $_{22}$Ti

5-102. *Refer to Section 5-17 and Appendix B.*

		Number of Electrons		
		s	*p*	*d*
(a) $_{15}$P	$1s^22s^22p^63s^23p^3$	6	9	0
(b) $_{36}$Kr	$1s^22s^22p^63s^23p^63d^{10}4s^24p^6$	8	18	10
(c) $_{28}$Ni	$1s^22s^22p^63s^23p^63d^84s^2$	8	12	8
(d) $_{30}$Zn	$1s^22s^22p^63s^23p^63d^{10}4s^2$	8	12	10
(e) $_{22}$Ti	$1s^22s^22p^63s^23p^63d^24s^2$	8	12	2

5-104. *Refer to Sections 5-15, 5-16 and 5-17, and Examples 5-8 and 5-9.*

(a) $_{19}$K $\qquad 1s^22s^22p^63s^23p^64s^1$

Electron	n	ℓ	m_ℓ	m_s
1	1	0	0	+1/2
2	1	0	0	−1/2
3	2	0	0	+1/2
4	2	0	0	−1/2
5	2	1	−1	+1/2
6	2	1	0	+1/2
7	2	1	+1	+1/2
8	2	1	−1	−1/2
9	2	1	0	−1/2
10	2	1	+1	−1/2
11	3	0	0	+1/2
12	3	0	0	−1/2
13	3	1	−1	+1/2
14	3	1	0	+1/2
15	3	1	+1	+1/2
16	3	1	−1	−1/2
17	3	1	0	−1/2
18	3	1	+1	−1/2
19	4	0	0	+1/2

(c) $_{20}$Ca $\qquad 1s^22s^22p^63s^23p^64s^2$

Electron	n	ℓ	m_ℓ	m_s
1	1	0	0	+1/2
2	1	0	0	−1/2
3	2	0	0	+1/2
4	2	0	0	−1/2
5	2	1	−1	+1/2
6	2	1	0	+1/2
7	2	1	+1	+1/2
8	2	1	−1	−1/2
9	2	1	0	−1/2
10	2	1	+1	−1/2
11	3	0	0	+1/2
12	3	0	0	−1/2
13	3	1	−1	+1/2
14	3	1	0	+1/2
15	3	1	+1	+1/2
16	3	1	−1	−1/2
17	3	1	0	−1/2
18	3	1	+1	−1/2
19	4	0	0	+1/2
20	4	0	0	−1/2

(b) $_8$O $\qquad 1s^22s^22p^4$

Electron	n	ℓ	m_ℓ	m_s
1	1	0	0	+1/2
2	1	0	0	−1/2
3	2	0	0	+1/2
4	2	0	0	−1/2
5	2	1	−1	+1/2
6	2	1	0	+1/2
7	2	1	+1	+1/2
8	2	1	−1	−1/2

5-106. *Refer to Sections 5-17 and 5-19, and Table 5-5.*

	ns	*np*			*ns*	*np*
1A	↑	_ _ _		5A	↑↓	↑ ↑ ↑
2A	↑↓	_ _ _		6A	↑↓	↑↓ ↑ ↑
3A	↑↓	↑ _ _		7A	↑↓	↑↓ ↑↓ ↑
4A	↑↓	↑ ↑ _		8A	↑↓	↑↓ ↑↓ ↑↓

74

5-108. *Refer to Sections 5-17 and 5-19, Examples 5-8 and 5-9, Table 5-5 and Appendix B.*

(a) $_{14}$Si   [Ne]$3s^23p^2$ The highest energy or "last" electron entered a $3p$ orbital. Therefore, $n = 3$, $\ell = 1$, and $m_\ell = -1$, **0 or +1.**

(b) $_{89}$Ac   [Rn]$7s^26d^1$ The highest energy or "last" electron went into a $6d$ orbital. Therefore, $n = 6$, $\ell = 2$, and $m_\ell = -2$, **-1, 0, +1, or +2.**

(c) $_{17}$Cl   [Ne]$3s^23p^5$ The highest energy or "last" electron entered a $3p$ orbital. Therefore, $n = 3$, $\ell = 1$, and $m_\ell = -1$, **0 or +1.**

(d) $_{59}$Pr   [Kr]$4d^{10}4f^35s^25p^66s^2$ The highest energy or "last" electron entered a $4f$ orbital. Therefore, $n = 4$, $\ell = 3$, and $m_\ell = -3$, **-2, -1, 0, +1, +2 or +3.**

5-110. *Refer to Sections 5-17 and 5-19, Table 5-5, and Appendix B.*

$A \equiv {_{29}}$Cu $1s^22s^22p^63s^23p^63d^{10}4s^1$

$B \equiv {_{40}}$Zr $1s^22s^22p^63s^23p^63d^{10}4s^24p^64d^25s^2$

$C \equiv {_{54}}$Xe $1s^22s^22p^63s^23p^63d^{10}4s^24p^64d^{10}5s^25p^6$

$D \equiv {_{83}}$Bi $1s^22s^22p^63s^23p^63d^{10}4s^24p^64d^{10}4f^{14}5s^25p^65d^{10}6s^26p^3$

$E \equiv {_4}$Be $1s^22s^2$

5-112. *Refer to Sections 5-17 and 5-19, Example 5-11 and Appendix B.*

Given below are the shorthand notations and number of unpaired electrons for the elements in their ground states.

	Shorthand	**Unpaired Electrons**
$_{11}$Na	[Ne]$3s^1$	1
$_{10}$Ne	$1s^22s^22p^6$	0
$_{13}$Al	$1s^22s^22p^63s^22p^1$	1
$_4$Be	$1s^22s^2$	0
$_{35}$Br	[Ar]$3d^{10}4s^24p^5$	1
$_{33}$As	[Ar] $3d^{10}4s^24p^3$	3
$_{22}$Ti	[Ar]$3d^24s^2$	2

5-114. *Refer to Sections 5-17 and 5-18, and Appendix B.*

An element is paramagnetic if it has any unpaired electrons. Given below are the shorthand notations and number of unpaired electrons present for the elements and ions in their ground states.

		Shorthand	**Unpaired Electrons**	**Paramagnetic**
(a)	$_9$F	$1s^22s^22p^5$	1	yes
(b)	$_{36}$Kr	$1s^22s^22p^63s^23p^63d^{10}4s^24p^6$	0	no
(c)	$_{10}$Ne$^+$	$1s^22s^22p^5$	1	yes
(d)	$_{26}$Fe	[Ar]$3d^64s^2$	4	yes
(e)	$_{17}$Cl$^-$	[Ne]$3s^23p^6$ or [Ar]	0	no

5-116. *Refer to Section 5-9 and Table 5-3.*

The atomic mass of chlorine is reported as 35.5 to three significant figures. No single atom of chlorine has that mass because the atomic mass of any element is the weighted average of all the isotopes, not the mass of any one atom. Chlorine is ~76% ^{35}Cl and ~24% ^{37}Cl.

5-118. *Refer to Section 5-14.*

The following statement is false: "If an atom has two electrons in its $2p_x$ orbital, one electron must be in one lobe of the orbital and the other electron must be in the other lobe." Heisenberg's Uncertainty Principle states that we cannot know exactly where the electrons are in any orbital.

5-120. *Refer to Section 5-16, and Figures 5-22, 5-24, 5-25 and 5-27.*

(a) $3p_x$ (Figure 5-24)

(b) $2s$ (Figures 5-22 and 5-27)

(c) $3d_{xy}$ (Figure 5-25)

(d) $3d_{z^2}$ (Figure 5-25)

5-122. *Refer to Section 5-15 and 5-17.*

An atom in its ground state containing 18 electrons (argon) has 3 s ($\ell = 0$) orbitals: $1s$, $2s$ and $3s$. These 3 orbitals hold 2 electrons each for a total of **6 electrons**.

5-124. *Refer to Section 5-17.*

For $_3$Li
(a) ground state: $1s^2 2s^1$
(b) excited state: $1s^2 2p^1$
(c) forbidden or impossible state: $1s^3$

5-126. *Refer to Section 5-8.*

(a),(b) There are four different HCl molecules which can be formed from the naturally-occurring hydrogen and chlorine isotopes:

Molecules	^{1}H^{35}Cl	^{1}H^{37}Cl	^{2}H^{35}Cl	^{2}H^{37}Cl
Approximate Masses	36 amu	38 amu	37 amu	39 amu

(c) From the relative abundance of the isotopes, the expected abundances of the molecules are in the following decreasing order:

$$^1H^{35}Cl > {}^1H^{37}Cl > {}^2H^{35}Cl > {}^2H^{37}Cl$$

5-128. *Refer to Sections 5-10 and 5-12.*

Plan: (1) Use $E = hc/\lambda$ to calculate the energy of the photon emitted by an excited sodium atom in joules.
(2) Convert energy (J) to mass (kg) using Einstein's equation, $E = mc^2$, where $1\ \text{J} = 1\ \text{kg·m}^2/\text{s}^2$.

(1) $\lambda = 589\ \text{nm} = 589 \times 10^{-9}\ \text{m} = 5.89 \times 10^{-7}\ \text{m}$

$$E \text{ per photon} = \frac{(6.63 \times 10^{-34}\ \text{J·s})(3.00 \times 10^8\ \text{m/s})}{(5.89 \times 10^{-7}\ \text{m})} = 3.38 \times 10^{-19}\ \text{J/photon}$$

(2) ? mass of one photon (kg) $= \dfrac{E}{c^2} = \dfrac{3.38 \times 10^{-19}\ \text{J}}{(3.00 \times 10^8\ \text{m/s})^2} = \mathbf{3.76 \times 10^{-36}\ kg}$

5-130. *Refer to Sections 5-10 and 5-12.*

Plan: (1) Use $E = hc/\lambda$ to calculate the energy loss in J/atom of barium.
(2) Convert J/atom into kJ/mol.

(1) $\lambda = 554\ \text{nm} = 554 \times 10^{-9}\ \text{m} = 5.54 \times 10^{-7}\ \text{m}$

$$E \text{ lost per atom} = \frac{(6.63 \times 10^{-34}\ \text{J·s})(3.00 \times 10^8\ \text{m/s})}{(5.54 \times 10^{-7}\ \text{m})} = 3.59 \times 10^{-19}\ \text{J/atom}$$

(2) ? kJ/mol Ba atoms $= 3.59 \times 10^{-19}\ \text{J/atom} \times \dfrac{6.02 \times 10^{23}\ \text{atoms}}{1\ \text{mol}} \times \dfrac{1\ \text{kJ}}{1000\ \text{J}} = \mathbf{216\ kJ/mol}$

5-132. *Refer to Section 5-10.*

We know: $c\ (\text{m/s}) = \lambda\ (\text{m}) \times \nu\ (\text{s}^{-1})$

Since $\nu = 89.5\ \text{MHz} = 8.95 \times 10^7\ \text{s}^{-1}$ and $c = 3.00 \times 10^8\ \text{m/s}$

$$\lambda\ (\text{m}) = \frac{c\ (\text{m/s})}{\nu\ (\text{s}^{-1})} = \frac{3.00 \times 10^8\ \text{m/s}}{8.95 \times 10^7\ \text{s}^{-1}} = \mathbf{3.35\ m}$$

5-134. *Refer to Section 5-5.*

(a) ? electrons per 1 mol N_2 molecules $= \dfrac{7\ \text{electrons}}{1\ \text{atom N}} \times \dfrac{2\ \text{atoms N}}{1\,\text{molecule } N_2} \times \dfrac{6.02 \times 10^{23}\ \text{molecules } N_2}{1\ \text{mol } N_2\ \text{molecules}}$

$$= \mathbf{8.43 \times 10^{24}\ electrons}$$

(b) ? electrons $= 30.0\ \text{g } H_2O \times \dfrac{1\ \text{mol } H_2O}{18.0\ \text{g } H_2O} \times \dfrac{6.02 \times 10^{23}\ \text{molecules } H_2O}{1\ \text{mol } H_2O} \times \dfrac{10\ \text{electrons}}{1\ \text{molecule } H_2O} = \mathbf{1.00 \times 10^{25}\ electrons}$

Note: there are 10 electrons in an H_2O molecule because there is 1 electron per hydrogen atom and 8 electrons in an oxygen atom, for a total of 10 electrons.

6 Chemical Periodicity

6-2. *Refer to Section 6-1*

The general order in which the shells are filled starting with the $n = 3$ shell is:

$$3s \quad 3p \quad 4s \quad 3d \quad 4p \quad \text{etc.}$$

Period 3 includes the elements whose outer electrons are in $3s$ or $3p$ subshells. The maximum number of electrons in these subshells is a total of 8. Hence Period 3 contains only 8 elements. Since the $3d$ subshell is higher in energy than the $4s$ subshell, it is not going to be filled until Period 4.

6-4. *Refer to Section 6-1 and Table 5-5.*

The atomic number of the yet-to-be discovered alkali metal in period 8 is 119. The last portion of its electron configuration after [Rn] (atomic number = 86) should be:

$$7s^2 \, 5f^{14} \, 6d^{10} \, 7p^6 \, 8s^1$$

6-6. *Refer to Section 6-1.*

(a) ns^2np^6 Group 8A (noble gases)

(b) ns^1 Group 1A (alkali metals)

(c) $ns^2(n{-}1)d^{1\text{-}10}$ d-transition elements

(d) ns^2np^1 Group 3A

6-8. *Refer to Section 6-1.*

(a) alkali metals	B	(h) actinides	D
(b) outer configuration of d^7s^2	H	(i) d-transition elements	E, H, K
(c) lanthanides	A	(j) noble gases	G
(d) p-block representative elements	C, F, G, I	(k) alkaline earth elements	J
(e) partially filled f-subshells	A, D		
(f) halogens	I		
(g) s-block representative elements	B, J		

6-10. *Refer to Section 6-5 and the Key Terms for Chapter 6.*

Isoelectronic species are species that have the same total number of electrons.

O^{2-}, F^-, Ne, Mg^{2+} and Al^{3+} have 10 electrons and therefore are isoelectronic species. However, Na has 11 electrons and is not isoelectronic with the others.

Electrons that are in filled sets of orbitals between the nucleus and outer shell electrons shield the outer shell electrons partially from the effect of the protons in the nucleus; this effect is called nuclear shielding.

As we move from left to right along a period, the outer shell electrons do experience a progressively stronger force of attraction to the nucleus due to the combination of an increase in the number of protons and a constant nuclear shielding by inner electrons. As a result the atomic radii decrease.

As we move down a group, the outer electrons are partially shielded from the attractive force of the nucleus by an increasing number of inner electrons. This effect is *partially* responsible for the observed increase in atomic radii going down a group.

6-14. *Refer to Sections 6-2 and 6-5, and Figures 6-1 and 6.4.*

Consider the element with atomic number 116 in Group 6A. Even though it has not been isolated, its atomic radius is expected to be somewhat larger than that of Po (1.68 Å), probably about 1.9 - 2.0 Å, since it lies just below Po on the periodic table. Its outer electrons would lie in the $n=7$ shell, which would be further away from the nucleus than Po's outermost electrons in the $n=6$ shell. The ionic radius of the element with a 2– charge, is expected to be somewhat larger than that of Po^{2-} and Te^{2-}.

6-16. *Refer to Section 6-2, Figure 6-1 and Example 6-1.*

Atomic radii increase from top to bottom within a group and from right to left within a period. Therefore, in order of increasing size, we have:

(a) $Be < Mg < Ca < Sr < Ba < Ra$

(b) $He < Ne < Ar < Kr < Xe < Rn$

(c) $Ar < Cl < S < P < Si < Al < Mg < Na$

(d) $N < B < Sb < Sr < Ba$

6-18. *Refer to Section 6-3.*

(a) The first ionization energy, IE_1, also called the first ionization potential, is the minimum amount of energy required to remove the most loosely bound electron from an isolated gaseous atom to form an ion with a 1+ charge.

$$X(g) + IE_1 \rightarrow X^+(g) + e^-$$

(b) The second ionization energy, IE_2, is the amount of energy required to remove a second electron from an isolated gaseous singly charged cation; i.e. to remove an electron from an ion with a 1+ charge to give an ion with a 2+ charge.

$$X^+(g) + IE_2 \rightarrow X^{2+}(g) + e^-$$

6-20. *Refer to Sections 6-2 and 6-3.*

As we move down a given group, the valence electrons are further and further away from the nucleus. The first ionization energies of the elements, which is the energy required to remove an electron from an isolated gaseous atom, decrease while the atomic radii increase.

Likewise, from left to right across a period, the forces of attraction between the outermost electron and the nucleus increase. Therefore the ionization energies increase while the atomic radii decrease. Refer to Figure 6-2 for the exceptions to the general trends for ionization energy.

Effective nuclear charge, Z_{eff}, felt by the outer electrons for the positively charged nucleus is less than the actual nuclear charge because there is repulsion between the outer electrons and the inner electrons, counteracting the attractive forces between the electrons and the protons in the nucleus. Z_{eff} increases going from left to right across a period, because the nuclear charge is increasing. The increase in effective nuclear charge causes the outermost electrons to be held more tightly making them harder to remove. Therefore, the first ionization energies generally increase from left to right across the periodic table.

6-24. *Refer to Section 6-3, Table 6-1, Figure 6-2 and Example 6-2.*

First ionization energies increase from left to right and bottom to top in the periodic table. However, there are exceptions: elements of Group 3A generally have lower first ionization energies than elements of Group 2A, and elements of Group 6A generally have lower first ionization energies than elements of Group 5A. Therefore, we obtain the following orders of increasing first ionization energies:

(a) $Fr < Cs < Rb < K < Na < Li$

(b) $At < I < Br < Cl < F$

(c) $Li < B < Be < C < O < N < F < Ne$

(d) $Cs < Ga < B < Br < H < F$

6-26. *Refer to Section 6-3 and Exercise 6-24 Solution.*

As we move from left to right across Period 2 of the periodic table, there is an increase in effective nuclear charge and a decrease in atomic radii. Outer valence electrons are held more tightly and first ionization energies *generally* increase. Therefore, as the atomic radii decrease, the first ionization energies increase. Refer to Figure 6-2 for the exceptions to the general trend for ionization energy.

6-28. *Refer to Section 6-3.*

It is difficult to prepare compounds containing Li^{2+} due to the immense amount of energy that is required to remove a second electron from an ion of lithium, i.e., there is a very large amount of energy (the second ionization energy) required for this reaction:

$$Li^+(g) + 7298 \text{ kJ/mol} \rightarrow Li^{2+}(g) + e^-$$

This energy is not likely to be repaid during compound formation. The reason for such a high second ionization energy for lithium is because the electron configuration of Li^+ is $1s^2$ which has a filled s orbital. It is the special stability of the filled s orbital which prevents the formation of Li^{2+} ions. Also, the formation of Li^{2+} requires 14 times more energy than the formation of Li^+ and so is much less likely.

On the other hand, Be^{2+} has the very stable electron configuration of $1s^2$, isoelectronic with the noble gas, He. Compounds with Be^{2+} ions are to be expected.

6-30. *Refer to Section 6-4, Figure 6-3, Table 6-2, and Example 6-3.*

The electron affinity of an element is defined as the amount of energy absorbed when an electron is added to an isolated gaseous atom to form an ion with a 1− charge.

In general, electron affinities become more negative from bottom to top and from left to right in the periodic table, but there are many exceptions. According to Table 6-2, the order of increasing negative values of electron affinity is:

(least negative EA) $P < S < Br < Cl$ (most negative EA)

6-32. *Refer to Section 6-4.*

Elements that gain electrons easily to form negative ions have very negative electron affinities. The halogens, with electronic configurations of $ns^2 np^5$, easily gain one electron to form stable ions with a filled set of p orbitals. These ions are isoelectronic with the noble gases and have noble gas electronic configurations, $ns^2 np^6$. Therefore, the halogens have the most negative electron affinities. This does not occur when a Group 6A element gains an electron.

6-34. *Refer to Section 6-4 and Table 6-2.*

			Electronic Configuration		
(a)	$O(g) + e^- \rightarrow O^-(g) + 141 \text{ kJ/mol}$	O	$1s^2\, 2s^2\, 2p^4$	O^-	$1s^2\, 2s^2\, 2p^5$
(b)	$Cl(g) + e^- \rightarrow Cl^-(g) + 349 \text{ kJ/mol}$	Cl	$[Ne]\, 3s^2\, 3p^5$	Cl^-	$[Ar]$
(c)	$Mg(g) + e^- + {\sim}0 \text{ kJ/mol} \rightarrow Mg^-(g)$	Mg	$[Ne]\, 3s^2$	Mg^-	$[Ne]\, 3s^2\, 3p^1$

6-36. *Refer to Section 6-5, Figure 6-4 and Example 6-4.*

(a) Within an isoelectronic series, ionic radii increase with decreasing atomic number. Therefore, in order of increasing ionic radii, we have
$$Ga^{3+} < Ca^{2+} < K^+$$

(b) Ionic radii increase down a group. So, $Be^{2+} < Mg^{2+} < Ca^{2+} < Ba^{2+}$

(c) $Al^{3+} < Sr^{2+} < K^+ < Rb^+$ (See Figure 6-4)

(d) $Ca^{2+} < K^+ < Rb^+$ (See Figure 6-4)

6-38. *Refer to Section 6-5, Figure 6-4, and Example 6-4.*

(a) In an isoelectronic series, ionic radii increase with decreasing atomic number because of decreasing nuclear charge. Therefore, in order of increasing ionic radii, we have
$$Cl^- < S^{2-} < P^{3-}$$

(b) Ionic radii increase down a group. So, $O^{2-} < S^{2-} < Se^{2-}$

(c) $S^{2-} < N^{3-} < Br^-$ and $S^{2-} < N^{3-} < P^{3-}$ but we don't know the size relationship between P^{3-} and Br^-. (See Figure 6-4)

(d) Ionic radii increase down a group. So, $Cl^- < Br^- < I^-$

6-40. *Refer to Section 6-5.*

The Fe^{2+} ion has 26 protons pulling on 24 electrons, whereas the Fe^{3+} ion has 26 protons pulling on 23 electrons. The electrons in the Fe^{3+} ion are more tightly held and therefore, Fe^{3+} is the smaller ion.

Likewise, the Sn^{2+} ion has 50 protons pulling on 48 electrons, whereas the Sn^{4+} ion has 50 protons attracting 46 electrons. The electrons in the Sn^{4+} ion are more tightly held and therefore, Sn^{4+} is the smaller ion.

6-42. *Refer to Section 6-6, Table 6-3 and Example 6-5.*

Electronegativities usually increase from left to right across periods and from bottom to top within groups. Exceptions are explained in Section 6-6.

(a) Pb < Sn < Ge < C   (b) Na < Mg < S < Cl   (c) Bi < Sb < P < N   (d) Ba < Sc < Si < Se < F

6-44. *Refer to Section 6-6 and Table 6-3.*

The electronegativity values of the metalloids are clustered between 1.8 to 2.1. These lie between the higher values of the nonmetals and the lower values of the metals.

6-46. *Refer to Section 6-2.*

atomic radius of F = 1/2 x bond length of F_2 = 1/2 x 1.42 Å = **0.71 Å**

atomic radius of Cl = 1/2 x bond length of Cl_2 = 1/2 x 1.98 Å = **0.99 Å**

predicted F-Cl bond length = (atomic radius of F) + (atomic radius of Cl) = 0.71 Å + 0.99 Å = **1.70 Å**
        (actual F-Cl bond length = 1.64 Å)

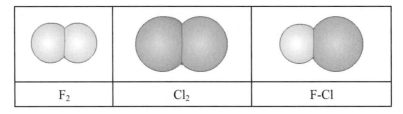

| F_2 | Cl_2 | F-Cl |

6-48. *Refer to Sections 6-3 and 6-4, and Tables 6-1 and 6-2.*

If we compare the values of the first ionization energy and electron affinity for the Period 3 elements, we have

	Na	Mg	Al	Si	P	S	Cl	Ar
First Ionization Energy (kJ/mol)	496	738	578	786	1012	1000	1251	1521
Electron Affinity (kJ/mol)	−53	(~0)	−43	−134	−72	−200	−349	0

The magnitude of the electron affinity values is less than that of the first ionization energies. It is much more difficult and hence more energy is required to remove an electron from a neutral gaseous atom, quantified by the first ionization energy, than to add an electron to a neutral gaseous atom, quantified by the electron affinity. In fact, many atoms actually release energy when an extra electron is added as denoted by the negative sign attached to the electron affinity value.

6-50. *Refer to Section 6-4 and Table 6-2.*

The electron affinities (in kJ/mol) of the halogens are: F (−328) > Cl (−349) < Br (−324) < I (−295).

The actual electron affinity of fluorine is less negative than expected from the trend set by the other halogens. This implies that it is more difficult to add an electron to fluorine than to chlorine. Although the variations in electron affinity down a group (family) are not so easily explained, one plausible reason is that because fluorine is small and the electron is added to the small $2p$ sublevel which is already crowded with five other electrons, the repulsion between the incoming electron and the others will cause a slightly lower value in electron affinity.

Elemental hydrogen exists as a colorless, odorless, tasteless, diatomic gas with the lowest atomic weight and density of any known substance. This flammable gas melts at $-259.14°C$ and boils at $-252.8°C$.

6-54. *Refer to Section 6-7 and Example 6-6.*

(a) Hydrogen gas reacts with the alkali metals and heavier alkaline earth metals to form ionic hydrides:

$$2Li(molten) + H_2(g) \rightarrow 2LiH(s)$$

(b) Hydrogen gas reacts with other nonmetals to form molecular (covalent) hydrides:

$$H_2(g) + Cl_2(g) \rightarrow 2HCl(g)$$

6-56. *Refer to Section 6-7 and Example 6-7.*

NaH, sodium hydride, is the product of hydrogen gas reacting with an active metal, sodium. A compound consisting of a metal and a nonmetal is ionic, hence this is an ionic hydride.

H_2S, hydrogen sulfide, is the product of hydrogen gas reacting with a nonmetal, sulfur. A compound consisting of two nonmetals is covalent, and therefore this is a molecular (covalent) hydride.

6-58. *Refer to Section 6-7.*

(a) H_2S hydrogen sulfide   (e) H_2Se hydrogen selenide

(b) HCl   hydrogen chloride   (f) MgH_2 magnesium hydride

(c) KH   potassium hydride   (g) AlH_3 aluminum hydride

(d) NH_3 ammonia

6-60. *Refer to Sections 6-7 and 6-8.*

H_2, hydrogen, is a colorless, odorless, tasteless, nonpolar, diamagnetic, diatomic gas with the lowest atomic weight and density of any known substance. It has low solubility in water and is very flammable. Hydrogen is prepared by reactions of metals with water, steam or various acids, electrolysis of water, the water gas reaction and thermal cracking of hydrocarbons. It combines with metals and nonmetals to form hydrides.

O_2, oxygen, is nearly colorless, odorless, tasteless, nonpolar, paramagnetic, diatomic gas. It is nonflammable but participates in all combustion reactions. It is prepared by cooling air until it changes to a liquid, then separating the gas components, electrolysis of water and thermal decomposition of certain oxygen-containing salts. Oxygen combines with almost all other elements to form oxides and can be converted to an allotropic form, ozone, O_3.

6-62. *Refer to Section 6-8 and Table 6-4.*

The elements that react with oxygen to form primarily normal oxides include:
     (a) Li,
     (d) Mg,
     (e) Zn and
     (f) Al.

6-64. *Refer to Section 6-8 and Example 6-10.*

(a) $2C(s) + O_2(g) \rightarrow 2CO(g)$ (O_2 is limited)

(b) $As_4(s) + 3O_2(g) \rightarrow 2As_2O_3(s)$ (O_2 is limited)

(c) $2Ge(s) + O_2(g) \rightarrow 2GeO(s)$ (O_2 is limited)

6-66. *Refer to Section 6-8 and Table 6-4.*

A normal oxide is a binary (two element) compound containing oxygen in the –2 oxidation state. BaO is an example of an ionic oxide and SO_2 is an example of a molecular (covalent) oxide.

A peroxide can be a binary ionic compound containing the O_2^{2-} ion, such as Na_2O_2, or a covalent compound, such as H_2O_2, with oxygen in the –1 oxidation state.

A superoxide is a binary ionic compound containing the O_2^- ion with oxygen in the –1/2 oxidation state, such as KO_2.

6-68. *Refer to Section 6-8 and Example 6-10.*

(a) $CO_2(g) + H_2O(\ell) \rightarrow H_2CO_3(aq)$ carbonic acid

(b) $SO_3(\ell) + H_2O(\ell) \rightarrow H_2SO_4(aq)$ sulfuric acid

(c) $SeO_3(s) + H_2O(\ell) \rightarrow H_2SeO_4(aq)$ selenic acid

(d) $N_2O_5(s) + H_2O(\ell) \rightarrow 2HNO_3(aq)$ nitric acid

(e) $Cl_2O_7(\ell) + H_2O(\ell) \rightarrow 2HClO_4(aq)$ perchloric acid

6-70. *Refer to Section 6-8.*

The acid anhydrides are:

(a) SO_3

(b) CO_2

(c) SO_2

(d) P_2O_5

(e) N_2O_3

6-72. *Refer to Section 6-8.*

Combustion is an oxidation-reduction reaction in which oxygen gas combines rapidly with oxidizable materials in highly exothermic reactions usually with a visible flame. A combustion reaction is a redox reaction. The oxygen atoms are being reduced since the oxidation number of oxygen is changed from 0 in O_2 to –2 in the products, usually CO_2 and H_2O when oxidizing hydrocarbons, while the other reactants have elements being oxidized.

6-74. *Refer to Section 6-8.*

(a) $2CH_4(g) + 3O_2(g) \rightarrow 2CO(g) + 4H_2O(g)$ (O_2 is limited)

(b) $2C_3H_8(g) + 7O_2(g) \rightarrow 6CO(g) + 8H_2O(g)$ (O_2 is limited)

Balanced reaction: $C_3H_8(g) + 5O_2(g) \rightarrow 3CO_2(g) + 4H_2O(g)$

However, this is a limiting reactant problem with propane, C_3H_8, being the limiting reactant since we are given 7 molecules of O_2 and not 5. When the reaction is complete, there will be 2 molecules of O_2 remaining unreacted.

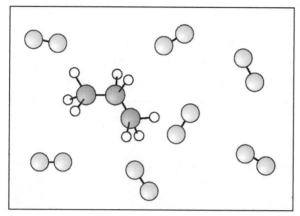

Before Reaction

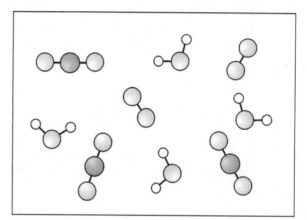

After Reaction

6-78. *Refer to Section 6-8.*

(a) $4C_6H_5NH_2(\ell) + 33O_2(g) \rightarrow 24CO_2(g) + 14H_2O(\ell) + 4NO(g)$

(b) $2C_2H_5SH(\ell) + 9O_2(g) \rightarrow 4CO_2(g) + 6H_2O(g) + 2SO_2(g)$

(c) $C_7H_{10}NO_2S(\ell) + 10O_2(g) \rightarrow 7CO_2(g) + 5H_2O(g) + NO(g) + SO_2(g)$

6-80. *Refer to Section 6-2 and Figure 6-1.*

Within a family or group, atomic radii increase from top to bottom as electrons are added to shells further and further from the nucleus. As we move from left to right across a period, atomic radii decrease. Therefore, the atomic radii are most likely associated with the following atoms and Ge must be **122 pm**, where 1 pm = 1 picometer = 1×10^{-12} m.

	Si - 118 pm	P -110 pm
Ga - 135 pm	Ge - 122 pm	As - 120 pm

6-82. *Refer to Section 6-2, and Figures 6-1 and 6-4.*

The electronic configurations for beryllium and magnesium:

$_4$Be $\quad 1s^2 2s^2 \qquad$ and $\qquad _{12}$Mg $\quad 1s^2 2s^2 2p^6 3s^2$

show that both elements are in Group 2A. Both are metals, exhibiting +2 oxidation number in all their compounds. They would be expected to form stable 2+ ions, with Mg having more metallic character than Be.

When we compare the sizes of their most stable ions, Be^{2+} and Mg^{2+}, we see that $Mg^{2+} > Be^{2+}$.

6-84. *Refer to Section 6-7 and Figure 6-1.*

Hydrogen gas obtained cheaply from the electrolysis of water: $2H_2O(\ell) \rightarrow 2H_2(g) + O_2(g)$
may be used to fuel automobiles in the future. The pollution would be greatly decreased, however, there may be problems with:

(1) safe storage and combustion of the gas which is very flammable,

(2) transferring the gas safely into the vehicle at neighborhood "hydrogen stations,"

(3) managing the high volume electrolysis plants so that the hydrologic cycle would not be overly disturbed, and

(4) pollution generated by the electricity required to electrolyze the water in the first place.

6-86. *Refer to Sections 6-3 and 5-10.*

Recall: For 1 atom, E (J/atom) $= h$ (J·s) x v (s^{-1})

For 1 mole of atoms, E (J/mol) $= hvN$ where N is Avogadro's Number

Solving for v, we have

$$v \text{ (s}^{-1}) = \frac{E}{hN} = \frac{419 \text{ kJ/mol x 1000 J/kJ}}{(6.63 \times 10^{-34} \text{ J·s})(6.02 \times 10^{23} \text{ mol}^{-1})} = \mathbf{1.05 \times 10^{15} \text{ s}^{-1}}$$

6-88. *Refer to Section 6-3 and Table 6-1.*

First Ionization Energy for Mg (kJ/mol) = 738 kJ/mol $\qquad$ So, $\qquad$ $Mg(g) + 738 \text{ kJ/mol} \rightarrow Mg^+(g) + e^-$

Second Ionization Energy for Mg (kJ/mol) = 1451 kJ/mol $\qquad$ So, $\underline{\quad Mg^+(g) + 1451 \text{ kJ/mol} \rightarrow Mg^{2+}(g) + e^- \quad}$

$$Mg(g) + 2189 \text{ kJ/mol} \rightarrow Mg^{2+}(g) + 2e^-$$

And so, 2189 kJ/mol of energy is required to produce 1 mole of gaseous Mg^{2+} ions from gaseous Mg atoms. So, using dimensional analysis:

$$? \text{ energy (kJ)} = 1.75 \text{ g Mg} \times \frac{1 \text{ mol Mg}}{24.305 \text{ g Mg}} \times \frac{2189 \text{ kJ}}{1 \text{mol Mg}} = \mathbf{158 \text{ kJ}}$$

7   Chemical Bonding

7-2.   *Refer to Sections 7-2 and 7-3.*

Ionic bonding results from electrostatic interactions between ions, which can be formed by the *transfer* of one or more electrons from a *metal* to a *nonmetal* or group of nonmetals. Covalent bonding, on the other hand, results from *sharing* one or more electron pairs between two *nonmetal* atoms.

(a) $Na + Cl_2$ ionic bonding, since Na is a metal and Cl is a nonmetal
(b) $C + O_2$ covalent bonding, since both C and O are nonmetals
(c) $N_2 + O_2$ covalent bonding, since both N and O are nonmetals
(d) $S + O_2$ covalent bonding, since both S and O are nonmetals

7-4.   *Refer to Section 7-1 and Table 7-1.*

(a) Lewis dot representations for the representative elements show only the valence electrons in the outermost occupied *s* and *p* orbitals. Paired and unpaired electrons are also indicated.

(b)   He: Si·   ·P·   :Ne: Mg·   ·Br:

7-6.   *Refer to Sections 7-2 and 7-3.*

In $NaClO_3$, there is ionic bonding occurring between the Na^+ ion and the ClO_3^- ion, and covalent bonding between the O and Cl atoms in the ClO_3^- ion.

7-8.   *Refer to Sections 7-2 and 7-3.*

In general, the bond between a metal and a nonmetal is ionic, whereas the bond between two nonmetals is covalent. In other words, the further apart across the periodic table the two elements are, the more likely they are to form an ionic bond.

(a) Ca (metal) and Cl (nonmetal)   ionic bond   (d) Na (metal) and I (nonmetal)   ionic bond

(b) P (nonmetal) and O (nonmetal)   covalent bond   (e) Si (metalloid) and Br (nonmetal)   covalent bond

(c) Br (nonmetal) and I (nonmetal)   covalent bond   (f) Ba (metal) and F (nonmetal)   ionic bond

7-10.   *Refer to Sections 7-2 and 7-3.*

In general, whenever a metal and a nonmetal are together in a compound, it is ionic. If the compound consists only of nonmetals, it is covalent. In other words, the further apart two elements are on the periodic table, the more likely they are to form an ionic compound.

(a) $Ca(NO_3)_2$ metal + nonmetals   ionic (within the NO_3^- ion, there are covalent bonds)
(b) H_2Se nonmetals   covalent
(c) KNO_3 metal + nonmetals   ionic (within the NO_3^- ion, there are covalent bonds)
(d) $CaCl_2$ metal + nonmetal   ionic

(e) H_2CO_3 nonmetals   covalent (H is not a metal)

(f) NCl_3 nonmetals   covalent

(g) Li_2O metal + nonmetal   ionic

(h) N_2H_4 nonmetals   covalent

(i) $SOCl_2$ nonmetals   covalent

7-12.   *Refer to Section 7-2 and Chapter 13.*

An ionic crystal is a solid characterized by a regular, ordered arrangement of ions in three-dimensional space. The specific geometrical arrangement of the ions is controlled by

(1) the compound formula, i.e., the ratio of cations to anions,
(2) the size of the ions and
(3) the conditions (temperature and pressure) under which the solid exists.

7-14.   *Refer to Section 7-2 and Table 7-2.*

(a) $Ca + Cl_2 \rightarrow \textbf{CaCl}_2$

(b) $Ba + Cl_2 \rightarrow \textbf{BaCl}_2$

(c) $Na + 1/2Cl_2 \rightarrow \textbf{NaCl}$

7-16.   *Refer to Section 7-2 and Appendix B.*

(a) Cr^{3+} [Ar] $3d^3$ (e) Cu^{2+} [Ar] $3d^9$

(b) Mn^{2+} [Ar] $3d^5$ (f) Sc^{2+} [Ar] $3d^1$

(c) Ag^+ [Kr] $4d^{10}$ (g) Fe^{2+} [Ar] $3d^6$

(d) Fe^{3+} [Ar] $3d^5$

7-18.   *Refer to Section 7-2.*

Stable binary ionic compounds are formed from ions that have noble gas configurations. None of the compounds meet this requirement. First of all, CO_4 is not an ionic compound at all because it is a covalent compound, made from 2 nonmetals. Even so, CO_4 is not stable because with O^{2-}, C would have an oxidation number of +8, which is very unlikely. Consider the following ionic compounds composed of a metal and nonmetals:

MgI $(Mg^+ + I^-)$ Al(OH)$_2$ $(Al^{2+} + 2OH^-)$ InF$_2$ $(In^{2+} + 2F^-)$

RbCl$_2$ $(Rb^{2+} + 2Cl^-)$ CsS $(Cs^{2+} + S^{2-})$ Be$_3$O $(3Be^+ + O^{3-})$.

Neither Mg^+, Al^{2+}, In^{2+}, Rb^{2+}, Cs^{2+}, Be^+ nor O^{3-} have noble gas configurations.

7-20.   *Refer to Section 7-2.*

(a) Cations with ....$3s^2\, 3p^6$ electronic configurations are isoelectronic with argon. Examples: K^+, Ca^{2+}

(b) Cations with ....$6s^2\, 6p^6$ electronic configurations are isoelectronic with radon. Examples: Fr^+, Ra^{2+}

7-22. *Refer to Section 7-3 and Figure 7-4.*

Figure 7-4 is a plot of potential energy versus the distance between 2 hydrogen atoms. The resulting function is the sum of two opposing forces: (1) the attractive force between the negatively charged electron of one H atom and the positively charged nucleus of the other H atom, and (2) the repulsive force between the two positively charged nuclei.

When the two atoms are relatively far apart, there is essentially no interaction at all between them; both the attractive and repulsive forces are about zero. As the two atoms get closer, the attractive forces dominate, and the potential energy decreases to a minimum at a distance of 0.74 Å, which is the H–H bond length. At distances less than 0.74 Å, the repulsive forces become more important and the energy increases sharply.

7-24. *Refer to Section 7-3.*

(a) A single covalent bond contains 2 shared electrons.

(b) A double covalent bond contains 4 shared electrons.

(c) A triple covalent bond contains 6 shared electrons.

7-26. *Refer to Section 7-4, and Tables 7-3 and 7-4.*

Here is the listing of the bond energies and bond lengths of C–O, C=O, and C≡O:

	Single bond C–O	Double bond C=O	Triple bond C≡O
Bond energy (kJ/mol)	358	732 (799 in CO_2)	1072
Bond length (Å)	1.43	1.22	1.13

We can easily see that as we go from single bond to double bond to triple bond between carbon and oxygen, the bond energies increase while the bond lengths decrease.

	Single bond C–C	Double bond C=C	Triple bond C≡C
Bond energy (kJ/mol)	346	602	835
Bond length (Å)	1.54	1.34	1.21

The same relationships are seen as we go from single bond to double bond to triple bond between carbon and carbon; the bond energies increase while the bond lengths decrease.

Overall, the bond energies are less for C–C bonds and the bond lengths are greater than for the corresponding C–O bonds

7-28. *Refer to Section 7-5.*

Lewis formulas are representations of molecules or ions which show

- the element symbols,
- the order in which the atoms are connected,
- the number of valence electrons linking the atoms together,
- the number of lone pairs of valence electrons not used for bonding,
- and the number and kind of bonds.

They do not show the shape of a chemical species.

H_2S H:Ö:H   H-Ö-H

$S = N - A$

$= [2 \times 2(\text{for H}) + 1 \times 8(\text{for O})] - [2 \times 1(\text{for H}) + 1 \times 6(\text{for O})]$

$= 12 - 8$

$= 4$ (there are 4 electrons shared in the molecule)

NH_3 H:N̈:H / Ḧ   H-N̈-H / H

$S = N - A$

$= [3 \times 2(\text{for H}) + 1 \times 8(\text{for N})] - [3 \times 1(\text{for H}) + 1 \times 5(\text{for N})]$

$= 14 - 8$

$= 6$ (there are 6 electrons shared in the molecule)

OH^- [:Ö:H]⁻   [:Ö-H]⁻

$S = N - A$

$= [1 \times 2(\text{for H}) + 1 \times 8(\text{for O})] - [1 \times 1(\text{for H}) + 1 \times 6(\text{for O}) + 1e^-]$

$= 10 - 8$

$= 2$ (there are 2 electrons shared in the diatomic ion)

F^- :F̈:⁻

(a) SCl_2 :Cl:S:Cl:

$S = N - A$

$= [1 \times 8(\text{for S}) + 2 \times 8(\text{for Cl})] - [1 \times 6(\text{for S}) + 2 \times 7(\text{for Cl})]$

$= 4$ (there are 4 electrons shared)

(b) AsF_3 :F:As:F: / :F:

$S = N - A$

$= [1 \times 8(\text{for As}) + 3 \times 8(\text{for F})] - [1 \times 5(\text{for As}) + 3 \times 7(\text{for F})]$

$= 6$ (there are 6 electrons shared)

(c) ICl :I:Cl:

$S = N - A$

$= [1 \times 8(\text{for I}) + 1 \times 8(\text{for Cl})] - [1 \times 7(\text{for I}) + 1 \times 7(\text{for Cl})]$

$= 2$ (there are 2 electrons shared)

(d) NCl_3 :Cl:N:Cl: / :Cl:

$S = N - A$

$= [1 \times 8(\text{for N}) + 3 \times 8(\text{for Cl})] - [1 \times 5(\text{for N}) + 3 \times 7(\text{for Cl})]$

$= 6$ (there are 6 electrons shared)

The number of valence electrons in a compound is the sum of the valence electrons of each atom in the compound. If the species is a positively-charged ion, one must subtract the charge on the ion to determine the total number of valence electrons. If the species is negatively-charged, one must add the charge on the ion to determine the number of valence electrons.

(a) H_2Se $2 \times 1(\text{for H}) + 1 \times 6(\text{for Se}) = 8$ valence electrons

(b) PCl_3 $1 \times 5(\text{for P}) + 3 \times 7(\text{for Cl}) = 26$ valence electrons

(c) $NOCl$ $1 \times 5(\text{for N}) + 1 \times 6(\text{for O}) + 1 \times 7(\text{for Cl}) = 18$ valence electrons

(d) OH^- $1 \times 6(\text{for O}) + 1 \times 1(\text{for H}) + 1 \ e^- = 8$ valence electrons

7-36. *Refer to Sections 7-5 and 7-6, Exercise 7-34, and Examples 7-1, 7-2, and 7-3.*

(a) H_2Se

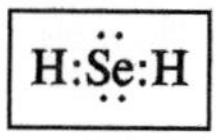

H:S̈e:H

$S = N - A$
$= [2 \times 2(\text{for H}) + 1 \times 8(\text{for Se})] - [2 \times 1(\text{for H}) + 1 \times 6(\text{for Se})]$
$= 4$ shared electrons

(b) PCl_3

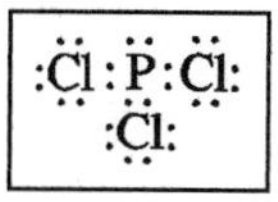

:C̈l:P̈:C̈l:
:C̈l:

$S = N - A$
$= [1 \times 8(\text{for P}) + 3 \times 8(\text{for Cl})] - [1 \times 5(\text{for P}) + 3 \times 7(\text{for Cl})]$
$= 6$ shared electrons

(c) $ClNO$

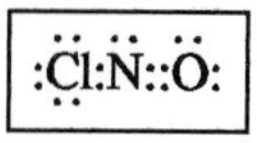

:C̈l:N::Ö:

$S = N - A$
$= [1 \times 8(\text{for Cl}) + 1 \times 8(\text{for N}) + 1 \times 8(\text{for O})]$
$\qquad - [1 \times 7(\text{for Cl}) + 1 \times 5(\text{for N}) + 1 \times 6(\text{for O})]$
$= 6$ shared electrons

(d) OH^-

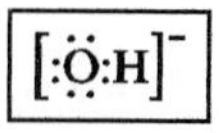

[:Ö:H]$^-$

$S = N - A$
$= [1 \times 2(\text{for H}) + 1 \times 8(\text{for O})] - [1 \times 1(\text{for H}) + 1 \times 6(\text{for O}) + 1e^-]$
$= 10 - 8$
$= 2$ shared electrons

7-38. *Refer to Sections 7-5 and 7-6, and Examples 7-1, 7-2 and 7-3.*

(a) H_2CO

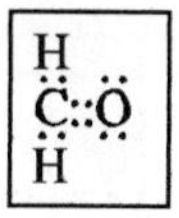

H
C̈::Ö
H

$S = N - A$
$= [2 \times 2(\text{for H}) + 1 \times 8(\text{for C}) + 1 \times 8(\text{for O})]$
$\qquad - [2 \times 1(\text{for H}) + 1 \times 4(\text{for C}) + 1 \times 6(\text{for O})]$
$= 8$ shared electrons

(b) ClF

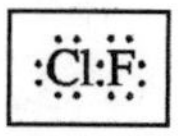

:C̈l:F̈:

$S = N - A$
$= [1 \times 8(\text{for Cl}) + 1 \times 8(\text{for F})] - [1 \times 7(\text{for Cl}) + 1 \times 7(\text{for F})]$
$= 2$ shared electrons

(c) BF_4^-

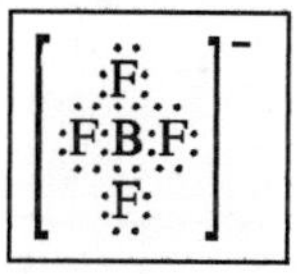

[:F̈:
:F̈:B:F̈:
:F̈:]$^-$

$S = N - A$
$= [1 \times 8(\text{for B}) + 4 \times 8(\text{for F})] - [1 \times 3(\text{for B}) + 4 \times 7(\text{for F}) + 1e^-]$
$= 8$ shared electrons

(d) PO_4^{3-}

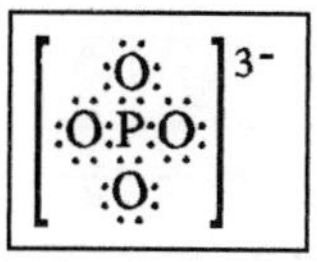

[:Ö:
:Ö:P:Ö:
:Ö:]$^{3-}$

$S = N - A$
$= [1 \times 8(\text{for P}) + 4 \times 8(\text{for O})] - [1 \times 5(\text{for P}) + 4 \times 6(\text{for O}) + 3e^-]$
$= 8$ shared electrons

(e) $HClO_3$

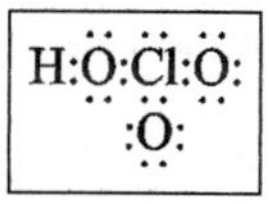

H:Ö:C̈l:Ö:
:Ö:

$S = N - A$
$= [1 \times 8(\text{for Cl}) + 3 \times 8(\text{for O}) + 1 \times 2(\text{for H})]$
$\qquad - [1 \times 7(\text{for Cl}) + 3 \times 6(\text{for O}) + 1 \times 1(\text{for H})]$
$= 8$ shared electrons

7-40. *Refer to Sections 7-5, 7-6 and 7-8.*

ClO_2

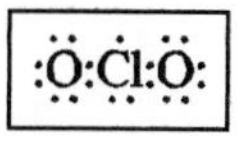

:Ö:C̈l:Ö:

The octet rule is not valid without modification (Section 7-8, Limitation C).
$A = 1 \times 7(\text{for Cl}) + 2 \times 6(\text{for O}) = 19$ (total number of valence electrons)

7-42. *Refer to Sections 7-5 and 7-6, and Examples 7-1 and 7-2.*

CCl_4 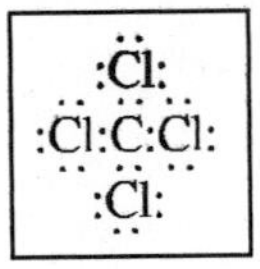

$S = N - A$
$= [1 \times 8(\text{for C}) + 4 \times 8(\text{for Cl})] - [1 \times 4(\text{for C}) + 4 \times 7(\text{for Cl})]$
$= 8$ shared electrons

SiF_4 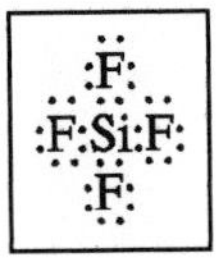

$S = N - A$
$= [1 \times 8(\text{for Si}) + 4 \times 8(\text{for F})] - [1 \times 4(\text{for Si}) + 4 \times 7(\text{for F})]$
$= 8$ shared electrons

PbI_4 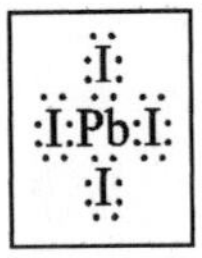

$S = N - A$
$= [1 \times 8(\text{for Pb}) + 4 \times 8(\text{for I})] - [1 \times 4(\text{for Pb}) + 4 \times 7(\text{for I})]$
$= 8$ shared electrons

All three compounds obey the octet rule and are formed from a 4A element bonded to four atoms of a 7A element and, therefore, look very similar.

7-44. *Refer to Section 7-7.*

The formal charge, FC = (Group No.) - [(No. of bonds) + (No. of unshared e^-)]

(a) 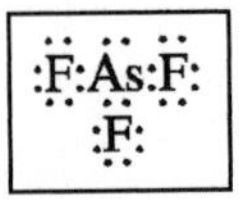

for As, FC = 5 - (3 + 2) = 0
for F, FC = 7 - (1 + 6) = 0

(b) 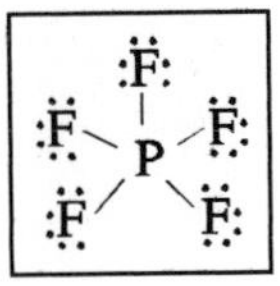

for P, FC = 5 - (5 + 0) = 0
for F, FC = 7 - (1 + 6) = 0

(c) 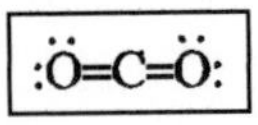

for C, FC = 4 - (4 + 0) = 0
for O, FC = 6 - (2 + 4) = 0

(d) 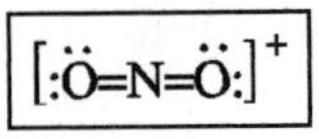

for N, FC = 5 - (4 + 0) = +1
for O, FC = 6 - (2 + 4) = 0

(e) 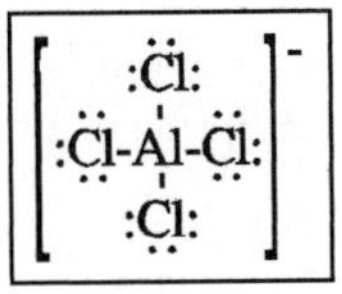

for Al, FC = 3 - (4 + 0) = -1
for Cl, FC = 7 - (1 + 6) = 0

7-46. *Refer to Sections 7-7, 7-8 and 7-9, and Chapter 8.*

Here are six resonance forms of the sulfate ion. The singly bonded oxygen atoms have a formal charge of –1; all the other atoms have a formal charge of 0.

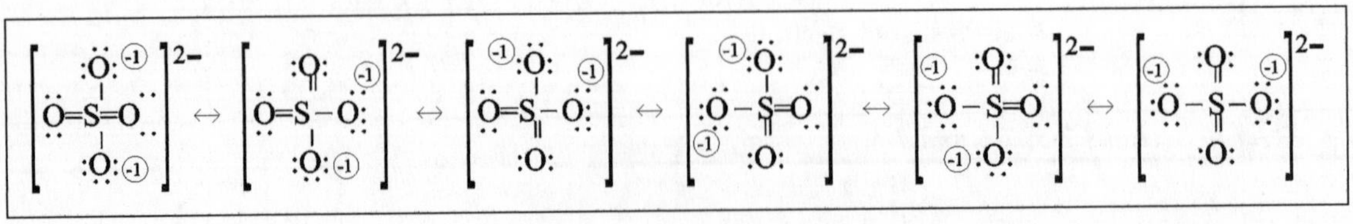

All the arrangements have the same stability. Since the ion is actually tetrahedral in shape and not square planar, all the bonds are equidistant from each other, even though it doesn't look that way on paper.

butane:

$$
\begin{array}{l}
\text{H H H H} \\
\text{H:C:C:C:C:H} \\
\text{H H H H}
\end{array}
\equiv
\begin{array}{l}
\text{H H H H} \\
\text{H:C:C-C:C:H} \\
\text{H H H H}
\end{array}
$$

$S = N - A$
$= [4 \times 8(\text{for C}) + 10 \times 2(\text{for H})]$
$\quad - [4 \times 4(\text{for C}) + 10 \times 1(\text{for H})]$
$= 26$ shared electrons

The indicated bond in butane is a nonpolar covalent single bond between two carbon atoms.

propane:

$$
\begin{array}{l}
\text{H H H} \\
\text{H:C:C:C:H} \\
\text{H H H}
\end{array}
\equiv
\begin{array}{l}
\text{H H H} \\
\text{H:C-C:C:H} \\
\text{H H H}
\end{array}
$$

$S = N - A$
$= [3 \times 8(\text{for C}) + 8 \times 2(\text{for H})]$
$\quad - [3 \times 4(\text{for C}) + 8 \times 1(\text{for H})]$
$= 20$ shared electrons

The indicated bond in propane is also a nonpolar covalent single bond between two carbon atoms.

(a) C_2F_4

$S = N - A$
$= [2 \times 8(\text{for C}) + 4 \times 8(\text{for F})] - [2 \times 4(\text{for C}) + 4 \times 7(\text{for F})]$
$= 12$ shared electrons

(b) CH_2CHCN

$S = N - A$
$= [3 \times 8(\text{for C}) + 3 \times 2(\text{for H}) + 1 \times 8(\text{for N})]$
$\quad - [3 \times 4(\text{for C}) + 3 \times 1(\text{for H}) + 1 \times 5(\text{for N})]$
$= 18$ shared electrons

Ozone, O_3 exhibits resonance, obeying the octet rule: $S = N - A = [3 \times 8(\text{for O})] - [3 \times 6(\text{for O})] = 6$ shared e^-

$$:\ddot{O}::\ddot{O}:\ddot{O}: \quad \leftrightarrow \quad :\ddot{O}:\ddot{O}::\ddot{O}:$$

As the number of electrons in a bond increases, the energy of the bond increases, and the length of the bond decreases. Therefore,

$$C–C > C=C > C\equiv C \qquad \text{in bond length}$$

From the discussion of resonance, the carbon-carbon bond length in the six-membered ring of toluene is intermediate in length between a single bond and a double bond. Therefore, this bond would be shorter than a regular single bond found between the CH_3 group and the carbon atom on the ring.

(a) NO_2^-

$$\left[:\ddot{O}:\ddot{N}::\ddot{O}:\right]^- \leftrightarrow \left[:\ddot{O}::\ddot{N}:\ddot{O}:\right]^-$$

$S = N - A$
$= [1 \times 8(\text{for N}) + 2 \times 8(\text{for O})]$
$\quad - [1 \times 5(\text{for N}) + 2 \times 6(\text{for O}) + 1e^-]$
$= 6$ shared electrons

(b) BrO_3^-

$$\left[\ddot{O}::Br::\ddot{O}\atop :\ddot{O}:\right]^- \leftrightarrow \left[\ddot{O}::Br:\ddot{O}:\atop :\ddot{O}:\right]^- \leftrightarrow \left[:\ddot{O}:Br::\ddot{O}\atop :\ddot{O}:\right]^-$$

The octet rule is not valid.
$A = 1 \times 7(\text{for Br}) + 3 \times 6(\text{for O}) + 1e^-$
  $= 26$ available electrons

(c) PO_4^{3-}

$$\left[\ddot{O}\atop \ddot{O}:P:\ddot{O}\atop :\ddot{O}:\right]^{3-} \leftrightarrow \left[\ddot{O}\atop :\ddot{O}:P:\ddot{O}\atop :\ddot{O}:\right]^{3-} \leftrightarrow \left[\ddot{O}\atop :\ddot{O}:P:\ddot{O}:\atop :\ddot{O}:\right]^{3-} \leftrightarrow \left[\ddot{O}\atop :\ddot{O}:P:\ddot{O}:\atop :\ddot{O}:\right]^{3-}$$

The octet rule is not valid.
$A = 1 \times 5(\text{for P}) + 4 \times 6(\text{for O}) + 3e^-$
  $= 32$ available electrons

7-58. *Refer to Sections 7-5, 7-6 and 7-8, and Examples 7-1, 7-4 and 7-5.*

(a) $BeBr_2$:Br:Be:Br: :Br-Be-Br:

The octet rule is not valid without modification (Section 7-8, Limitation A).
$A = 1 \times 2(\text{for Be}) + 2 \times 7(\text{for Br}) = 16$ (total number of valence electrons)

(b) BBr_3

The octet rule is not valid without modification (Section 7-8, Limitation B).
$A = 1 \times 3(\text{for B}) + 3 \times 7(\text{for Br}) = 24$
  (total number of valence electrons)

(c) NCl_3 :Cl:N:Cl: :Cl-N-Cl:

$S = N - A$
  $= [1 \times 8(\text{for N}) + 3 \times 8(\text{for Cl})] - [1 \times 5(\text{for N}) + 3 \times 7(\text{for Cl})]$
  $= 6$ shared electrons

(d) $AlCl_3$

The octet rule is not valid without modification (Section 7-8, Limitation B).
$A = 1 \times 3(\text{for Al}) + 3 \times 7(\text{for Cl}) = 24$ (total number of valence electrons)

Compounds (a), (b) and (d) have a central atom that disobeys the octet rule with a share in less than an octet of valence electrons.

7-60. *Refer to Sections 7-5, 7-6 and 7-8, and Examples 7-1 and 7-5.*

(a) CH_2Cl_2

$S = N - A$
  $= [1 \times 8(\text{for C}) + 2 \times 2(\text{for H}) + 2 \times 8(\text{for Cl})]$
      $- [1 \times 4(\text{for C}) + 2 \times 1(\text{for H}) + 2 \times 7(\text{for Cl})]$
  $= 8$ shared electrons

(b) BF_3

The octet rule is not valid without modification (Section 7-8, Limitation B).
$A = 1 \times 3(\text{for B}) + 3 \times 7(\text{for Br}) = 24$ (total no. of valence electrons)

(c) BCl_4^-

$S = N - A$
  $= [1 \times 8(\text{for B}) + 4 \times 8(\text{for Cl})]$
      $- [1 \times 3(\text{for B}) + 4 \times 7(\text{for Cl}) + 1e^-]$
  $= 8$ shared electrons

(d) AlF_4^-

$S = N - A$
  $= [1 \times 8(\text{for Al}) + 4 \times 8(\text{for F})]$
      $- [1 \times 3(\text{for Al}) + 4 \times 7(\text{for F}) + 1e^-]$
  $= 8$ shared electrons

Only **Compound (b)** has a central atom that disobeys the octet rule with a share in less than an octet of valence electrons.

(a) PF_6^-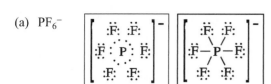

The octet rule is not valid without modification (Section 7-8, Limitation D).

$A = 1 \times 5(\text{for P}) + 6 \times 7(\text{for F}) + 1e^- = 48$
    (total number of valence electrons)

(b) $AsCl_5$

The octet rule is not valid without modification (Section 7-8, Limitation D).

$A = 1 \times 5(\text{for As}) + 5 \times 7(\text{for Cl}) = 40$
    (total number of valence electrons)

(c) ICl_3

The octet rule is not valid without modification (Section 7-8, Limitation D).

$A = 1 \times 7(\text{for I}) + 3 \times 7(\text{for Cl}) = 28$
    (total number of valence electrons)

(d) C_2H_6

$S = N - A$
  $= [2 \times 8(\text{for C}) + 6 \times 2(\text{for H})]$
      $- [2 \times 4(\text{for C}) + 6 \times 1(\text{for H})]$
  $= 14$ shared electrons

Compounds (a), (b) and (c) have central atoms that disobey the octet rule with a share in more than an octet of valence electrons.

7-64. *Refer to Sections 7-5 and 7-6.*

(a) $ElBr_3$

El is located in Group 5A because it brings 5 valence electrons to the compound. The example shown is NBr_3.

(b) ElO_2

El is located in Group 4A because it brings 4 valence electrons to the compound. The example shown is CO_2.

(c) ElH_4^+

El is located in Group 5A because it brings 5 valence electrons to the ion. One electron was lost giving the ion a 1+ charge. The example shown is NH_4^+.

(d) ElH_3^+

El is located in Group 6A because it brings 6 valence electrons to the ion. One electron was lost giving the ion a 1+ charge. The example shown is the hydronium ion, H_3O^+.

7-66. *Refer to Sections 7-5, 7-6, 7-8 and 7-9, and Example 7-9.*

(a) SO_2 exhibits resonance and obeys the octet rule:

$$\ddot{\text{O}}::\ddot{\text{S}}:\ddot{\text{O}}: \quad \leftrightarrow \quad :\ddot{\text{O}}:\ddot{\text{S}}::\ddot{\text{O}}$$

$S = N - A = [1 \times 8(\text{for S}) + 2 \times 8(\text{for O})] - [1 \times 6(\text{for S}) + 2 \times 6(\text{for O})] = 6$ shared electrons

(b) NO_2 exhibits resonance, but it violates the octet rule because the compound contains an odd number of valence electrons, 17 (Section 7-8, Limitation C).

$$:\ddot{O}::N:\ddot{O}: \quad \leftrightarrow \quad :\ddot{O}:N::\ddot{O}: \quad \leftrightarrow \quad \cdot\ddot{O}:N::\ddot{O}: \quad \leftrightarrow \quad :\ddot{O}::N:\ddot{O}\cdot$$

$A = 1 \times 5(\text{for N}) + 2 \times 6(\text{for O}) = 17$ (total number of valence electrons)

(c) CO exhibits resonance. It is known from experiments that the C-O bond in CO is intermediate between a typical double and triple bond length. Only one resonance structure obeys the octet rule.

$$:C:::O: \quad \leftrightarrow \quad :C::\ddot{O}$$

$S = N - A$
$\quad = [1 \times 8(\text{for C}) + 1 \times 8(\text{for O})] - [1 \times 4(\text{for C}) + 1 \times 6(\text{for O})]$
$\quad = 16 - 10$
$\quad = 6$ shared electrons

(d) O_3 exhibits resonance and obeys the octet rule.

$$:\ddot{O}::\ddot{O}:\ddot{O}: \quad \leftrightarrow \quad :\ddot{O}:\ddot{O}::\ddot{O}:$$

$S = N - A = [3 \times 8(\text{for O})] - [3 \times 6(\text{for O})] = 24 - 18 = 6$ shared electrons

(e) SO_3 exhibits resonance and obeys the octet rule.

$$\ddot{O}::S:\ddot{O}: \quad \leftrightarrow \quad :\ddot{O}:S:\ddot{O}: \quad \leftrightarrow \quad :\ddot{O}:S::\ddot{O}$$
$$:\ddot{O}: :\ddot{O}: :\ddot{O}:$$

$S = N - A = [1 \times 8(\text{for S}) + 3 \times 8(\text{for O})] - [1 \times 6(\text{for S}) + 3 \times 6(\text{for O})] = 8$ shared electrons

(f) $(NH_4)_2SO_4$ is an ionic solid composed of covalently bonded polyatomic ions, both of which obey the octet rule:

NH_4^+

$$\left[\begin{matrix} H \\ H:N:H \\ H \end{matrix} \right]^+$$

$S = N - A$
$\quad = [4 \times 2(\text{for H}) + 1 \times 8(\text{for N})] - [4 \times 1(\text{for H}) + 1 \times 5(\text{for N}) - 1\ e^-]$
$\quad = 16 - 8$
$\quad = 8$ shared electrons

SO_4^{2-}

$$\left[\begin{matrix} :\ddot{O}: \\ :\ddot{O}:S:\ddot{O}: \\ :\ddot{O}: \end{matrix} \right]^{2-}$$

$S = N - A$
$\quad = [4 \times 8(\text{for O}) + 1 \times 8(\text{for S})] - [4 \times 6(\text{for O}) + 1 \times 6(\text{for S}) + 2\ e^-]$
$\quad = 40 - 32$
$\quad = 8$ shared electrons

(See Exercise 7-46 Solution for more appropriate Lewis formulas due to formal charge examination.)

7-68.   *Refer to Section 7-10 and Example 7-10.*

An HCl molecule is a heteronuclear diatomic molecule composed of H (EN = 2.1) and Cl (EN = 3.0). Because the electronegativities of the elements are different, the pull on the electrons in the covalent bond between them is unequal. Hence HCl is a polar molecule. A homonuclear diatomic molecule contains a nonpolar bond, since the electron pair between the two atoms is shared equally. Cl_2 is an example of a homonuclear diatomic molecule.

7-70.   *Refer to Section 7-10 and Example 7-10.*

A covalent bond between two atoms is polar when the two atoms have different electronegativities. The greater the difference in electronegativity, the more polar is the bond.

Electronegativity is defined as the tendency of an atom to attract electrons to itself in a chemical bond. Electrons are more attracted to the fluorine atom in the C–F bond than to the bromine atom in the C–Br bond because the F atom has a higher electronegativity value (EN = 4.0) than the bromine atom (EN = 2.8).

(a) $\overset{\delta+}{C} - \overset{\delta-}{O}$ ($\Delta EN = 1.0$) is more polar than $\overset{\delta+}{C} - \overset{\delta-}{N}$ ($\Delta EN = 0.5$)

(b) Both the C – S ($\Delta EN = 0$) bond and the N – Cl ($\Delta EN = 0$) bond are non-polar bonds and as such, have no dipole moment.

(c) $\overset{\delta+}{P} - \overset{\delta-}{N}$ ($\Delta EN = 0.9$) is more polar than P – H ($\Delta EN = 0.0$)

(d) $\overset{\delta+}{B} - \overset{\delta-}{I}$ ($\Delta EN = 0.5$) is more polar than $\overset{\delta+}{B} - \overset{\delta-}{H}$ ($\Delta EN = 0.1$)

(a) The two pairs of elements most likely to form ionic bonds are (1) Ba (metal) and F (nonmetal) and (2) K (metal) and O (nonmetal).

(b) We know that bond polarity increases with increasing $\Delta(EN)$, the difference in electronegativity between 2 atoms that are bonded together.

Bond	I – H	C – F	N – F
EN	2.5 2.1	2.5 4.0	3.0 4.0
$\Delta(EN)$	0.4	1.5	1.0

Therefore, the least polar bond is I–H and the most polar bond is C–F.

The use of ΔEN alone to distinguish between ionic and polar covalent bonds will lead to the mis-labeling of some bonds, especially when the elements, H and F are involved.

Position on the periodic table can also be used as an indicator:

metal + nonmetal $\rightarrow$ ionic bond, and the compound is generally a solid and melts at high temperatures, and
nonmetal + nonmetal $\rightarrow$ covalent bond, and the compound is generally a liquid or gas at room temperature.

However, there are also many exceptions, especially when Be is involved.

		ΔEN	Bonding Type
(a)	K (metal, EN = 0.9) and O (nonmetal, EN = 3.5)	2.6	ionic (K_2O, decomposes at $350^{\circ}C$)
(b)	Br (nonmetal, EN = 2.8) and I (nonmetal, EN = 2.5)	0.3	polar covalent (BrI, m.p. $42^{\circ}C$)
(c)	Na (metal, EN = 1.0) and H (nonmetal, EN = 2.1)	1.1	ionic (NaH, m.p. $800^{\circ}C$)
(d)	O (nonmetal, EN = 3.5) and O (nonmetal, EN = 3.5)	0.0	nonpolar covalent (O_2, gas)
(e)	H (nonmetal, EN = 2.1) and O (nonmetal, EN = 3.5)	1.4	polar covalent (H_2O is a liquid)

The bond with the greater "ionic character" is the bond between atoms with the greater difference in electronegativity.

(a) Na–Cl ($\Delta EN = 2.0$) has more ionic character than Mg–Cl ($\Delta EN = 1.8$)

(b) Ca–S ($\Delta EN = 1.5$) has more ionic character than Fe–S ($\Delta EN = 0.8$)

(c) Al–Br ($\Delta EN = 1.3$) has more ionic character than O–Br ($\Delta EN = 0.7$)

(d) Ra–H ($\Delta EN = 1.1$) has more ionic character than C–H ($\Delta EN = 0.4$)

A molecule or polyatomic ion for which two or more Lewis formulas with the same arrangements of atoms can be drawn to describe the bonding is said to exhibit resonance. The two structures given here do not have the same arrangement of atoms, and hence are not resonance structures.

A chemical bond exhibiting 100% "covalent character" and 0% "ionic character" occurs between identical nonmetals atoms in which the difference in electronegativity (ΔEN) is zero. An example is the H–H bond. There are atoms with essentially the same electronegativity, e.g. N and Cl both have an electronegativity equal to 3.0 to 2 significant figures, so the N–Cl bond would exhibit close to 100% "covalent character" and 0% "ionic character".

Electrostatic charge potential (ECP) plots integrate dipole moment, electronegativity, and partial charges. It is a visual representation of the **relative polarity** of a molecule. Both Cl_2 and F_2 are nonpolar molecules with nonpolar bonds. However, Cl_2 is larger than F_2, so it is easy to see which is which.

From Table 7-4, there are indications that there is a trend of longer bond lengths as we move down a group.

VIA Group:	H–O (0.94 Å) < H–S (1.32 Å)	looking at X–O vs. X–S bond lengths
	C–O (1.43 Å) < C–S (1.81 Å)	
	N–O (1.36 Å) < N–S (1.74 Å)	
	F–O (1.30 Å) < F–S (1.68 Å)	
	S–O (1.70 Å) < S–S (2.08 Å)	
	O–O (1.32 Å) < O–S (1.70 Å)	
VIIA Group:	H–H (0.74 Å) < H–F (0.92 Å)	looking at X–H vs. X–F bond lengths
	C–H (1.10 Å) < C–F (1.41 Å)	
	N–H (0.98 Å) < N–F (1.34 Å)	
	O–H (0.94 Å) < O–F (1.30 Å)	
	F–H (0.92 Å) < F–F (1.28 Å)	
	S–H (1.32 Å) < S–F (1.68 Å)	

Looking at these examples, in every case X–O < X–S in bond length and X–H < X–F in bond length.

7-90. *Refer to Sections 7-10 and 7-11.*

(a) Ca_3N_2 calcium nitride (ionic)

(b) Al_2O_3 aluminum oxide (ionic)

(c) K_2Se potassium selenide (ionic)

(d) $SrBr_2$ strontium bromide (ionic)

7-92. *Refer to Sections 4-2, 4-3, 7-5, 7-6 and 7-9.*

(a) formula unit: $HCN(aq) + NaOH(aq) \rightarrow NaCN(aq) + H_2O(\ell)$

    total ionic: $HCN(aq) + Na^+(aq) + OH^-(aq) \rightarrow Na^+(aq) + CN^-(aq) + H_2O(\ell)$

    net ionic: $HCN(aq) + OH^-(aq) \rightarrow CN^-(aq) + H_2O(\ell)$

| H:C::N: | Na⁺ | [:Ö:H]⁻ | [:C::N:]⁻ | H-Ö: / H |

(b) formula unit: $HCl(aq) + NaOH(aq) \rightarrow NaCl(aq) + H_2O(\ell)$

    total ionic: $H^+(aq) + Cl^-(aq) + Na^+(aq) + OH^-(aq) \rightarrow Na^+(aq) + Cl^-(aq) + H_2O(\ell)$

    net ionic: $H^+(aq) + OH^-(aq) \rightarrow H_2O(\ell)$

| H⁺ | [:C̈l:]⁻ | Na⁺ | [:Ö:H]⁻ | H:Ö: / H |

(c) formula unit: $CaCl_2(aq) + Na_2CO_3(aq) \rightarrow 2NaCl(aq) + CaCO_3(s)$

    total ionic: $Ca^{2+}(aq) + 2Cl^-(aq) + 2Na^+(aq) + CO_3^{2-}(aq) \rightarrow 2Na^+(aq) + 2Cl^-(aq) + CaCO_3(s)$

    net ionic: $Ca^{2+}(aq) + CO_3^{2-}(aq) \rightarrow CaCO_3(s)$

| Ca²⁺ | [:C̈l:]⁻ | Na⁺ | [:Ö-C=Ö: / :Ö:]²⁻ | Ca²⁺, [:Ö-C=Ö: / :Ö:]²⁻ |

Note: Only one of the three resonance structures of the carbonate ion is shown.

7-94. *Refer to Sections 7-5 and 7-6.*

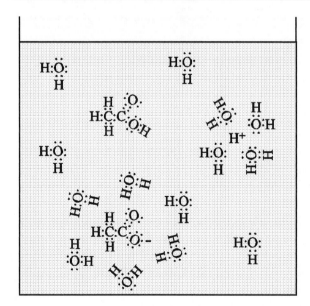

There are three solute species present due to the partial ionization of acetic acid, CH_3COOH:

$$CH_3COOH(aq) \rightleftarrows CH_3COO^-(aq) + H^+(aq)$$

The solvent species, H_2O, is a very polar molecule. The water molecules arrange themselves around the ions so that the slightly positive ends of the water molecules point toward the negative ions, and the slightly negative ends of the water molecules point toward the positive ions.

$$\overset{-\delta}{H:\ddot{O}:} \\ {}_{+\delta}\ H$$

In the actual solution, because acetic acid is a weak acid, there are very many more acetic acid molecules, CH_3COOH, than there are acetate ions, CH_3COO^-, or hydrogen ions, H^+.

8 Molecular Structure and Covalent Bonding Theories

8-2. *Refer to Sections 7-10 and 8-8.*

(a) "Bonding pair" is a term that refers to a pair of electrons that is shared between two nuclei in a covalent bond, while the term "lone pair" refers to an unshared pair of electrons that is associated with a single nucleus.

(b) Lone pairs of electrons occupy more space than bonding pairs. This fact was determined experimentally from measurements of bond angles of many molecules and polyatomic ions. An explanation for this is the fact that a lone pair has only one atom exerting strong attractive forces on it, and it exists closer to the nucleus than bonding pairs.

(c) The relative magnitudes of the repulsive forces between pairs of electrons on an atom are as follows:

$$bp/bp < lp/bp \ll lp/lp$$

where lp refers to lone pairs and bp refers to bonding pairs of valence shell electrons.

8-4. *Refer to Section 8-2.*

When VSEPR theory is used to predict molecular geometries, double and triple bonds are treated identically to single bonds: as a single electron group, i.e. as a single place where you can find electrons.

A single unshared nonbonding electron is also counted as one electron group.

8-6. *Refer to Sections 8-2, 8-11 and 8-15, and Tables 8-3 and 8-4.*

Three possible arrangements of AB_2U_3 with 2 atoms of B and 3 lone pairs around the central A atom are:

(1) 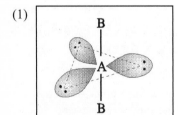 (2) 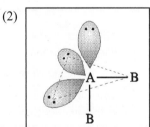 (3)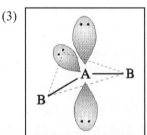

According to VSEPR theory, the most stable arrangement of the three lone pairs of electrons would be in the equatorial position, as shown in (1), where they would be less crowded. Therefore, a linear structure is the correct molecular geometry of the molecule.

8-8. *Refer to Sections 8-4 and 8-5, and Table 8-2.*

(a) The number of electron groups (also called the regions of high electron density) on an atom is equal to the number of its pure atomic orbitals that hybridize.

(b) The number of atomic orbitals that hybridize equals the number of hybrid orbitals formed.

(a) *sp*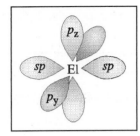

(b) *sp²*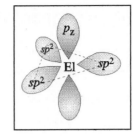

(c) *sp³*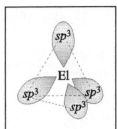

(d) *sp³d*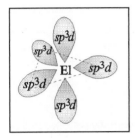

(e) *sp³d²*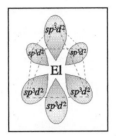

(a) ABU₅ *sp³d²*

(b) AB₂U₄ *sp³d²*

(c) AB₃ *sp²*

(d) AB₃U₂ *sp³d*

(e) AB₅ *sp³d*

From the sketch, we can see that two *p* orbitals and one *s* orbital are being hybridized, forming three new *sp²* hybrid orbitals. See the following figure in Section 8-6. The three *sp²* hybrid orbitals are arranged as so:

(a) *sp*   180°

(b) *sp²*   120°

(c) *sp³*   109.5°

(d) *sp³d*   90°, 120°, 180°

(e) *sp³d²*   90°, 180°

The hybridization at the central atoms in linear nonpolar Group 2A (also called IIA) compounds, like $BeCl_2$, is *sp*.

The molecular or ionic geometry is identical to its electronic geometry when there are no lone pairs on the central atom.

The central atom is an atom that is bonded to more than one other atom.

		Central Atom
(a)	HCO_3^-	C
		O (can also be thought of as a central atom when bonded to both the carbon and hydrogen)
(b)	SiO_2	Si
(c)	SO_3	S
(d)	$Al(OH)_4^-$	Al
(e)	$BeBr_2$	Be
(f)	$(CH_3)_4Pb$	Pb

(a) $CdCl_2$

:Cl:Cd:Cl:

This molecule (type AB_2) does not obey the octet rule without modification.

Available electrons, A = 2 x 7(for Cl) + 1 x 2(for Cd) = 16 (total number of valence e^-)

The Lewis formula predicts 2 electron groups around the central Be atom and a linear electronic geometry. There are no lone pairs on the Cd atom, so the molecular geometry is the same as the electronic geometry: linear (Section 8-5).

(b) $SnCl_4$

:Cl:
:Cl:Sn:Cl:
:Cl:

Sn is a 4A element and has 4 valence electrons.

$S = N - A$ = [4 x 8(for Cl) + 1 x 8(for Sn)] - [4 x 7(for Cl) + 1 x 4(for Sn)]
= 8 shared electrons

The Lewis formula for the molecule (type AB_4) predicts 4 electron groups around the central Sn atom and a tetrahedral electronic geometry. Since there are no lone pairs on Sn, the molecular geometry is also tetrahedral (Section 8-7).

(c) BrF_3

:F:Br:F:
:F:

This molecule (type AB_3U_2) does not obey the octet rule without modification.

Available electrons, A = 3 x 7(for F) + 1 x 7(for Br) = 28 (total number of valence e^-)

The Lewis formula predicts 5 electron groups around the central Br atom and a trigonal bipyramidal electronic geometry. There are two lone pairs on the Br atom, so the molecular geometry is T-shaped (Section 8-11).

(d) SbF_6^-

$$\left[\begin{array}{c} :F: \\ :F\overset{|}{\underset{|}{Sb}}F: \\ :F: F: \end{array} \right]^-$$

This polyatomic ion (type AB_6), like (c), does not obey the octet rule without modification since 12 electrons must be shared to form 6 Sb-F bonds. Sb is a 5A element, but the charge on the ion gives an extra electron which participates in bonding. The Lewis formula predicts 6 electron groups around the central Sb atom and an octahedral electronic geometry. There are no lone pairs on the Sb atom, so the ionic geometry is the same as the electronic geometry (Section 8-12).

(a) $CdCl_2$ The ideal bond angles would be 180° since the molecule is linear.

    $SnCl_4$ The ideal bond angles would be 109.5° since the structure is tetrahedral.

    BrF_3 The ideal bond angles would be those for a trigonal bipyramidal electronic geometry (type AB_3U_2): 90° and 180°. One F atom and the 2 lone pairs on the Br atom are separated by 120°.

    SbF_6^- The ideal bond angles would be those for an octahedron, 90° and 180°.

(b) These bond angles differ from the actual bond angles only for BrF_3, since this species has lone pairs of electrons on the central atom. Lone pairs of electrons require more space than bonding pairs of electrons: the F–Br–F bond angles in BrF_3 are slightly reduced from the ideal case.

8-28. *Refer to Sections 8-7, 8-8 and 8-9.*

(a) $H^+ + H_2O \rightarrow H_3O^+$ (b) $NH_3 + H^+ \rightarrow NH_4^+$

	Lewis Formula	Electronic Geometry	Molecular (Ionic) Geometry		Lewis Formula	Electronic Geometry	Molecular (Ionic) Geometry
H_2O	H-Ö: / H	tetrahedral	angular (bent)	NH_3	H-N̈-H / H	tetrahedral	trigonal pyramidal
H_3O^+	[H-Ö-H / H]$^+$	tetrahedral	trigonal pyramidal	NH_4^+	[H / H-N-H / H]$^+$	tetrahedral	tetrahedral

---

8-30. *Refer to Tables 8-3 and 8-4, and the Sections as stated.*

(a) H_3O^+

$S = N - A = [3 \times 2(\text{for H}) + 1 \times 8(\text{for O})] - [3 \times 1(\text{for H}) + 1 \times 6(\text{for O}) - 1e^-]$
= 6 shared electrons

The Lewis formula for the ion (type AB_3U) predicts 4 electron groups around O including 1 lone pair of electrons. The electronic geometry is tetrahedral and the ionic geometry is trigonal pyramidal (Section 8-8).

(b) PCl_6^-

The ion (type AB_6) does not obey the octet rule.

$A = 6 \times 7(\text{for Cl}) + 1 \times 5(\text{for P}) + 1\,e^- = 48$ (total number of valence electrons)

The Lewis formula shows 6 electron groups around the central P atom. The electronic geometry and the ionic geometry are both octahedral because there are no lone pairs of electrons on P (Section 8-12).

(c) PCl_4^-

The ion (type AB_4U) does not obey the octet rule.

$A = 4 \times 7(\text{for Cl}) + 1 \times 5(\text{for P}) + 1\,e^- = 34$ (total number of valence electrons)

The Lewis formula shows 5 electron groups around the central P atom and its electronic geometry is trigonal bipyramidal. The ionic geometry is a seesaw due to the presence of 1 lone pair of electrons on the central P atom (Section 8-11).

(d) $SbCl_4^+$

$S = N - A = [4 \times 8(\text{for Cl}) + 1 \times 8(\text{for Sb})] - [4 \times 7(\text{for Cl}) + 1 \times 5(\text{for Sb}) - 1e^-]$
$= 8$ shared electrons

The Lewis formula for the ion (type AB_4) predicts 4 electron groups around the central Sb atom and a tetrahedral electronic geometry. Since there are no lone pairs on Sb, the ionic geometry is also tetrahedral (Section 8-7).

(a)

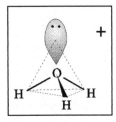

(b)

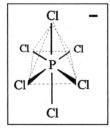

(c)

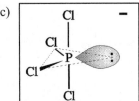

(d)

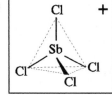

8-32. *Refer to Sections 8-7, 8-11 and 8-12.*

(a) SiF_4

$S = N - A = [4 \times 8(\text{for F}) + 1 \times 8(\text{for Si})] - [4 \times 7(\text{for F}) + 1 \times 4(\text{for Si})]$
$= 8$ shared electrons

The Lewis formula for the molecule (type AB_4) predicts 4 electron groups around the central Si atom with no lone pairs of electrons. The molecular geometry is the same as the electronic geometry: tetrahedral.

SF_4

The molecule (type AB_4U) does not obey the octet rule.

$A = 4 \times 7(\text{for F}) + 1 \times 6(\text{for S}) = 34$ (total number of valence electrons)

The Lewis formula shows 5 electron groups around the central S atom and its electronic geometry is trigonal bipyramidal. The molecular geometry is a seesaw due to the presence of 1 lone pair of electrons on the central S atom.

XeF_4

This molecule (type AB_4U_2) does not obey the octet rule.

$A = 4 \times 7(\text{for F}) + 1 \times 8(\text{for Xe}) = 36$ (total number of valence electrons)

The Lewis formula predicts 6 electron groups around the central Xe atom and its electronic geometry is octahedral. The molecular geometry is square planar due to the presence of 2 lone pairs of electrons on the central Xe atom.

(b) It is obvious that the molecular geometries of SiF_4, SF_4 and XeF_4 are not the same even though their molecular formulas are similar. The differences are due to the different number of valence electrons that must be accommodated by the central atom as lone pairs of electrons.

(a) I_3^-

The ion (type AB_2U_3) does not obey the octet rule.

$A = 3 \times 7(\text{for I}) + 1\ e^- = 22$ (total number of valence electrons)

The Lewis formula predicts 5 electron groups around the central I atom and a trigonal bipyramidal electronic geometry. This ionic geometry is probably linear.

(b) $TeCl_4$

The molecule (type AB_4U) does not obey the octet rule.

$A = 4 \times 7(\text{for Cl}) + 1 \times 6(\text{for Te}) = 34$ (total number of valence electrons)

The Lewis formula predicts 5 electron groups around Te with 1 lone pair of electrons. The electronic geometry is trigonal bipyramidal and the molecular geometry is a seesaw (Section 8-11).

(c) XeO_3

$S = N - A\ = [3 \times 8(\text{for O}) + 1 \times 8(\text{for Xe})] - [3 \times 6(\text{for O}) + 1 \times 8(\text{for Xe})]$
$= 6$ shared electrons

The Lewis formula for the molecule (type AB_3U) predicts 4 electron groups around Xe including 1 lone pair of electrons. The electronic geometry is tetrahedral and the molecular geometry is trigonal pyramidal (Section 8-8).

(d) BrNO

:Br:N::Ö:

$S = N - A\ = [1 \times 8(\text{for Br}) + 1 \times 8(\text{for N}) + 1 \times 8(\text{for O})]$
$- [1 \times 7(\text{for Br}) + 1 \times 5(\text{for N}) + 1 \times 6(\text{for O})]$
$= 6$ shared electrons

The Lewis formula for the molecule (type AB_2U) predicts 3 electron groups around the central N atom including 1 lone pair of electrons. The electronic geometry is trigonal planar and the molecular geometry is angular or bent (Table 8-3).

(e) $ClNO_2$

$S = N - A\ = [1 \times 8(\text{for Cl}) + 1 \times 8(\text{for N}) + 2 \times 8(\text{for O})]$
$- [1 \times 7(\text{for Cl}) + 1 \times 5(\text{for N}) + 2 \times 6(\text{for O})]$
$= 8$ shared electrons

The Lewis formula for the molecule (type AB_3) predicts 3 electron groups around the central N atom. Only 1 of the two resonance structures is shown. The electronic and molecular geometries are the same, trigonal planar, because there are no lone pairs of electrons on the N atom (Section 8-6).

(f) Cl_2SO

$S = N - A\ = [2 \times 8(\text{for Cl}) + 1 \times 8(\text{for S}) + 1 \times 8(\text{for O})]$
$- [2 \times 7(\text{for Cl}) + 1 \times 6(\text{for S}) + 1 \times 6(\text{for O})]$
$= 6$ shared electrons

The Lewis formula for the molecule (type AB_3U) predicts 4 electron groups around the central S atom including 1 lone pair of electrons. The electronic geometry is tetrahedral and the molecular geometry is trigonal pyramidal (Section 8-8).

The magnitude of bond polarity depends on the difference in electronegativity (ΔEN) between the two atoms involved. For the N–O bond, ΔEN = 3.5 - 3.0 = 0.5. For the H–O bond, ΔEN = 3.5 - 2.1 = 1.4. Since the H–O bond has the greater ΔEN, the H–O bond is predicted to be more polar than the N–O bond.

(a) CdI_2

This molecule (type AB_2) has a linear electronic and molecular geometry. The Cd–I bonds are polar. Since the molecule is symmetric, the bond dipoles cancel to give a nonpolar molecule (Section 8-5).

(b) BCl_3

This molecule (type AB_3) has a trigonal planar electronic geometry and trigonal planar molecular geometry. The B–Cl bonds are polar, but since the molecule is symmetrical, the bond dipoles cancel to give a nonpolar molecule (Section 8-6).

(c) $AsCl_3$

This molecule (type AB_3U) has a tetrahedral electronic geometry and a pyramidal molecular geometry. Cl (EN = 3.0) is more electronegative than As (EN = 2.1). The polar As–Cl bond dipoles oppose the effect of the lone pair. The molecule is only slightly polar (Section 8-8).

(d) H_2O

This molecule (type AB_2U_2) has a tetrahedral electronic geometry and an angular molecular geometry. Oxygen (EN = 25.5) is more electronegative than H (EN = 2.1). The O–H bond dipole reinforces the effect of the two lone pairs of electrons and so, H_2O is very polar (Section 8-9).

(e) SF_6

This molecule (type AB_6) has an octahedral electronic and molecular geometry. The S-F bonds are polar, but the molecule is symmetrical. The S–F bond dipoles cancel to give a nonpolar molecule (Section 8-12).

PF_2Cl_3

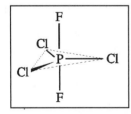

The Lewis formula for this molecule (type AB_5) predicts 5 electron groups with no lone pairs of electrons on the central P atom. The molecular geometry is the same as the electronic geometry: trigonal bipyramidal. For the molecule to be nonpolar (dipole moment = 0), the P–F and P–Cl bonds must be symmetrically arranged, i.e., the three Cl atoms must be at the equatorial positions and the two F atoms must be at the axial positions.

PCl_3

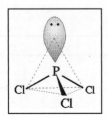

The P–Cl bond in phosphorus trichloride is a polar bond because the difference in electronegativity, ΔEN, between the two atoms is not zero.

$$\Delta EN = 3.0 \text{ (for Cl)} - 2.1 \text{ (for P)} = 0.9$$

This molecule (type AB_3U) has a tetrahedral electronic geometry and a trigonal pyramidal molecular geometry. The polar P–Cl bond dipoles oppose the effect of the lone pair. The molecule is polar.

(a) CS_2

This molecule (type AB_2) has a linear electronic and molecular geometry. The C–S bonds are nonpolar, since $\Delta EN = 2.5$ (for S) - 2.5 (for C) = 0. Moreover, the molecule is symmetric giving a nonpolar molecule (Section 8-5).

(b) AlF_3

This molecule (type AB_3) has a trigonal planar electronic geometry and trigonal planar molecular geometry. The Al–F bonds are polar, but since the molecule is symmetrical, the bond dipoles cancel to give a nonpolar molecule (Section 8-6).

(c) H_2S

This molecule (type AB_2U_2) has a tetrahedral electronic geometry and an angular molecular geometry. Sulfur (EN = 2.5) is more electronegative than H (EN = 2.1). The S–H bond dipole reinforces the effect of the two lone pairs of electrons and so, H_2S is very polar (Section 8-9).

(d) SnF_2

This molecule (type AB_2U) has a trigonal planar electronic geometry and an angular molecular geometry. F (EN = 4.0) is more electronegative than Sn (EN = 1.8). The Sn–F bond dipoles lessens the effect of the lone pair of electrons and so, SnF_2 is polar, but not extremely so (Section 8-9).

8-46. *Refer to Sections 8-1 and 8-4.*

The orbital overlap model of covalent bonding, described in *valence bond* (VB) *theory*, starts by describing covalent bonding as electron pair sharing that results from the overlap of orbitals from two atoms. VB theory describes the atomic orbitals that overlap to produce the bonding that generates molecules or ions that have a certain geometry. It is also assumed that each unshared pair of electrons occupies a separate orbital. Many times, pure atomic orbitals do not have the correct energies or orientations to describe where the electrons are when atoms are bonded to other atoms. So a process called hybridization is used, in which atomic orbitals are "mixed" together to form new hybrid orbitals which do have a lower total energy and are in the appropriate spatial arrangement to explain the molecular and ionic geometries that are observed.

8-48. *Refer to Table 8-2, Section 8-4 and Exercise 8-10 Solution.*

(a) NCl_3 sp^3 (b) $AlCl_3$ sp^2 (c) CF_4 sp^3

(d) SF_6 sp^3d^2 (e) IO_4^- sp^3

8-50. *Refer to Table 8-2.*

	Lewis Formula	Hybridization	Electronic Geometry	Molecular Geometry
(i) $CHCl_3$		sp^3	tetrahedral	tetrahedral (distorted)

	Lewis Formula	Hybridization	Electronic Geometry	Molecular Geometry
(ii) CH_2Cl_2	 :Cl: H:C:Cl: H	sp^3	tetrahedral	tetrahedral (distorted)
(iii) NF_3	 :F:N:F: :F:	sp^3	tetrahedral	trigonal pyramidal (1 lone pair of e^-)
(iv) PO_4^{3-}	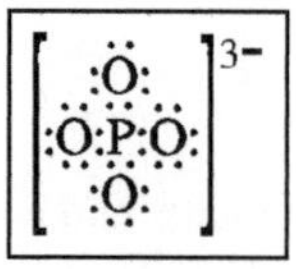 [:O: :O:P:O: :O:]$^{3-}$	sp^3	tetrahedral	tetrahedral
(v) IF_6^+	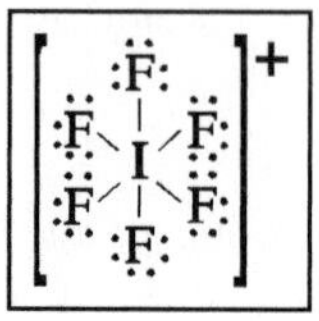 [:F: :F: :F: I :F: :F: :F:]$^+$	sp^3d^2	octahedral	octahedral
(vi) SiF_6^{2-}	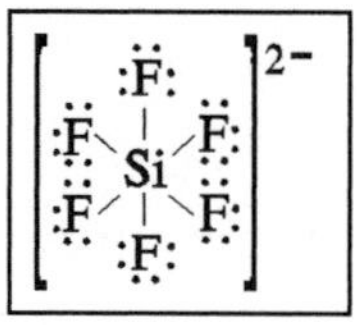 [:F: :F: :F: Si :F: :F: :F:]$^{2-}$	sp^3d^2	octahedral	octahedral

---

8-52   *Refer to Sections 8-13.*

$CH_3CCl=CH_2$

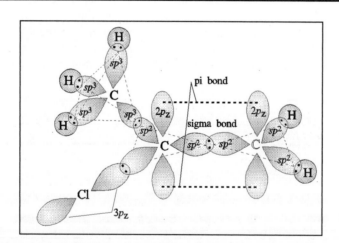

8-54.   *Refer to Table 8-2 and Sections 8-13 and 8-14.*

(a)

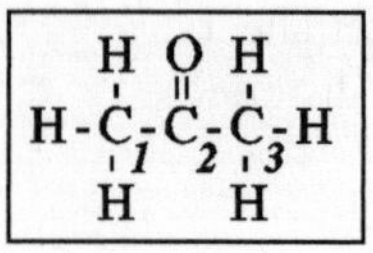

H O H
H–C–C–C–H
    1 2 3
H   H

C_1 sp^3 hybridization (4 electron groups)
C_2 sp^2 hybridization (3 electron groups)
C_3 sp^3 hybridization (4 electron groups)

(b)

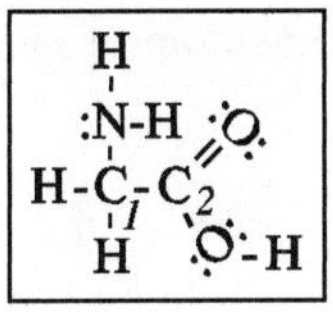

H
|
:N-H :O.
| ‖
H-C₁-C₂
| |
H   .O.-H

C_1 sp^3 hybridization (4 electron groups)
C_2 sp^2 hybridization (3 electron groups)

(c)

H   H
 \ /
  C-C
 / \\
H-C   C-N
 \\ / \\
  C=C   O
 / \   O
H   H

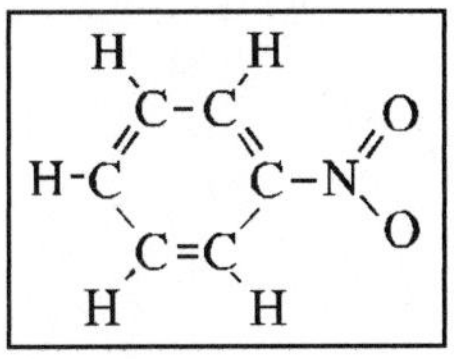

C   (all C atoms) sp^2 hybridization (3 electron groups)

(d)

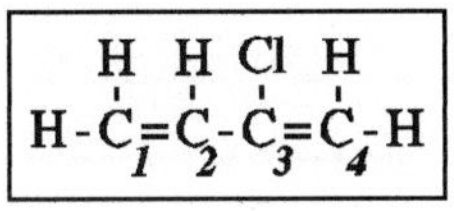

H H Cl H
| | | |
H-C=C-C=C-H
  1 2 3 4

C_1 sp^2 hybridization (3 electron groups)
C_2 sp^2 hybridization (3 electron groups)
C_3 sp^2 hybridization (3 electron groups)
C_4 sp^2 hybridization (3 electron groups)

(e)

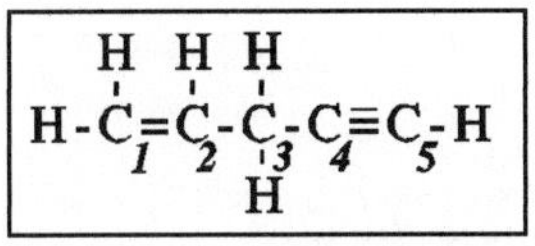

H H H
| | |
H-C=C-C-C≡C-H
  1 2 |3 4 5
        H

C_1 sp^2 hybridization (3 electron groups)
C_2 sp^2 hybridization (3 electron groups)
C_3 sp^3 hybridization (4 electron groups)
C_4 sp hybridization (2 electron groups)
C_5 sp hybridization (2 electron groups)

8-56.   *Refer to the Figures in Chapter 8*

(a) *s-s* overlap   (b) *s-p* overlap 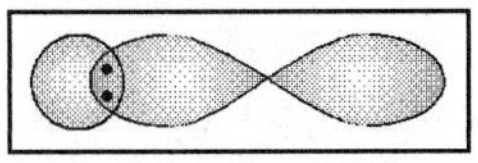   (c) *p-p* overlap along bond axis   (d) *p-p* overlap perpendicular to bond axis

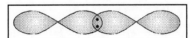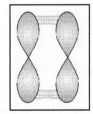

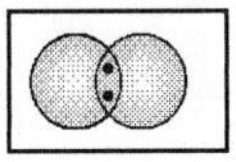

8-58.   *Refer to Sections 8-13 and 8-14, and Figures 8-6 and 8-8.*

(a) The molecule contains 10 single bonds (10 sigma bonds) and 1 double bond (1 sigma bond, 1 pi bond) for a total of 11 sigma bonds and 1 pi bond.

(b) The molecule contains 4 single bonds (4 sigma bonds) and 2 double bonds (2 sigma bonds, 2 pi bonds) for a total of 6 sigma bonds and 2 pi bonds.

(c) The molecule contains 9 single bonds (9 sigma bonds) and 1 double bond (1 sigma bond, 1 pi bond) for a total of 10 sigma bonds and 1 pi bond.

(d) The molecule contains 5 single bonds (5 sigma bonds), 1 double bond (1 sigma bond and 1 pi bond) and 1 triple bond (1 sigma bonds, 2 pi bonds) for a total of 7 sigma bonds and 3 pi bonds.

8-60.   *Refer to Section 8-14 and Figure 8-8b.*

C_2^{2-} $[:C ::C:]^{2-}$ $S = N - A$ = [2 x 8(for C)] - [2 x 4(for C) + 2e^-] = 6 shared electrons

There are 2 electron groups around each C, so each C atom is *sp* hybridized and the bond angle is 180°.

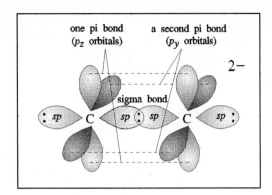

The sigma bond results from the head-on overlap of the *sp* hybrid orbitals located on each C atom.

One pi bond results from the side-on overlap of the $2p_y$ orbitals located on each C atom. The other pi bond results from the overlap of the $2p_z$ orbitals on each C atom.

8-62. *Refer to Sections 8-13 and 8-14.*

(a) butane

C_4H_{10}

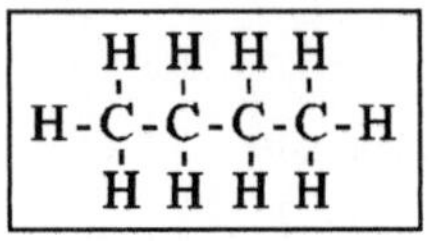

C_1, C_2, C_3, C_4 sp^3 hybridized with bond angles of 109.5°

(b) propene

$H_2C=CHCH_3$

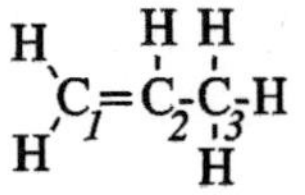

C_1, C_2 sp^2 hybridized with bond angles of 120°
C_3 sp^3 hybridized with bond angles of 109.5°

(c) 1-butyne

$HC\equiv CCH_2CH_3$

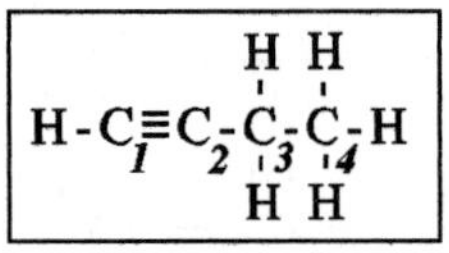

C_1, C_2 sp hybridized with bond angles of 180°.
C_3, C_4 sp^3 hybridized with bond angles of 109.5°.

(d) acetaldehyde

CH_3CHO

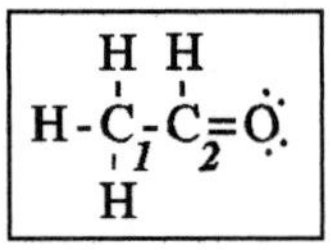

C_1 sp^3 hybridized with bond angles of 109.5°.
C_2 sp^2 hybridized with bond angles of 120°.

(a) butane, $CH_3CH_2CH_2CH_3$

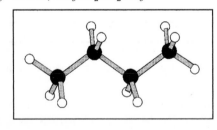

(b) propene, CH_2CHCH_3

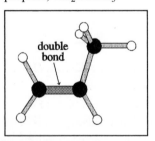

(c) 1-butyne, $CHCCH_2CH_3$

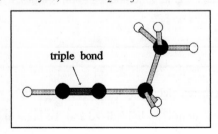

(d) acetaldehyde, CH_3CHO

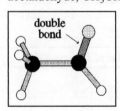

(i) NH_3

    (a) The N atom is sp^3 hybridized because the molecule (type AB_3U) has 4 electron groups around the central N atom.

    (b) Three of the sp^3 hybrid orbitals overlap with the $1s$ orbitals of the H atoms to form sigma bonds, leaving the fourth hybrid orbital to contain 1 lone pair of electrons on the N.

(ii) NH_4^+

    (a) The N atom is sp^3 hybridized because the ion (type AB_4) has 4 electron groups around the central N atom.

    (b) Each of the four sp^3 hybrid orbitals overlap with the $1s$ orbital of an H atom to form a sigma bond.

(iii) N_2H_2

    (a) Each N atom is sp^2 hybridized because there are 3 electron groups around each N atom (a single bond, a double bond and a lone pair of electrons).

    (b) Each N-H single bond is formed from the overlap of an sp^2 hybrid orbital of N with a $1s$ orbital of H. The double bond is the result of the side-on overlap of the unhybridized $2p$ orbitals on the N atoms, yielding a pi bond, and the head-on overlap of two sp^2 hybrid orbitals, yielding a sigma bond. The third sp^2 hybrid orbital on each N atom is used to accommodate the lone pair of electrons.

(iv) HCN

    (a) The N atom is sp hybridized because there are 2 electron groups around the N atom.

    (b) A lone pair of electrons occupies one sp hybrid orbital on N. The other sp hybrid orbital on N overlaps head-on with an sp hybrid orbital on C forming a sigma bond. The unhybridized $2p$ orbitals on C and N overlap side-on to form 2 pi bonds. The one sigma bond with the two pi bonds creates a C–N triple bond. The H–C bond is formed from the overlap of a $1s$ H orbital with an sp hybrid orbital on C.

(v) H_2NNH_2

    (a) Both N atoms are sp^3 hybridized because there are 4 electron groups around each N atom.

    (b) Each N–H single bond is formed from the overlap of an sp^3 hybrid orbital of N with a $1s$ orbital of H. The single N–N bond is the result of the head-on overlap of an sp^3 hybrid orbital on each N atom, yielding a sigma bond. The fourth sp^3 hybrid orbital on each N atom is used to accommodate the lone pair of electrons.

(i) H_2CO

    (a) The carbon atom is sp^2 hybridized.

    (b) Since there are no lone pairs of electrons on C, the electronic and molecular geometries are trigonal planar. The molecule is planar (flat) with H–C–H and H–C–O bond angles of 120º.

(ii) HCN

    (a) The carbon atom is sp hybridized.

    (b) The molecular geometry is linear and the H–C–N bond angle is 180º.

(iii) $CH_3CH_2CH_3$

    (a) All three carbon atoms are sp^3 hybridized.

    (b) The molecular geometry about each carbon atom is tetrahedral with H–C–H, C–C–C and H–C–C bond angles of 109.5º.

(iv) H_2C_2O

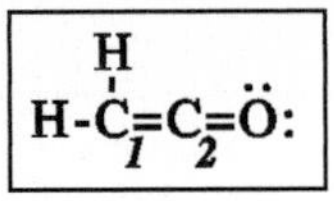

$$\begin{array}{c} H \\ | \\ H-C_1=C_2=\ddot{O}: \end{array}$$

(a) Carbon-1 is sp^2 hybridized and carbon-2 is sp hybridized.

(b) The molecular geometry at carbon-1 is trigonal planar with H–C–H and H–C–C bond angles of 120°. The molecular geometry at carbon-2 is linear with C–C–O bond angle of 180°.

8-68. *Refer to Table 8-4.*

	Lewis Formula	Hybridization	Molecular (or Ionic) Geometry
(i) IF	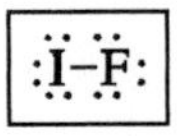 :Ï–F̈:	sp^3	linear
(ii) IF_3	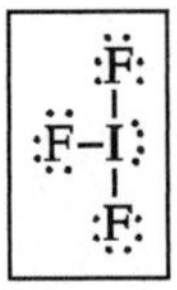	sp^3d	T-shaped
(iii) IF_4^-	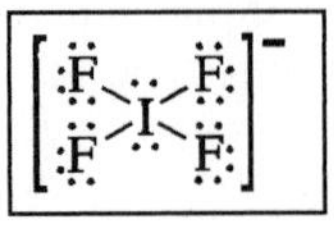	sp^3d^2	square planar
(iv) IF_5	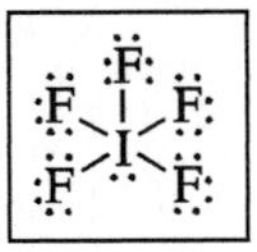	sp^3d^2	square pyramidal

---

8-70. *Refer to Sections 8-8 and 8-9.*

hydroxylamine

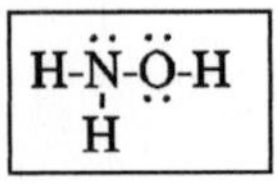

Since there are 4 electron groups around both the N and O atoms, the ideal bond angles about these atoms should be 109.5°. The lone pair of electrons on the N atom would make the H–N–H and the H–N–O bond angles less than 109.5°. The two lone pairs of electrons on the O atom should make the N–O–H bond angle even smaller. In fact, the observed bond angles are 107° for H–N–H and H–N–O, and 102° for N–O–H.

8-72. *Refer to Exercise 8-2 Solution and Section 8-8.*

Experimental evidence demonstrates that ·CH_3, the methyl free radical, has bond angles of about 120°, indicating a trigonal planar arrangement. This fact indicates that a single unpaired electron does not have much, if any, repulsive force. The species can be treated as if the central carbon atom had only 3 electron groups, sp^2 hybridization and a resulting trigonal planar geometry.

The methyl carbanion, :CH_3^-, has bond angles close to that in a tetrahedral arrangement of atoms, 109°, indicating 4 electron groups around the central C atom and sp^3 hybridization. The lone pair of electrons exerts significant repulsive force on the electrons in bonding orbitals and must be counted as an electron group.

8-74. *Refer to Table 8-2.*

(a) NO_2^+

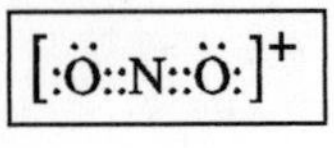

$[:\ddot{O}::N::\ddot{O}:]^+$

$S = N - A$ = [1 x 8(for N) + 2 x 8(for O)] - [1 x 5(for N) + 2 x 6(for O) - 1e^-]
= 8 shared electrons

The Lewis formula of NO_2^+ (type AB_2) shows 2 electron groups around the central N atom. The hybridization of N is sp.

NO$_2^-$

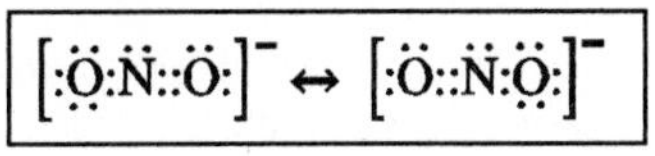

$S = N - A = [1 \times 8(\text{for N}) + 2 \times 8(\text{for O})]$
$\qquad\qquad - [1 \times 5(\text{for N}) + 2 \times 6(\text{for O}) + 1e^-]$
$\qquad = 6$ shared electrons

The Lewis resonance structures of NO$_2^-$ (type AB$_2$U) show 3 electron groups around the central N atom. The hybridization of N is sp^2.

(b) The bond angle for the linear ion, NO$_2^+$, is 180°. The ideal bond angle for the angular ion, NO$_2^-$, is 120°. The actual O–N–O bond angle is slightly less than 120°, because a lone pair of electrons requires more room than a bonding pair.

8-76. *Refer to Table 8-4.*

(a) PF$_5$

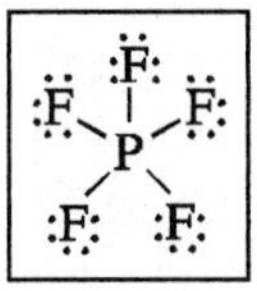

The Lewis formulas predict sp^3d hybridization of P in PF$_5$ (5 electron groups), changing to sp^3d^2 hybridization of P in PF$_6^-$ (6 electron groups).

PF$_6^-$

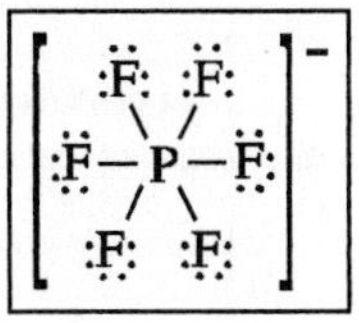

(b) CO

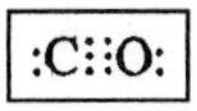

:C::O:

The Lewis formulas predict sp hybridization of C in both molecules since both C atoms have 2 electron groups around them.

CO$_2$

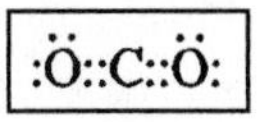

:O::C::O:

(c) AlI$_3$

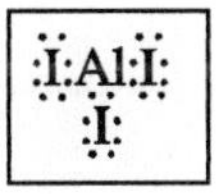

:I:Al:I:
　:I:

The Lewis formulas predict sp^2 hybridization of Al in AlI$_3$ (3 electron groups) changing to sp^3 hybridization of Al in AlI$_4^-$
(4 electron groups).

AlI$_4^-$

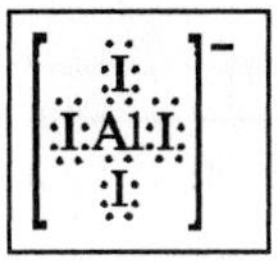

(d) NH$_3$

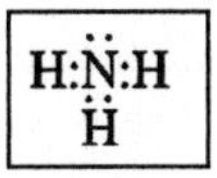

H:N:H
　H

The Lewis formulas predict sp^3 hybridization of N in NH$_3$ (4 electron groups) and sp^2 hybridization of B in BF$_3$ (3 electron groups). The product of the reaction, H$_3$N:BF$_3$, contains both N and B atoms with sp^3 hybridization.

BF$_3$

:F:B:F:
　:F:

H$_3$N:BF$_3$

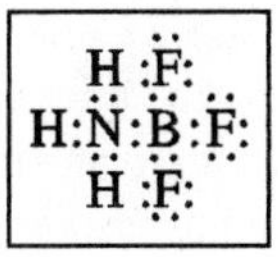

Using Valence Bond (VB) theory, the central atoms of the molecules with formulas AB_2U_2 and AB_3U should undergo sp^3 hybridized with predicted bond angles of 109.5°. If no hybridization occurs, bonds would be formed by the use of p orbitals. Since the p orbitals are oriented at 90° from each other, the bond angles would be 90°. Note that hybridization is only invoked if the actual molecular geometry data indicate that it is necessary.

The actual B–A–B bond angles for molecules of some representative elements are:

H_2O	104.5°	NH_3	106.7°
H_2S	92.2°	PH_3	93.7°
H_2Se	91.0°	AsH_3	91.8°
H_2Te	89.5°	SbH_3	91.3°

It is not necessary to invoke hybridization for the larger elements (Period 3 and greater). The above data show it is only the molecules containing the smaller Period 2 elements (O and N) that have bond angles approaching those for sp^3 hybridization, 109.5°. The other molecules formed from the larger elements have bond angles closer to 90°, so the application of the hybridization concept is not necessary to explain their molecular geometries.

Molecule or Ion	Electronic Geometry	Molecular Geometry	Hybridization of S Atom
SO_2	trigonal planar	bent or angular	sp^2
SCl_2	tetrahedral	bent or angular	sp^3
SO_3	trigonal planar	trigonal planar	sp^2
SO_3^{2-}	tetrahedral	trigonal pyramidal	sp^3
SF_4	trigonal bipyramidal	see-saw	sp^3d
SO_4^{2-}	tetrahedral	tetrahedral	sp^3
SF_5^+	trigonal bipyramidal	trigonal bipyramidal	sp^3d
SF_6	octahedral	octahedral	sp^3d^2

(a) CO_3^{2-}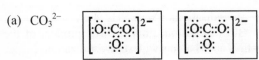

Here are two Lewis formulas for the carbonate ion, CO_3^{2-}. They do represent two resonance forms of this ion. Note: there are a total of 3 resonance forms for this ion.

(b) HSO_4^-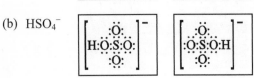

These two Lewis formulas for the hydrogen sulfate ion, HSO_4^-, do NOT represent two resonance forms of this ion. According to the definition of resonance, resonance is exhibited by a molecule or polyatomic ion for which two or more Lewis formulas with the same arrangements of atoms can be drawn to describe the bonding. The two structures given here do not have the same arrangement of atoms, and hence are not resonance structures.

8-84. *Refer to Sections 7-6, 8-6, 8-7 and 8-13.*

Evidence can be presented to show that rotation around a C–C single bond happens readily, but rotation around a C=C double bond does not. Consider the compound, CH_2ClCH_2Cl. No matter how this compound is synthesized, there is only one compound that is made with that formula. However, when CHClCHCl is prepared, there are two different compounds made with that formula. We call these two compounds geometric isomers. One is labeled "cis" and the other is "trans." They have different physical and chemical properties. If there were free rotation around a double bond, this could not happen.

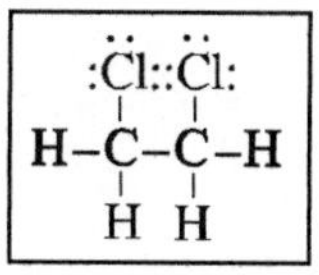

CH_2ClCH_2Cl

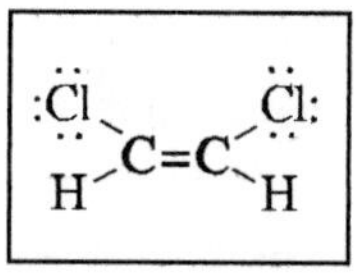

cis-CHClCHCl
(bp = 333.2 K)

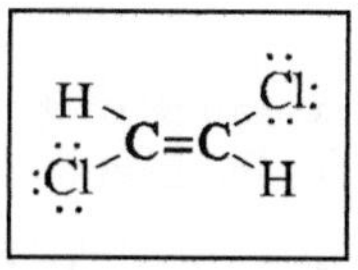

trans-CHClCHCl
(bp = 320.7 K)

8-86. *Refer to Table 8-4.*

C_3O_2 $S = N - A = [2 \times 8(\text{for O}) + 3 \times 8(\text{for C})] - [2 \times 6(\text{for O}) + 3 \times 4(\text{for C})] = 16$ shared electrons

The Lewis formula can be represented as:

$:\overset{..}{O}::C::C::C::\overset{..}{O}:$ or $:\overset{..}{O}=C=C=C=\overset{..}{O}:$

The Lewis formula predicts 2 electron groups around each C atom, resulting in the linear structure of C_3O_2.

8-88. *Refer to Section 7-4.*

The correct Lewis formula is:

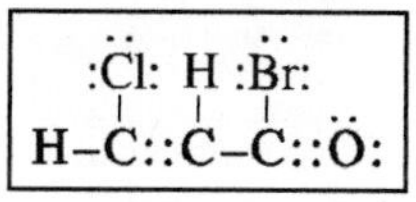

115

9 Molecular Orbitals in Chemical Bonding

9-2.   *Refer to the Introduction to Chapter 9 and Section 9-1.*

A molecular orbital (MO) is an orbital resulting from the overlap and combination of atomic orbitals on different atoms. An MO and the electrons in it belong to the molecule as a whole. Molecular orbitals calculations are used to develop

(1) mathematical representations of the orbital shapes, and
(2) energy level diagrams for the molecules.

The mathematical pictures called "electron density maps" are used to determine molecular structures and the energy level diagrams are used to determine the energies of bond formation and to interpret spectroscopy data.

9-4.   *Refer to Section 8-4.*

A set of hybridized atomic orbitals holds the same maximum number of electrons as the set of atomic orbitals from which the hybridized atomic orbitals were formed. A hybridized atomic orbital can hold a maximum of 2 electrons having opposite spin.

9-6.   *Refer to Section 9-1, and Figures 9-2 and 9-3.*

A σ (sigma) and a σ* (sigma star) molecular orbital result from the head-on overlap and the subsequent combination of atomic orbitals on adjacent atoms. Both molecular orbitals are cylindrically symmetrical about the axis linking the two atoms. There is a high electron density in the region between the atoms for the bonding σ orbital, promoting bonding and stabilizing the system. However, the electron density between the atoms approaches zero at the nodal plane for the anti-bonding σ* orbital, which destabilizes the system. As an example, see below for the overlap of 2 $1s$ orbitals (from Figure 9-2). Figure 9-3 shows the bonding and antibonding sigma orbitals formed by the combining head-on of two p orbitals.

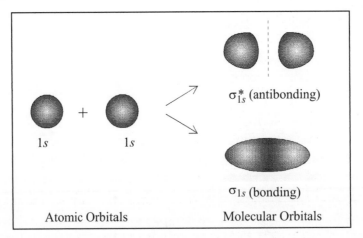

9-8.   *Refer to Section 9-2.*

Rules for placing electrons in molecular orbitals:

(1) Choose the correct molecular orbital energy level diagram.

(2) Determine the *total* number of electrons in the molecule or ion.

(3) Put the electrons into the energy level diagram as follows:

- Each electron is placed in the lowest possible energy level.

- A maximum of 2 electrons can be placed in an orbital, but they must have opposite spins (Pauli Exclusion Principle).

- Electrons must occupy all the orbitals of a given energy level singly with the same spin before pairing begins (Hund's Rule).

9-10.   *Refer to Sections 9-1 and 9-6, and Figures 9-10 and 9-11.*

(a) Atomic orbitals are *pure* orbitals that have not mixed with other orbitals in the same atom, molecule or ion. Molecular orbitals are orbitals resulting from the overlapping and mixing of atomic orbitals from *all* the atoms in a molecule. A molecular orbital belongs to the molecule as a whole and not simply to a particular atom in the molecule.

For example, each H atom has a $1s$ atomic orbital, but in the H_2 molecule, the σ_{1s} molecular orbital belongs to the entire H_2 molecule.

(b) A bonding molecular orbital (MO) results when two atomic orbitals overlap in phase. The energy of the bonding MO is always lower than the original atomic orbitals and is therefore more stable. An antibonding MO results when two atomic orbitals overlap out of phase; its energy is higher than the original atomic orbitals. Refer to Figure 9-2.

(c) A sigma (σ) molecular orbital results from the head-on overlap of two atomic orbitals. A pi (π) molecular orbital results from the side-on overlap of atomic orbitals. See Figures 9-2, 9-3 and 9-4.

(d) A localized molecular orbital is an MO which is associated with one or two particular atoms in a molecule. The bonding orbitals are localized between the atoms that are bonded together, the nonbonding orbitals are localized on one particular atom. Delocalized molecular orbitals, on the other hand, cover the entire molecule. These orbitals cannot easily be labeled as specific bonds. See Figures 9-10 and 9-11.

9-12.   *Refer to Section 9-1.*

Bonding molecular orbitals (MOs) have energies that are lower than those of the original atomic orbitals. Antibonding molecular orbitals have energies that are higher than those of the original atomic orbitals. Therefore, bonding MOs are more stable than the original atomic orbitals, whereas, antibonding MOs are less stable than the original atomic orbitals.

9-14.   *Refer to the Key Terms for Chapter 9.*

(a) A homonuclear molecule or ion is a species containing only one type of element. For example, O_2, O_3 and O_2^+ are homonuclear.

(b) A heteronuclear molecule or ion is a species containing different elements. For example, HF, CN^- and H_2O are heteronuclear.

(c) A diatomic molecule or ion is a species containing two atoms. It can either be homonuclear (O_2, O_2^-) or heteronuclear (HF, CN^-).

9-16. *Refer to Sections 9-2, 9-3, 9-4, Table 9-1, Example 9-1, Figure 9-5 and Exercise 9-15.*

Recall: Bond Order $= \dfrac{\text{No. bonding electrons } - \text{ No. antibonding electrons}}{2}$

		No. Bonding Electrons	No. Antibonding Electrons	Bond Order
(a)	$Be_2 \quad \sigma_{1s}^{2} \; \sigma_{1s}^{*\,2} \; \sigma_{2s}^{2} \; \sigma_{2s}^{*\,2}$	4	4	0
	$Be_2^{+} \quad \sigma_{1s}^{2} \; \sigma_{1s}^{*\,2} \; \sigma_{2s}^{2} \; \sigma_{2s}^{*\,1}$	4	3	0.5
	$Be_2^{-} \quad \sigma_{1s}^{2} \; \sigma_{1s}^{*\,2} \; \sigma_{2s}^{2} \; \sigma_{2s}^{*\,2} \; \pi_{2p_y}^{1}$	5	4	0.5
(b)	$B_2 \quad \sigma_{1s}^{2} \; \sigma_{1s}^{*\,2} \; \sigma_{2s}^{2} \; \sigma_{2s}^{*\,2} \; \pi_{2p_y}^{1} \; \pi_{2p_z}^{1}$	6	4	1
	$B_2^{+} \quad \sigma_{1s}^{2} \; \sigma_{1s}^{*\,2} \; \sigma_{2s}^{2} \; \sigma_{2s}^{*\,2} \; \pi_{2p_y}^{1}$	5	4	0.5
	$B_2^{-} \quad \sigma_{1s}^{2} \; \sigma_{1s}^{*\,2} \; \sigma_{2s}^{2} \; \sigma_{2s}^{*\,2} \; \pi_{2p_y}^{2} \; \pi_{2p_z}^{1}$	7	4	1.5

9-18. *Refer to Section 9-3 and Exercise 9-16 Solution.*

In Molecular Orbital Theory, the greater the bond order, the more stable is the molecule or ion. Therefore, we predict:

Unstable: Be_2

Somewhat stable: Be_2^+, Be_2^-, B_2^+

Stable: B_2, B_2^-

This means that although Be_2 is unstable, both its 1+ cation and 1− anion are somewhat stable. It also shows that the stability of the boron species is in the order: $B_2^+ < B_2 < B_2^-$.

All these predictions are generally correct. The fact that many of these supposed stable species are not observed in nature is because they are chemically very reactive and are observed only at high temperatures and reduced pressures. Therefore, the chemical reactivity of a species is also instrumental when considering the survival probability of the species with stable chemical bonding.

9-20. *Refer to Sections 9-2, 9-3, 9-4, Table 9-1, and Example 9-1.*

Note: Bond Order $= \dfrac{\text{No. bonding electrons } - \text{ No. antibonding electrons}}{2}$

		No. Bonding Electrons	No. Antibonding Electrons	Bond Order
(a)	$Ne_2^{+} \quad \sigma_{1s}^{2} \; \sigma_{1s}^{*\,2} \; \sigma_{2s}^{2} \; \sigma_{2s}^{*\,2} \; \sigma_{2p}^{2} \; \pi_{2p_y}^{2} \; \pi_{2p_z}^{2} \; \pi_{2p_y}^{*\,2} \; \pi_{2p_z}^{*\,2} \; \sigma_{2p}^{*\,1}$	2	1	0.5
(b)	$Ne_2 \quad \sigma_{1s}^{2} \; \sigma_{1s}^{*\,2} \; \sigma_{2s}^{2} \; \sigma_{2s}^{*\,2} \; \sigma_{2p}^{2} \; \pi_{2p_y}^{2} \; \pi_{2p_z}^{2} \; \pi_{2p_y}^{*\,2} \; \pi_{2p_z}^{*\,2} \; \sigma_{2p}^{*\,2}$	2	2	0
(c)	$Ne_2^{2+} \quad \sigma_{1s}^{2} \; \sigma_{1s}^{*\,2} \; \sigma_{2s}^{2} \; \sigma_{2s}^{*\,2} \; \sigma_{2p}^{2} \; \pi_{2p_y}^{2} \; \pi_{2p_z}^{2} \; \pi_{2p_y}^{*\,2} \; \pi_{2p_z}^{*\,2}$	2	0	1

The ions Ne_2^+ and Ne_2^{2+} have non-zero bond orders and would exist, but Ne_2 has a bond order of zero and would not exist.

9-22. *Refer to Sections 9-2, 9-3, 9-4, Table 9-1, and Example 9-1.*

Recall: $\text{Bond Order} = \dfrac{\text{No. bonding electrons - No. antibonding electrons}}{2}$

		No. Bonding Electrons	No. Antibonding Electrons	Bond Order
(a) Li_2	$\sigma_{1s}^2\ \sigma_{1s}^{*2}\ \sigma_{2s}^2$	4	2	1
(b) Li_2^+	$\sigma_{1s}^2\ \sigma_{1s}^{*2}\ \sigma_{2s}^1$	3	2	0.5
(c) O_2^{2-}	$\sigma_{1s}^2\ \sigma_{1s}^{*2}\ \sigma_{2s}^2\ \sigma_{2s}^{*2}\ \sigma_{2p}^2\ \pi_{2p_y}^2\ \pi_{2p_z}^2\ \pi_{2p_y}^{*2}\ \pi_{2p_z}^{*2}$	10	8	1

All these species would exist since their bond order is greater than zero.

9-24. *Refer to Section 9-3 and Exercise 9-23.*

Recall: $\text{Bond Order} = \dfrac{\text{No. bonding electrons - No. antibonding electrons}}{2}$

(a) X_2 bond order $= \dfrac{10 - 4}{2} = 3$

(b) X_2 bond order $= \dfrac{10 - 6}{2} = 2$

(c) X_2^- bond order $= \dfrac{10 - 5}{2} = 2.5$

9-26. *Refer to Sections 9-3 and 9-4, Example 9-1, and Exercise 9-25.*

In Exercise 9-25, the following electron configurations were obtained:

F_2 $\sigma_{1s}^2\ \sigma_{1s}^{*2}\ \sigma_{2s}^2\ \sigma_{2s}^{*2}\ \sigma_{2p}^2\ \pi_{2p_y}^2\ \pi_{2p_z}^2\ \pi_{2p_y}^{*2}\ \pi_{2p_z}^{*2}$

F_2^- $\sigma_{1s}^2\ \sigma_{1s}^{*2}\ \sigma_{2s}^2\ \sigma_{2s}^{*2}\ \sigma_{2p}^2\ \pi_{2p_y}^2\ \pi_{2p_z}^2\ \pi_{2p_y}^{*2}\ \pi_{2p_z}^{*2}\ \sigma_{2p}^{*1}$

F_2^+ $\sigma_{1s}^2\ \sigma_{1s}^{*2}\ \sigma_{2s}^2\ \sigma_{2s}^{*2}\ \sigma_{2p}^2\ \pi_{2p_y}^2\ \pi_{2p_z}^2\ \pi_{2p_y}^{*2}\ \pi_{2p_z}^{*1}$

C_2 $\sigma_{1s}^2\ \sigma_{1s}^{*2}\ \sigma_{2s}^2\ \sigma_{2s}^{*2}\ \pi_{2p_y}^2\ \pi_{2p_z}^2$

C_2^+ $\sigma_{1s}^2\ \sigma_{1s}^{*2}\ \sigma_{2s}^2\ \sigma_{2s}^{*2}\ \pi_{2p_y}^2\ \pi_{2p_z}^1$

C_2^- $\sigma_{1s}^2\ \sigma_{1s}^{*2}\ \sigma_{2s}^2\ \sigma_{2s}^{*2}\ \pi_{2p_y}^2\ \pi_{2p_z}^2\ \sigma_{2p}^1$

(a) Recall: $\text{Bond Order} = \dfrac{\text{No. bonding electrons - No. antibonding electrons}}{2}$

	F_2	F_2^-	F_2^+	C_2	C_2^+	C_2^-
Bonding e^-	10	10	10	8	7	9
Antibonding e^-	8	9	7	4	4	4
Bond order	1	0.5	1.5	2	1.5	2.5
(b) Unpaired e^-	0	1	1	0	1	1
Paramagnetic or Diamagnetic	D	P	P	D	P	P

(c) Based on bond orders, we have the following order of stability:

Somewhat stable: F_2^-

Stable: F_2, F_2^+, C_2, C_2^+, and C_2^-

Order of stability: $F_2^- < F_2 < F_2^+$ and $C_2^+ < C_2 < C_2^-$

9-28. *Refer to Sections 9-3 and 9-4.*

(a) N_2 $\quad \sigma_{1s}^2 \ \sigma_{1s}^{*\,2} \ \sigma_{2s}^2 \ \sigma_{2s}^{*\,2} \ \pi_{2p_y}^2 \ \pi_{2p_z}^2 \ \sigma_{2p}^2$

N_2^- $\quad \sigma_{1s}^2 \ \sigma_{1s}^{*\,2} \ \sigma_{2s}^2 \ \sigma_{2s}^{*\,2} \ \pi_{2p_y}^2 \ \pi_{2p_z}^2 \ \sigma_{2p}^2 \ \pi_{2p_y}^{*\,1}$

N_2^+ $\quad \sigma_{1s}^2 \ \sigma_{1s}^{*\,2} \ \sigma_{2s}^2 \ \sigma_{2s}^{*\,2} \ \pi_{2p_y}^2 \ \pi_{2p_z}^2 \ \sigma_{2p}^1$

(b) Recall: Bond Order $= \dfrac{\text{No. bonding electrons } - \text{ No. antibonding electrons}}{2}$

N_2 $\qquad$ bond order $= \dfrac{10 - 4}{2} = 3$

N_2^- $\qquad$ bond order $= \dfrac{10 - 5}{2} = 2.5$

N_2^+ $\qquad$ bond order $= \dfrac{9 - 4}{2} = 2.5$

(c) The species with the greatest bond order has the greatest bond energy and the shortest bond length.

Therefore, the N_2 molecule with a bond order of 3 has the shortest bond.

Both N_2^- and N_2^+ ions with bond orders of 2.5, should have slightly longer bond lengths.

in bond length: $\quad N_2 < N_2^-$ and N_2^+

NO⁺

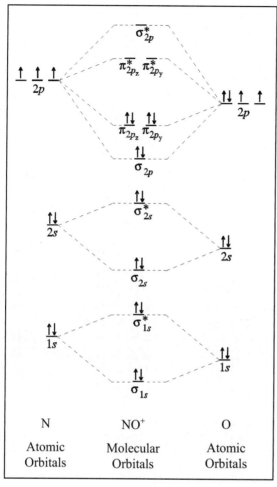

N
Atomic
Orbitals

NO⁺
Molecular
Orbitals

O
Atomic
Orbitals

The bond order of NO^+ is

$$\text{bond order} = \frac{\text{bonding } e^- - \text{antibonding } e^-}{2}$$

$$= \frac{10 - 4}{2}$$

$$= 3$$

There are no lone pairs of electrons, so the molecular ion is **diamagnetic**.

The NO^+ ion is **very stable**, since its bond order is high.

Recall: Bond Order $= \dfrac{\text{No. bonding electrons } - \text{ No. antibonding electrons}}{2}$

CN $\quad \sigma_{1s}^{2}\ \sigma_{1s}^{*\,2}\ \sigma_{2s}^{2}\ \sigma_{2s}^{*\,2}\ \pi_{2p_y}^{2}\ \pi_{2p_z}^{2}\ \sigma_{2p}^{1}$

CN⁻ $\quad \sigma_{1s}^{2}\ \sigma_{1s}^{*\,2}\ \sigma_{2s}^{2}\ \sigma_{2s}^{*\,2}\ \pi_{2p_y}^{2}\ \pi_{2p_z}^{2}\ \sigma_{2p}^{2}$

CN²⁻ $\quad \sigma_{1s}^{2}\ \sigma_{1s}^{*\,2}\ \sigma_{2s}^{2}\ \sigma_{2s}^{*\,2}\ \pi_{2p_y}^{2}\ \pi_{2p_z}^{2}\ \sigma_{2p}^{2}\ \pi_{2p_y}^{*\,1}$

Bond order calculations:

bond order of CN $= \dfrac{9-4}{2} = 2.5$ $\qquad$ bond order of CN⁻ $= \dfrac{10-4}{2} = 3$ $\qquad$ bond order of CN²⁻ $= \dfrac{10-5}{2} = 2.5$

The most stable species is **CN⁻** since it has the largest bond order.

9-34. *Refer to Section 9-5 and the Molecular Orbital Diagram before Exercise 9-29.*

(a) NF

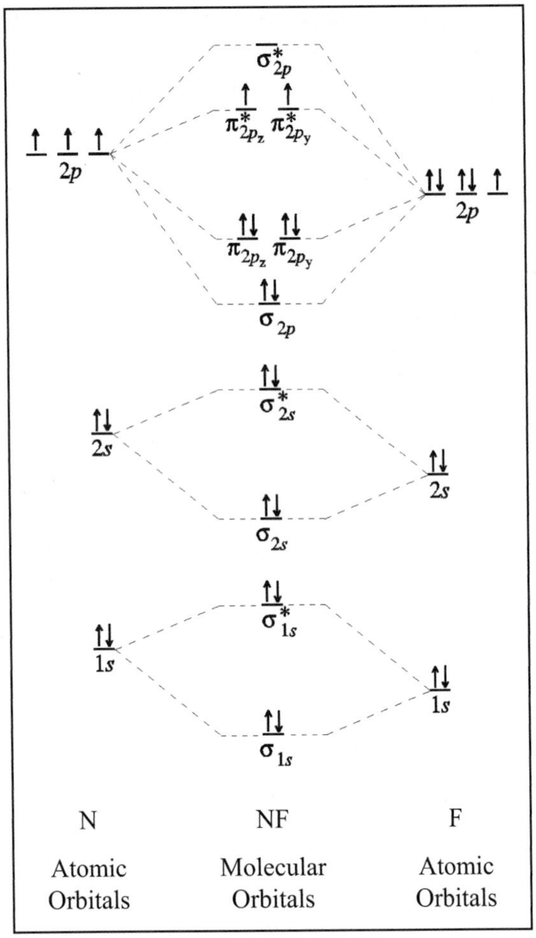

NF+

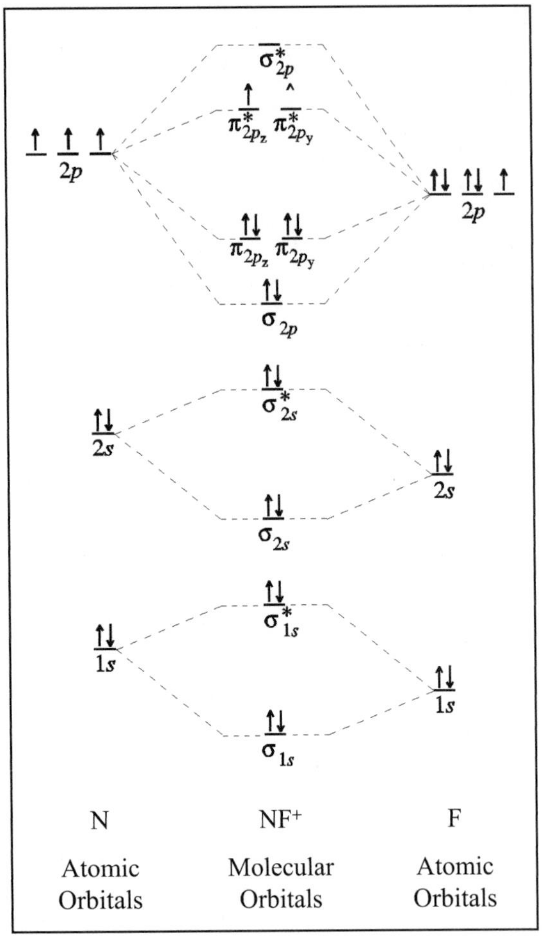

(b) NF $\sigma_{1s}^{2}\ \sigma_{1s}^{*2}\ \sigma_{2s}^{2}\ \sigma_{2s}^{*2}\ \sigma_{2p}^{2}\ \pi_{2p_z}^{2}\ \pi_{2p_y}^{2}\ \pi_{2p_z}^{*1}\ \pi_{2p_y}^{*1}$

NF^+ $\sigma_{1s}^{2}\ \sigma_{1s}^{*2}\ \sigma_{2s}^{2}\ \sigma_{2s}^{*2}\ \sigma_{2p}^{2}\ \pi_{2p_z}^{2}\ \pi_{2p_y}^{2}\ \pi_{2p_z}^{*1}$

(c) Recall: $\text{Bond Order} = \dfrac{\text{No. bonding electrons - No. antibonding electrons}}{2}$

	NF	NF+
Bonding e^-	10	10
Antibonding e^-	6	5
Bond order	2	2.5
Unpaired e^-	2	1
(d) Paramagnetic or Diamagnetic	P	P

Both NF and NF^+ are stable, but NF^+ is slightly more stable than NF due to its higher bond order.

BC

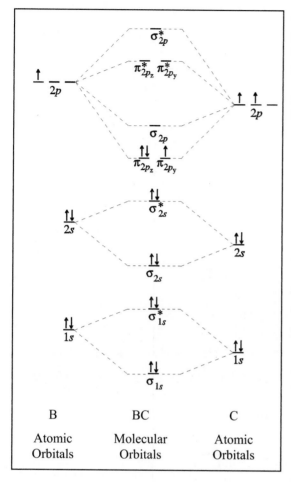

B	BC	C
Atomic Orbitals	Molecular Orbitals	Atomic Orbitals

To decrease the strength of the B–C bond, an electron is removed from the BC molecule, forming the BC$^+$ ion. One of the electrons from the bonding orbital, π_{2p_x} is removed.

The bond order decreases from 1.5 for BC to 1.0 for BC$^+$. Likewise, the bond strength also decreases.

9-38. *Refer to Section 9-6 and Figure 9-10.*

(a) SO_2

:Ö::S̈:   ↔   :Ö:S̈:
    :Ö: :Ö:

(b) O_3

:Ö::Ö:Ö: ↔   :Ö:Ö::Ö:

(c) HCO_2^-

$\left[\text{H:C:Ö:}\right]^-$ ↔ $\left[\text{H:C::Ö:}\right]^-$
      :Ö: :Ö:

Molecular Orbital Descriptions:

(a)

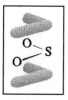

(b)

(c)

(a) O_2^{2+} $\sigma_{1s}^2 \; \sigma_{1s}^{*\,2} \; \sigma_{2s}^2 \; \sigma_{2s}^{*\,2} \; \sigma_{2p}^2 \; \pi_{2p_y}^2 \; \pi_{2p_z}^2$ bond order $= \dfrac{10 - 4}{2} = 3$

The ion is diamagnetic and very stable.

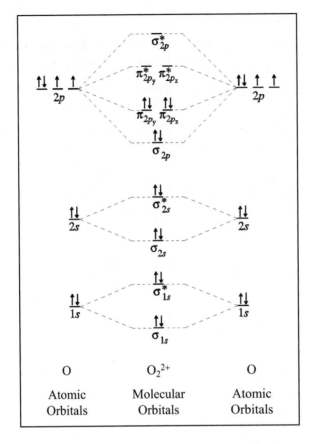

(b) HO^- $1s^2 \; 2s^2 \; \sigma_{sp}^2 \; 2p_x^2 \; 2p_y^2$ bond order $= \dfrac{2 - 0}{2} = 1$

The ion is diamagnetic and stable.

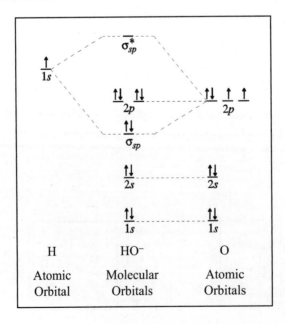

(c) HF $1s^2\ 2s^2\ \sigma_{sp}^2\ 2p_x^2\ 2p_y^2$

bond order $= \dfrac{2-0}{2} = 1$

The HF molecule is diamagnetic and stable.

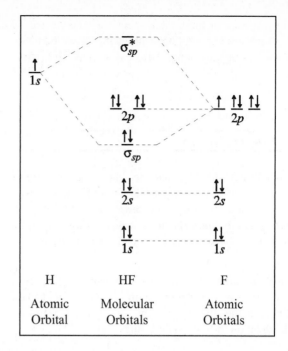

H	HF	F
Atomic Orbital	Molecular Orbitals	Atomic Orbitals

9-42.   *Refer to Section 9-4, Figure 9-5, and Table 9-1.*

If a diatomic species has 20 or fewer electrons, its bond order cannot be greater than three. The difference between the number of bonding electrons and the number of antibonding electrons is never more than 6 according to Figure 9-5. Since bond order equals (no. bonding electrons - antibonding electrons)/2, the maximum bond order can never be more than three.

Bond order must be a value that is divisible by 0.5 because the difference between the number of bonding electrons and antibonding electrons is always a whole number. When the difference is divided by 2 to determine bond order, the answer must be a multiple of 0.5.

9-44.   *Refer to Section 9-4, Figure 9-5, and Table 9-1.*

The homonuclear diatomic molecules of the second row of the periodic table are predicted by MO theory to (1) be paramagnetic: B_2 and O_2, (2) have a bond order of 1: Li_2, B_2 and F_2, (3) have a bond order of 2: C_2, O_2, and (4) have the highest bond order: N_2 with bond order of 3.

9-46.   *Refer to Section 9-4, Figure 9-7 and Exercise 9-30 Solution.*

To explain how the bonds in N–O, N–O$^+$ and N–O$^-$ compare, let's first find their electron configurations and bond order:

NO (15 e$^-$) $\sigma_{1s}^2\ \sigma_{1s}^{*\,2}\ \sigma_{2s}^2\ \sigma_{2s}^{*\,2}\ \sigma_{2p}^2\ \pi_{2p_z}^2\ \pi_{2p_y}^2\ \pi_{2p_z}^{*\,1}$ bond order = (10 - 5)/2 = 2.5

NO$^+$ (14 e$^-$) $\sigma_{1s}^2\ \sigma_{1s}^{*\,2}\ \sigma_{2s}^2\ \sigma_{2s}^{*\,2}\ \sigma_{2p}^2\ \pi_{2p_z}^2\ \pi_{2p_y}^2$ bond order = (10 - 4)/2 = 3

NO$^-$ (16 e$^-$) $\sigma_{1s}^2\ \sigma_{1s}^{*\,2}\ \sigma_{2s}^2\ \sigma_{2s}^{*\,2}\ \sigma_{2p}^2\ \pi_{2p_z}^2\ \pi_{2p_y}^2\ \pi_{2p_z}^{*\,1}\ \pi_{2p_y}^{*\,1}$ bond order = (10 - 6)/2 = 2

The atoms in each of these three species are connected by a strong bond, since their bond orders are large, so they should definitely exist. As bond order increases, bond strength increases and bond length decreases. Therefore, in order of increasing bond length (decreasing bond strength), we have:

in bond length: $NO^+ < NO < NO^-$

9-48.   *Refer to Section 9-4.*

(a) N and P are both VA elements but N_2 is much more stable than P_2 because N is a smaller atom than P and therefore can effectively participate in pi bonding, whereas P cannot. The $3p$ orbitals of a P atom do not overlap side-on in a pi bond with the corresponding $3p$ orbitals of another P atom nearly as well as the corresponding $2p$ orbitals of the much smaller nitrogen atoms. Therefore, as explained by valence bond theory, P forms 3 sigma bonds instead and acquires an octet of electrons using sp^3 hybridization to form P_4 molecules.

(b) O and S are both VIA elements. However, O_2 and O_3 are much more stable than S_2 because O is a smaller atom than S and therefore can effectively participate in pi bonding, whereas S cannot. The $3p$ orbitals of a S atom do not overlap side-on with the corresponding $3p$ orbitals of another S atom nearly as well as the corresponding $2p$ orbitals of the smaller O atoms. As a result, S forms 2 sigma bonds and obtains its octet of electrons using sp^3 hybridization to form S_8 molecules.

10 Reactions in Aqueous Solutions I: Acids, Bases, and Salts

10-2. *Refer to Section 10-2.*

(a) According to Arrhenius, an acid is a substance that contains hydrogen and produces hydrogen ions in aqueous solution. A base is a substance that contains the OH group and produces hydroxide ions in aqueous solution. Neutralization is the reaction between hydrogen ions and hydroxide ions yielding water molecules.

(b) acid: $HBr(aq) \rightarrow H^+(aq) + Br^-(aq)$

base: $Ba(OH)_2(aq) \rightarrow Ba^{2+}(aq) + 2OH^-(aq)$

neutralization: $2HBr(aq) + Ba(OH)_2(aq) \rightarrow BaBr_2(aq) + 2H_2O(\ell)$

10-4. *Refer to Section 4-2.*

To distinguish between strong electrolytes, weak electrolytes and nonelectrolytes, prepare equimolar aqueous solutions of the compounds and test their electrical conductivity. If a compound's solution conducts electricity well, it is a strong electrolyte; if its solution conducts electricity poorly, it is a weak electrolyte. A solution of a nonelectrolyte does not conduct electricity at all.

strong electrolyte: Na_2SO_4, $HClO_4$

weak electrolyte: HCN, CH_3COOH, HF, $HCOOH$, NH_3

nonelectrolyte: CH_3OH

10-6. *Refer to Section 10-3.*

A hydrated hydrogen ion containing only one water of hydration: $H^+(H_2O)$ or H_3O^+
This ion is also called the hydronium ion.

10-8. *Refer to Section 10-3 and Exercise 10-6 Solution.*

The statement, "The hydrated hydrogen ion should always be represented as H_3O^+," has two main flaws. First, the true extent of hydration of the H^+ in many solutions is unknown. Secondly, when balancing equations, it is generally much easier to use H^+ rather than H_3O^+.

10-10. *Refer to Section 10-4 and Example 10-1.*

In accordance with Brønsted-Lowry terminology,

(a) acid: a proton donor, e.g. HCl, NH_4^+, H_2O, H_3O^+

(b) conjugate base: a species that is produced when an acid donates a proton,

e.g. Cl^- is the conjugate base of HCl OH^- is the conjugate base of H_2O
         NH_3 is the conjugate base of NH_4^+ H_2O is the conjugate base of H_3O^+

(c) base: a proton acceptor, e.g., NH_3, H_2O, OH^-

(d) conjugate acid: a species that is produced when a base accepts a proton,

e.g.   HCl is the conjugate acid of Cl^- H_2O is the conjugate acid of OH^-
        NH_4^+ is the conjugate acid of NH_3 H_3O^+ is the conjugate acid of H_2O

(e) conjugate acid-base pair: two species with formulas that differ only by a proton, e.g., HCl and Cl^-, NH_4^+ and NH_3, $H_2PO_4^-$ and HPO_4^{2-}. The species with the extra proton is the conjugate acid, whereas the other is the conjugate base.

10-12.   Refer to Section 10-4.

(a) In dilute aqueous solution, ammonia is a Brønsted-Lowry base and water is a Brønsted-Lowry acid. The reaction produces an ammonium ion (the conjugate acid of ammonia) and a hydroxide ion (the conjugate base of water).

$$NH_3(aq) + H_2O(\ell) \rightleftarrows NH_4^+(aq) + OH^-(aq)$$

(b) In the gaseous state, ammonia also behaves as a Brønsted-Lowry base when reacting with gaseous hydrogen chloride. It takes a proton from the gaseous acid and produces the ionic salt, ammonium chloride.

$$NH_3(g) + HCl(g) \rightleftarrows NH_4Cl(s)$$

10-14.   Refer to Section 10-4.

These species are Brønsted-Lowry bases in water (none are Arrhenius bases) since they are proton acceptors and OH^- ions are produced: NH_3, HS^-, $HCOO^-$ and O^{2-}.

$$NH_3 + H_2O \rightleftarrows NH_4^+ + OH^-$$ $$HS^- + H_2O \rightleftarrows H_2S + OH^-$$
$$HCOO^- + H_2O \rightleftarrows HCOOH + OH^-$$ $$O^{2-} + H_2O \rightleftarrows 2OH^-$$

10-16.   Refer to Section 10-4 and Example 10-1.

(a) NH_4^+ + CH_3COO^- → NH_3 + CH_3COOH
    acid$_1$ base$_2$ base$_1$ acid$_2$

(c) $HClO_4$ + $[H_2NNH_3]^+$ → ClO_4^- + $[H_3NNH_3]^{2+}$
    acid$_1$ base$_2$ base$_1$ acid$_2$

(b) S^{2-} + H_2SO_4 → HS^- + HSO_4^-
    base$_1$ acid$_2$ acid$_1$ base$_2$

(d) NH_2^- + H_2O → NH_3 + OH^-
    base$_1$ acid$_2$ acid$_1$ base$_2$

10-18.   Refer to Section 10-4 and Example 10-1.

Acid	H_2O	HS^-	HBr	PH_4^+	$HOCH_3$
Conjugate Base	OH^-	S^{2-}	Br^-	PH_3	CH_3O^-

10-20.   Refer to Section 10-4 and Example 10-1.

	Brønsted-Lowry Acids	Brønsted-Lowry Bases
(a)	H_2O, H_2S	HS^-, OH^-
(b)	H_2CO_3, H_2SO_4	HSO_4^-, HCO_3^-
(c)	CH_3COOH, HNO_2	NO_2^-, CH_3COO^-
(d)	H_2O, NH_3	NH_2^-, OH^-

128

(a) $H_2CO_3 + H_2O \rightleftarrows H_3O^+ + HCO_3^-$
    acid$_1$ base$_2$ acid$_2$ base$_1$

(c) $H_3PO_4 + S^{2-} \rightleftarrows HS^- + H_2PO_4^-$
    acid$_1$ base$_2$ acid$_2$ base$_1$

(b) $HSO_4^- + H_2O \rightleftarrows H_3O^+ + SO_4^{2-}$
    acid$_1$ base$_2$ acid$_2$ base$_1$

(d) $HS^- + OH^- \rightleftarrows H_2O + S^{2-}$
    acid$_1$ base$_2$ acid$_2$ base$_1$

10-24. *Refer to Section 10-4 and Example 10-1.*

(a) $H_2SO_4 + H_2O \rightleftarrows HSO_4^- + H_3O^+$
    acid$_1$ base$_2$ base$_1$ acid$_2$

(b) $H_2SO_3 + H_2O \rightleftarrows HSO_3^- + H_3O^+$
    acid$_1$ base$_2$ base$_1$ acid$_2$

$HSO_4^- + H_2O \rightleftarrows SO_4^{2-} + H_3O^+$
    acid$_1$ base$_2$ base$_1$ acid$_2$

$HSO_3^- + H_2O \rightleftarrows SO_3^{2-} + H_3O^+$
    acid$_1$ base$_2$ base$_1$ acid$_2$

10-26. *Refer to Section 10-1.*

Aqueous solutions of strong soluble bases (1) have a bitter taste, (2) have a slippery feeling, (3) change the colors of many acid-base indicators, (4) react with protic acids to form salts and water, and (5) conduct an electrical current since they contain ions.

Aqueous ammonia reacts with protic acids to form salts (and no water), but exhibits all the other traits to a lesser degree since it is a weak base exhibiting limited ionization and providing a lower OH^- concentration.

10-28. *Refer to Section 10-6 and Table 10-1.*

Amphoteric metal or metalloid hydroxides can act either as an acid or as a base. Such compounds include $Be(OH)_2$, $Zn(OH)_2$, $As(OH)_3$ and $Sb(OH)_3$.

(a) As acids:
$$Be(OH)_2 + 2OH^- \rightarrow [Be(OH)_4]^{2-}$$
$$Zn(OH)_2 + 2OH^- \rightarrow [Zn(OH)_4]^{2-}$$
$$As(OH)_3 + OH^- \rightarrow [As(OH)_4]^-$$
$$Sb(OH)_3 + OH^- \rightarrow [Sb(OH)_4]^-$$

(b) As bases:
$$Be(OH)_2 + 2H^+ \rightarrow Be^{2+} + 2H_2O$$
$$Zn(OH)_2 + 2H^+ \rightarrow Zn^{2+} + 2H_2O$$
$$As(OH)_3 + 3H^+ \rightarrow As^{3+} + 3H_2O$$
$$Sb(OH)_3 + 3H^+ \rightarrow Sb^{3+} + 3H_2O$$

10-30. *Refer to Section 10-7 and the Key Terms at the end of Chapter 10.*

Consider the following reaction: $Rb_2O(s) + H_2O(\ell) \rightarrow 2Rb^+(aq) + 2OH^-(aq)$

The leveling effect is the effect by which all acids stronger than the acid that is characteristic of the solvent react with the solvent to produce that acid. For example, in water, the strongest acid and base that can exist in water is H^+ and OH^-, respectively. In our example, the oxide ion (O^{2-}) is a stronger base than OH^-. When K_2O dissolves in water, the O^{2-} cannot exist in the solution. O^{2-} immediately removes a proton from H_2O, forming OH^-:

$$O^{2-}(aq) + H_2O(\ell) \rightarrow 2OH^-(aq).$$

Because any base stronger than OH^- will undergo a similar reaction, we say that water does not allow us to discriminate between these bases with regard to strength. Therefore, water is called a leveling solvent.

10-32. *Refer to Sections 10-4 and 10-7.*

Base strength refers to the relative tendency to produce OH^- ions in aqueous solution by (1) the dissociation of soluble metal hydroxides or (2) by ionization reactions with water using Arrhenius theory. A more general definition, applying Brønsted-Lowry theory, is that base strength is a measure of the relative tendency to accept a proton from any acid.

Acid strength refers to the relative tendency of a substance to produce H_3O^+ ions in aqueous solutions, according to Arrhenius. Using Brønsted-Lowry theory, acid strength is a measure of the relative tendency of a species to donate a proton.

10-34. *Refer to Section 10-4.*

(a) A binary protic acid is a covalent compound consisting of hydrogen atom(s) and one other element. The compound can act as a proton donor.

(b)

hydrofluoric acid	$HF(aq)$
hydrosulfuric acid	$H_2S(aq)$
hydrobromic acid	$HBr(aq)$
hydroselenic acid	$H_2Se(aq)$

10-36. *Refer to Section 10-7.*

In order of decreasing acidity:

(a) $H_2Se > H_2S > H_2O$ (b) $HI > HBr > HCl > HF$ (c) $H_2S > HS^- > S^{2-}$ (S^{2-} is not an acid)

10-38. *Refer to Section 10-4.*

Brønsted-Lowry bases: NH_3, H_2O and $:H^-$ in ionic NaH Brønsted-Lowry acids: H_2O, HF

The hydrides, BeH_2, BH_3 and CH_4 are *generally* considered as neither Brønsted-Lowry acids nor Brønsted-Lowry bases.

10-40. *Refer to Section 10-7.*

Ternary acids, including nitric and sulfuric acids, can be described as hydroxyl compounds of nonmetals since they contain 1 or more -O-H groups attached to the central nonmetal atom. For example,

$$HNO_3 \equiv NO_2(OH) \qquad HClO_4 \equiv ClO_2(OH)$$

10-42. *Refer to Section 10-7.*

Consider phosphoric acid, H_3PO_4 (a triprotic acid), and phosphorus acid, H_3PO_3 (a diprotic acid). Two structures that show this acidic behavior are:

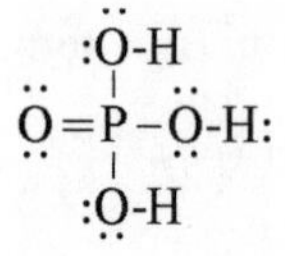

phosphoric acid

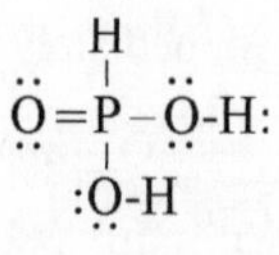

phosphorus acid

Each H atoms attached to an O atom is an acidic hydrogen and can be released by the acid to a base. Phosphoric acid has 3 such H atoms and thus is triprotic. Although phosphorus acid has 3 H atoms, only 2 H atoms are bonded to an O atom, the other one is attached directly to the P atom.. Thus phosphorus acid is only diprotic.

10-44. *Refer to Section 10-7.*

(a) Acid strengths of most ternary acids containing different elements in the same oxidation state from the same group in the periodic table increase with increasing electronegativity of the central element.

(b) In order of increasing acid strength:

(1) $H_3PO_4 < HNO_3$ (2) $H_3AsO_4 < H_3PO_4$ (3) $H_2TeO_4 < H_2SeO_4$ (4) $HIO_3 < HBrO_3 < HClO_3$

10-46. *Refer to Section 10-8.*

Acid-base reactions are called neutralization reactions because the reaction of an acid with a base generally produces a salt with little or no acid-base character and, in many cases, water.

10-48. *Refer to Section 4-2.*

The electrolytes are NH_4Cl, HI, RaF_2, $Zn(CH_3COO)_2$, $Cu(NO_3)_2$, CH_3COOH, KOH, $KHCO_3$, $NaClO_4$ and $La_2(SO_4)_3$.

The nonelectrolytes are C_6H_6, $C_{12}H_{22}O_{11}$ (table sugar) and I_2.

10-50. *Refer to Section 10-8 and Examples 10-2 and 10-3.*

(a) formula unit:

$$HNO_2(aq) + LiOH(aq) \rightarrow LiNO_2(aq) + H_2O(\ell)$$

nitrous   lithium   lithium
acid   hydroxide   nitrite

total ionic:

$$HNO_2\,(aq) + Li^+(aq) + OH^-(aq) \rightarrow Li^+(aq) + NO_2^-(aq) + H_2O(\ell)$$

net ionic:

$$HNO_2\,(aq) + OH^-(aq) \rightarrow NO_2^-(aq) + H_2O(\ell)$$

(b) formula unit:

$$H_2SO_4(aq) + 2KOH(aq) \rightarrow K_2SO_4(aq) + 2H_2O(\ell)$$

sulfuric   potassium   potassium
acid   hydroxide   sulfate

total ionic:

$$2H^+(aq) + SO_4^{2-}(aq) + 2K^+(aq) + 2OH^-(aq) \rightarrow 2K^+(aq) + SO_4^{2-}(aq) + 2H_2O(\ell)$$

net ionic:

$$2H^+(aq) + 2OH^-(aq) \rightarrow 2H_2O(\ell)$$

therefore,

$$H^+(aq) + OH^-(aq) \rightarrow H_2O(\ell)$$

(c) formula unit:

$$HCl(aq) + NH_3(aq) \rightarrow NH_4Cl(aq)$$

hydrochloric ammonia   ammonium
acid   chloride

total ionic:

$$H^+(aq) + Cl^-(aq) + NH_3(aq) \rightarrow NH_4^+(aq) + Cl^-(aq)$$

net ionic:

$$H^+(aq) + NH_3(aq) \rightarrow NH_4^+(aq)$$

(d) formula unit:

$$CH_3COOH(aq) + NaOH(aq) \rightarrow NaCH_3COO(aq) + H_2O(\ell)$$

acetic   sodium   sodium
acid   hydroxide   acetate

total ionic:

$$CH_3COOH(aq) + Na^+(aq) + OH^-(aq) \rightarrow Na^+(aq) + CH_3COO^-(aq) + H_2O(\ell)$$

net ionic:

$$CH_3COOH(aq) + OH^-(aq) \rightarrow CH_3COO^-(aq) + H_2O(\ell)$$

(e) formula unit:

$$HI(aq) + NaOH(aq) \rightarrow NaI(aq) + H_2O(\ell)$$

hydroiodic   sodium   sodium
acid   hydroxide   iodide

total ionic:

$$H^+(aq) + I^-(aq) + Na^+(aq) + OH^-(aq) \rightarrow Na^+(aq) + I^-(aq) + H_2O(\ell)$$

net ionic:

$$H^+(aq) + OH^-(aq) \rightarrow H_2O(\ell)$$

10-52. *Refer to Section 10-8 and Examples 10-2 and 10-3.*

(a) formula unit:

$$2HClO_4(aq) + Ba(OH)_2(aq) \rightarrow Ba(ClO_4)_2(aq) + 2H_2O(\ell)$$

perchloric   barium   barium
acid   hydroxide   perchlorate

total ionic:

$$2H^+(aq) + 2ClO_4^-(aq) + Ba^{2+}(aq) + 2OH^-(aq) \rightarrow Ba^{2+}(aq) + 2ClO_4^-(aq) + 2H_2O(\ell)$$

net ionic:

$$2H^+(aq) + 2OH^-(aq) \rightarrow 2H_2O(\ell)$$

therefore,

$$H^+(aq) + OH^-(aq) \rightarrow H_2O(\ell)$$

(b) formula unit:

$$2HBr(aq) + Ca(OH)_2(aq) \rightarrow CaBr_2(aq) + 2H_2O(\ell)$$

hydrobromic   calcium   calcium
acid   hydroxide   bromide

total ionic:

$$2H^+(aq) + 2Br^-(aq) + Ca^{2+}(aq) + 2OH^-(aq) \rightarrow Ca^{2+}(aq) + 2Br^-(aq) + 2H_2O(\ell)$$

net ionic:

$$2H^+(aq) + 2OH^-(aq) \rightarrow 2H_2O(\ell)$$

therefore,

$$H^+(aq) + OH^-(aq) \rightarrow H_2O(\ell)$$

(c) formula unit:

$$H_2SO_4(aq) + 2NH_3(aq) \rightarrow (NH_4)_2SO_4(aq)$$

sulfuric   ammonia   ammonium
acid   sulfate

total ionic:

$$2H^+(aq) + SO_4^{2-}(aq) + 2NH_3(aq) \rightarrow 2NH_4^+(aq) + SO_4^{2-}(aq)$$

net ionic:

$$2H^+(aq) + 2NH_3(aq) \rightarrow 2NH_4^+(aq)$$

therefore,

$$H^+(aq) + NH_3(aq) \rightarrow NH_4^+(aq)$$

(d) formula unit:

$$3H_2SO_4(aq) + 2Fe(OH)_3(s) \rightarrow Fe_2(SO_4)_3(aq) + 6H_2O(\ell)$$

sulfuric   iron(III)   iron(III)
acid   hydroxide   sulfate

total ionic:

$$6H^+(aq) + 3SO_4^{2-}(aq) + 2Fe(OH)_3(s) \rightarrow 2Fe^{3+}(aq) + 3SO_4^{2-}(aq) + 6H_2O(\ell)$$

net ionic:

$$6H^+(aq) + 2Fe(OH)_3(s) \rightarrow 2Fe^{3+}(aq) + 6H_2O(\ell)$$

therefore,

$$3H^+(aq) + Fe(OH)_3(s) \rightarrow Fe^{3+}(aq) + 3H_2O(\ell)$$

(e) formula unit:

$$H_2SO_4(aq) + Ca(OH)_2(aq) \rightarrow CaSO_4(s) + 2H_2O(\ell)$$

sulfuric   calcium   calcium
acid   hydroxide   sulfate

total ionic:

$$2H^+(aq) + SO_4^{2-}(aq) + Ca^{2+}(aq) + 2OH^-(aq) \rightarrow CaSO_4(s) + 2H_2O(\ell)$$

net ionic: same as the total ionic equation

10-54. *Refer to Section 10-8 and Example 10-4.*

(a) formula unit:

$$2HNO_3(aq) + Pb(OH)_2(s) \rightarrow Pb(NO_3)_2(aq) + 2H_2O(\ell)$$

total ionic:

$$2H^+(aq) + 2NO_3^-(aq) + Pb(OH)_2(s) \rightarrow Pb^{2+}(aq) + 2NO_3^-(aq) + 2H_2O(\ell)$$

net ionic:

$$2H^+(aq) + Pb(OH)_2(s) \rightarrow Pb^{2+}(aq) + 2H_2O(\ell)$$

(b) formula unit:

$$3HCl(aq) + Al(OH)_3(s) \rightarrow AlCl_3(aq) + 3H_2O(\ell)$$

total ionic:

$$3H^+(aq) + 3Cl^-(aq) + Al(OH)_3(s) \rightarrow Al^{3+}(aq) + 3Cl^-(aq) + 3H_2O(\ell)$$

net ionic:

$$3H^+(aq) + Al(OH)_3(s) \rightarrow Al^{3+}(aq) + 3H_2O(\ell)$$

(c) formula unit:

$$H_2SO_4(aq) + 2NH_3(aq) \rightarrow (NH_4)_2SO_4(aq)$$

total ionic:

$$2H^+(aq) + SO_4^{2-}(aq) + 2NH_3(aq) \rightarrow 2NH_4^+(aq) + SO_4^{2-}(aq)$$

net ionic:

$$2H^+(aq) + 2NH_3(aq) \rightarrow 2NH_4^+(aq)$$

therefore,

$$H^+(aq) + NH_3(aq) \rightarrow NH_4^+(aq)$$

(d) formula unit:

$$2HClO_4(aq) + Ca(OH)_2(aq) \rightarrow Ca(ClO_4)_2(aq) + 2H_2O(\ell)$$

total ionic:

$$2H^+(aq) + 2ClO_4^-(aq) + Ca^{2+}(aq) + 2OH^-(aq) \rightarrow Ca^{2+}(aq) + 2ClO_4^-(aq) + 2H_2O(\ell)$$

net ionic:

$$2H^+(aq) + 2OH^-(aq) \rightarrow 2H_2O(\ell)$$

therefore,

$$H^+(aq) + OH^-(aq) \rightarrow H_2O(\ell)$$

(e) formula unit:

$$3H_2SO_4(aq) + 2Al(OH)_3(s) \rightarrow Al_2(SO_4)_3(aq) + 6H_2O(\ell)$$

total ionic:

$$6H^+(aq) + 3SO_4^{2-}(aq) + 2Al(OH)_3(s) \rightarrow 2Al^{3+}(aq) + 3SO_4^{2-}(aq) + 6H_2O(\ell)$$

net ionic:

$$6H^+(aq) + 2Al(OH)_3(s) \rightarrow 2Al^{3+}(aq) + 6H_2O(\ell)$$

therefore,

$$3H^+(aq) + Al(OH)_3(s) \rightarrow Al^{3+}(aq) + 3H_2O(\ell)$$

10-56. *Refer to Section 10-8 and Example 10-4.*

(a) The salts are (1) $KMnO_4$, (2) $CaSO_4$, (4) SnF_2 and (5) K_3PO_4. Compound (3) P_4O_{10} is an oxide, not a salt.

(b) Acid-base equations that account for salt formation are:

(1) $KOH(aq) + HMnO_4(aq) \rightarrow KMnO_4(aq) + H_2O(\ell)$

(2) $Ca(OH)_2(aq) + H_2SO_4(aq) \rightarrow CaSO_4(aq) + 2H_2O(\ell)$

(4) $Sn(OH)_2(s) + 2HF(aq) \rightarrow SnF_2(s) + 2H_2O(\ell)$

(5) $3KOH(aq) + H_3PO_4(aq) \rightarrow K_3PO_4(aq) + 3H_2O(\ell)$

An acidic salt is a salt that contains an ionizable hydrogen atom. It is the product formed when less than a stoichiometric amount of base reacts with a polyprotic acid:

$$H_2SO_3(aq) + NaOH(aq) \rightarrow NaHSO_3(aq) + H_2O(\ell)$$

$$H_2CO_3(aq) + NaOH(aq) \rightarrow NaHCO_3(aq) + H_2O(\ell)$$

$$H_3PO_4(aq) + KOH(aq) \rightarrow KH_2PO_4(aq) + H_2O(\ell)$$

$$H_3PO_4(aq) + 2NaOH(aq) \rightarrow Na_2HPO_4(aq) + 2H_2O(\ell)$$

$$H_2S(aq) + NaOH(aq) \rightarrow NaHS(aq) + H_2O(\ell)$$

(a) $HNO_3 + NH_3 \rightarrow NH_4NO_3$

(b) $H_3PO_4 + NH_3 \rightarrow NH_4H_2PO_4$

(c) $H_3PO_4 + 2NH_3 \rightarrow (NH_4)_2HPO_4$

(d) $H_3PO_4 + 3NH_3 \rightarrow (NH_4)_3PO_4$

(e) $H_2SO_4 + 2NH_3 \rightarrow (NH_4)_2SO_4$

A basic salt is a salt containing an ionizable OH group and can therefore neutralize acids.

(a),(b)

$$HCl(aq) \ + \ Ca(OH)_2(aq) \ \rightarrow \ Ca(OH)Cl(aq) \ + \ H_2O(\ell)$$
1 mol   1 mol

$$HCl(aq) \ + \ Al(OH)_3(s) \ \rightarrow \ Al(OH)_2Cl(aq) \ + \ H_2O(\ell)$$
1 mol   1 mol

$$2HCl(aq) \ + \ Al(OH)_3(s) \ \rightarrow \ Al(OH)Cl_2(aq) \ + \ 2H_2O(\ell)$$
2 mol   1 mol

(a) for $Cr(OH)_3$ formula unit: $Cr(OH)_3(s) + 3HNO_3(aq) \rightarrow Cr(NO_3)_3(aq) + 3H_2O(\ell)$

total ionic: $Cr(OH)_3(s) + 3H^+(aq) + 3NO_3^-(aq) \rightarrow Cr^{3+}(aq) + 3NO_3^-(aq) + 3H_2O(\ell)$

net ionic: $Cr(OH)_3(s) + 3H^+(aq) \rightarrow Cr^{3+}(aq) + 3H_2O(\ell)$

for $Pb(OH)_2$ formula unit: $Pb(OH)_2(s) + 2HNO_3(aq) \rightarrow Pb(NO_3)_2(aq) + 2H_2O(\ell)$

total ionic: $Pb(OH)_2(s) + 2H^+(aq) + 2NO_3^-(aq) \rightarrow Pb^{2+}(aq) + 2NO_3^-(aq) + 2H_2O(\ell)$

net ionic: $Pb(OH)_2(s) + 2H^+(aq) \rightarrow Pb^{2+}(aq) + 2H_2O(\ell)$

(b) for $Cr(OH)_3$ formula unit: $Cr(OH)_3(s) + KOH(aq) \rightarrow K[Cr(OH)_4](aq)$

total ionic: $Cr(OH)_3(s) + K^+(aq) + OH^-(aq) \rightarrow K^+(aq) + [Cr(OH)_4]^-(aq)$

net ionic: $Cr(OH)_3(s) + OH^-(aq) \rightarrow [Cr(OH)_4]^-(aq)$

for $Pb(OH)_2$ formula unit: $Pb(OH)_2(s) + 2KOH(aq) \rightarrow K_2[Pb(OH)_4](aq)$

total ionic: $Pb(OH)_2(s) + 2K^+(aq) + 2OH^-(aq) \rightarrow 2K^+(aq) + [Pb(OH)_4]^{2-}(aq)$

net ionic: $Pb(OH)_2(s) + 2OH^-(aq) \rightarrow [Pb(OH)_4]^{2-}(aq)$

10-66. *Refer to Section 10-7 and Appendix F.*

Ionization of citric acid, $C_6H_8O_7$ or $C_3H_5O(COOH)_3$:

$$C_3H_5O(COOH)_3(aq) \rightleftarrows H^+(aq) + C_3H_5O(COO)(COOH)_2^-(aq)$$

$$C_3H_5O(COO)(COOH)_2^-(aq) \rightleftarrows H^+(aq) + C_3H_5O(COO)_2(COOH)^{2-}(aq)$$

$$C_3H_5O(COO)_2(COOH)^{2-}(aq) \rightleftarrows H^+(aq) + C_3H_5O(COO)_3^{3-}(aq)$$

10-68. *Refer to Section 10-10.*

(a)

H:Ö: +   H:Ö: →   H:Ö:H⁺   +   H:Ö:⁻
H   H   H
base   acid   acid   base

(b)

H:Cl: +   H:Ö: →   :Cl:⁻   +   H:Ö:H⁺
           H   H
acid   base   base   acid

(c)

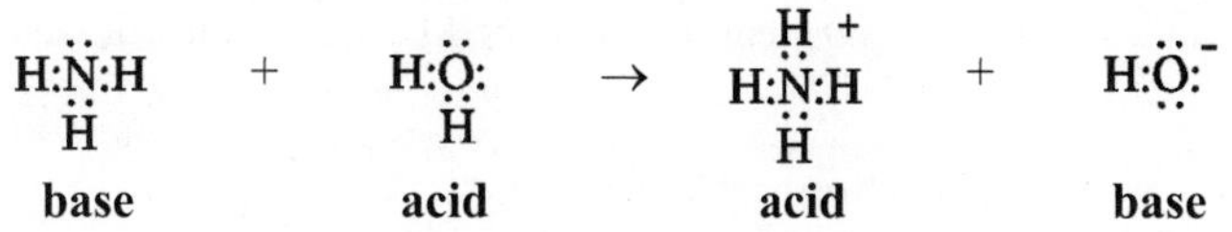

H:N:H   +   H:Ö: →   H:N:H   +   H:Ö:⁻
H   H   H
base   acid   acid   base

(d)

H:N:H   +   H:Br: →   [H:N:H]⁺   :Br:⁻
H   H
base   acid   acid   base

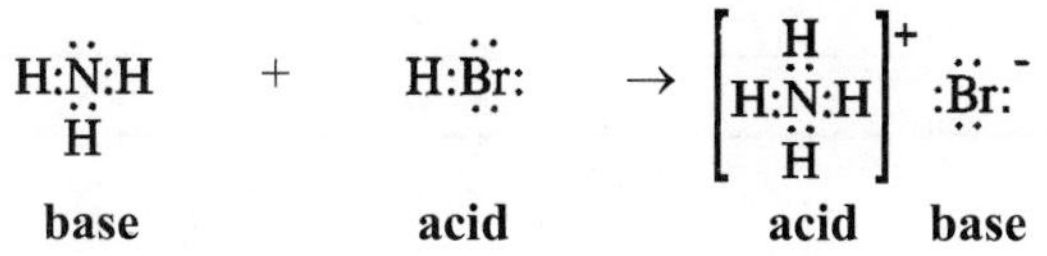

10-70. *Refer to Sections 10-4 and 10-10.*

The advantages of Brønsted-Lowry theory over Arrhenius theory are twofold. (1) Brønsted-Lowry theory extends the definitions of acids and bases so that they can be used in nonaqueous systems and (2) this theory also recognizes that other constituents besides the hydroxide ion can have basic properties.

A limitation of the Brønsted-Lowry theory is that it does not permit the acid-base classification to include those systems which do not involve proton transfer. This limitation is met by the Lewis theory.

10-72. *Refer to Section 10-10.*

(a)

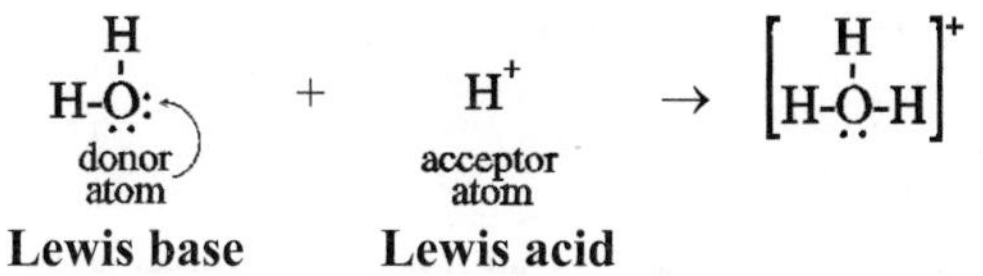

H
|
H-Ö: +   H⁺   →   [H-Ö-H]⁺
donor   acceptor
atom   atom
Lewis base   **Lewis acid**

(b)

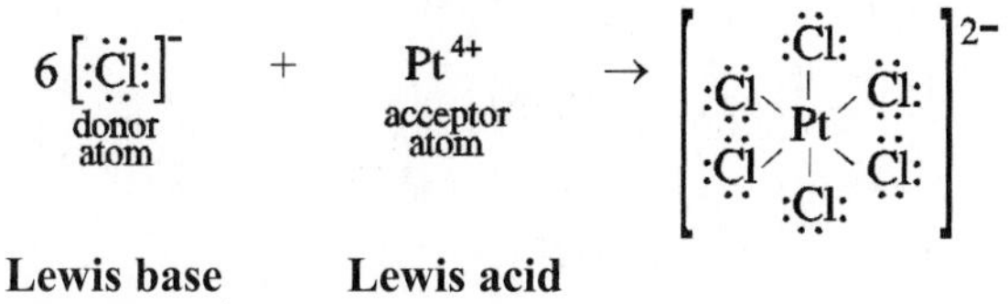

$6[:Cl:]^-$ + Pt^{4+} →   [Cl₆Pt]²⁻
donor   acceptor
atom   atom
Lewis base   **Lewis acid**

(a) HF + SbF_5 → $H(SbF_6)$
    Lewis   Lewis
    base   acid

(b) HF + BF_3 → $H(BF_4)$
    Lewis   Lewis
    base   acid

(b) In $H(SbF_6)$, H is bonded to the SbF_6^- ion through an ionic bond. In $H(BF_4)$, the H is bonded to the BF_4^- ion through an ionic bond.

(a) Hydrogen sulfide, $H_2S(g)$, can be prepared by combining elemental sulfur with hydrogen gas.

$$S_8(s) + 8H_2(g) \rightarrow 8H_2S(g)$$

(b) Hydrogen chloride, $HCl(g)$, can be prepared in small quantities by dropping concentrated nonvolatile acids such as phosphoric acid, onto an appropriate salt such as $NaCl(s)$.

$$H_3PO_4(\ell) + NaCl(s) \rightarrow HCl(g) + NaH_2PO_4(s)$$

(c) An aqueous solution of the weak acid acetic acid, $CH_3COOH(aq)$, can be produced by using sulfuric acid and an acetate salt:

$$H_2SO_4(aq) + Ca(CH_3COO)_2(aq) \rightarrow 2CH_3COOH(aq) + CaSO_4(aq)$$

(a) acidic oxides: CO_2, SO_2, SO_3

(b) amphoteric oxides: Al_2O_3, Ga_2O_3, SnO_2

(c) basic oxides: Na_2O, K_2O, CaO, BaO

(a) H_2S Arrhenius acid, Brønsted-Lowry acid

(b) $PO(OH)_3 \equiv H_3PO_4$ Arrhenius acid, Brønsted-Lowry acid

(c) $H_2CaO_2 \equiv Ca(OH)_2$ Arrhenius base, Brønsted-Lowry base

(d) $ClO_3(OH) \equiv HClO_4$ Arrhenius acid, Brønsted-Lowry acid

(e) $Sb(OH)_3$ amphoteric hydroxide (can act as either an acid or base)

(a)

Acid	Conjugate base		Base	Conjugate Acid
H_3PO_4	$H_2PO_4^-$		HSO_4^-	H_2SO_4
NH_4^+	NH_3		PH_3	PH_4^+
OH^-	O^{2-}		PO_4^{3-}	HPO_4^{2-}

(b) We know that the weaker a base, the stronger is its conjugate acid. Therefore, given that NO_2^- is a stronger base than NO_3^-, then HNO_3 (the conjugate acid of NO_3^-) is a stronger acid than HNO_2 (the conjugate acid of NO_2^-).

10-84. *Refer to Section 10-8 and Figure 4-2.*

In a conductivity experiment, the indicator light bulb glows brightly when the electrodes are placed in an aqueous solution containing a high concentration of ions, such as can be found in aqueous solutions of strong electrolytes (strong acids, strong soluble bases and soluble salts). The bulb will only glow dimly in the presence of weak electrolytes because there are few ions present to conduct electricity through the solution.

(a) $NaOH(aq)$ and $HCl(aq)$ are both strong electrolytes. The light bulb glows brightly for these solutions. The neutralization reaction between NaOH and HCl that results when the two solutions are mixed can be represented as follows:

formula unit: $$NaOH(aq) + HCl(aq) \rightarrow NaCl(aq) + H_2O(\ell)$$

dissociation of the product, $NaCl(aq)$: $$NaCl(aq) \rightarrow Na^+(aq) + Cl^-(aq)$$

Even though ions are lost as the reaction proceeds, due to H^+ and OH^- ions combining to form water, there are still plenty of Na^+ and Cl^- ions remaining in the solution to cause the indicator bulb to glow brightly, but not quite as brightly as the initial solution.

(b) $NH_3(aq)$ and $CH_3COOH(aq)$ are both weak electrolytes and the light bulb will only glow dimly for these solutions. The neutralization reaction between the weak base, NH_3, and the weak acid, CH_3COOH, is as follows:

formula unit: $$NH_3(aq) + CH_3COOH(aq) \rightarrow NH_4CH_3COO(aq)$$

dissociation of the product, $NH_4CH_3COO(aq)$: $$NH_4CH_3COO(aq) \rightarrow NH_4^+(aq) + CH_3COO^-(aq)$$

As can readily be seen, the product formed is a soluble salt, the strong electrolyte, NH_4CH_3COO, which dissociates completely into NH_4^+ and CH_3COO^- ions. The indicator bulb glows brightly in this solution.

10-86. *Refer to Sections 10-4 and 10-8, and the Key Terms for Chapter 10.*

Solubility refers to the extent to which a substance will dissolve in a solvent. Molecular substances that do dissolve in water may or may not ionize into ions If they do ionize, they may or may not ionize completely.

$HCl(aq)$, a soluble gas, ionizes almost totally into its ions in aqueous solution, whereas glucose, a soluble molecular solid, does not ionize at all. Weak acids, such as HF, are soluble in water but ionize only slightly into ions.

10-88. *Refer to Section 10-7.*

(a) Hydrochloric acid is a strong acid and ionizes completely in aqueous solution:
$$HCl(aq) \rightarrow H^+(aq) + Cl^-(aq)$$
It is best represented by Diagram (d).

(b) Hydrofluoric acid is a weak acid and only ionizes slightly in aqueous solution:
$$HF(aq) \rightleftarrows F^-(aq) + H^+(aq)$$
It is best represented by Diagram (b).

137

10-90. *Refer to Section 10-5.*

On planet Baseacidopolous, ammonia is the primary solvent: $NH_3 + NH_3 \rightleftarrows NH_2^- + NH_4^+$

(a) The cation that would indicate that a compound is an acid is NH_4^+.

(b) The anion that would indicate that a compound is a base is NH_2^-.

(c) NaCl could be a salt on this planet. It would be formed as follows: $NaNH_2 + NH_4Cl \rightarrow NaCl + 2NH_3$

10-92. *Refer to Sections 10-4, 10-5, 10-6 and 10-10.*

Autoionization of PCl_5:

$$PCl_5 \quad + \quad PCl_5 \quad \rightarrow \quad PCl_4^+ \quad + \quad PCl_6^-$$

| trigonal bipyramidal | trigonal bipyramidal | tetrahedral | octahedral |

10-94. *Refer to Section 10-9.*

(a) $CH_3COOH(aq) + NaHCO_3(aq) \rightarrow NaCH_3COO(aq) + CO_2(g) + H_2O(\ell)$

The "fizz" is caused by the gaseous product, carbon dioxide, escaping from the solution.

(b) $CH_3CH(OH)COOH(aq) + NaHCO_3(aq) \rightarrow NaCH_3CH(OH)COO(aq) + CO_2(g) + H_2O(\ell)$

"Quick" bread "rises" during baking due to the reaction between baking soda and lactic acid found in the added milk. The resulting carbon dioxide gas bubbles are caught in the bread dough, giving the bread more volume. Yeast breads "rise" due to carbon dioxide bubbles released in the fermentation of sugar by yeast.

11 Reactions in Aqueous Solutions II: Calculations

11-2. *Refer to Sections 3-6 and 11-1.*

Molarity is defined as the number of moles of solute per 1 liter of solution and has units of mol/L. If we multiply molarity by unity $= 10^{-3}/10^{-3}$,

$$\text{Molarity, } M\left(\frac{\text{mol}}{\text{L}}\right) = \frac{\text{mol solute}}{\text{L soln}} \times \frac{10^{-3}}{10^{-3}} = \frac{\text{mmol solute}}{\text{mL soln}}$$

11-4. *Refer to Section 3-6 and Example 3-18.*

Plan: (1) Calculate the moles of $MgSO_4$ present in the solution.
        (2) Calculate the molarity.

(1) $? \text{ mol } MgSO_4 = \dfrac{44.4 \text{ g } MgSO_4}{120.4 \text{ g/mol}} = 0.369 \text{ mol } MgSO_4$

(2) $? \, M \, MgSO_4 = \dfrac{0.369 \text{ mol } MgSO_4}{3.00 \text{ L}} = \textbf{0.123 } \textbf{\textit{M}} \textbf{ MgSO}_4 \textbf{ soln}$

Dimensional Analysis:

$? \, M \, MgSO_4 = \dfrac{44.4 \text{ g } MgSO_4}{3.00 \text{ L}} \times \dfrac{1 \text{ mol } MgSO_4}{120.4 \text{ g } MgSO_4} = \textbf{0.123 } \textbf{\textit{M}} \textbf{ MgSO}_4$

11-6. *Refer to Section 3-6 and Example 3-20.*

Assume a 1 liter solution of 39.77% H_2SO_4 with a density of 1.305 g/mL   (density = specific gravity x 1.00 g/mL)

$? \text{ g } H_2SO_4 \text{ soln in 1 L soln} = \dfrac{1.305 \text{ g soln}}{1 \text{ mL soln}} \times 1000 \text{ mL soln} = 1305 \text{ g soln}$

$? \text{ g } H_2SO_4 \text{ in 1 L soln} = 1305 \text{ g soln} \times \dfrac{39.77 \text{ g } H_2SO_4}{100 \text{ g soln}} = 519.0 \text{ g } H_2SO_4$

$? \text{ mol } H_2SO_4 \text{ in 1 L soln} = \dfrac{519.0 \text{ g } H_2SO_4}{98.08 \text{ g/mol}} = 5.292 \text{ mol } H_2SO_4$

Therefore, $? \, M \, H_2SO_4 = \textbf{5.292 } \textbf{\textit{M}}$

11-8. *Refer to Section 11-1 and Example 11-1.*

This is a possible limiting reactant problem.

Plan: (1) Calculate the number of moles of HCl and NaOH.
        (2) Determine the limiting reactant, if there is one.
        (3) Calculate the moles of NaCl formed.
        (4) Determine the molarity of NaCl in the solution.

Balanced equation: $HCl(aq) + NaOH(aq) \rightarrow NaCl(aq) + H_2O(\ell)$

(1) $? \text{ mol HCl} = 3.35 \, M \text{ HCl} \times 0.225 \text{ L} = 0.754 \text{ mol HCl}$
     $? \text{ mol NaOH} = 1.77 \, M \text{ NaOH} \times 0.426 \text{ L} = 0.754 \text{ mol NaOH}$

(2) This is not a problem with a single limiting reactant since we have stoichiometric amounts of HCl and NaOH. Our final solution is a salt solution with no excess acid or base.

(3) ? mol NaCl = mol HCl = mol NaOH = 0.754 mol NaCl

(4) $? M \text{NaCl} = \dfrac{\text{mol NaCl}}{\text{total volume}} = \dfrac{0.754 \text{ mol}}{(0.225 \text{ L} + 0.426 \text{ L})} = \textbf{1.16 } \boldsymbol{M} \textbf{ NaCl}$

11-10. *Refer to Section 11-1 and Example 11-1.*

This is a possible limiting reactant problem.

Plan: (1) Calculate the number of moles of HI and KOH.
(2) Determine the limiting reactant, if there is one.
(3) Calculate the moles of KI formed.
(4) Determine the molarity of KI in the solution.

Balanced equation: $\text{HI}(aq) + \text{KOH}(aq) \rightarrow \text{KI}(aq) + \text{H}_2\text{O}(\ell)$

(1) ? mol HI = 8.99 M HI x 0.0555 L = 0.499 mol HI

? mol KOH = 14.1 M KOH x 0.0354 L = 0.499 mol KOH

(2) This is not a problem with a single limiting reactant since we have stoichiometric amounts of HI and KOH. Our final solution is a salt solution with no excess acid or base.

(3) ? mol KI = mol HI = mol KOH = 0.499 mol KI

(4) $? M \text{KI} = \dfrac{\text{mol KI}}{\text{total volume}} = \dfrac{0.499 \text{ mol}}{(0.0555 \text{ L} + 0.0354 \text{ L})} = \textbf{5.49 } \boldsymbol{M} \textbf{ KI}$

Note: The answers to each step have been rounded to the correct number of significant figures. However, the entire number has been kept in the calculator and used throughout the entire calculation to minimize rounding errors.

11-12. *Refer to Section 11-1 and Examples 11-2 and 11-4.*

This is a possible limiting reactant problem.

Plan: (1) Calculate the number of millimoles of HI and Ba(OH)$_2$.
(2) Determine the limiting reactant, if there is one.
(3) Calculate the millimoles of BaI$_2$ formed.
(4) Determine the molarity of BaI$_2$ in the solution.

Balanced equation: $2\text{HI}(aq) + \text{Ba(OH)}_2(aq) \rightarrow \text{BaI}_2(aq) + 2\text{H}_2\text{O}(\ell)$

(1) ? mmol HI = 0.0650 M HI x 19.4 mL = 1.26 mmol HI

? mmol Ba(OH)$_2$ = 0.135 M Ba(OH)$_2$ x 7.50 mL = 1.01 mmol Ba(OH)$_2$

(2)   Required ratio = $\dfrac{2 \text{ mmol HI}}{1 \text{ mmol Ba(OH)}_2} = 2$ Available ratio = $\dfrac{1.26 \text{ mmol HI}}{1.01 \text{ mmol Ba(OH)}_2} = 1.25$

Available ratio < required ratio; HI is the limiting reactant.

(3) ? mmol BaI$_2$ = 1.26 mmol HI x $\dfrac{1 \text{ mmol BaI}_2}{2 \text{ mmol HI}}$ = 0.630 mmol BaI$_2$

(4) $? M \text{BaI}_2 = \dfrac{\text{mmol BaI}_2}{\text{total volume in mL}} = \dfrac{0.630 \text{ mmol}}{(19.4 \text{ mL} + 7.50 \text{ mL})} = \textbf{0.0234 } \boldsymbol{M} \textbf{ BaI}_2$

Note: When doing these calculations, do not round off your answers until the end. The answers here are rounded after each step to illustrate the concept of significant figures.

Balanced equation: $H_3PO_4(aq) + 3NaOH(aq) \rightarrow Na_3PO_4(aq) + 3H_2O(\ell)$

Plan: (1) Calculate the moles of H_3PO_4 and NaOH.
           (2) Determine the limiting reactant, if there is one.
           (3) Calculate the moles of Na_3PO_4 formed.
           (4) Determine the molarity of the salt in the solution.
           (5) Determine the moles and concentration of excess reactant in the solution.

(1) ? mol H_3PO_4 = 5.52 M H_3PO_4 x 0.250 L = 1.38 mol H_3PO_4

     ? mol NaOH = 5.52 M NaOH x 0.775 L = 4.28 mol NaOH

(2) In the balanced equation, H_3PO_4 reacts with NaOH in a 1:3 mole ratio.

     mol H_3PO_4:mol NaOH = 1.38 mol:4.28 mol = 1:3.10

     We do not have stoichiometric amounts of both reactants; this is a limiting reactant problem. We have less H_3PO_4 than is necessary to react with all of the NaOH, so H_3PO_4 is the limiting reactant and NaOH is in excess. The amount of salt formed is set then by the amount of H_3PO_4.

(3) ? mol Na_3PO_4 = mol H_3PO_4 = 1.38 mol Na_3PO_4

(4) ? M $Na_3PO_4 = \dfrac{1.38 \text{ mol } Na_3PO_4}{(0.250 \text{ L} + 0.775 \text{ L})} = \mathbf{1.35}$ $\boldsymbol{M}$ $\mathbf{Na_3PO_4}$

(5) The moles of NaOH consumed are determined from the amount of limiting reactant, H_3PO_4. The moles and molarity of NaOH remaining is determined by subtraction.

     ? mol NaOH consumed = 3 x 1.38 mol H_3PO_4 = 4.14 mol NaOH

     ? excess mol NaOH = total mol NaOH - mol NaOH consumed by H_3PO_4 = 4.28 mol – 4.14 mol = 0.14 mol

     ? M NaOH $= \dfrac{0.14 \text{ mol NaOH}}{(0.250 \text{ L} + 0.775 \text{ L})} = \mathbf{0.137}$ $\boldsymbol{M}$ **NaOH in excess**

Note: When doing these calculations, do not round off your answers until the end. The answers here are rounded after each step to illustrate the concept of significant figures.

Assume a 1 liter solution of 5.11% CH_3COOH.

? g CH_3COOH soln in 1 L soln $= \dfrac{1.007 \text{ g soln}}{1 \text{ mL soln}}$ x 1000 mL soln = 1007 g soln

? g CH_3COOH in 1 L soln = 1007 g soln x $\dfrac{5.11 \text{ g } CH_3COOH}{100 \text{ g soln}}$ = 51.5 g CH_3COOH

? mol CH_3COOH in 1 L soln $= \dfrac{51.5 \text{ g } CH_3COOH}{60.1 \text{ g/mol}}$ = 0.857 mol CH_3COOH

Therefore, ? M CH_3COOH = **0.857** $\boldsymbol{M}$

(a) Balanced equation: $3NaOH(aq) + H_3PO_4(aq) \rightarrow Na_3PO_4(aq) + 3H_2O(\ell)$

    Plan: (1) Calculate the moles of NaOH and H_3PO_4 required to form 1 mole of Na_3PO_4.
              (2) Find the volumes of each solution.

      (1) ? mol NaOH = mol Na_3PO_4 x $\dfrac{3 \text{ mol NaOH}}{1 \text{ mol } Na_3PO_4}$ = 1.00 mol x 3 = 3.00 mol NaOH

          ? mol H_3PO_4 = mol Na_3PO_4 = 1.00 mol H_3PO_4

(2) $? \text{ L NaOH soln} = \dfrac{3.00 \text{ mol NaOH}}{3.25 \; M \text{ NaOH}} = \textbf{0.923 L NaOH soln}$

$? \text{ L H}_3\text{PO}_4 \text{ soln} = \dfrac{1.00 \text{ mol H}_3\text{PO}_4}{4.50 \; M \text{ H}_3\text{PO}_4} = \textbf{0.222 L H}_3\textbf{PO}_4 \textbf{ soln}$

(b) Balanced equation: $2\text{NaOH}(aq) + \text{H}_3\text{PO}_4(aq) \rightarrow \text{Na}_2\text{HPO}_4(aq) + 2\text{H}_2\text{O}(\ell)$

    Plan: (1) Calculate the moles of NaOH and H_3PO_4 required to form 1 mole of Na_2HPO_4.

                 (2) Find the volumes of each solution.

(1) $? \text{ mol NaOH} = \text{mol Na}_2\text{HPO}_4 \times \dfrac{2 \text{ mol NaOH}}{1 \text{ mol Na}_2\text{HPO}_4} = 1.00 \text{ mol} \times 2 = 2.00 \text{ mol NaOH}$

$? \text{ mol H}_3\text{PO}_4 = \text{mol Na}_2\text{HPO}_4 = 1.00 \text{ mol H}_3\text{PO}_4$

(2) $? \text{ L NaOH soln} = \dfrac{2.00 \text{ mol NaOH}}{3.25 \; M \text{ NaOH}} = \textbf{0.615 L NaOH soln}$

$? \text{ L H}_3\text{PO}_4 \text{ soln} = \dfrac{1.00 \text{ mol H}_3\text{PO}_4}{4.50 \; M \text{ H}_3\text{PO}_4} = \textbf{0.222 L H}_3\textbf{PO}_4 \textbf{ soln}$

11-20.   *Refer to Sections 11-2 and 11-3.*

A standard solution of NaOH cannot be prepared directly because the solid is hydroscopic and absorbs moisture and CO_2 from the air.

Step 1: Weigh out an amount of solid NaOH and dissolve it in water to obtain a solution with the approximate concentration.

Step 2: Weigh out an appropriate amount of an acidic material, suitable for use as a primary standard, such as potassium hydrogen phthalate (KHP).

Step 3: Titrate the KHP sample with the NaOH solution and calculate the molarity of the NaOH solution using the fact that KHP and NaOH react in a 1:1 stoichiometric ratio.

The prepared NaOH solution is called a secondary standard because its concentration is determined by titration against a primary standard.

11-22.   *Refer to Section 11-3 and Example 11-8.*

(a) Potassium hydrogen phthalate (KHP) is the acidic salt, $\text{KC}_6\text{H}_4(\text{COO})(\text{COOH})$.

(b) KHP is used as a primary standard for the standardization of strong bases.

11-24.   *Refer to Section 11-3.*

Balanced equation: $2\text{CH}_3\text{COOH} + \text{Ba(OH)}_2 \rightarrow \text{Ba(CH}_3\text{COO)}_2 + 2\text{H}_2\text{O}$

Plan: $\underset{(1)}{M, \text{ L Ba(OH)}_2 \text{ soln}} \Rightarrow \underset{(2)}{\text{mol Ba(OH)}_2} \Rightarrow \underset{(3)}{\text{mol CH}_3\text{COOH}} \Rightarrow V \text{ CH}_3\text{COOH soln}$

(1) $? \text{ mol Ba(OH)}_2 = 0.105 \; M \times 0.02158 \text{ L} = 0.00227 \text{ mol Ba(OH)}_2$

(2) $? \text{ mol CH}_3\text{COOH} = 0.00227 \text{ mol Ba(OH)}_2 \times \dfrac{2 \text{ mol CH}_3\text{COOH}}{1 \text{ mol Ba(OH)}_2} = 0.00453 \text{ mol CH}_3\text{COOH}$

(3) $? \text{ L CH}_3\text{COOH soln} = \dfrac{0.00453 \text{ mol CH}_3\text{COOH}}{0.145 \; M \text{ CH}_3\text{COOH}} = \textbf{0.0313 L or 31.3 mL of CH}_3\textbf{COOH soln}$

Dimensional Analysis:

? L CH_3COOH soln $= 0.02158$ L $Ba(OH)_2$ soln $\times \dfrac{0.105 \text{ mol } Ba(OH)_2}{1 \text{ L } Ba(OH)_2 \text{ soln}} \times \dfrac{2 \text{ mol } CH_3COOH}{1 \text{ mol } Ba(OH)_2} \times \dfrac{1 \text{ L } CH_3COOH \text{ soln}}{0.145 \text{ mol } CH_3COOH}$

$\qquad\qquad$ **$= 0.0313$ L or 31.3 mL of CH_3COOH soln**

11-26. *Refer to Section 11-3 and Example 11-7.*

Balanced equation: $2HNO_3(aq) + Na_2CO_3(s) \rightarrow 2NaNO_3(aq) + CO_2(g) + H_2O(\ell)$

$\qquad\qquad\qquad\quad$ (1) $\qquad\qquad$ (2) $\qquad\qquad$ (3)

Plan: g $Na_2CO_3 \Rightarrow$ mol $Na_2CO_3 \Rightarrow$ mol $HNO_3 \Rightarrow M\, HNO_3$ soln

(1) ? mol $Na_2CO_3 = 0.2040$ g $Na_2CO_3 \times \dfrac{1 \text{ mol } Na_2CO_3}{106.0 \text{ g } Na_2CO_3} = 1.925 \times 10^{-3}$ mol Na_2CO_3

(2) ? mol $HNO_3 = 1.925 \times 10^{-3}$ mol $Na_2CO_3 \times \dfrac{2 \text{ mol } HNO_3}{1 \text{ mol } Na_2CO_3} = 3.849 \times 10^{-3}$ mol HNO_3

(3) ? $M\, HNO_3$ soln $= \dfrac{3.849 \times 10^{-3} \text{ mol } HNO_3}{0.01572 \text{ L } HNO_3} =$ **$0.2449\, M\, HNO_3$**

Dimensional Analysis (start by setting up the ratio of mass to volume, then convert to moles and do stochiometry):

? $M\, HNO_3$ soln $= \dfrac{0.2040 \text{ g } Na_2CO_3}{0.01572 \text{ L } HNO_3 \text{ soln}} \times \dfrac{1 \text{ mol } Na_2CO_3}{106.0 \text{ g } Na_2CO_3} \times \dfrac{2 \text{ mol } HNO_3}{1 \text{ mol } Na_2CO_3} =$ **$0.2449\, M\, HNO_3$**

11-28. *Refer to Section 11-3 and Example 11-8.*

Balanced equation: $NaOH + KHP \rightarrow NaKP + H_2O$

$\qquad\qquad\qquad$ (1) $\qquad\qquad\quad$ (2) $\qquad\qquad$ (3)

Plan: g KHP $\Rightarrow$ mmol KHP $\Rightarrow$ mmol NaOH $\Rightarrow M\, NaOH$

(1) ? mmol KHP $= \dfrac{0.5536 \text{ g KHP}}{204.2 \text{ g/mol}} \times \dfrac{1000 \text{ mmol}}{1 \text{ mol}} = 2.711$ mmol KHP

(2) ? mmol NaOH $=$ mmol KHP $= 2.711$ mmol NaOH

(3) ? $M\, NaOH = \dfrac{2.711 \text{ mmol NaOH}}{(37.26 \text{ mL} - 0.23 \text{ mL})} =$ **$0.07321\, M\, NaOH$**

11-30. *Refer to Section 11-3 and Example 11-7.*

Balanced equation: $2HCl(aq) + CaCO_3(s) \rightarrow CaCl_2(aq) + CO_2(g) + H_2O(\ell)$

$\qquad\qquad\qquad$ (1) $\qquad\qquad\quad$ (2) $\qquad\qquad$ (3)

Plan: g $CaCO_3 \Rightarrow$ mol $CaCO_3 \Rightarrow$ mol HCl $\Rightarrow$ L HCl soln

(1) ? mol $CaCO_3 = 0.900$ g $CaCO_3 \times \dfrac{1 \text{ mol } CaCO_3}{100.1 \text{ g } CaCO_3} = 0.00899$ mol $CaCO_3$

(2) ? mol HCl $= 0.00899$ mol $CaCO_3 \times \dfrac{2 \text{ mol HCl}}{1 \text{ mol } CaCO_3} = 0.0180$ mol HCl

(3) ? L HCl soln $= \dfrac{0.0180 \text{ mol HCl}}{1.0\, M\, HCl} =$ **0.018 L or 18 mL HCl**

Dimensional Analysis:

? L HCl soln $= 0.900$ g $CaCO_3 \times \dfrac{1 \text{ mol } CaCO_3}{100.1 \text{ g } CaCO_3} \times \dfrac{2 \text{ mol HCl}}{1 \text{ mol } CaCO_3} \times \dfrac{1.0 \text{ L soln}}{1.0 \text{ mol HCl}} =$ **0.018 L or 18 mL HCl**

(a) An ideal primary standard:

    (1) does not react with or absorb water vapor, oxygen or carbon dioxide,
    (2) reacts according to a single known reaction,
    (3) is available in high purity,
    (4) has a high formula weight,
    (5) is soluble in the solvent of interest,
    (6) is nontoxic,
    (7) is inexpensive, and
    (8) is environmentally friendly.

(b) The significance of each factor is given below.

    (1) The compound must be weighed accurately and must not undergo composition change due to reaction with atmospheric components.
    (2) The reaction must be one of known stoichiometry with no side reactions.
    (3) Solutions of precisely known concentration must be prepared by directly weighing the primary standard.
    (4) The high formula weight is necessary to minimize the effect of weighing errors.
    (5),(6),(7),(8) The significance is self-explanatory.

11-34. *Refer to Section 11-3 and Example 11-9.*

Balanced equation: $(COOH)_2 + 2NaOH \rightarrow Na_2(COO)_2 + 2H_2O$

Plan: M, L NaOH $\overset{(1)}{\Rightarrow}$ mol NaOH $\overset{(2)}{\Rightarrow}$ mol $(COOH)_2$ $\overset{(3)}{\Rightarrow}$ mol $(COOH)_2 \cdot 2H_2O$ $\overset{(4)}{\Rightarrow}$ g $(COOH)_2 \cdot 2H_2O$ $\overset{(5)}{\Rightarrow}$ % purity

(1) ? mol NaOH = 0.298 M NaOH x 0.01916 L = 0.00571 mol NaOH

(2) ? mol $(COOH)_2$ = mol NaOH x $\dfrac{1 \text{ mol } (COOH)_2}{2 \text{ mol NaOH}}$ = 0.00571 mol NaOH x 1/2 = 0.00285 mol $(COOH)_2$

(3) ? mol $(COOH)_2 \cdot 2H_2O$ = mol $(COOH)_2$ = 0.00285 mol $(COOH)_2 \cdot 2H_2O$

(4) ? g $(COOH)_2 \cdot 2H_2O$ = 0.00285 mol $(COOH)_2 \cdot 2H_2O$ x 126 g/mol = 0.360 g $(COOH)_2 \cdot 2H_2O$

(5) ? % $(COOH)_2 \cdot 2H_2O$ = $\dfrac{\text{g }(COOH)_2 \cdot 2H_2O}{\text{g sample}}$ x 100 = $\dfrac{0.360 \text{ g}}{1.00 \text{ g}}$ x 100 = **36.0% $(COOH)_2 \cdot 2H_2O$**

11-36. *Refer to Section 11-3.*

Balanced equation: $2HCl + CaCO_3 \rightarrow CaCl_2 + CO_2 + H_2O$

Plan: M, L HCl $\overset{(1)}{\Rightarrow}$ mol HCl $\overset{(2)}{\Rightarrow}$ mol $CaCO_3$ $\overset{(3)}{\Rightarrow}$ g $CaCO_3$

(1) ? mol HCl = 0.112 M HCl x 0.0268 L = 0.00300 mol HCl

(2) ? mol $CaCO_3$ = 0.00300 mol HCl x (1 mol $CaCO_3$/2 mol HCl) = 0.00150 mol $CaCO_3$

(3) ? g $CaCO_3$ = 0.00150 mol $CaCO_3$ x 100.1 g/mol = **0.150 g $CaCO_3$**

Dimensional Analysis:

? g $CaCO_3$ = 0.0268 L HCl soln x $\dfrac{0.112 \text{ mol HCl}}{1 \text{ L HCl soln}}$ x $\dfrac{1 \text{ mol } CaCO_3}{2 \text{ mol HCl}}$ x $\dfrac{100.1 \text{ g } CaCO_3}{1 \text{ mol } CaCO_3}$ = **0.150 g $CaCO_3$**

Balanced equation: $HNO_3 + NaOH \rightarrow NaNO_3 + H_2O$

Plan: M, L HNO_3 soln $\overset{(1)}{\Rightarrow}$ mol HNO_3 $\overset{(2)}{\Rightarrow}$ mol NaOH $\overset{(3)}{\Rightarrow}$ M NaOH soln

(1) ? mol HNO_3 = 0.0342 M x 0.0375 L = 1.28 x 10^{-3} mol HNO_3

(2) ? mol NaOH = 1.28 x 10^{-3} mol HNO_3 x (1 mol NaOH/1 mol HNO_3) = 1.28 x 10^{-3} mol NaOH

(3) ? M NaOH $= \dfrac{\text{mol NaOH}}{\text{L NaOH}} = \dfrac{1.28 \times 10^{-3} \text{ mol}}{0.0414 \text{ L}} =$ **0.0310 M NaOH soln**

11-40. *Refer to Sections 3-6 and 11-1, and Examples 3-18 and 11-3.*

(1) Plan: g H_3PO_4 $\overset{(1)}{\Rightarrow}$ mol H_3PO_4 $\overset{(2)}{\Rightarrow}$ M H_3PO_4 soln

(1) ? mol H_3PO_4 $= \dfrac{\text{g } H_3PO_4}{\text{FW } H_3PO_4} = \dfrac{0.978 \text{ g}}{97.99 \text{ g/mol}} =$ 0.00998 mol H_3PO_4

(2) ? M H_3PO_4 $= \dfrac{\text{mol } H_3PO_4}{\text{L soln}} = \dfrac{0.00998 \text{ mol}}{0.185 \text{ L}} =$ **0.0539 M H_3PO_4**

Dimensional Analysis – Find the moles of H_3PO_4, then divide moles by liters to calculate the molarity:

? mol H_3PO_4 = 0.978 g H_3PO_4 x $\dfrac{1 \text{ mol } H_3PO_4}{97.99 \text{ g } H_3PO_4}$ = 0.0998 mol H_3PO_4

? M H_3PO_4 $= \dfrac{\text{mol } H_3PO_4}{\text{L soln}} = \dfrac{0.00998 \text{ mol}}{0.185 \text{ L}} =$ **0.0539 M H_3PO_4**

(2) Balanced equation: $H_3PO_4 + 3NaOH \rightarrow Na_3PO_4 + 3H_2O$

? mL H_3PO_4 soln = 11.58 mL NaOH soln x $\dfrac{0.454 \text{ mol NaOH}}{1000 \text{ mL NaOH soln}}$ x $\dfrac{1 \text{ mol } H_3PO_4}{3 \text{ mol NaOH}}$ x $\dfrac{1000 \text{ mL } H_3PO_4 \text{ soln}}{0.0539 \text{ mol } H_3PO_4}$

= **32.5 mL of H_3PO_4 soln**

11-42. *Refer to Sections 3-6 and 11-1, and Examples 3-18 and 11-3.*

(1) ? M H_3AsO_4 $= \dfrac{\text{mol } H_3AsO_4}{\text{L soln}} = \dfrac{(18.6 \text{ g } H_3AsO_4)/(142 \text{ g/mol})}{0.475 \text{ L}} =$ **0.276 M H_3AsO_4**

(2) Balanced equation: $H_3AsO_4 + 3NaOH \rightarrow Na_3AsO_4 + 3H_2O$

? mL H_3AsO_4 soln = 11.58 mL NaOH soln x $\dfrac{0.454 \text{ mol NaOH}}{1000 \text{ mL NaOH soln}}$ x $\dfrac{1 \text{ mol } H_3PO_4}{3 \text{ mol NaOH}}$ x $\dfrac{1000 \text{ mL } H_3AsO_4 \text{ soln}}{0.276 \text{ mol } H_3AsO_4}$

= **6.35 mL H_3AsO_4 soln**

11-44. *Refer to Section 11-3 and Example 11-7.*

Balanced equation: $2HCl + Na_2CO_3 \rightarrow 2NaCl + CO_2 + H_2O$

Plan: g Na_2CO_3 $\overset{(1)}{\Rightarrow}$ mol Na_2CO_3 $\overset{(2)}{\Rightarrow}$ mol HCl $\overset{(3)}{\Rightarrow}$ M HCl

? mol HCl = 0.483 g Na_2CO_3 x $\dfrac{1 \text{ mol } Na_2CO_3}{106.0 \text{ g } Na_2CO_3}$ x $\dfrac{2 \text{ mol HCl}}{1 \text{ mol } Na_2CO_3}$ = 0.00911 mol HCl

? M HCl $= \dfrac{0.00911 \text{ mol HCl}}{0.0391 \text{ L soln}} =$ **0.233 M HCl**

Balanced equations: $Mg(OH)_2(s) + 2HCl(aq) \rightarrow MgCl_2(aq) + 2H_2O(\ell)$

$HCl(aq) + NaOH(aq) \rightarrow NaCl(aq) + H_2O(\ell)$

This is an example of the method of back titration, in which more acid (HCl) is added than is necessary to stoichiometrically react with the base $(Mg(OH)_2)$, in order to be certain that all the base has reacted. One then titrates the excess acid with a standardized base solution (NaOH) and in a series of calculations, determines the amount of unknown base $(Mg(OH)_2)$.

Plan: (1) Calculate the total moles of HCl that were added to the tablet.
(2) Calculate the moles of HCl in excess, which are equal to the moles of NaOH added.
(3) Calculate the moles of HCl that reacted with $Mg(OH)_2$ in the tablet = (1) - (2).
(4) Determine the mass of $Mg(OH)_2$ that reacted with the HCl.
(5) Determine the mass % of $Mg(OH)_2$ in the tablet.

(1) ? mol HCl added = 0.953 M HCl x 0.02500 L = 0.0238 mol HCl

(2) ? mol HCl in excess = mol NaOH titrated = 0.602 M NaOH x 0.01229 L = 0.00740 mol HCl

(3) ? mol HCl reacted = 0.0238 mol HCl - 0.00740 mol HCl = 0.0164 mol HCl

(4) ? g $Mg(OH)_2$ = 0.0164 mol HCl x $\dfrac{1 \text{ mol } Mg(OH)_2}{2 \text{ mol HCl}}$ x $\dfrac{58.32 \text{ g } Mg(OH)_2}{1 \text{ mol } Mg(OH)_2}$ = 0.478 g $Mg(OH)_2$

(5) ? % $Mg(OH)_2$ = $\dfrac{0.478 \text{ g } Mg(OH)_2}{1.462 \text{ g tablet}}$ x 100 = **32.7%**

(a) $Fe(s) + 2HCl(aq) \rightarrow FeCl_2(aq) + H_2(g)$

(b) $2Cr(s) + 3H_2SO_4(aq) \rightarrow Cr_2(SO_4)_3(aq) + 3H_2(g)$

(c) $Sn(s) + 4HNO_3(aq) \rightarrow SnO_2(s) + 4NO_2(g) + 2H_2O(\ell)$

Determining the net ionic equation by balancing the oxidation-reduction reaction:

skeletal equation:	$Cu(s) + HNO_3(aq) \rightarrow Cu^{2+}(aq) + NO(g)$
ox. half-rxn:	$Cu(s) \rightarrow Cu^{2+}(aq)$
balanced ox. half-rxn:	$Cu(s) \rightarrow Cu^{2+}(aq) + 2e^-$
red. half-rxn:	$NO_3^-(aq) \rightarrow NO(g)$ since HNO_3 is a strong acid
balanced red. half-rxn:	$3e^- + 4H^+(aq) + NO_3^-(aq) \rightarrow NO(g) + 2H_2O(\ell)$

Now, we balance the electron transfer and add the half-reactions term-by-term and cancel electrons:

oxidation: $\qquad\qquad 3[Cu(s) \rightarrow Cu^{2+}(aq) + 2e^-]$

reduction: $\qquad\qquad \underline{2[3e^- + 4H^+(aq) + NO_3^-(aq) \rightarrow NO(g) + 2H_2O(\ell)]}$

balanced net ionic eq.: $\quad 3Cu(s) + 8H^+(aq) + 2NO_3^-(aq) \rightarrow 3Cu^{2+}(aq) + 2NO(g) + 4H_2O(\ell)$

The formula unit equation is obtained by recognizing that there is no net charge in a solution, so all the cations are paired with anions to neutralize the charge. In this case the Cu^{2+} and H^+ are paired with NO_3^-. This does mean that some NO_3^- did not react but remained as spectator ions.

balanced formula unit eq.: $\qquad 3Cu(s) + 8HNO_3(aq) \rightarrow 3Cu(NO_3)_2(aq) + 2NO(g) + 4H_2O(\ell)$

(a) skeletal equation: $\qquad MnO_4^-(aq) + Br^-(aq) \rightarrow Mn^{2+}(aq) + Br_2(\ell)$

ox. half-rxn: $\qquad\qquad\qquad\qquad\quad Br^-(aq) \rightarrow Br_2(\ell)$

balanced ox. half-rxn: $\qquad\qquad\quad 2Br^-(aq) \rightarrow Br_2(\ell) + 2e^-$

red. half-rxn: $\qquad\qquad\qquad MnO_4^-(aq) \rightarrow Mn^{2+}(aq)$

balanced red. half-rxn: $\quad 5e^- + 8H^+(aq) + MnO_4^-(aq) \rightarrow Mn^{2+}(aq) + 4H_2O(\ell)$

Now, we balance the electron transfer and add the half-reactions term-by-term and cancel electrons:

oxidation: $\qquad\qquad\qquad\qquad 5[2Br^-(aq) \rightarrow Br_2(\ell) + 2e^-]$

reduction: $\qquad\qquad 2[5e^- + 8H^+(aq) + MnO_4^-(aq) \rightarrow Mn^{2+}(aq) + 4H_2O(\ell)]$

balanced.: $\qquad 16H^+(aq) + 2MnO_4^-(aq) + 10Br^-(aq) \rightarrow 2Mn^{2+}(aq) + 5Br_2(\ell) + 8H_2O(\ell)$

Br is oxidized from -1 (in Br^-) to 0 (in Br_2), therefore, Br^- is the reducing agent.
Mn is reduced from +7 (in MnO_4^-) to +2 (in Mn^{2+}), therefore, MnO_4^- is the oxidizing agent.

(b) skeletal equation: $\qquad Cr_2O_7^{2-}(aq) + I^-(aq) \rightarrow Cr^{3+}(aq) + I_2(s)$

ox. half-rxn: $\qquad\qquad\qquad\qquad\quad I^-(aq) \rightarrow I_2(s)$

balanced ox. half-rxn: $\qquad\qquad\quad 2I^-(aq) \rightarrow I_2(s) + 2e^-$

red. half-rxn: $\qquad\qquad\quad Cr_2O_7^{2-}(aq) \rightarrow Cr^{3+}(aq)$

balanced red. half-rxn: $\quad 6e^- + 14H^+(aq) + Cr_2O_7^{2-}(aq) \rightarrow 2Cr^{3+}(aq) + 7H_2O(\ell)$

Now, we balance the electron transfer and add the half-reactions term-by-term and cancel electrons:

oxidation: $\qquad\qquad\qquad\qquad 3[2I^-(aq) \rightarrow I_2(s) + 2e^-]$

reduction: $\qquad\qquad [6e^- + 14H^+(aq) + Cr_2O_7^{2-}(aq) \rightarrow 2Cr^{3+}(aq) + 7H_2O(\ell)]$

balanced.: $\qquad 14H^+(aq) + Cr_2O_7^{2-}(aq) + 6I^-(aq) \rightarrow 2Cr^{3+}(aq) + 3I_2(s) + 7H_2O(\ell)$

I is oxidized from -1 (in I^-) to 0 (in I_2), therefore, I^- is the reducing agent.
Cr is reduced from +6 (in $Cr_2O_7^{2-}$) to +3 (in Cr^{3+}), therefore, $Cr_2O_7^{2-}$ is the oxidizing agent.

(c) oxidation: $\qquad\qquad\qquad 5[H_2O(\ell) + SO_3^{2-}(aq) \rightarrow SO_4^{2-}(aq) + 2H^+(aq) + 2e^-]$

reduction: $\qquad\qquad 2[5e^- + 8H^+(aq) + MnO_4^-(aq) \rightarrow Mn^{2+}(aq) + 4H_2O(\ell)]$

balanced.: $\qquad 2MnO_4^-(aq) + 5SO_3^{2-}(aq) + 6H^+(aq) \rightarrow 2Mn^{2+}(aq) + 5SO_4^{2-}(aq) + 3H_2O(\ell)$

S is oxidized from +4 (in SO_3^{2-}) to +6 (in SO_4^{2-}), therefore, SO_3^{2--} is the reducing agent.
Mn is reduced from +7 (in MnO_4^-) to +2 (in Mn^{2+}), therefore, MnO_4^- is the oxidizing agent.

(d) oxidation: $\qquad\qquad\qquad 6[Fe^{2+}(aq) \rightarrow Fe^{3+}(aq) + e^-]$

reduction: $\qquad\qquad [6e^- + 14H^+(aq) + Cr_2O_7^{2-}(aq) \rightarrow 2Cr^{3+}(aq) + 7H_2O(\ell)]$

balanced.: $\qquad Cr_2O_7^{2-}(aq) + 6Fe^{2+}(aq) + 14H^+(aq) \rightarrow 2Cr^{3+}(aq) + 6Fe^{3+}(aq) + 7H_2O(\ell)$

Fe is oxidized from +2 (in Fe^{2+}) to +3 (in Fe^{3+}), therefore, Fe^{2+} is the reducing agent.
Cr is reduced from +6 (in $Cr_2O_7^{2-}$) to +3 (in Cr^{3+}), therefore, $Cr_2O_7^{2-}$ is the oxidizing agent.

11-54. *Refer to Sections 11-4 and 11-5, and Example 11-13.*

The net ionic equation is obtained by balancing the oxidation-reduction reaction. First we determine the half-reactions:

skeletal equation: $\qquad Al(s) + OH^-(aq) + H_2O(\ell) \rightarrow [Al(OH)_4]^-(aq) + H_2(g)$

ox. half-rxn: $\qquad\qquad\qquad\qquad Al(s) \rightarrow [Al(OH)_4]^-(aq)$

balanced ox. half-rxn: $\qquad Al(s) + 4OH^-(aq) \rightarrow [Al(OH)_4]^-(aq) + 3e^-$

red. half-rxn: $\qquad\qquad\qquad\qquad H_2O(\ell) \rightarrow H_2(g)$

balanced red. half-rxn:
$$2e^- + 2H_2O(\ell) \rightarrow H_2(g) + 2OH^-(aq)$$

Now, we balance the electron transfer and add the half-reactions term-by-term and cancel electrons:

oxidation: $\qquad 2[Al(s) + 4OH^-(aq) \rightarrow [Al(OH)_4]^-(aq) + 3e^-]$

reduction: $\qquad 3[2e^- + 2H_2O(\ell) \rightarrow H_2(g) + 2OH^-(aq)]$

balanced: $\qquad 6e^- + 2Al(s) + 8OH^-(aq) + 6H_2O(\ell) \rightarrow 2[Al(OH)_4]^-(aq) + 3H_2(g) + 6OH^-(aq) + 6e^-$

simplifying: $\qquad 2Al(s) + 2OH^-(aq) + 6H_2O(\ell) \rightarrow 2[Al(OH)_4]^-(aq) + 3H_2(g)$

The formula unit equation is obtained by recognizing that there is no net charge in a solution, so all the cations are paired with anions to neutralize the charge. In this case the anions are paired with Na^+ ions from the NaOH, giving:

formula unit equation: $\qquad 2Al(s) + 2NaOH(aq) + 6H_2O(\ell) \rightarrow 2NaAl(OH)_4(aq) + 3H_2(g)$

11-56. *Refer to Sections 11-4 and 11-5, and Example 11-11.*

(a) To balance the redox reaction, first we determine the oxidation and the reduction half-reactions:

skeletal equation: $\qquad Fe^{2+}(aq) + MnO_4^-(aq) \rightarrow Fe^{3+}(aq) + Mn^{2+}(aq)$

ox. half-rxn: $\qquad Fe^{2+}(aq) \rightarrow Fe^{3+}(aq)$

balanced ox. half-rxn: $\qquad Fe^{2+}(aq) \rightarrow Fe^{3+}(aq) + e^-$

red. half-rxn: $\qquad MnO_4^-(aq) \rightarrow Mn^{2+}(aq)$

balanced red. half-rxn: $\qquad 5e^- + 8H^+(aq) + MnO_4^-(aq) \rightarrow Mn^{2+}(aq) + 4H_2O(\ell)$

Now, we balance the electron transfer and add the half-reactions term-by-term and cancel electrons:

oxidation: $\qquad 5[Fe^{2+}(aq) \rightarrow Fe^{3+}(aq) + e^-]$

reduction: $\qquad 5e^- + MnO_4^-(aq) + 8H^+(aq) \rightarrow Mn^{2+}(aq) + 4H_2O(\ell)$

balanced: $\qquad 5Fe^{2+}(aq) + MnO_4^-(aq) + 8H^+(aq) \rightarrow 5Fe^{3+}(aq) + Mn^{2+}(aq) + 4H_2O(\ell)$

(b) skeletal equation: $\qquad Br_2(\ell) + SO_2(g) \rightarrow Br^-(aq) + SO_4^{2-}(aq)$

ox. half-rxn: $\qquad SO_2(g) \rightarrow SO_4^{2-}(aq)$

balanced ox. half-rxn: $\qquad 2H_2O(\ell) + SO_2(g) \rightarrow SO_4^{2-}(aq) + 4H^+(aq) + 2e^-$

red. half-rxn: $\qquad Br_2(\ell) \rightarrow Br^-(aq)$

balanced red. half-rxn: $\qquad 2e^- + Br_2(\ell) \rightarrow 2Br^-(aq)$

The electron transfer is already balanced and we can write:

oxidation: $\qquad 2H_2O(\ell) + SO_2(g) \rightarrow SO_4^{2-}(aq) + 4H^+(aq) + 2e^-$

reduction: $\qquad 2e^- + Br_2(\ell) \rightarrow 2Br^-(aq)$

balanced: $\qquad Br_2(\ell) + SO_2(g) + 2H_2O(\ell) \rightarrow 2Br^-(aq) + SO_4^{2-}(aq) + 4H^+(aq)$

(c) oxidation: $\qquad Cu(s) \rightarrow Cu^{2+}(aq) + 2e^-$

reduction: $\qquad 2[1e^- + 2H^+(aq) + NO_3^-(aq) \rightarrow NO_2(g) + H_2O(\ell)]$

balanced: $\qquad Cu(s) + 2NO_3^-(aq) + 4H^+(aq) \rightarrow Cu^{2+}(aq) + 2NO_2(g) + 2H_2O(\ell)$

(d) oxidation: $\qquad 2Cl^-(aq) \rightarrow Cl_2(g) + 2e^-$

reduction: $\qquad 2e^- + 4H^+(aq) + 2Cl^-(aq) + PbO_2(s) \rightarrow PbCl_2(s) + 2H_2O(\ell)$

balanced: $\qquad PbO_2(s) + 4Cl^-(aq) + 4H^+(aq) \rightarrow PbCl_2(s) + Cl_2(g) + 2H_2O(\ell)$

(e) oxidation: $\qquad 5[Zn(s) \rightarrow Zn^{2+}(aq) + 2e^-]$

reduction: $\qquad 10e^- + 12H^+(aq) + 2NO_3^-(aq) \rightarrow N_2(g) + 6H_2O(\ell)$

balanced: $\qquad 5Zn(s) + 2NO_3^-(aq) + 12H^+(aq) \rightarrow 5Zn^{2+}(aq) + N_2(g) + 6H_2O(\ell)$

(a) The net ionic equation is obtained by balancing the oxidation-reduction reaction. First we determine the half-reactions. Then we balance the electron transfer and add the half-reactions term-by-term and cancel electrons:

skeletal equation:	$MnO_4^-(aq) + C_2O_4^{2-}(aq) \rightarrow Mn^{2+}(aq) + CO_2(g)$
ox. half-rxn:	$C_2O_4^{2-}(aq) \rightarrow CO_2(g)$
balanced ox. half-rxn:	$C_2O_4^{2-}(aq) \rightarrow 2\,CO_2(g) + 2e^-$
red. half-rxn:	$MnO_4^-(aq) \rightarrow Mn^{2+}(aq)$
balanced red. half-rxn:	$5e^- + 8H^+(aq) + MnO_4^-(aq) \rightarrow Mn^{2+}(aq) + 4H_2O(\ell)$

oxidation:	$5[C_2O_4^{2-}(aq) \rightarrow 2\,CO_2(g) + 2e^-]$
reduction:	$2[5e^- + MnO_4^-(aq) + 8H^+(aq) \rightarrow Mn^{2+}(aq) + 4H_2O(\ell)]$
balanced:	$2MnO_4^-(aq) + 5C_2O_4^{2-}(aq) + 16H^+(aq) \rightarrow 2Mn^{2+}(aq) + 10\,CO_2(g) + 8H_2O(\ell)$

Balanced formula unit equation (cations = K^+, anions Cl^-):

$$2KMnO_4(aq) + 5K_2C_2O_4(aq) + 16HCl(aq) \rightarrow 2MnCl_2(aq) + 12KCl(aq) + 10\,CO_2(g) + 8H_2O(\ell)$$

(b)
skeletal equation:	$Zn(s) + NO_3^-(aq) \rightarrow Zn^{2+}(aq) + NH_4^+(aq)$
ox. half-rxn:	$Zn(s) \rightarrow Zn^{2+}(aq)$
balanced ox. half-rxn:	$Zn(s) \rightarrow Zn^{2+}(aq) + 2e^-$
red. half-rxn:	$NO_3^-(aq) \rightarrow NH_4^+(aq)$
balanced red. half-rxn:	$8e^- + 10H^+(aq) + NO_3^-(aq) \rightarrow NH_4^+(aq) + 3H_2O(\ell)$

oxidation:	$4[Zn(s) \rightarrow Zn^{2+}(aq) + 2e^-]$
reduction:	$1[8e^- + 10H^+(aq) + NO_3^-(aq) \rightarrow NH_4^+(aq) + 3H_2O(\ell)]$
balanced:	$4Zn(s) + NO_3^-(aq) + 10H^+(aq) \rightarrow 4Zn^{2+}(aq) + NH_4^+(aq) + 3H_2O(\ell)$

Balanced formula unit equation (cations = H^+, anions = unreacted NO_3^-):

$$4Zn(s) + 10\,HNO_3(aq) \rightarrow 4Zn(NO_3)_2(aq) + NH_4NO_3(aq) + 3H_2O(\ell)$$

(a) The net ionic equation is obtained by balancing the oxidation-reduction reaction. First we determine the half-reactions. Then we balance the electron transfer and add the half-reactions term-by-term and cancel electrons:

Balanced net ionic equation:

oxidation:	$Zn(s) \rightarrow Zn^{2+}(aq) + 2e^-$
reduction:	$2e^- + Cu^{2+}(aq) \rightarrow Cu(s)$
balanced:	$Zn(s) + Cu^{2+}(aq) \rightarrow Zn^{2+}(aq) + Cu(s)$

Balanced formula unit equation (anion = SO_4^{2-}): $Zn(s) + CuSO_4(aq) \rightarrow ZnSO_4(aq) + Cu(s)$

(b) Balanced net ionic equation:

oxidation:	$2[Cr(s) \rightarrow Cr^{3+}(aq) + 3e^-]$
reduction:	$3[2e^- + 2H^+(aq) \rightarrow H_2(g)]$
balanced:	$2Cr(s) + 6H^+(aq) \rightarrow 2Cr^{3+}(aq) + 3H_2(g)$

Balanced formula unit equation (anion = SO_4^{2-}): $2Cr(s) + 3H_2SO_4(aq) \rightarrow Cr_2(SO_4)_3(aq) + 3H_2(g)$

11-62. *Refer to Section 11-6 and Example 11-14.*

Balanced <u>net</u> <u>ionic</u> equation is:

$$5Fe^{2+}(aq) + MnO_4^-(aq) + 8H^+(aq) \rightarrow 5Fe^{3+}(aq) + Mn^{2+}(aq) + 4H_2O(\ell)$$

Note: This exercise uses $KMnO_4$ and $FeSO_4$. These are both soluble salts which dissociate into their ions. The K^+ and SO_4^{2-} ions are spectator ions and are omitted from the balanced net ionic equation.

Plan: M, mL $FeSO_4$ soln $\overset{(1)}{\Rightarrow}$ mmol $FeSO_4$ $\overset{(2)}{\Rightarrow}$ mmol $KMnO_4$ $\overset{(3)}{\Rightarrow}$ mL $KMnO_4$

(1) ? mmol $FeSO_4 = 0.150\ M \times 25.0\ mL = 3.75$ mmol $FeSO_4$

(2) ? mmol $KMnO_4 = 3.75$ mmol $FeSO_4 \times \dfrac{1\ mmol\ KMnO_4}{5\ mmol\ FeSO_4} = 0.750$ mmol $KMnO_4$

(3) ? mL $KMnO_4 = \dfrac{0.750\ mmol\ KMnO_4}{0.213\ M\ KMnO_4} = \mathbf{3.52\ mL\ KMnO_4\ soln}$ \qquad (since molarity can have units: mmol/mL)

Alternative: Dimensional Analysis (Each step above is a separate unit factor):

? mL $KMnO_4 = 25.0$ mL $FeSO_4 \times \dfrac{0.150\ mmol\ FeSO_4}{1\ mL\ FeSO_4} \times \dfrac{1\ mmol\ KMnO_4}{5\ mmol\ FeSO_4} \times \dfrac{1\ mL\ KMnO_4}{0.213\ mmol\ KMnO_4}$

$= \mathbf{3.52\ mL\ KMnO_4\ soln}$

11-64. *Refer to Section 11-6 and Example 11-14.*

Balanced <u>net</u> <u>ionic</u> equation - you need to balance this before you can begin:

$$2MnO_4^-(aq) + 16H^+(aq) + 10\ I^-(aq) \rightarrow 2Mn^{2+}(aq) + 5I_2(s) + 8H_2O(\ell)$$

Note: This exercise uses $KMnO_4$ and KI, which are both soluble salts that dissociate into their ions. The K^+ ions are spectator ions and are omitted from the balanced net ionic equation.

Plan: M, L KI soln $\overset{(1)}{\Rightarrow}$ mol KI $\overset{(2)}{\Rightarrow}$ mol $KMnO_4$ $\overset{(3)}{\Rightarrow}$ V $KMnO_4$ soln

(1) ? mol $KI = 0.150\ M \times 0.0270\ L = 0.00405$ mol KI

(2) ? mol $KMnO_4 = 0.00405$ mol $KI \times \dfrac{1\ mol\ KMnO_4}{5\ mol\ KI} = 0.000810$ mol $KMnO_4$

Note: The ratio of mol $KMnO_4$ to mol KI was simplified from 2/10 to 1/5.

(3) ? L $KMnO_4$ soln $= \dfrac{0.000810\ mol\ KMnO_4}{0.190\ M\ KMnO_4} = \mathbf{0.00426\ L\ or\ 4.26\ mL\ KMnO_4\ soln}$

Alternative: Dimensional Analysis (Each step above is a separate unit factor):

? mL $KMnO_4 = 0.0270$ L $KI \times \dfrac{0.150\ mol\ KI}{1\ L\ KI} \times \dfrac{1\ mol\ KMnO_4}{5\ mol\ KI} \times \dfrac{1\ L\ KMnO_4}{0.190\ mol\ KMnO_4}$

$= \mathbf{0.00426\ L\ or\ 4.26\ mL\ KMnO_4\ soln}$

11-66. *Refer to Section 11-6 and Example 11-15.*

(a) Balanced equation: $2Na_2S_2O_3 + I_2 \rightarrow Na_2S_4O_6 + 2NaI$

Plan: M, L $Na_2S_2O_3$ soln $\overset{(1)}{\Rightarrow}$ mol $Na_2S_2O_3$ $\overset{(2)}{\Rightarrow}$ mol I_2 $\overset{(3)}{\Rightarrow}$ M I_2 soln

(1) ? mol $Na_2S_2O_3 = 0.1442\ M \times 0.03700\ L = 0.005335$ mol $Na_2S_2O_3$

(2) ? mol $I_2 = 0.005335$ mol $Na_2S_2O_3 \times \dfrac{1\ \text{mol}\ I_2}{2\ \text{mol}\ Na_2S_2O_3} = 0.002668$ mol I_2

(3) ? $M\ I_2 = \dfrac{\text{mol}\ I_2}{L\ I_2} = \dfrac{0.002668\ \text{mol}\ I_2}{0.02885\ L\ I_2} = \mathbf{0.09247\ \textit{M}\ I_2}$

(b) Balanced equation: $As_2O_3 + 5H_2O + 2I_2 \rightarrow 2H_3AsO_4 + 4HI$

Plan: M, L I_2 soln $\overset{(1)}{\Rightarrow}$ mol I_2 $\overset{(2)}{\Rightarrow}$ mol As_2O_3 $\overset{(3)}{\Rightarrow}$ g As_2O_3

(1) ? mol $I_2 = 0.09247\ M \times 0.03532\ L = 0.003266$ mol I_2

(2) ? mol $As_2O_3 = 0.003266$ mol $I_2 \times \dfrac{1\ \text{mol}\ As_2O_3}{2\ \text{mol}\ I_2} = 0.001633$ mol As_2O_3

(3) ? g $As_2O_3 = 0.001633$ mol $As_2O_3 \times 197.8$ g/mol $= \mathbf{0.3230\ g\ As_2O_3}$

Alternative: Dimensional Analysis (Each step above is a separate unit factor):

? g $As_2O_3 = 0.03532\ L\ I_2 \times \dfrac{0.09247\ \text{mol}\ I_2}{1\ L\ I_2} \times \dfrac{1\ \text{mol}\ As_2O_3}{2\ \text{mol}\ I_2} \times \dfrac{197.8\ \text{g}\ As_2O_3}{1\ \text{mol}\ As_2O_3}$

   $= \mathbf{0.3230\ g\ As_2O_3}$

11-68. *Refer to Sections 11-6 and Example 11-14.*

Balanced <u>net ionic</u> equation - you need to balance this before you can begin:

   $2NO_3^-(aq) + 3S^{2-}(aq) + 8H^+(aq) \rightarrow 2NO(g) + 3S(s) + 4H_2O(\ell)$

Plan: M, L S^{2-} soln $\overset{(1)}{\Rightarrow}$ mol S^{2-} $\overset{(2)}{\Rightarrow}$ mol NO_3^- $\overset{(3)}{\Rightarrow}$ V NO_3^- soln

(1) ? mol $S^{2-} = 0.75\ M \times 0.035\ L = 0.026$ mol S^{2-}

(2) ? mol $NO_3^- = 0.026$ mol $S^{2-} \times \dfrac{2\ \text{mol}\ NO_3^-}{3\ \text{mol}\ S^{2-}} = 0.018$ mol NO_3^-

(3) ? L NO_3^- soln $= \dfrac{0.018\ \text{mol}\ NO_3^-}{5.0\ M\ NO_3^-} = \mathbf{0.0035\ L\ or\ 3.5\ mL\ NO_3^-\ soln}$

Note: when doing calculations step-wise, it is critical that you keep the entire number in your calculator and don't round between steps. Otherwise, major rounding errors can develop. Rounding was done in this manual at every step only to illustrate the concept of significant figures.

Dimensional Analysis:

? L $NO_3^- = 0.035\ L\ S^{2-}$ soln $\times \dfrac{0.75\ \text{mol}\ H_2S}{1\ L\ H_2S\ \text{soln}} \times \dfrac{2\ \text{mol}\ NO_3^-}{3\ \text{mol}\ S^{2-}} \times \dfrac{1\ L\ NO_3^-\ \text{soln}}{5.0\ \text{mol}\ NO_3^-}$

   $= \mathbf{0.0035\ L\ or\ 3.5\ mL\ NO_3^-\ soln}$

Plan: $g\ KMnO_4 \overset{(1)}{\Rightarrow} mol\ KMnO_4 \overset{(2)}{\Rightarrow} M\ KMnO_4$

(1) $?\ mol\ KMnO_4 = \dfrac{14.6\ g\ KMnO_4}{158\ g/mol} = 0.0924\ mol\ KMnO_4$

(2) $?\ M\ KMnO_4 = \dfrac{0.0924\ mol\ KMnO_4}{0.750\ L} = \textbf{0.123 } \boldsymbol{M}\ \textbf{KMnO}_4$

The balanced half-reaction involving the reduction of MnO_4^- to MnO_4^{2-} requires 1 electron:

$$e^- + MnO_4^-(aq) \rightarrow MnO_4^{2-}(aq)$$

This fact is irrelevant since the molarity of a solution is *independent* of the number of electrons involved in the reaction. The molarity depends only on the moles of solute and the liters of solution.

11-72. *Refer to Section 11-3 and Example 11-7.*

Balanced equation: $2HCl + Na_2CO_3 \rightarrow 2NaCl + H_2O + CO_2$

Plan: $g\ Na_2CO_3 \overset{(1)}{\Rightarrow} mol\ Na_2CO_3 \overset{(2)}{\Rightarrow} mol\ HCl \overset{(3)}{\Rightarrow} M\ HCl$

(1) $?\ mol\ Na_2CO_3 = \dfrac{0.4811\ g\ Na_2CO_3}{106.0\ g/mol} = 0.004539\ mol\ Na_2CO_3$

(2) $?\ mol\ HCl = 0.004539\ mol\ Na_2CO_3 \times (2\ mol\ HCl/1\ mol\ Na_2CO_3) = 0.009077\ mol\ HCl$

(3) $?\ M\ HCl = \dfrac{0.009077\ mol\ HCl}{0.03275\ L\ soln} = \textbf{0.2772 } \boldsymbol{M}\ \textbf{HCl}$

11-74. *Refer to Section 11-2 and Examples 11-5 and 11-6.*

Balanced equation: $HCl + NaOH \rightarrow NaCl + H_2O$

(1) Plan: $M,\ mL\ NaOH \overset{(1)}{\Rightarrow} mmol\ NaOH \overset{(2)}{\Rightarrow} mmol\ HCl$

$?\ mmol\ HCl = 25.5\ mL\ NaOH \times \dfrac{0.298\ mmol\ NaOH}{1\ mL\ NaOH} \times \dfrac{1\ mmol\ HCl}{1\ mmol\ NaOH} = \textbf{7.60 mmol HCl}$

(2) $?\ mL\ HCl = 7.60\ mmol\ HCl \times \dfrac{1\ mL\ HCl}{0.606\ mmol\ HCl} = \textbf{12.5 mL HCl}$

11-76. *Refer to Section 11-1 and Example 11-3.*

Balanced equation: $2HCl + Ca(OH)_2 \rightarrow CaCl_2 + 2H_2O$

Plan: $g\ Ca(OH)_2 \overset{(1)}{\Rightarrow} mol\ Ca(OH)_2 \overset{(2)}{\Rightarrow} mol\ HCl \overset{(3)}{\Rightarrow} V\ HCl$

(1) $?\ mol\ Ca(OH)_2 = \dfrac{1.79\ g\ Ca(OH)_2}{74.1\ g/mol} = 0.0242\ mol\ Ca(OH)_2$

(2) $?\ mol\ HCl = 0.0242\ mol\ Ca(OH)_2 \times (2\ mol\ HCl/1\ mol\ Ca(OH)_2) = 0.0483\ mol\ HCl$

(3) $?\ L\ HCl = \dfrac{0.0483\ mol\ HCl}{0.1153\ M\ HCl} = \textbf{0.419 L or 419 mL HCl soln}$

Dimensional Analysis:

$$? \text{ L HCl} = 1.79 \text{ g Ca(OH)}_2 \times \frac{1 \text{ mol Ca(OH)}_2}{74.1 \text{ g Ca(OH)}_2} \times \frac{2 \text{ mol HCl}}{1 \text{ mol Ca(OH)}_2} \times \frac{1 \text{ L HCl soln}}{0.1153 \text{ mol HCl}}$$

= 0.419 L or 419 mL HCl soln

11-78.   *Refer to Section 11-1 and Example 11-3.*

Balanced equation: $H_2SO_4 + 2KOH \rightarrow K_2SO_4 + 2H_2O$

Plan: M, L KOH $\overset{(1)}{\Rightarrow}$ mol KOH $\overset{(2)}{\Rightarrow}$ mol H_2SO_4 $\overset{(3)}{\Rightarrow}$ $V\,H_2SO_4$

(1) ? mol KOH = 0.322 M x 0.0344 L = 0.0111 mol KOH

(2) ? mol H_2SO_4 = 0.0111 mol KOH x (1 mol H_2SO_4/2 mol KOH) = 0.00554 mol H_2SO_4

(3) ? L $H_2SO_4 = \dfrac{0.00554 \text{ mol } H_2SO_4}{0.296 \, M\, H_2SO_4} =$ **0.0187 L or 18.7 mL H_2SO_4 soln**

Dimensional Analysis:

$$? \text{ L } H_2SO_4 = 0.0344 \text{ L KOH soln} \times \frac{0.322 \text{ mol KOH}}{1 \text{ L KOH soln}} \times \frac{1 \text{ mol } H_2SO_4}{2 \text{ mol KOH}} \times \frac{1 \text{ L } H_2SO_4 \text{ soln}}{0.296 \text{ mol } H_2SO_4}$$

= 0.0187 L or 18.7 mL H_2SO_4 soln

11-80.   *Refer to Sections 11-1 and Example 11-3.*

Balanced equation: $H_2SO_4 + 2NaOH \rightarrow Na_2SO_4 + 2H_2O$

Plan: M, L H_2SO_4 $\overset{(1)}{\Rightarrow}$ mol H_2SO_4 $\overset{(2)}{\Rightarrow}$ mol NaOH $\overset{(3)}{\Rightarrow}$ V NaOH

(1) ? mol H_2SO_4 = 0.1023 M x 0.03438 L = 0.003517 mol H_2SO_4

(2) ? mol NaOH = 0.003517 mol H_2SO_4 x (2 mol NaOH/1 mol H_2SO_4) = 0.007034 mol NaOH

(3) ? L NaOH = $\dfrac{0.007034 \text{ mol NaOH}}{0.1945 \, M \text{ NaOH}} =$ **0.03617 L or 36.17 mL NaOH soln**

Dimensional Analysis:

$$? \text{ L NaOH} = 0.03438 \text{ L } H_2SO_4 \text{ soln} \times \frac{0.1023 \text{ mol } H_2SO_4}{1 \text{ L } H_2SO_4 \text{ soln}} \times \frac{2 \text{ mol NaOH}}{1 \text{ mol } H_2SO_4} \times \frac{1 \text{ L NaOH soln}}{0.1945 \text{ mol NaOH}}$$

= 0.03617 L or 36.17 mL NaOH soln

11-82.   *Refer to Sections 11-1 and 11-6.*

(a) Balanced equation: HI + NaOH $\rightarrow$ NaI + H_2O

Plan: M, L NaOH soln $\overset{(1)}{\Rightarrow}$ mol NaOH $\overset{(2)}{\Rightarrow}$ mol HI $\overset{(3)}{\Rightarrow}$ V HI soln

(1) ? mol NaOH = 0.100 M x 0.0250 L = 0.00250 mol NaOH

(2) ? mol HI = mol NaOH = 0.00250 mol HI

(3) ? L HI soln = $\dfrac{0.00250 \text{ mol HI}}{0.245 \, M \text{ HI}} =$ **0.0102 L or 10.2 mL of HI soln**

Dimensional Analysis:

? L HI soln = 0.0250 L NaOH soln $\times \dfrac{0.100 \text{ mol NaOH}}{1 \text{ L NaOH soln}} \times \dfrac{1 \text{ mol HI}}{1 \text{ mol NaOH}} \times \dfrac{1 \text{ L HI soln}}{0.245 \text{ mol HI}}$ = **0.0102 L of HI soln**

(b) Balanced net ionic equation: $Ag^+ + I^- \rightarrow AgI$

Plan: $\underset{(1)}{g \text{ AgNO}_3} \Rightarrow \underset{(2)}{\text{mol AgNO}_3} \Rightarrow \underset{(3)}{\text{mol HI}} \Rightarrow V \text{ HI soln}$

(1) ? mol $AgNO_3 = \dfrac{0.503 \text{ g AgNO}_3}{169.9 \text{ g/mol}} = 0.00296 \text{ mol AgNO}_3$

(2) ? mol HI = mol $AgNO_3$ = 0.00296 mol HI

(3) ? L HI soln $= \dfrac{0.00296 \text{ mol HI}}{0.245 \, M \text{ HI}} =$ **0.0121 L or 12.1 mL of HI soln**

(c) Balanced equation: $2Cu^{2+} + 4I^- \rightarrow 2CuI + I_2$

Plan: $\underset{(1)}{g \text{ CuSO}_4} \Rightarrow \underset{(2)}{\text{mol CuSO}_4} \Rightarrow \underset{(3)}{\text{mol HI}} \Rightarrow L \text{ HI soln}$

(1) ? mol $CuSO_4 = \dfrac{0.621 \text{ g CuSO}_4}{159.6 \text{ g/mol}} = 0.00389 \text{ mol CuSO}_4$

(2) ? mol HI = 0.00389 mol $CuSO_4 \times (4 \text{ mol HI}/2 \text{ mol CuSO}_4) = 0.00778 \text{ mol HI}$

(3) ? L HI $= \dfrac{0.00778 \text{ mol HI}}{0.245 \, M \text{ HI}} =$ **0.0318 L or 31.8 mL of HI soln**

11-84. *Refer to Sections 4-4 and 4-7.*

An antioxidant is a compound that opposes oxidation or inhibits reactions promoted by oxygen or peroxides. Such a compound is ascorbic acid, $H_2C_6H_6O_6$, also called Vitamin C, which can undergo a decomposition reaction as follows:

Oxidation Numbers: $\quad\quad \overset{+1}{} \; \overset{+\frac{2}{3}}{} \overset{+1}{} \; \overset{-2}{} \quad\quad \overset{+1}{} \; \overset{+1}{} \; \overset{-2}{} \quad\quad\quad \overset{0}{}$

$$H_2C_6H_6O_6 \rightarrow C_6H_6O_6 + H_2$$

Vitamin C is both oxidized (C = +2/3 → C = +1) and reduced (H = +1 → H = 0) in this reaction.

11-86. *Refer to Section 11-6 and Example 11-15.*

Balanced equation: $Cr_2O_7{}^{2-} + 6Fe^{2+} + 14H^+ \rightarrow 2Cr^{3+} + 6Fe^{3+} + 7H_2O$

Plan: $\underset{(1)}{M, L \text{ Na}_2\text{Cr}_2\text{O}_7} \Rightarrow \text{mol Na}_2\text{Cr}_2\text{O}_7 \, (= \text{mol Cr}_2\text{O}_7{}^{2-}) \Rightarrow \underset{(2)}{\text{mol Fe}^{2+}} \Rightarrow \underset{(3)}{g \text{ Fe}^{2+}} (= g \text{ Fe}) \Rightarrow \underset{(4)}{\%\text{Fe}}$

(1) ? mol $Cr_2O_7{}^{2-}$ = 42.96 mL $Na_2Cr_2O_7 \times \dfrac{1.000 \text{ L Na}_2\text{Cr}_2\text{O}_7}{1000 \text{ mL Na}_2\text{Cr}_2\text{O}_7} \times \dfrac{0.02130 \text{ mol Na}_2\text{Cr}_2\text{O}_7}{1.000 \text{ L Na}_2\text{Cr}_2\text{O}_7} \times \dfrac{1 \text{ mol Cr}_2\text{O}_7{}^{2-}}{1 \text{ mol Na}_2\text{Cr}_2\text{O}_7}$

$= 9.150 \times 10^{-4} \text{ mol Cr}_2\text{O}_7{}^{2-}$

(2) ? mol Fe^{2+} = $9.150 \times 10^{-4} \text{ mol Cr}_2\text{O}_7{}^{2-} \times \dfrac{6 \text{ mol Fe}^{2+}}{1 \text{ mol Cr}_2\text{O}_7{}^{2-}} = 5.490 \times 10^{-3} \text{ mol Fe}^{2+}$

(3) ? g Fe = $5.490 \times 10^{-3} \text{ mol Fe}^{2+} \times \dfrac{1 \text{ mol Fe}}{1 \text{ mol Fe}^{2+}} \times \dfrac{55.85 \text{ g Fe}}{1 \text{ mol Fe}} = 0.3066 \text{ g Fe}$

(4) ? %Fe $= \dfrac{0.3066 \text{ g Fe}}{0.5166 \text{ g sample}} \times 100 =$ **59.36% Fe**

Pure limonite, $2Fe_2O_3 \cdot 3H_2O$ (FW = 373.4 g/mol), is 59.83% Fe by mass.

$$\% \text{ Fe by mass} = \frac{\text{AW Fe x 4}}{\text{FW}} \times 100\% = \frac{223.4 \text{ g}}{373.4 \text{ g}} \times 100\% = 59.83\% \text{ Fe}$$

If the percentage of Fe in the limonite ore had been calculated to be greater than 59.83%, one might conclude that there were other components in the dissolved ore solution in addition to Fe^{2+} that were capable of reducing $Cr_2O_7^{2-}$ to Cr^{3+}, (assuming of course that the analytical data were correct).

Therefore, the volume of $Na_2Cr_2O_7$ necessary to reach the equivalence point would increase, and the amount of Fe present would appear to be larger than it really was.

11-88. *Refer to Section 11-1 and Example 11-2.*

This is a possible limiting reactant problem.

Plan: (1) Calculate the number of moles of $AgNO_3$ and $CaCl_2$.
(2) Determine the limiting reactant, if there is one.
(3) Calculate the moles of AgCl formed.
(4) Determine the mass of AgCl produced.

Balanced equation: $2AgNO_3(aq) + CaCl_2(aq) \rightarrow 2AgCl(s) + Ca(NO_3)_2(aq)$

(1) ? mol $AgNO_3$ = 6.0 M $AgNO_3$ x 0.095 L = 0.57 mol $AgNO_3$
? mol $CaCl_2$ = 6.0 M $CaCl_2$ x 0.040 L = 0.24 mol $CaCl_2$

(2) In the balanced equation, $AgNO_3$ reacts with $CaCl_2$ in a 2:1 mole ratio.

mol $AgNO_3$:mol $CaCl_2$ = 0.57 mol:0.24 mol = 2.4:1

We do not have stoichiometric amounts of both reactants; this is a limiting reactant problem. We have more $AgNO_3$ than is necessary to react with all of the $CaCl_2$, so $CaCl_2$ is the limiting reactant and $AgNO_3$ is in excess. The amount of salt formed is set then by the amount of $CaCl_2$.

(3) ? mol AgCl = 0.24 mol $CaCl_2$ x (2 mol AgCl/1 mol $CaCl_2$) = 0.48 mol AgCl

(4) ? g AgCl = 0.48 mol AgCl x $\dfrac{143 \text{ g AgCl}}{1 \text{ mol AgCl}}$ = **69 g AgCl**

11-90. *Refer to Section 11-2.*

Balanced equation: $NaAl(OH)_2CO_3 + 4HCl \rightarrow NaCl + AlCl_3 + CO_2 + 3H_2O$

Plan: (1) Calculate the mmoles of HCl in your stomach acid.
(2) Calculate the mmoles of $NaAl(OH)_2CO_3$ in one antacid tablet.
(3) Calculate the mmoles of HCl that can be neutralized by the antacid tablet.

(1) ? mmol HCl in stomach = 0.10 M HCl x 800. mL = **80. mmol HCl in stomach**

(2) ? mmol $NaAl(OH)_2CO_3$ = $\dfrac{334 \text{ mg } NaAl(OH)_2CO_3}{144 \text{ mg/mmol}}$ = 2.32 mmol $NaAl(OH)_2CO_3$

(3) ? mmol neutralized HCl = 2.32 mmol $NaAl(OH)_2CO_3$ x $\dfrac{4 \text{ mmol HCl}}{1 \text{ mmol } NaAl(OH)_2CO_3}$ = **9.28 mmol HCl**

The number of **mmoles of HCl in your stomach** is roughly nine times greater than the number of mmoles of HCl that can be neutralized by a single antacid tablet. However, about 2 tablets are sufficient to neutralize the excess HCl in the stomach by reducing its concentration down to the normal 8.0×10^{-2} M level.

11-92. *Refer to Section 4-7.*

Balanced equation: $SiO_2(s) + 6HF(aq) \rightarrow H_2SiF_6(aq) + 2H_2O(\ell)$

The etching of glass, SiO_2, by hydrofluoric acid, HF, is **not** an oxidation-reduction reaction, since no element in the reaction is undergoing a change in oxidation number.

11-94. *Refer to Section 4-7.*

If we consider the two electrostatic charge potential plots for potassium hydrogen phthalate (KHP), the only difference between them is the location of the hydrogen in the –COOH group. The figure on the left is more stable because the hydrogen is oriented in such a way as to promote intramolecular H-bonding within the compound. The figure on the right has two oxygen atoms with partial negative charges adjacent to each other; this is not as stable a configuration.

12 Gases and the Kinetic Molecular Theory

12-2. *Refer to Sections 12-1 and 12-2.*

All gases are (a) transparent to light. Some gases are (b) colorless and (e) odorless. However, no gas (c) is unable to pass through filter paper, (d) is more difficult to compress than water and (f) settles on standing.

12-4. *Refer to Section 12-3 and Figure 12-1.*

A manometer is a device employing the change in liquid levels to measure gas pressure differences between a standard and an unknown system. For example, a typical mercury manometer consists of a glass tube partially filled with mercury. One arm is open to the atmosphere and the other is connected to a container of gas. When the pressure of the gas in the container is greater than atmospheric pressure, the level of the mercury in the open side will be higher and

$$P_{gas} = P_{atm} + \Delta h$$

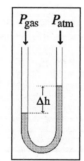 where Δh is the difference in mercury levels

However, when the pressure of the gas is less than atmospheric pressure, the level of the mercury in the closed side will be higher, and

$$P_{gas} = P_{atm} - \Delta h$$

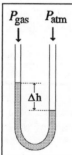 where Δh is the difference in mercury levels

---

12-6. *Refer to Section 12-3 and Appendix C.*

(a) $? \text{ psi} = 752 \text{ torr} \times \dfrac{1 \text{ atm}}{760 \text{ torr}} \times \dfrac{14.70 \text{ psi}}{1 \text{ atm}} = \textbf{14.5 psi}$

(b) $? \text{ cm Hg} = 752 \text{ torr} \times \dfrac{1 \text{ mm Hg}}{1 \text{ torr}} \times \dfrac{1 \text{ cm Hg}}{10 \text{ mm Hg}} = \textbf{75.2 cm Hg}$

(c) $? \text{ inches Hg} = 752 \text{ torr} \times \dfrac{1 \text{ mm Hg}}{1 \text{ torr}} \times \dfrac{1 \text{ cm Hg}}{10 \text{ mm Hg}} \times \dfrac{1 \text{ in Hg}}{2.54 \text{ cm Hg}} = \textbf{29.6 in Hg}$

(d) $? \text{ kPa} = 752 \text{ torr} \times \dfrac{1 \text{ atm}}{760 \text{ torr}} \times \dfrac{1.013 \times 10^5 \text{ Pa}}{1 \text{ atm}} \times \dfrac{1 \text{ kPa}}{1000 \text{ Pa}} = \textbf{100. kPa}$

(e)　$? \text{ atm} = 752 \text{ torr} \times \dfrac{1 \text{ atm}}{760 \text{ torr}} = \textbf{0.989 atm}$

(f)　$? \text{ ft H}_2\text{O} = 752 \text{ torr} \times \dfrac{1 \text{ mm Hg}}{1 \text{ torr}} \times \dfrac{1 \text{ cm Hg}}{10 \text{ mm Hg}} \times \dfrac{1 \text{ in Hg}}{2.54 \text{ cm Hg}} \times \dfrac{1 \text{ ft Hg}}{12 \text{ in Hg}} \times \dfrac{13.59 \text{ ft H}_2\text{O}}{1.00 \text{ ft Hg}} = \textbf{33.5 ft H}_2\textbf{O}$

> Note: The final unit factor uses the relative densities of water and mercury at 25°C. Since the density of water is only 1/13.59 that of mercury, a 13.59 ft column of H_2O has the same mass as a 1.00 ft column of mercury.

12-8.　*Refer to Sections 12-1 and 12-2.*

(a) The material is not a gas. If the container did hold a gas and was opened to the atmosphere, the material would expand without limit.

(b) The material discharging from the smokestack is not a gas, but a colloidal mixture that light cannot penetrate.

(c) The material is not a gas because its density, 8.2 g/mL, is far too great.

(d) The material is a gas for two reasons. (1) It is much less dense than fresh water since it rises rapidly to the surface. (2) At 30 ft below the water's surface the material is exposed to 2 atm pressure: 1 atm (760 mm Hg) atmospheric pressure and 1 atm (76 cm Hg) of water pressure. As the pressure on the material decreased as the material rises to the surface, its volume increased. This is an illustration of Boyle's Law (Section 12-4).

(e) The material may be a gas, but insufficient information is given.

(f) The material is definitely a gas.

12-10.　*Refer to Section 12-3.*

Since 1 atm = 14.7 psi,

$? \text{ psi} = 150. \text{ atm} \times \dfrac{14.7 \text{ psi}}{1 \text{ atm}} = \textbf{2.20 x 10}^3 \textbf{ psi}$

12-12.　*Refer to Section 12-4 and Figures 12-3 and 12-4.*

(a) Boyle studied the effect of changing pressure on a volume of a known mass of gas at constant temperature. Boyle's Law states: at a given temperature, the product of pressure and volume of a definite mass of gas is constant.

(b) When the mathematical relationship, XY = constant, is plotted on the X-Y axes, a hyperbola results. Boyle's Law can be stated as

$$\text{pressure x volume} = \text{constant} \qquad (\text{at constant } n, T)$$

resulting in the graph shown in Figure 12-4. Since pressure and volume can never have negative values, the other branch of the hyperbola is omitted.

12-14.　*Refer to Section 12-4 and Examples 12-1 and 12-2.*

Boyle's Law states:　$P_1V_1 = P_2V_2$　　　at constant n and T

Substituting,　　$P_2 = \dfrac{P_1V_1}{V_2} = \dfrac{2.00 \text{ atm x } 300. \text{ mL}}{575 \text{ mL}} = \textbf{1.04 atm}$

12-16. *Refer to Section 12-4 and Examples 12-1 and 12-2.*

Recall Boyle's Law: $P_1V_1 = P_2V_2$ at constant n and T

(a) Given: $P_1 = 48.6$ torr $\times \dfrac{1 \text{ atm}}{760 \text{ torr}} = 6.39 \times 10^{-2}$ atm $V_1 = 35.0$ L

            $P_2 = ?$ $V_2 = 150.$ mL $\times \dfrac{1 \text{ L}}{1000 \text{ mL}} = 0.150$ L

$P_2 = \dfrac{P_1V_1}{V_2} = \dfrac{6.39 \times 10^{-2} \text{ atm} \times 35.0 \text{ L}}{0.150 \text{ L}} = \mathbf{14.9 \text{ atm}}$

(b) Given: $P_1 = 6.39 \times 10^{-2}$ atm $V_1 = 35.0$ L
                $P_2 = 10.0$ atm $V_2 = ?$

$V_2 = \dfrac{P_1V_1}{P_2} = \dfrac{6.39 \times 10^{-2} \text{ atm} \times 35.0 \text{ L}}{10.0 \text{ atm}} = \mathbf{0.224 \text{ L}}$

12-18. *Refer to Section 12-4.*

Plan: (1) Use Boyle's Law to find the maximum volume occupied by the gas at 1.1 atm.
           (2) After subtracting out the volume of the cylinder, divide the remaining volume by the volume of each balloon to get the number of balloons.

(1) Recall Boyle's Law: $P_1V_1 = P_2V_2$ at constant n and T

    Given: $P_1 = 165$ atm $V_1 = 11$ L
               $P_2 = 1.1$ atm $V_2 = ?$

    Solving, $V_2 = \dfrac{P_1V_1}{P_2} = \dfrac{165 \text{ atm} \times 11 \text{ L}}{1.1 \text{ atm}} = 1650$ L (good to 2 significant figures)

(2) This volume of gas is distributed between the balloons and the "empty" cylinder.

         N $\times V_{\text{balloon}} = V_2 - V_{\text{cylinder}}$ where N = number of balloons (a whole number)
         N $\times 2.5$ L $= 1650$ L - 11 L

             N $= \dfrac{1650 \text{ L} - 11 \text{ L}}{2.5 \text{ L}} = \mathbf{660 \text{ balloons}}$ (2 significant figures)

12-20. *Refer to Section 12-5 and Figure 12-5.*

(a) An "absolute temperature scale" is a scale in which properties such as gas volume change linearly with temperature while the origin of the scale is set at absolute zero. The Kelvin scale is a typical example of it.

(b) Boyle, in his experiments, noticed that temperature affected gas volume. About 1800, Charles and Gay-Lussac found that the rate of gas expansion with increased temperature was constant at constant pressure. Later, Lord Kelvin noticed that for a series of constant pressure systems, volume decreased as temperature decreased and the extrapolation of these different T-V lines back to zero volume yielded a common intercept, -273.15°C on the temperature axis. He defined this temperature as absolute zero. The relationship between the Celsius and Kelvin temperature scales is

$$K = °C + 273.15°.$$

(c) Absolute zero may be thought of as the limit of thermal contraction for an ideal gas. In other words, an ideal gas would have zero volume at absolute zero temperature. Theoretically, it is also the temperature at which molecular motion ceases.

(a) Experiments have shown that at constant pressure, the volume of a definite mass of gas is directly proportional to its absolute temperature (in K).

(b) This is known as Charles's Law and is expressed as V/T = constant at constant n and P. Therefore, for a sample of gas when volume is plotted against temperature, a straight line results. See line A in the graph above.

12-24. *Refer to Section 12-5 and Figure 12-5.*

In the graph above, we see that for the real gas B, the volume drops to nearly zero at about 50°C. This is because the gas must have liquefied at that temperature, i.e., its boiling point must be about 50°C. Since liquids are much more dense than gases, the volume would have decreased greatly at that temperature.

From the graph, we can read that at 100°C (373 K), the volume of the gas is 13 mL and at 400°C (673 K), the volume is about 22 mL. Charles's Law states that for an ideal gas, V/T = constant at constant n and P.

$$\begin{array}{lll} \text{At 373 K} & V/T = 0.035 \text{ mL/K} \\ \text{At 673 K} & V/T = 0.033 \text{ mL/K} \end{array}$$

Since V/T is approximately the same within the reading error of the graph, the gas does behave ideally above 50°C.

It is expected that most real gases would exhibit similar ideal behavior above their liquefaction points, i.e., their boiling points.

12-26. *Refer to Section 12-5.*

This is a Charles's Law calculation: $\dfrac{V_1}{T_1} = \dfrac{V_2}{T_2}$ at constant n and P

Given: $V_1 = 0.78$ L $T_1 = 26°C + 273° = 299$ K
         $V_2 = ?$ L $T_2 = 21°C + 273° = 294$ K

$$V_2 = \frac{V_1 T_2}{T_1} = \frac{0.78 \text{ L} \times 294 \text{ K}}{299 \text{ K}} = \textbf{0.77 L}$$

12-28. *Refer to Section 12-5.*

(a) Recall Charles's Law: $\dfrac{V_1}{T_1} = \dfrac{V_2}{T_2}$ at constant n and P

Given: $V_1 = 1.400$ L $T_1 = 0.0°C + 273.15° = 273.2$ K
             $V_2 = ?$ $T_2 = 8.0°C + 273.15° = 281.2$ K

$$V_2 = \frac{V_1 T_2}{T_1} = \frac{1.400 \text{ L} \times 281.2 \text{ K}}{273.2 \text{ K}} = \textbf{1.441 L}$$

(b) The volume change corresponding to the temperature change from 0.0°C to 8.0°C is (1.441 - 1.400) L = 0.041 L or 41 mL or 41 cm^3. When the cross-sectional area of the graduated arm is 1.0 cm^2, the difference in height (cm) is equivalent to the difference in volume (cm^3). Hence, the height will increase by **41 cm**.

(c) To improve the thermometer's sensitivity (measured in Δheight/°C) for the same volume change, the cross-sectional area of the graduated arm should be decreased. This will cause the height difference to increase. Also, a larger volume of gas could be used.

12-30. *Refer to Section 12-5 and Figure 12-5.*

Recall Charles's Law: $\dfrac{V_1}{T_1} = \dfrac{V_2}{T_2}$ at constant n and P

Dry ice: volume at $-78.5°C$: $\quad V_2 = \dfrac{V_1 T_2}{T_1} = \dfrac{5.00\ \text{L} \times (-78.5°C + 273.15°)}{25.0°C + 273.15°} = \textbf{3.26 L}$

Liquid N_2: volume at $-195.8°C$: $\quad V_2 = \dfrac{V_1 T_2}{T_1} = \dfrac{5.00\ \text{L} \times (-195.8°C + 273.15°)}{25.0°C + 273.15°} = \textbf{1.30 L}$

Liquid He: volume at $-268.9°C$: $\quad V_2 = \dfrac{V_1 T_2}{T_1} = \dfrac{5.00\ \text{L} \times (-268.9°C + 273.15°)}{25.0°C + 273.15°} = \textbf{0.0713 L}$

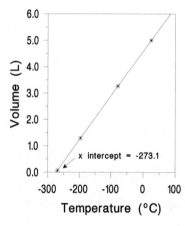

If the line is extrapolated to the x axis, the x intercept is the temperature at which zero volume is theoretically reached. That temperature is $-273.1°C$, also known as absolute zero, 0.0 K.

12-32. *Refer to Sections 12-4 and 12-5, and Figures 12-4 and 12-5.*

(a) $P \times V = \text{constant}$

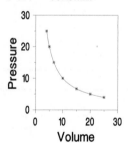

(b) $P = \text{constant} \times 1/V$

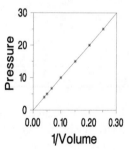

(c) $V = \text{constant} \times T \quad \text{or} \quad V/T = \text{constant}$

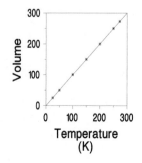

(d) $P = \text{constant} \times T \quad \text{or} \quad P/T = \text{constant}$

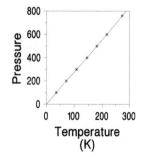

The graphs were obtained by plotting the hypothetical data given below. It is assumed that for (a) and (b), n and T are constant; for (c), n and P are constant; and for (d), n and V are constant.

(a),(b)	P	V	$P \times V$	$1/V$
	4.00	25	100	0.0400
	5.00	20.0	100	0.0500
	6.67	15.0	100	0.0667
	10.0	10.0	100	0.100
	15.0	6.67	100	0.150
	20.0	5.00	100	0.200
	25.0	4.00	100	0.250

(c)	V	$T(K)$	V/T
	273	273	1.00
	250	250	1.00
	200	200	1.00
	150	150	1.00
	100	100	1.00
	50	50	1.00
	25	25	1.00

(d)	P	$T(K)$	P/T
	760	273	2.78
	600	216	2.78
	500	180	2.78
	400	144	2.78
	300	108	2.78
	200	71.8	2.78
	100	35.9	2.78

12-34. *Refer to Section 12-7, and Examples 12-4 and 12-5.*

Given: $T_1 = 26°C + 273.15° = 299$ K $V_1 = 385$ mL $P_1 = 670.$ torr
   $T_2 = ?$ $V_2 = 440.$ mL $P_2 = 940.$ torr

Combined Gas Law: $\dfrac{P_1 V_1}{T_1} = \dfrac{P_2 V_2}{T_2}$ at constant n

$$\frac{670.\ \text{torr} \times 385\ \text{mL}}{299\ \text{K}} = \frac{940.\ \text{torr} \times 440.\ \text{mL}}{T_2}$$

$$T_2 = \textbf{479 K or 206°C}$$

12-36. *Refer to Sections 12-4, 12-5 and 12-7.*

Boyle's Law and Charles's Law can both be derived from the Combined Gas Law: $\dfrac{P_1 V_1}{T_1} = \dfrac{P_2 V_2}{T_2}$ at constant n

(1) When applying Boyle's Law, we are working at constant temperature and with a constant number of moles of gas. In this case, $T_1 = T_2$, and the Combined Gas Law simplifies to

   Boyle's Law: $P_1 V_1 = P_2 V_2$ at constant n and T

(2) When applying Charles's Law, we are working at constant pressure and with a constant number of moles of gas. In this case, $P_1 = P_2$, and the Combined Gas Law simplifies to

   Charles's Law: $\dfrac{V_1}{T_1} = \dfrac{V_2}{T_2}$ at constant n and P

12-38. *Refer to Section 12-8.*

(a) Avogadro's Law states that, at the same temperature and pressure, equal volumes of all gases contain the same number of molecules. This means that equal number of moles of any gas take up equal volumes as long as the temperature and pressure are the same.

(b) The standard molar volume is the volume occupied by 1 mole of an ideal gas under standard conditions. It is 22.4 L/mol at STP, where STP is defined as being 1 atmosphere pressure (760 torr) and 0.00°C (273.15 K).

(c) The term "standard molar volume" can also apply to liquids and solids as well as gases. It is defined as the volume that one mole of substance occupies at standard conditions, usually 1 atmosphere pressure (760 torr) and 0.00°C (273.15 K). Note that there is much less variation of volume with temperature and pressure with liquids and solids, as there is with gases.

(d) Yes, there are other temperature and pressure conditions at which 1 mole of any ideal gas would occupy 22.4 L. The law we need here that is applied to situations where the number of moles of gas and the volume stay constant can be derived from the Combined Gas Law:

$$\frac{P_1 V_1}{T_1} = \frac{P_2 V_2}{T_2} \quad \text{at constant } n$$

We want the volume to remain constant, so the equation simplifies to $\dfrac{P_1}{T_1} = \dfrac{P_2}{T_2}$ at constant n and V

We know that 1 mole of any ideal gas occupies 22.4 L at 1 atmosphere and 273.15 K, so plugging in, we can determine that any time the pressure to temperature ratio is 1 atm/273.15 K = 0.00366 atm/K, the volume will be 22.4 L.

12-40.   *Refer to Section 12-9, and Example 12-11.*

Plan: (1) Find the volume of 1.00 mole of EDB vapor using the ideal gas equation.
    (2) Calculate the density, D (g/L), of the EDB vapor.

(1) ? V EDB (in L) $= \dfrac{nRT}{P} = \dfrac{(1.00 \text{ mol})(0.0821 \text{ L·atm/mol·K})(165°C + 273°)}{1.00 \text{ atm}} = 36.0$ L EDB

(2) Density (g/L) $= \dfrac{\text{mass of vapor (g)}}{\text{volume of vapor (L)}} = \dfrac{188 \text{ g}}{36.0 \text{ L}} =$ **5.23 g/L**

12-42.   *Refer to Section 12-8, and Examples 12-6 and 12-12.*

Recall: at STP, 1 mol of gas having a mass equal to its molecular weight occupies 22.4 L. Therefore at STP for a gas that behaves ideally,

$$\text{Density (g/L)} = \dfrac{\text{molecular weight (g/mol)}}{22.4 \text{ L/mol}}$$

(a) Plan: Using the above formula, calculate the molecular weights of the 2 unknown gases and identify them.

Cylinder #1: $D = 3.74$ g/L

MW (g/mol) $= D$ (g/L) x 22.4 L/mol = 3.74 g/L x 22.4 L/mol = 83.8 g/mol

Therefore, the gas must be **krypton, Kr**.

Cylinder #2: $D = 0.900$ g/L

MW (g/mol) $= D$ (g/L) x 22.4 L/mol = 0.900 g/L x 22.4 L/mol = 20.2 g/mol

Therefore, the gas must be **neon, Ne**.

(b) It is possible using the tools we now have to identify the gases if the density had been made at a different temperature and pressure than STP.

Plan: (1) Use the Combined Gas Law to calculate the volume that 1 L of the gas would have occupied at STP.
    (2) Calculate the new density at STP, which is equal to the same number of grams of gas divided by the newly calculated volume at STP.
    (3) Calculate the MW of the gas as above.

Note: However, it is standard procedure to solve this problem using the variation of the ideal gas law. See Example 12-12.

12-44.   *Refer to Section 12-9.*

(a) An "ideal gas" is a hypothetical gas that follows all of the postulates of the kinetic molecular theory. It also obeys exactly all of the gas laws.

(b) The ideal gas equation, also called the ideal gas law, is the relationship, $PV = nRT$.

(c) The ideal gas law is derived by combining Boyle's Law, Charles's Law and Avogadro's Law, obtaining

$$V \propto \dfrac{nT}{P} \qquad \text{with no restrictions}$$

(d) The symbol for the proportionality constant for the conversion of the above proportion to an equality is R. The formula obtained is

$$V = R\left(\frac{nT}{P}\right)$$

which can be rearranged to give

$$PV = nRT.$$

The value of R is obtained by experimentally measuring a complete set of P, V, n and T values, then solving for R by substituting into the ideal gas law.

12-46.   *Refer to Section 12-9 and Example 12-8.*

Recall the ideal gas law: $PV = nRT$

$$P = \frac{nRT}{V} = \frac{(2.54 \text{ mol})(0.0821 \text{ L·atm/mol·K})(45°C + 273°)}{7.75 \text{ L}} = \textbf{8.56 atm}$$

12-48.   *Refer to Section 12-9, Table 1-7, and Example 12-9.*

Plan: (a) (1) Calculate the moles of Cl_2 involved.
          (2) Determine the volume (in L and ft³) of Cl_2 at 750. torr and 18°C using $PV = nRT$.
       (b) Determine the length (ft) of the Cl_2 cloud knowing that V (ft³) = length (ft) x width (ft) x depth (ft).

(a) (1) ? mol Cl_2 = 565 tons x $\dfrac{2000 \text{ lb}}{1 \text{ ton}}$ x $\dfrac{453.6 \text{ g}}{1 \text{ lb}}$ x $\dfrac{1 \text{ mol}}{70.9 \text{ g}}$ = 7.23 x 10⁶ mol Cl_2

   (2) ? V Cl_2 (in L) = $\dfrac{nRT}{P}$ = $\dfrac{(7.23 \times 10^6 \text{ mol})(0.0821 \text{ L·atm/mol·K})(18.0°C + 273°)}{(750./760.) \text{ atm}}$ = **1.75 x 10⁸ L Cl_2**

   ? V Cl_2 (in ft³) = 1.75 x 10⁸ L x $\dfrac{1 \text{ ft}^3}{28.32 \text{ L}}$ = **6.18 x 10⁶ ft³ Cl_2**

(b) ? length of Cl_2 cloud (ft) = $\dfrac{V \text{ (ft}^3)}{\text{width (ft) x depth (ft)}}$ = $\dfrac{6.18 \times 10^6 \text{ ft}^3}{(0.500 \text{ mi x 5280 ft/mi) x 60.0 ft}}$

$$= \textbf{39.0 ft} \text{ (to 3 significant figures)}$$

12-50.   *Refer to Section 12-10 and Example 12-12.*

Plan: (1) Use the ideal gas law, $PV = nRT$, to calculate the number of moles of ethane in the container at STP.
       (2) Determine the experimental molecular weight of ethane and compare it to the theoretical value.

(1) $n = \dfrac{PV}{RT} = \dfrac{(1 \text{ atm})(0.185 \text{ L})}{(0.0821 \text{ L·atm/mol·K})(273 \text{ K})}$ = 8.25 x 10⁻³ mol C_2H_6

(2) MW $C_2H_6 = \dfrac{0.244 \text{ g } C_2H_6}{8.25 \times 10^{-3} \text{ mol}}$ = **29.6 g/mol**

The actual molecular weight of ethane, C_2H_6, is 30.1 g/mol.

Percent error = $\dfrac{\text{actual MW - experimental MW}}{\text{actual MW}}$ x 100% = $\dfrac{30.1 - 29.6}{30.1}$ x 100% = **2%**

Possible sources of error which would result in a slightly low experimental molecular weight include:
   (a) the container volume is slightly less than 185 mL,
   (b) the mass of ethane is slightly more than 0.244 g, and
   (c) ethane deviates slightly from ideal behavior under STP conditions (Refer to methane, CH_4, in Table 12-5.)

12-52. *Refer to Section 12-10 and Example 12-12.*

Plan: (1) Use the ideal gas law, $PV = nRT$, to find the number of moles of gas.
          (2) Calculate the molecular weight of the gas.

(1) $n = \dfrac{PV}{RT} = \dfrac{[(745/760)\text{ atm}](0.00413\text{ L})}{(0.0821\text{ L·atm/mol·K})(23°C + 273°)} = 1.67 \times 10^{-4}$ mol

(2) MW (g/mol) $= \dfrac{0.00500\text{ g gas}}{1.67 \times 10^{-4}\text{ mol}} = \textbf{30.0 g/mol}$

Within experimental error, the gaseous hydrocarbon could be **ethane (C_2H_6)** with a molecular weight of 30.1 g/mol.

12-54. *Refer to Section 12-10 and Example 12-13.*

Plan: (1) Find the volume, V, of the container. In this case, you cannot assume that the flask is 250. mL.
          (2) Determine the number of moles of gas, using $PV = nRT$.
          (3) Determine the mass of the gas in the flask.
          (4) Calculate the molecular weight of the gas.

(1) ? V, volume of container $=$ volume of water in container

$$= \frac{\text{mass of water (g)}}{\text{density of water (g/mL)}} \qquad \left(\text{since } D = \frac{\text{mass}}{\text{volume}}\right)$$

$$= \frac{\text{mass of flask filled with water - empty flask}}{\text{density of water}}$$

$$= \frac{327.4\text{ g} - 65.347\text{ g}}{0.997\text{ g/mL}}$$

$$= 263\text{ mL}$$

(2) $n = \dfrac{PV}{RT} = \dfrac{[(743.3/760)\text{ atm}](0.263\text{ L})}{(0.0821\text{ L·atm/mol·K})(99.8°C + 273°)} = 0.00840$ mol

(3) mass of gas $=$ mass of condensed liquid $=$ mass of flask and condensed liquid - mass of empty flask
$$= 65.739\text{ g} - 65.347\text{ g}$$
$$= 0.392\text{ g}$$

(4) MW (g/mol) $= \dfrac{0.392\text{ g gas}}{0.00840\text{ mol}} = 46.7$ g/mol or **46.7 amu/molecule**

12-56. *Refer to Section 12-11.*

(a) The partial pressure of a gas is the pressure it exerts in a mixture of gases. It is equal to the pressure the gas would exert if it were alone in the container at the same temperature.

(b) Dalton's Law states that the total pressure exerted by a mixture of ideal gases is the sum of the partial pressures of those gases.
$$P_{total} = P_A + P_B + P_C \; \dots \qquad\qquad \text{at constant } V, T$$

12-58. *Refer to Section 12-11 and Examples 12-17 and 12-18.*

From Dalton's Law of Partial Pressures,

$P_{total} = \dfrac{n_{total}RT}{V}$ where $n_{total} = n_{CHCl_3} + n_{CH_4} = \dfrac{3.23\text{ g }CHCl_3}{119.4\text{ g/mol}} + \dfrac{1.22\text{ g }CH_4}{16.04\text{ g/mol}}$

$$= 0.0271\text{ mol} + 0.0761\text{ mol}$$

$$= 0.1031\text{ mol (4 significant figures due to rules of addition)}$$

$$P_{total} = \frac{(0.1031 \text{ mol gas})(0.0821 \text{ L·atm/mol·K})(275°C + 273°)}{0.0500 \text{ L}} = \mathbf{92.8 \text{ atm}}$$

$$P_{CHCl_3} = \left(\frac{n_{CHCl_3}}{n_{CHCl_3} + n_{CH_4}}\right)P_{total} = \left(\frac{0.0271 \text{ mol}}{0.0271 \text{ mol} + 0.0761 \text{ mol}}\right)92.8 \text{ atm} = \mathbf{24.4 \text{ atm}}$$

Alternative method: use $P_{CHCl_3} = \dfrac{n_{CHCl_3}RT}{V}$

12-60.   Refer to Section 12-11 and Example 12-16.

mole fraction of He $X_{He} = \dfrac{P_{He}}{P_{total}} = \dfrac{0.467 \text{ atm He}}{0.467 \text{ atm He} + 0.317 \text{ atm Ar} + 0.277 \text{ atm Xe}} = \dfrac{0.467 \text{ atm}}{1.061 \text{ atm}} = \mathbf{0.440}$

mole fraction of Ar $X_{Ar} = \dfrac{P_{Ar}}{P_{total}} = \dfrac{0.317 \text{ atm}}{1.061 \text{ atm}} = \mathbf{0.299}$

mole fraction of Xe $X_{Xe} = \dfrac{P_{Xe}}{P_{total}} = \dfrac{0.277 \text{ atm}}{1.061 \text{ atm}} = \mathbf{0.261}$ Note: $X_{He} + X_{Ar} + X_{Xe} = 1$

12-62.   Refer to Section 12-11 and Examples 12-15 and 12-17.

(a) Boyle's Law states that $P_1V_1 = P_2V_2$.

For each gas, $P_2 = \dfrac{P_1V_1}{V_2} = \dfrac{1.50 \text{ atm} \times 2.50 \text{ L}}{1.00 \text{ L}} = 3.75 \text{ atm}$

Dalton's Law of Partial Pressures states that $P_{total} = P_1 + P_2 + P_3 + \dots$ at constant V, T

Therefore, $P_{total} = P_{O_2} + P_{N_2} + P_{He} = 3.75 \text{ atm} + 3.75 \text{ atm} + 3.75 \text{ atm} = \mathbf{11.25 \text{ atm}}$

(b) partial pressure of O_2, $P_{O_2} = \mathbf{3.75 \text{ atm}}$

(c) partial pressure of N_2, $P_{N_2} = $ partial pressure of He, $P_{He} = \mathbf{3.75 \text{ atm}}$

12-64.   Refer to Section 12-11, Figure 12-7, Table 12-4 and Example 12-19.

Plan: (1) Calculate the partial pressure of nitrogen in the container at 25°C and 750. torr.
         (2) Use the Combined Gas Law to calculate the volume of gas ($N_2 + H_2O$) at the new conditions.

(1) $P_{N_2} = P_{atm} - P_{H_2O} = 750. \text{ torr} - 24 \text{ torr} = 726 \text{ torr}$

(2) Combined Gas Law: $\dfrac{P_1V_1}{T_1} = \dfrac{P_2V_2}{T_2}$

$$\frac{760 \text{ torr} \times 469 \text{ mL}}{273 \text{ K}} = \frac{726 \text{ torr} \times V_2}{25°C + 273°}$$

$$V_2 = \mathbf{536 \text{ mL}}$$

12-66.   Refer to Section 12-11 and Example 12-18.

(a) Recall Boyle's Law: $P_1V_1 = P_2V_2$ at constant n, T

The total volume of the flasks = 2.00 L + 4.00 L = 6.00 L

for He: 6.00 atm $\times$ 4.00 L $= P_2 \times$ 6.00 L   for N_2:   3.00 atm $\times$ 2.00 L $= P_2 \times$ 6.00 L

$P_2 = \mathbf{4.00 \text{ atm}}$ $P_2 = \mathbf{1.00 \text{ atm}}$

(b) According to Dalton's Law of Partial Pressures, $P_{total} = P_1 + P_2 + \ldots$.
Therefore,

$P_{total} = P_{He} + P_{N_2} = 4.00 \text{ atm} + 1.00 \text{ atm} = \textbf{5.00 atm}$

(c) mole fraction of He, $X_{He} = \dfrac{P_{He}}{P_{total}} = \dfrac{4.00 \text{ atm}}{5.00 \text{ atm}} = \textbf{0.800}$

12-68.   *Refer to Section 12-12 and Example 12-21.*

Balanced equation: $2NaN_3(s) \rightarrow 2Na(s) + 3N_2(g)$

Plan: $V N_2 \overset{(1)}{\Rightarrow} \text{mol } N_2 \overset{(2)}{\Rightarrow} \text{mol } NaN_3 \overset{(3)}{\Rightarrow} \text{g } NaN_3$

(1) ? mol $N_2 = n = \dfrac{PV}{RT} = \dfrac{(1.40 \text{ atm})(30.0 \text{ L})}{(0.0821 \text{ L·atm/mol·K})(25°C + 273°)} = 1.72 \text{ mol } N_2$

(2) ? mol $NaN_3 = \dfrac{2 \text{ mol } NaN_3}{3 \text{ mol } N_2} \times 1.72 \text{ mol } N_2 = 1.14 \text{ mol } NaN_3$

(3) ? g $NaN_3 = 1.14 \text{ mol } NaN_3 \times 65.0 \text{ g/mol} = \textbf{74.4 g NaN}_3$

Note: When doing an extensive calculation, keep all the numbers in your calculator. The answers here are rounded off to the appropriate number of significant figures after each step only to illustrate the concept of significant figures.

12-70.   *Refer to Section 12-12.*

Balanced equation: $S_8(g) + 8O_2(g) \rightarrow 8SO_2(g)$

In reality, above 444°C, sulfur boils to give a vapor containing a mixture of S_8, S_6, S_4 and S_2 molecules. However, we are assuming at this temperature that the sulfur exists primarily as S_8.

Plan: $V S_8 \overset{(1)}{\Rightarrow} \text{mol } S_8 \overset{(2)}{\Rightarrow} \text{mol } SO_2 \overset{(3)}{\Rightarrow} \text{g } SO_2$

(1) ? mol $S_8 = n = \dfrac{PV}{RT} = \dfrac{(1.00 \text{ atm})(1.00 \text{ L})}{(0.0821 \text{ L·atm/mol·K})(600.°C + 273°)} = 0.0140 \text{ mol } S_8$

(2) ? mol $SO_2 = \dfrac{8 \text{ mol } SO_2}{1 \text{ mol } S_8} \times 0.0140 \text{ mol } S_8 = 0.112 \text{ mol } SO_2$

(3) ? g $SO_2 = 0.112 \text{ mol } SO_2 \times 64.07 \text{ g/mol} = \textbf{7.15 g SO}_2$

12-72.   *Refer to Section 12-12 and Example 12-20.*

Balanced equation: $2KClO_3(s) \rightarrow 2KCl(s) + 3O_2(g)$

Plan: $V_{actual} O_2 \overset{(1)}{\Rightarrow} V_{theoretical} O_2 \overset{(2)}{\Rightarrow} \text{mol } O_2 \overset{(3)}{\Rightarrow} \text{mol } KClO_3 \overset{(4)}{\Rightarrow} \text{g } KClO_3$

(1) In order to fill four 250. mL bottles, 1.00 L O_2 is actually required. However, more than 1.00 L O_2 must be produced since some O_2 will be lost in the process. If 25% of O_2 will be wasted, the percentage yield of the process is 75%. The theoretical amount of O_2 that must be produced can be calculated:

Recall: percentage yield $= \dfrac{\text{actual yield}}{\text{theoretical yield}} \times 100\%$.   Therefore,

theoretical volume of O_2 needed $= \dfrac{\text{actual volume of } O_2 \text{ needed}}{75\%} \times 100\% = \dfrac{1.00 \text{ L}}{75\%} \times 100\% = 1.33 \text{ L } O_2$

(2) $? \text{ mol } O_2 = n = \dfrac{PV}{RT} = \dfrac{[(762/760) \text{ atm}](1.33 \text{ L})}{(0.0821 \text{ L·atm/mol·K})(25°C + 273°)} = 0.0545 \text{ mol } O_2$

(3) $? \text{ mol } KClO_3 = 0.0545 \text{ mol } O_2 \times (2 \text{ mol } KClO_3/3 \text{ mol } O_2) = 0.0363 \text{ mol } KClO_3$

(4) $? \text{ g } KClO_3 = 0.0363 \text{ mol } KClO_3 \times 122.6 \text{ g/mol} = \textbf{4.45 g } KClO_3$

12-74. *Refer to Section 12-12.*

Balanced equation: $N_2(g) + 3H_2(g) \rightarrow 2NH_3(g)$

This is a limiting reactant problem. Due to Gay-Lussac's Law, we can work directly in volumes instead of moles.

(1) Compare the required ratio to the available ratio of reactants to find the limiting reactant.

$\text{Required ratio} = \dfrac{1 \text{ volume } N_2}{3 \text{ volumes } H_2} = 0.333 \qquad \text{Available ratio} = \dfrac{3.00 \text{ L } N_2}{7.00 \text{ L } H_2} = 0.429$

Available ratio > required ratio; H_2 is the limiting reactant.

(2) $? \text{ L } NH_3 = 7.00 \text{ L } H_2 \times \dfrac{2 \text{ L } NH_3}{3 \text{ L } H_2} = \textbf{4.67 L } NH_3$

12-76. *Refer to Section 12-12, and Examples 12-20 and 12-21.*

Balanced equation: $2KNO_3(s) \rightarrow 2KNO_2(s) + O_2(g)$

Recall that 1 mole of ideal gas at STP occupies 22.4 L.

$$\text{Plan:} \quad \overset{(1)}{V_{STP} \, O_2} \Rightarrow \overset{(2)}{\text{mol } O_2} \Rightarrow \overset{(3)}{\text{mol } KNO_3} \Rightarrow \text{g } KNO_3$$

Method 1:

(1) $? \text{ mol } O_2 = \dfrac{21.1 \text{ L}_{STP} \, O_2}{22.4 \text{ L}_{STP}/\text{mol}} = 0.942 \text{ mol } O_2$

(2) $? \text{ mol } KNO_3 = 0.942 \text{ mol } O_2 \times (2 \text{ mol } KNO_3/1 \text{ mol } O_2) = 1.88 \text{ mol } KNO_3$

(3) $? \text{ g } KNO_3 = 1.88 \text{ mol } KNO_3 \times 101 \text{ g/mol} = \textbf{190. g } KNO_3$

Method 2: Dimensional Analysis

$? \text{ g } KNO_3 = 21.1 \text{ L}_{STP} \, O_2 \times \dfrac{1 \text{ mol } O_2}{22.4 \text{ L}_{STP} \, O_2} \times \dfrac{2 \text{ mol } KNO_3}{1 \text{ mol } O_2} \times \dfrac{101 \text{ g } KNO_3}{1 \text{ mol } KNO_3} = \textbf{190. g } KNO_3$

12-78. *Refer to Section 12-12.*

$$\text{Plan:} \quad \overset{(1)}{V \, SO_2} \Rightarrow \overset{(2)}{\text{mol } SO_2} \Rightarrow \overset{(3)}{\text{mol } S} \Rightarrow \overset{(4)}{\text{g } S} \Rightarrow \%S \text{ by mass}$$

(1) $? \text{ mol } SO_2 = n = \dfrac{PV}{RT} = \dfrac{[(755/760) \text{ atm}](1.177 \text{ L})}{(0.0821 \text{ L·atm/mol·K})(35.0°C + 273°)} = 0.0462 \text{ mol } SO_2$

(2) $? \text{ mol } S = \text{mol } SO_2 = 0.0462 \text{ mol } S$

(3) $? \text{ g } S = 0.0462 \text{ mol } S \times 32.066 \text{ g/mol} = 1.48 \text{ g } S$

(4) $? \%S \text{ by mass} = \dfrac{\text{g } S}{\text{g sample}} \times 100 = \dfrac{1.48 \text{ g}}{5.913 \text{ g}} \times 100 = \textbf{25.1\% S by mass}$

Balanced equations: A: $2C_8H_{18} + 25O_2 \rightarrow 16CO_2 + 18H_2O$

B: $2C_8H_{18} + 17O_2 \rightarrow 16CO + 18H_2O$

(a) Plan: CO concentration $\overset{(1)}{\Rightarrow}$ g CO $\overset{(2)}{\Rightarrow}$ mol CO $\overset{(3)}{\Rightarrow}$ mol C_8H_{18} (from Reaction B) $\overset{(4)}{\Rightarrow}$ mol C_8H_{18} (total)

$\overset{(5)}{\Rightarrow}$ g C_8H_{18} (total) $\overset{(6)}{\Rightarrow}$ $V\,C_8H_{18}$

Method 1:

(1) ? g CO = concentration (g/m³) x volume (m³) = 2.00 g/m³ x 97.5 m³ = 195 g CO produced

(2) ? mol CO $= \dfrac{195 \text{ g CO}}{28.0 \text{ g/mol}} = 6.96$ mol CO

(3) ? mol C_8H_{18} (Reaction B) = (2/16) x 6.96 mol CO = 0.871 mol C_8H_{18} (Reaction B)

(4) ? mol C_8H_{18} (total) $= \dfrac{0.871 \text{ mol } C_8H_{18} \text{ (Reaction B)}}{0.050} = 17.4$ mol C_8H_{18} (total)

since only 5.0% of the total amount of C_8H_{18} burned in the engine produced CO.

(5) ? g C_8H_{18} (total) = 17.4 mol C_8H_{18} x 114 g/mol = 1980 g C_8H_{18} (total)

(6) ? $V\,C_8H_{18} = \dfrac{\text{mass (g)}}{\text{Density (g/mL)}} = \dfrac{1980 \text{ g } C_8H_{18}}{0.702 \text{ g/mL}} = 2830$ mL or **2.83 L C_8H_{18}** $(\text{g/mL}) = \dfrac{\text{mass (g)}}{\text{volume (mL)}}$

Method 2: Dimensional Analysis

? L C_8H_{18} = 97.5 m³ x $\dfrac{2.00 \text{ g CO}}{1 \text{ m}^3}$ x $\dfrac{1 \text{ mol CO}}{28.0 \text{ g CO}}$ x $\dfrac{2 \text{ mol } C_8H_{18} \text{ (Reaction B)}}{16 \text{ mol CO}}$ x $\dfrac{1.00 \text{ mol } C_8H_{18} \text{ (total)}}{0.050 \text{ mol } C_8H_{18} \text{ (Reaction B)}}$

x $\dfrac{114 \text{ g } C_8H_{18} \text{ (total)}}{1 \text{ mol } C_8H_{18} \text{ (total)}}$ x $\dfrac{1 \text{ mL } C_8H_{18}}{0.702 \text{ g } C_8H_{18}}$ x $\dfrac{1 \text{ L } C_8H_{18}}{1000 \text{ mL } C_8H_{18}}$ = **2.83 L C_8H_{18}**

(b) fuel rate $\left(\dfrac{\text{L}}{\text{min}}\right) = \dfrac{\text{volume of fuel burned (L)}}{\text{time (min)}}$

therefore,   time (min) $= \dfrac{\text{volume of fuel burned (L)}}{\text{fuel rate (L/min)}} = \dfrac{2.83 \text{ L}}{0.0631 \text{ L/min}} =$ **44.8 min**

(a) Since 1 Å = 1 x 10⁻¹⁰ m = 1 x 10⁻⁸ cm; and 1 cm³ = 1 mL

Molecular radius: $r = 2.00$ Å x $\dfrac{1 \times 10^{-8} \text{ cm}}{1 \text{ Å}} = 2.00 \times 10^{-8}$ cm

Molecular volume: $V = \dfrac{4}{3}\pi r^3$

$= \dfrac{4}{3}$ x 3.1416 x (2.00 x 10⁻⁸ cm)³

= **3.35 x 10⁻²³ cm³ or 3.35 x 10⁻²³ mL**

(b) Molecular volume for 1 mol molecules:

$$? \, V = 1.00 \text{ mol} \times \frac{6.02 \times 10^{23} \text{ molecules}}{1.00 \text{ mol}} \times \frac{3.35 \times 10^{-23} \text{ mL}}{1 \text{ molecule}} = \textbf{20.2 mL} \text{ or } \textbf{0.0202 L}$$

(c) fraction of volume occupied by 1.00 moles of molecules $= \dfrac{0.0202 \text{ L}}{22.4 \text{ L}} = \textbf{9.02} \times \textbf{10}^{-4}$ ($\equiv \textbf{0.0902\%}$)

(d) This result verifies the first statement of the Kinetic Molecular Theory of Ideal Gases: the molecular volume of the gas molecules is indeed negligible in comparison to the volume occupied by them.

12-84. *Refer to Section 12-13 and Exercise 12-86 Solution.*

According to the Kinetic-Molecular Theory, all gas molecules have the same average kinetic energy ($= 1/2 \, m\bar{u}^2$) at the same temperature, where $\bar{u}$ is the average velocity. Hence, at the same T:

$$1/2(m_{SiH_4})(\bar{u}_{SiH_4})^2 = 1/2 \, (m_{CH_4})(\bar{u}_{CH_4})^2$$

$$\frac{\bar{u}_{CH_4}}{\bar{u}_{SiH_4}} = \sqrt{\frac{m_{SiH_4}}{m_{CH_4}}} = \sqrt{\frac{MW_{SiH_4}}{MW_{CH_4}}} = \sqrt{\frac{32}{16}} = 1.4$$

SiH_4 is heavier than CH_4; however, both molecules have the same average kinetic energy. This is due to the fact that methane molecules have an average speed which is 1.4 times faster than that of silane molecules.

12-86. *Refer to Section 12-13.*

(a) The third assumption of the Kinetic-Molecular Theory states that the average kinetic energy of gaseous molecules is directly proportional to the absolute temperature of the sample.

Average kinetic energy $= 1/2m\bar{u}^2 \propto T$ where $\quad m$ = mass (g)

$\bar{u}$ = average molecular speed (m/s)

T = absolute temperature (K)

We see that the average molecular speed is directly proportional to the square root of the absolute temperature.

(b) $\dfrac{\text{rms speed of } N_2 \text{ molecules at } 100°C}{\text{rms speed of } N_2 \text{ molecules at } 0°C} = \sqrt{\dfrac{100°C + 273°}{0°C + 273°}} = \textbf{1.17}$

12-88. *Refer to Section 12-13.*

According to Kinetic-Molecular Theory, the pressure exerted by a gas upon the walls of its container is caused by gas molecules hitting the walls. Pressure depends on

(1) the number of molecules hitting the walls per unit time and
(2) how vigorously the molecules hit the walls.

(a) When a gaseous sample in a fixed volume is heated, the pressure increases. Recall that the average kinetic energy is directly proportional to the absolute temperature. As temperature increases, so does the energy and also the velocity of the molecules. At higher temperatures, there is an increase in the force and number of collisions and the pressure increases.

(b) When the volume of a gaseous sample is reduced at constant temperature, the pressure increases. When the volume is reduced, more molecules hit a given area on the walls per unit time, increasing the pressure. This is Boyle's Law.

For H_2, F_2 and HF under the same conditions, H_2 would behave the most ideally, because for such small nonpolar molecules, the dispersion forces would be small and therefore the intermolecular attractions would be negligible. The behavior of HF, on the other hand, would deviate the most from ideality, because even though HF is smaller than F_2, it is very polar and its molecules exhibit great attraction for one another.

12-92. *Refer to Section 12-15.*

(a) The effect of molecular volume on the properties of a gas becomes more important when a gas is compressed at constant temperature.

(b) Molecular volume also becomes more important when more gas molecules are added to a system.

(c) When the temperature of the gas is raised at constant pressure, the volume expands. At a larger occupied volume, the effect of molecular volume on the properties of a gas becomes less significant.

12-94. *Refer to Section 12-15 and Example 12-23.*

(a) Assuming CCl_4 obeys the ideal gas law: $PV = nRT$

$$P = \frac{nRT}{V} = \frac{(1.00 \text{ mol})(0.0821 \text{ L·atm/mol·K})(77.0°C + 273°)}{35.0 \text{ L}} = \textbf{0.821 atm}$$

(b) Assuming CCl_4 obeys the van der Waals equation: $\left(P + \frac{n^2a}{V^2}\right)(V - nb) = nRT$

for CCl_4, $a = 20.39 \text{ L}^2\text{·atm/mol}^2$, $b = 0.1383 \text{ L/mol}$

$$\left[P + \frac{(1.00 \text{ mol})^2(20.39 \text{ L}^2\text{·atm/mol}^2)}{(35.0 \text{ L})^2}\right]\left[35.0 \text{ L} - (1.00 \text{ mol})\left(0.1383 \frac{\text{L}}{\text{mol}}\right)\right]$$

$$= (1.00 \text{ mol})(0.0821 \text{ L·atm/mol·K})(77°C + 273°)$$

$$[P + 0.0166 \text{ atm}][34.9 \text{ L}] = 28.7 \text{ L·atm}$$

$$P + 0.0166 \text{ atm} = 0.822 \text{ atm}$$

$$P = \textbf{0.805 atm}$$

(c) (1) Assuming CCl_4 obeys the ideal gas law: $PV = nRT$

$$P = \frac{nRT}{V} = \frac{(3.10 \text{ mol})(0.0821 \text{ L·atm/mol·K})(135°C + 273°)}{5.75 \text{ L}} = \textbf{18.1 atm}$$

(2) Assuming CCl_4 obeys the van der Waals equation: $\left(P + \frac{n^2a}{V^2}\right)(V - nb) = nRT$

for CCl_4, $a = 20.39 \text{ L}^2\text{·atm/mol}^2$, $b = 0.1383 \text{ L/mol}$

$$\left[P + \frac{(3.10 \text{ mol})^2(20.39 \text{ L}^2\text{·atm/mol}^2)}{(5.75 \text{ L})^2}\right]\left[5.75 \text{ L} - (3.10 \text{ mol})\left(0.1383 \frac{\text{L}}{\text{mol}}\right)\right]$$

$$= (3.10 \text{ mol})(0.0821 \text{ L·atm/mol·K})(135°C + 273°)$$

$$[P + 5.93 \text{ atm}][5.32 \text{ L}] = 104 \text{ L·atm}$$

$$P + 5.93 \text{ atm} = 19.5 \text{ atm}$$

$$P = \textbf{13.6 atm}$$

The molar volume of compounds is the volume that one mole of compound will occupy at a particular temperature. It can be calculated by dividing the atomic weight, AW, of an element by its density.

Molar volume of liquid iron at 1600.°C:

$$\text{Molar volume (cm}^3/\text{mol)} = \frac{AW}{D} = \frac{55.85 \text{ g/mol}}{6.88 \text{ g/cm}^3} = \textbf{8.12 cm}^3/\textbf{mol}$$

Alternatively, by dimensional analysis: $\text{molar volume (cm}^3/\text{mol)} = \frac{55.85 \text{ g}}{1 \text{ mol Fe}} \times \frac{1 \text{ cm}^3}{6.88 \text{ g}} = \textbf{8.12 cm}^3/\textbf{mol}$

Molar volume of solid iron at 20°C:

$$\text{Molar volume (cm}^3/\text{mol)} = \frac{AW}{D} = \frac{55.85 \text{ g/mol}}{7.86 \text{ g/cm}^3} = \textbf{7.11 cm}^3/\textbf{mol}$$

Alternatively, by dimensional analysis: $\text{molar volume (cm}^3/\text{mol)} = \frac{55.85 \text{ g}}{1 \text{ mol Fe}} \times \frac{1 \text{ cm}^3}{7.86 \text{ g}} = \textbf{7.11 cm}^3/\textbf{mol}$

Most substances have a slightly smaller molar volume as a solid than as a liquid; substances generally decrease in volume when they freeze. Water is the notable exception; it expands (increases in molar volume) upon freezing. That is why closed containers holding water and other aqueous solutions will break in freezing weather. Although iron is much more dense overall than the compounds in Table 12-1, and its molar volumes are much less than those of benzene and carbon tetrachloride, iron still behaves normally. The molar volume of liquid iron is still greater than that of solid iron.

Balanced equation: $2H_2O(\ell) \rightarrow 2H_2(g) + O_2(g)$

Plan: $V O_2 \overset{(1)}{\Rightarrow} \text{mol } O_2 \overset{(2)}{\Rightarrow} \text{mol } H_2O \overset{(3)}{\Rightarrow} \text{g } H_2O$

(1) $? \text{ mol } O_2 = n = \frac{PV}{RT} = \frac{(1.00 \text{ atm})(0.075 \text{ L})}{(0.0821 \text{ L·atm/mol·K})(25°C + 273°)} = 0.0031 \text{ mol } O_2$

(2) $? \text{ mol } H_2O = \frac{2 \text{ mol } H_2O}{1 \text{ mol } O_2} \times 0.0031 \text{ mol } N_2 = 0.0061 \text{ mol } H_2O$

(3) $? \text{ g } H_2O = 0.0061 \text{ mol } H_2O \times 18.0 \text{ g/mol} = \textbf{0.11 g } \textbf{H}_2\textbf{O}$ (2 significant figures)

Note: When doing an extensive calculation, keep all the numbers in your calculator. The answers here are rounded off to the appropriate number of significant figures after each step only to illustrate the concept of significant figures.

Recall Charles' Law: $\frac{V_1}{T_1} = \frac{V_2}{T_2}$ at constant n and P

Given: $V_1 = 175 \text{ m}^3$ $T_1 = 10.°C + 273° = 283 \text{ K}$
     $V_2 = ?$ $T_2 = 18°C + 273° = 291 \text{ K}$

$V_2 = \frac{V_1 T_2}{T_1} = \frac{175 \text{ m}^3 \times 291 \text{ K}}{283 \text{ K}} = 180. \text{ m}^3$

Therefore, **5 m³** (= 180. m³ - 175 m³) of air had been forced out of the cabin.

$? \text{ L air forced from cabin} = 5 \text{ m}^3 \times \frac{(100 \text{ cm})^3}{(1 \text{ m})^3} \times \frac{1 \text{ mL}}{1 \text{ cm}^3} \times \frac{1 \text{ L}}{1000 \text{ mL}} = \textbf{5000 L air}$

Plan: (1) Use the ideal gas law, $PV = nRT$, to find the moles of Freon-12.
        (2) Calculate the molecular weight of Freon-12.

(1) $n = \dfrac{PV}{RT} = \dfrac{[(790./760)\ \text{atm}](8.29\ \text{L})}{(0.0821\ \text{L·atm/mol·K})(200.°C + 273°)} = 0.222\ \text{mol Freon-12}$

(2) MW (g/mol) $= \dfrac{26.8\ \text{g Freon-12}}{0.222\ \text{mol}} = \textbf{121 g/mol}$

(1) Thought Process #1: Consider the ideal gas law, $PV = nRT$. We know that

Boyle's Law: P is inversely proportional to V at constant n and T.
Charles's Law: V is directly proportional to T at constant P and n.
Avogadro's Law: V is directly proportional to n at constant P and T.

So, what we see is that variables on the same side of the ideal gas law equality are inversely proportional to each other and variables on opposite sides of the equality are directly proportional to each other.

Therefore, we can deduce that P is directly proportional T at constant n and V.

(2) Thought Process #2: Another way to figure out the relationship is to say that at constant n and V,

$$P = \frac{nRT}{V} = \text{constant} \times T \quad \text{and } P \text{ is directly proportional } T \text{ since } n, R \text{ and } V \text{ are all constants}$$

We have a formula that is similar to Charles's Law: $\dfrac{P_1}{T_1} = \dfrac{P_2}{T_2}$ at constant n and V

A non-laboratory application of this law is that in a closed constant volume container holding a gas, the pressure increases with increasing temperature. In a fire, supposedly empty sealed containers containing air or vapor can become lethal weapons if the container is incapable of handling the increased stress provided by increased pressure. This fact would be invaluable to fire fighters.

(a) Plan: (1) Determine the actual partial pressure of H_2O vapor before and after air conditioning.
             (2) Calculate the moles and mass of water present before and after air conditioning using the ideal gas law, $PV = nRT$.
             (3) Determine the mass of water removed by the air conditioning process.

(1) From Appendix E,   vapor pressure of water at 33°C = 37.7 torr
                            vapor pressure of water at 25°C = 23.8 torr

Given: relative humidity $= \dfrac{\text{actual partial pressure of } H_2O \text{ vapor}}{\text{partial pressure of } H_2O \text{ vapor if saturated}}$

before air conditioning at 33.0°C: $P_{H_2O,\text{actual}}$ = relative humidity $\times P_{H_2O,\text{sat}}$
                                                        = 0.800 $\times$ 37.7 torr
                                                        = 30.2 torr

after air conditioning at 25.0°C: $P_{H_2O,\text{actual}}$ = relative humidity $\times P_{H_2O,\text{sat}}$
                                                        = 0.150 $\times$ 23.8 torr
                                                        = 3.57 torr

173

(2) $? V_{house}$ (L) $= 245\ m^3 \times \dfrac{(100\ cm)^3}{(1\ m)^3} \times \dfrac{1\ mL}{1\ cm^3} \times \dfrac{1\ L}{1000\ mL} = 2.45 \times 10^5\ L$

before air conditioning: $\quad n = \dfrac{PV}{RT} = \dfrac{[(30.2/760)\ atm](2.45 \times 10^5\ L)}{(0.0821\ L \cdot atm/mol \cdot K)(33.0°C + 273°)} = 388\ mol\ H_2O$

$? g\ H_2O = 388\ mol \times 18.0\ g/mol = 6980\ g\ H_2O$

after air conditioning: $\quad n = \dfrac{PV}{RT} = \dfrac{[(3.57/760)\ atm](2.45 \times 10^5\ L)}{(0.0821\ L \cdot atm/mol \cdot K)(25.0°C + 273°)} = 47.0\ mol\ H_2O$

$? g\ H_2O = 47.0\ mol \times 18.0\ g/mol = 846\ g\ H_2O$

(3) The mass of water removed $= 6980\ g - 846\ g = \textbf{6130 g H}_2\textbf{O}$

(b) $? mL\ H_2O\ at\ 25°C = \dfrac{6130\ g\ H_2O}{0.997\ g/cm^3} = \textbf{6150 cm}^3\ \textbf{H}_2\textbf{O}$ $\qquad$ since Density (g/mL) $= \dfrac{mass\ (g)}{volume\ (mL)}$

12-108. *Refer to Section 12-15, Example 12-23 and Table 12-5.*

(1) Assuming NH_3 obeys the ideal gas law: $PV = nRT$

$$P = \dfrac{nRT}{V} = \dfrac{(4.00\ mol)(0.0821\ L \cdot atm/mol \cdot K)(100°C + 273°)}{25.0\ L} = \textbf{4.90 atm}$$

(2) Assuming NH_3 obeys the van der Waals equation: $\quad \left(P + \dfrac{n^2a}{V^2}\right)(V-nb) = nRT$

$\qquad\qquad\qquad\qquad\qquad$ for NH_3, $\qquad a = 4.17\ L^2 \cdot atm/mol^2,\ b = 0.0371\ L/mol$

$$\left[P + \dfrac{(4.00\ mol)^2(4.17\ L^2 \cdot atm/mol^2)}{(25.0\ L)^2}\right]\left[25.0\ L - (4.00\ mol)\left(0.0371\ \dfrac{L}{mol}\right)\right]$$

$$= (4.00\ mol)(0.0821\ L \cdot atm/mol \cdot K)(100°C + 273°)$$

Simplifying,
$$[P + 0.107\ atm][24.9\ L] = 122\ L \cdot atm$$
$$P + 0.116\ atm = 4.90\ atm$$
$$P = \textbf{4.78 atm}$$

(3) $\%$ difference $= \dfrac{P_{ideal} - P_{real}}{P_{ideal}} \times 100 = \dfrac{4.90 - 4.78}{4.90} \times 100 = \textbf{2.4\%}$ (2 significant figures)

12-110. *Refer to Section 12-10 and Examples 12-13 and 12-14.*

Plan: (1) Find the empirical formula for cyanogen.
 (2) Calculate the molecular weight of cyanogen, using the ideal gas law, $PV = nRT$.
 (3) Determine the molecular formula.

(1) Assume 100 g of cyanogen.

$? mol\ C = \dfrac{46.2\ g\ C}{12.0\ g/mol} = 3.85\ mol\ C$ $\qquad\qquad$ Ratio $= \dfrac{3.85}{3.84} = 1.00$

$? mol\ N = \dfrac{53.8\ g\ N}{14.0\ g/mol} = 3.84\ mol\ N$ $\qquad\qquad$ Ratio $= \dfrac{3.84}{3.84} = 1.00$

The empirical formula for cyanogen is C_1N_1 or CN (formula weight $= 26.0\ g/mol$)

(2) $?$ mol cyanogen $= n = \dfrac{PV}{RT} = \dfrac{[(750./760) \text{ atm}](0.476 \text{ L})}{(0.0821 \text{ L·atm/mol·K})(25°C + 273°)} = 0.0192$ mol

$MW \text{ (g/mol)} = \dfrac{1.00 \text{ g}}{0.0192 \text{ mol}} = 52.1$ g/mol

(3) let $n = \dfrac{\text{molecular weight}}{\text{simplest formula weight}} = \dfrac{52.1 \text{ g/mol}}{26.0 \text{ g/mol}} = 2$ Therefore, the true molecular formula for cyanogen is $\mathbf{C_2N_2}$.

12-112. *Refer to Section 12-11.*

Plan: (1) Determine the partial pressure of acetic acid, CH_3COOH, and carbon dioxide, CO_2, in the mixture.
(2) Determine the moles of each gas in the mixture using the ideal gas law.
(3) Calculate the mass of each gas and the total mass of the sample.

(1) $P_{\text{acetic acid}} = 400.$ torr
$P_{\text{carbon dioxide}} = P_{\text{total}} - P_{\text{acetic acid}} = 760.$ torr $- 400.$ torr $= 360.$ torr

(2) n of acetic acid $= \dfrac{P_{\text{acetic acid}}V}{RT} = \dfrac{[(400./760.) \text{ atm}](0.500 \text{ L})}{(0.0821 \text{ L·atm/mol·K})(16.0°C + 273°)} = 0.0111$ mol CH_3COOH

n of carbon dioxide $= \dfrac{P_{\text{carbon dioxide}}V}{RT} = \dfrac{[(360./760.) \text{ atm}](0.500 \text{ L})}{(0.0821 \text{ L·atm/mol·K})(16.0°C + 273°)} = 0.00998$ mol CO_2

(3) $?$ g $CH_3COOH = 0.0111$ mol $CH_3COOH \times \dfrac{60.05 \text{ g } CH_3COOH}{1 \text{ mol } CH_3COOH} = 0.667$ g CH_3COOH

$?$ g $CO_2 = 0.00998$ mol $CO_2 \times \dfrac{44.01 \text{ g } CO_2}{1 \text{ mol } CO_2} = 0.439$ g CO_2

The total mass of the sample $= 0.667$ g $CH_3COOH + 0.439$ g $CO_2 = \mathbf{1.106 \text{ g}}$

12-114. *Refer to Section 12-13 and Figure 12-12.*

Use your imagination in this drawing and have fun. You might make the gas cylinders narrower to accentuate the decrease in volume between the molecules. To show changes in kinetic energy, recall that as the kinetic energy of the gas molecules decrease with temperature, so does the molecules' average velocity. You could put comic strip-like lines off each molecule indicating the speed and direction the molecules are traveling - longer lines for molecules moving faster and shorter lines for the slower molecules, such as:

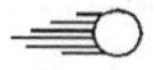

12-116. *Refer to Section 12-12.*

According to Gay-Lussac's Law, we can think directly in volumes instead of moles or molecules if the temperature and pressure are constant during a reaction.

(a) In the figure denoting the initial conditions, there are 6 reactant molecules in 2.5 L. After the reaction, the volume is only 1.25 L. Therefore, since the volume halved, the number of product molecules must have also halved. The only two answers involving 3 molecules are (i) and (ii). Answer (i) is wrong, because mass, i.e., the number of atoms before and after the reaction, was not conserved. Therefore, the answer must be **(ii)**.

(b) Balanced equation: $2AB_2(g) \rightarrow A_2B_4(g)$ where A are the orange atoms and B are the yellow atoms

In Chapter 3, we discussed percent by mass as a concentration unit: % by mass $= \dfrac{\text{g solute}}{\text{g solution}} \times 100\%$

We can similarly equate mole percent and mole fraction: mole percent = mole fraction $\times$ 100%

12-120. *Refer to Sections 12-9 and 12-12.*

Since $PV = nRT$; $P = \dfrac{nRT}{V}$. Therefore, at constant T and V, $P \propto n$.

As a result, at constant T and V, pressure can be used to measure the relative amount of compounds.

Balanced equation: $2H_2(g) + O_2(g) \rightarrow 2H_2O(g)$

Compare the required ratio of reactants (using moles) to the available ratio of reactants using partial pressures to find the limiting reactant.

Required ratio $= \dfrac{2 \text{ moles } H_2}{1 \text{ mole } O_2} = 2.00$ Available ratio $= \dfrac{0.588 \text{ atm } H_2}{0.302 \text{ atm } O_2} = 1.95$

Available ratio < required ratio; **H_2 is the limiting reactant**.

12-122. *Refer to Section 12-12.*

Balanced equations: $C(s) + O_2(g) \rightarrow CO_2(g)$

$CO_2(g) + 2NaOH(aq) \rightarrow Na_2CO_3(aq) + H_2O(\ell)$

$HCl(aq) + NaOH(aq) \rightarrow NaCl(aq) + H_2O(\ell)$

This is an example of the method of back titration, in which there is stoichiometrically more base, NaOH, present than is necessary to react with CO_2 dissolved in the water. One then titrates the excess NaOH with a standardized HCl solution and in a series of calculations, calculates the initial moles of CO_2. In this problem, the calculations are taken even further to determine the volume of O_2.

Plan: (1) Calculate the total moles of NaOH present.

(2) Calculate the moles of NaOH in excess after the CO_2 was bubbled in, which are equal to the moles of HCl reacted.

(3) Calculate the total moles of NaOH that reacted with CO_2 = (1) - (2).

(4) Determine the moles of CO_2.

(5) Determine the moles and volume of O_2 involved in the first equation.

(1) ? mol NaOH_{total} = 0.437 M NaOH x 3.50 L = 1.53 mol NaOH

(2) ? mol NaOH_{excess} = 1.71 L HCl soln x $\dfrac{0.350 \text{ mol HCl}}{1 \text{ L HCl}}$ x $\dfrac{1 \text{ mol NaOH}}{1 \text{ mol HCl}}$ = 0.598 mol NaOH

(3) ? mol $\text{NaOH}_{reacted}$ = 1.53 mol NaOH - 0.60 mol NaOH = 0.93 mol NaOH

(4) ? mol CO_2 = 0.93 mol NaOH x $\dfrac{1 \text{ mol } CO_2}{2 \text{ mol NaOH}}$ = 0.46 mol CO_2

(5) ? mol O_2 = 0.46 mol CO_2 x $\dfrac{1 \text{ mol } O_2}{1 \text{ mol } CO_2}$ = 0.46 mol O_2

$? V O_2 = \dfrac{nRT}{P} = \dfrac{[(0.46 \text{ mol } O_2)(0.0821 \text{ L·atm/mol·K})(20.°C + 273°)]}{8.6 \text{ atm}} = \textbf{1.3 L } O_2$

Balanced equation: $Mg^{2+}(aq) + SiO_2(s,dispersed) + 2HCO_3^-(aq) \rightarrow MgSiO_3(s) + 2CO_2(g) + H_2O(\ell)$

Plan: $L\ CO_2 \overset{(1)}{\Rightarrow} mol\ CO_2 \overset{(2)}{\Rightarrow} mol\ MgSiO_3 \overset{(3)}{\Rightarrow} g\ MgSiO_3$

(1) $?\ mol\ CO_2 = n = \dfrac{PV}{RT} = \dfrac{[(775/760.)\ atm](100.\ L)}{(0.0821\ L \cdot atm/mol \cdot K)(30.°C + 273°)} = 4.10\ mol\ CO_2$

(2) $?\ mol\ MgSiO_3 = 4.10\ mol\ CO_2 \times (1\ mol\ MgSiO_3/2\ mol\ CO_2) = 2.05\ mol\ MgSiO_3$

(3) $?\ g\ MgSiO_3 = 2.05\ mol\ MgSiO_3 \times 100.4\ g/mol =$ **206 g MgSiO$_3$**

12-126. *Refer to Section 12-12.*

(a) $?\ molarity\ SO_2\ (mol/L) = \dfrac{0.135\ mg\ SO_2}{1\ L} \times \dfrac{1\ g}{1000\ mg} \times \dfrac{1\ mol\ SO_2}{64.1\ g\ SO_2} =$ **2.11 x 10^{-6} M SO$_2$**

(b) (1) We must first determine the average FW of an air molecule. From Table 12-2, using the average composition of air, we can calculate:

FW_{air} = weighted average of the major gases' formula weights

= (fractional composition of N_2)(FW N_2) + (fractional composition of O_2)(FW O_2)
+ (fractional composition of Ar)(FW Ar) + (fractional composition of CO_2)(FW CO_2)

= (0.7809)(28.01 g/mol) + (0.2094)(32.00 g/mol) + (0.0093)(39.95 g/mol) + (0.0003)(44.0 g/mol)

= 29.0 g/mol   (this is close to 29.2 g/mol given in Exercise 12-41)

(2) Using the density of the air, we can determine how many moles of air are present in 1 liter of air

$?\ mol/L\ of\ air = \dfrac{1\ mol\ air}{29.0\ g\ air} \times \dfrac{1.29\ g\ air}{1\ L\ air} = 0.0445$ mol of air molecules per liter

(3) We can now calculate the mole fraction of SO_2 in one liter of air:

$?\ mole\ fraction\ SO_2 = \dfrac{mol\ SO_2}{mol\ air + mol\ SO_2} = \dfrac{2.11 \times 10^{-6}\ mol\ SO_2}{0.0445\ mol\ air + 2.11 \times 10^{-6}\ mol\ SO_2} =$ **4.74 x 10^{-5}**

13 Liquids and Solids

13-2. *Refer to Sections 13-2 and the Key Terms for Chapter 13.*

Hydrogen bonding is an especially strong dipole-dipole interaction between molecules in which one contains H in a highly polarized bond and the other contains a lone pair of electrons. The energy of a hydrogen bond is 4 - 5 times larger than a normal dipole-dipole interaction and roughly 10% of a covalent bond. It only occurs in systems where a hydrogen atom is directly bonded to a small, highly electronegative atom, such as N, O or F.

13-4. *Refer to Section 13-2 and Example 13-1.*

Permanent dipole-dipole forces can be found acting between the polar molecules of (c) NO and (d) SeF_4.

13-6. *Refer to Section 13-2, Example 13-1, and Exercise 13-4.*

Dispersion forces are the only important intermolecular forces of attraction operating between the nonpolar molecules of (a) molecular $AlBr_3$ and (b) PCl_5.

13-8. *Refer to Section 13-2, Example 13-1, and Solution to Exercise 13-12.*

The substances exhibiting strong hydrogen bonding in the liquid and solid states are (a) CH_3OH, (d) $(CH_3)_2NH$ and (e) CH_3NH_2.

13-10. *Refer to Section 13-2, Table 13-3 and Example 13-1.*

(a) The physical properties of ethyl alcohol (ethanol), $C_2H_6O \equiv CH_3CH_2OH$, are influenced mainly by hydrogen bonding since there is an H atom directly bonded to an O atom, but are also affected by dispersion forces like any other molecule.

(b) Phosphene, PH_3, is a polar molecule. Refer to Exercise 13-14 for its structure. The intermolecular forces existing between the molecules are dispersion forces and dipole-dipole forces. One commonly says that dipole-dipole forces are more important. However, when one looks at Table 13-3, since PH_3 is larger than NH_3 and has no H-bonding, one can deduce that dispersion forces are more important than dipole-dipole interactions.

(c) Sulfur hexafluoride, SF_6, is a nonpolar molecule and therefore has only dispersion forces acting between its molecules.

13-12. *Refer to Section 13-2.*

Hydrogen bonding, usually occurring between molecules having an H atom directly bonded to a F, O, or N atom, is very strong compared with other dipole-dipole interactions. H bonding results from the attractions between the $\delta+$ atoms of one molecule (the H atoms) and the $\delta-$ atoms (usually F, O, and N atoms) of another molecule. The small sizes of the F, O, and N atoms, combined with their high electronegativities, concentrate the electrons in the molecule around the $\delta-$ atoms. This causes the H atom to behave somewhat like a bare proton. There is then a very strong attraction between the $\delta+$ H atom and a lone pair of electrons on a F, O, or N atom on another molecule.

The normal melting points and boiling points generally increase as the intermolecular forces between the molecules in the compounds increase.

silane, SiH$_4$

$$\begin{array}{c} H \\ H\!:\!\overset{\displaystyle H}{\underset{\displaystyle \cdot\cdot}{Si}}\!:\!H \\ \ddot{H} \end{array}$$

SiH$_4$ is a nonpolar covalent molecule and has only dispersion forces acting between the molecules. PH$_3$ is a polar covalent molecule with dispersion forces and dipole-dipole interactions in operation. Since both molecules are about the same size as evidenced by having similar molecular weights, their dispersion forces are about the same. Therefore, the melting and boiling points of PH$_3$ (−133°C and −88°C) are predicted to be higher than those of SiH$_4$ (−185°C and −112°C) and they are.

phosphine, PH$_3$

$$\begin{array}{c} H\!:\!\overset{\cdot\cdot}{P}\!:\!H \\ \ddot{H} \end{array}$$

(a)   ammonia, NH$_3$

$$\begin{array}{c} H\text{-}\overset{\cdot\cdot}{N}\text{-}H \\ | \\ H \end{array}$$

NH$_3$ has hydrogen bonds operating between its molecules, but not PH$_3$. NH$_3$ contains a hydrogen atom directly bonded to the small, highly electronegative atom, N.

phosphine, PH$_3$

$$\begin{array}{c} H\text{-}\overset{\cdot\cdot}{P}\text{-}H \\ | \\ H \end{array}$$

(b)   ethylene, C$_2$H$_4$

$$\begin{array}{cc} H & H \\ \!\!\diagdown & \diagup\!\! \\ C=C \\ \!\!\diagup & \diagdown\!\! \\ H & H \end{array}$$

Hydrazine, N$_2$H$_4$, has hydrogen bonding, but not ethylene, C$_2$H$_4$, since N$_2$H$_4$ contains the small highly electronegative element, N, which is directly bonded to hydrogen atoms and has 2 lone pairs of electrons.

hydrazine, N$_2$H$_4$

$$\begin{array}{c} H\text{-}\overset{\cdot\cdot}{N}\text{-}\overset{\cdot\cdot}{N}\text{-}H \\ | \quad | \\ H \ H \end{array}$$

(c)   hydrogen fluoride, HF

$$H\text{-}\overset{\cdot\cdot}{\underset{\cdot\cdot}{F}}\!:$$

Hydrogen fluoride, HF, has hydrogen bonding, but not hydrogen chloride, HCl, since in HF, the hydrogen atom is directly bonded to the small highly electronegative element, F.

hydrogen chloride, HCl

$$H\text{-}\overset{\cdot\cdot}{\underset{\cdot\cdot}{Cl}}\!:$$

The difference in appearance of the water surface between a glass and a plastic 10-mL graduated cylinder is due to the comparative attractiveness water molecules have for themselves (cohesive forces) and for the solid surfaces (adhesive forces). In the case of the glass cylinder, polar water molecules are strongly attracted to the polar glass, resulting in stronger adhesive forces than cohesive forces. In the case of the plastic cylinder, the adhesive and cohesive forces are comparable, resulting in the flat meniscus.

Acetic acid dimer
     (with 2 hydrogen bonds)

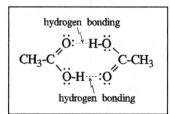

13-22. *Refer to Section 13-2 and Figure 13-5.*

(a) Ne, Ar, and Kr are nonpolar noble gases. Their boiling points are determined solely by dispersion forces, which in turn are dependent on an atom's size and polarizability. In order of increasing boiling point (i.e., increasing size),

$$Ne (-246°C) < Ar (-186°C) < Kr (-152°C)$$

(b) All three compounds, NH_3, H_2O and HF exhibit hydrogen bonding. Figure 13-5 gives the order of increasing boiling points as:

$$NH_3 (-33°C) < HF (20°C) < H_2O (100°C)$$

Since the electronegativities of the elements follow the order, F > O > N, the charge separation for the bonds in these three molecules should follow the order, F-H > O-H > N-H. The same order also should be followed when *each* of these hydrogen atoms forms hydrogen bonding with a lone pair of electrons on a neighboring molecule.

The reason why H_2O has a higher boiling point than HF is because there is a larger number of H-bonds in the H_2O system. The limiting reactant for the formation of H-bonds in the NH_3 system is the number of lone pairs of electrons, while that in the HF system is the number of H atoms. In the H_2O system, all of the H atoms and lone pairs can participate in H-bond formation. In either the NH_3 or the HF system, 1 mol of the molecules could only give a maximum of 1 mole of H-bonds. But, in 1 mole of H_2O, a maximum of 2 moles of H-bonds could be obtained. This explains why the *total* H-bonding force is much higher in H_2O than in HF, as shown by the higher boiling point.

13-24. *Refer to Section 13-8.*

(a) The normal boiling point is the temperature at which the vapor pressure of a liquid is exactly equal to one atmosphere (760 torr) pressure.

(b) When the boiling point of a liquid is measured, the atmospheric pressure over the liquid must be specified since the boiling point of a liquid is the temperature at which its vapor pressure is exactly equal to the applied pressure. As the atmospheric pressure decreases, so does the boiling point of a liquid.

13-26. *Refer to Section 13-6 and 13-8.*

Evaporation is the process by which molecules at the *surface* of the liquid escape and go into the gas phase at a temperature below its boiling point. During boiling, bubbles of vapor begin to form beneath the surface in the *bulk* of the liquid, rise to the surface and burst, releasing vapor into the air. At the boiling point, the vapor pressure equals the external pressure.

As temperature increases, the average kinetic energy of the molecules increase. A greater fraction of molecules then has the necessary energy to more readily overcome the attractions of neighboring molecules to escape into the vapor phase, so the rate of evaporation increases.

13-28. *Refer to Section 13-4 and Figure 13-8.*

Surface tension is a measure of the inward intermolecular forces of attraction among liquid particles that must be overcome to expand the surface area. A measurable tension in the surface of the liquid is caused by an imbalance set up between the intermolecular forces operating at the surface and those operating within the liquid. As the temperature increases, the increased kinetic energy and greater movement of the particles in the liquid tend to counteract the intermolecular forces, resulting in the lowering of surface tension.

There are similarities between the intermolecular attractions used to describe on a molecular level (1) viscosity, (2) surface tension, (3) the rate of evaporation and resulting vapor pressure of a liquid. For compounds in the liquid phase that have *strong* intermolecular forces of attraction operating between its molecules:

(1) the molecules cannot slide easily past each other, and the liquid has a *high* viscosity;

(2) the molecules at the surface of a liquid have an extra strong attraction for each other, and the liquid has a *high* surface tension, and

(3) the molecules are attracted strongly to each other in the liquid phase and do not evaporate as easily, effecting a *low* rate of evaporation and a *low* vapor pressure.

The weaker the intermolecular forces between molecules in the liquid phase, the higher is the vapor pressure. Generally, hydrogen bonding and the cumulative dispersion forces in larger molecules are the most significant factors (see Table 13-3). In order of decreasing vapor pressure,

(a) $BiCl_3 > BiBr_3$ $BiCl_3$ is smaller than $BiBr_3$, has weaker dispersion forces and thus has a higher vapor pressure at the same temperature.

(b) $CO > CO_2$ CO is smaller, has weaker dispersion forces and a higher vapor pressure than CO_2. Note from Table 13-3 that even though CO is polar, its dipole moment is extremely low and its dipole-dipole interaction energy is nearly zero.

(c) $N_2 > NO$ Both are small molecules of comparable size, and therefore have comparable dispersion forces. NO however is polar and has permanent dipole-dipole interactions, whereas N_2 is nonpolar and has only dispersion forces. Therefore N_2 has the higher vapor pressure.

(d) $HCOOCH_3 > CH_3COOH$ Methyl formate ($HCOOCH_3$) and acetic acid (CH_3COOH) have the same molecular formula, $C_2H_4O_2$, and are about the same size, so both have similar dispersion forces. CH_3COOH can form hydrogen bonds, whereas $HCOOCH_3$ cannot. Therefore $HCOOCH_3$ has weaker intermolecular interactions and has a higher vapor pressure than CH_3COOH.

To confirm the above reasoning, use boiling points given below to indicate strengths of intermolecular forces. Remember, the lower the boiling point is, the weaker the intermolecular forces present in the liquid, and the higher the vapor pressure at a specific temperature.

Compound	B.P. (°C)	Compound	B.P. (°C)
$BiCl_3$	447	N_2	−195.8
$BiBr_3$	453	NO	−151.8
CO	−191.5	CH_3COOH	117.9
CO_2	−78.5 (sublimes)	$HCOOCH_3$	31.5

The order of increasing boiling points corresponds with the order of increasing temperatures when their vapor pressures are constant, e.g. 100 torr:

$$butane \ < \ diethyl \ ether \ < \ 1\text{-butanol}$$

13-36. *Refer to Section 13-8.*

Vapor pressure curve for Cl_2O_7

The boiling point of Cl_2O_7 at 125 torr from the graph is about 34°C.

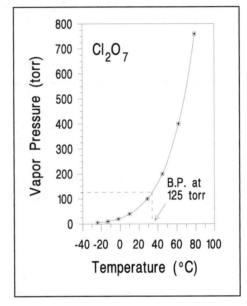

t (°C)	VP (torr)
−24	5
−13	10
−2	20
10	40
29	100
45	200
62	400
79	760

13-38. *Refer to Sections 13-7 and 12-11.*

Using Dalton's Law of Partial Pressures,

$$P_{total} = P_{air} + P_{water \ vapor}$$

Therefore, $P_{air} = P_{total} - P_{water \ vapor}$

$$= 633.5 \ torr - 289.1 \ torr$$

$$= \textbf{344.4 torr}$$

13-40. *Refer to Section 13-9.*

(a) at 37°C, the heat of vaporization of water = 2.41 kJ/g

at 37°C, ΔH°_{vap} (kJ/mol) $= 2.41 \ \dfrac{kJ}{g} \ \text{x} \ \dfrac{18.0 \ g}{1 \ mol} = \textbf{43.4 kJ/mol}$

(b) The heat of vaporization at a certain temperature is the amount of heat required to change 1 gram of liquid to 1 gram of vapor at that temperature. The heat of vaporization is greater at 37°C than at 100°C because the average kinetic energy of the molecules is lower at the lower temperature. Therefore, more energy must be added per unit mass of the liquid to break the intermolecular forces between the molecules at the lower temperature.

Vapor pressure curve for $C_2H_4F_2$

The boiling point of $C_2H_4F_2$ at 200. torr from the graph is about −1°C.

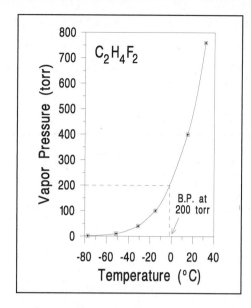

t (°C)	VP (torr)
−77.2	1
−51.2	10
−31.1	40
−15.0	100
14.8	400
31.7	760

13-44. *Refer to Section 13-6 and the Key Terms for Chapter 13.*

A dynamic equilibrium is a situation in which two (or more) processes occur at the same rate so that no net change occurs. This is the kind of equilibrium that is established between two physical states of matter, e.g., between a liquid and its vapor, in which the rate of evaporation is equal to the rate of condensation in a closed container:

$$\text{liquid} \rightleftarrows \text{vapor.}$$

On a molecular level at equilibrium, molecules of liquid are escaping into the vapor and vapor molecules are condensing into the liquid. This is not a static situation. No net change occurs because the rates are the same. This can be demonstrated by observing a glass of water. In time, the water will evaporate and the water volume will decrease. By covering the glass with a glass or clear plastic plate, droplets of condensed water will be observed on the underside of the plate and the water volume will remain the same.

13-46. *Refer to Section 13-11.*

Plan: (1) Calculate the amount of heat (in J) required to melt the zinc sample.

(2) Determine $\Delta H^{\circ}_{fusion}$ at its melting point.

(1) ? heat absorbed = Rate (J/s) × time (s) = 9.84 J/s × (3.60 min × 60 s/min) = 2.13 × 10³ J

(2) $\Delta H^{\circ}_{fusion}\left(\dfrac{J}{mol}\right) = \dfrac{\text{heat required to melt Zn sample at m.p.}}{\text{mol Zn}} = \dfrac{2.13 \times 10^3 \text{ J}}{21.8 \text{ g}/65.4 \text{ g/mol}} = \mathbf{6.38 \times 10^3 \text{ J/mol}}$

13-48. *Refer to Sections 13-9 and 13-11, and Example 13-2.*

This exercise requires 3 separate calculations:

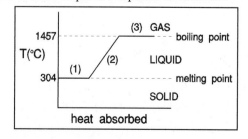

$$\underset{304°C}{\text{solid}} \overset{(1)}{\Rightarrow} \underset{304°C}{\text{liquid}} \overset{(2)}{\Rightarrow} \underset{1457°C}{\text{liquid}} \overset{(3)}{\Rightarrow} \underset{1457°C}{\text{gas}}$$

(1) heat required = mass x heat of fusion = (204 g)(21 J/g) = 4.3 x 10³ J

(2) heat required = mass x specific heat(ℓ) x Δt = (204 g)(0.13 J/g·°C)(1457°C - 304°C) = 3.1 x 10⁴ J

(3) heat required = mass x heat of vaporization = (204 g)(795 J/g) = 1.62 x 10⁵ J

Therefore, the total heat required = (1) + (2) + (3) = **1.97 x 10⁵ J**

Note: Even though the individual calculations were each rounded to the correct number of significant figures, the final answer was obtained by rounding only at the very end.

13-50. *Refer to Sections 13-9 and: 13-11.*

This exercise involves 5 separate calculations:

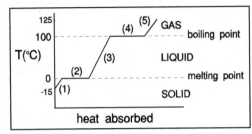

$$\begin{array}{ccccccccc}
 & (1) & & (2) & & (3) & & (4) & & (5) \\
\text{ice} & \Rightarrow & \text{ice} & \Rightarrow & \text{water} & \Rightarrow & \text{water} & \Rightarrow & \text{steam} & \Rightarrow & \text{steam} \\
-15.0°C & & 0.0°C & & 0.0°C & & 100.0°C & & 100.0°C & & 125.0°C
\end{array}$$

(1) heat required = mass x specific heat (s) x Δt = (70.0 g)(2.09 J/g·°C)(0.0°C - (-15.0°C)) = 2.19 x 10³ J

(2) heat required = mass x heat of fusion = (70.0 g)(334 J/g) = 2.34 x 10⁴ J

(3) heat required = mass x specific heat (ℓ) x Δt = (70.0 g)(4.184 J/g·°C)(100.0°C - 0.0°C) = 2.93 x 10⁴ J

(4) heat required = mass x heat of vaporization = (70.0 g)(2260 J/g) = 1.58 x 10⁵ J

(5) heat required = mass x specific heat (g) x Δt = (70.0 g)(2.03 J/g·°C)(125.0°C - 100.0°C) = 3.55 x 10³ J

Therefore, the total heat required = (1) + (2) + (3) + (4) + (5) = **2.17 x 10⁵ J**

Note: Even though the individual calculations were each rounded to the correct number of significant figures, the final answer was obtained by rounding only at the very end.

13-52. *Refer to Section 1-13 and the Solution to Exercise 1-60.*

Recall: When two substances are brought into contact with each other, the heat lost by one substance is equal in absolute value to the heat gained by the other.

$$|\text{heat gained by cold water}| = |\text{heat lost by hot water}|$$
$$|\text{mass x specific heat}(\ell) \text{ x } \Delta t|_{\text{cold}} = |\text{mass x specific heat}(\ell) \text{ x } \Delta t|_{\text{hot}}$$
$$|\text{mass x } \Delta t|_{\text{cold}} = |\text{mass x } \Delta t|_{\text{hot}}$$
$$(550. \text{ g})(t_{\text{final}} - 30.0°C) = (275 \text{ g})(100.°C - t_{\text{final}})$$
$$(550. \text{ x } t_{\text{final}}) - 16500 = 27500 - (275 \text{ x } t_{\text{final}})$$
$$825 \text{ x } t_{\text{final}} = 44000 \quad (3 \text{ significant figures})$$
$$t_{\text{final}} = \textbf{53.3°C}$$

13-54. *Refer to Section 13-9.*

Plan: (1) Determine if the final phase will be liquid or gas.

(2) Calculate the final temperature, t_{final} (°C).

(1) The amount of heat required to change the liquid water to steam at 100°C is

heat required = |mass x specific heat (ℓ) x Δt|$_{\text{water}}$ + |mass x heat of vaporization|
= (180. g)(4.184 J/g·°C)(100.°C - 0.°C) + (180. g)(2.26 x 10³ J/g)
= 4.82 x 10⁵ J

The amount of heat released when the steam changes to liquid water at 100°C is

heat released = |mass × specific heat (g) × Δt|$_{steam}$ + |mass × heat of vaporization|

$$= (18.0 \text{ g})(2.03 \text{ J/g·°C})(110.°C - 100.°C) + (18.0 \text{ g})(2.26 \times 10^3 \text{ J/g})$$

$$= 4.10 \times 10^4 \text{ J}$$

The heat required to convert water to steam is greater than the heat released when the steam condenses to water. Therefore, when the two systems are mixed, the liquid water will cause all of the steam to condense and the final temperature will be between 0°C and 100°C.

(2) |amount of heat gained by water| = |amount of heat lost by steam|

|mass × specific heat (ℓ) × Δt| = |mass × specific heat (g) × Δt| + |mass × heat of vaporization|

+ |mass × specific heat (ℓ) × Δt|$_{water\ from\ steam}$

$(180. \text{ g})(4.184 \text{ J/g·°C})(t_{final} - 0.°C)$ = $(18.0 \text{ g})(2.03 \text{ J/g·°C})(110.°C - 100.°C) + (18.0 \text{ g})(2.26 \times 10^3 \text{ J/g})$

$+ (18.0 \text{ g})(4.184 \text{ J/g·°C})(100.°C - t_{final})$

$(753 \times t_{final}) - 0$ = $(370 \text{ J}) + (4.07 \times 10^4 \text{ J}) + (7.53 \times 10^3 \text{ J}) - (75.3 \times t_{final})$

$828 \times t_{final}$ = 4.86×10^4

t_{final} = **58.7°C**

13-56.   *Refer to Sections 13-9 and 13-11.*

(a) This part of the exercise requires 2 separate calculations:

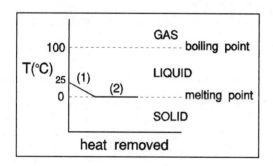

$$\begin{array}{ccccc} & (1) & & (2) & \\ \text{water} & \Rightarrow & \text{water} & \Rightarrow & \text{ice} \\ 25.0°C & & 0.0°C & & 0.0°C \end{array}$$

(1) heat removed = mass × specific heat (ℓ) × Δt = $(14.0 \text{ g})(4.184 \text{ J/g·°C})(25.0°C - 0.0°C) = 1.46 \times 10^3 \text{ J}$

(2) heat removed = mass × heat of fusion = $(14.0 \text{ g})(334 \text{ J/g}) = 4.68 \times 10^3 \text{ J}$

Therefore, the total heat removed = (1) + (2) = **6.14 × 10³ J**

(b) The mass of water at 100.0°C that could be cooled to 23.5°C by removing 6.14 × 10³ J of heat from the water sample is:

heat removed from water in (a) = mass$_{water}$ × specific heat (ℓ) × Δt

$6.14 \times 10^3 \text{ J}$ = $(mass_{water})(4.184 \text{ J/g·°C})(100.0°C - 23.5°C)$

= $(mass_{water})(3.20 \times 10^2 \text{ J/g})$

mass$_{water}$ = **19.2 g**

13-58. *Refer to Section 13-9 and Example 13-4.*

We must use the Clausius-Clapeyron equation to find the vapor pressure at 80.00°C:

$$\ln\left(\frac{P_2}{P_1}\right) = \frac{\Delta H_{vap}}{R}\left(\frac{1}{T_1} - \frac{1}{T_2}\right)$$

where
ΔH_{vap} = molar heat of vaporization (J/mol)
P_1 = 760.00 torr
P_2 = vapor pressure at 80.00°C
R = 8.314 J/mol·K
T_1 = normal boiling point of liquid (K)
T_2 = 80.00°C + 273.15° = 353.15 K

(1) for water, H_2O: ΔH_{vap} = 40,656 J/mol
T_1 = 100.00°C + 273.15° = 373.15 K

Substituting,

$$\ln\left(\frac{P_2}{760.00\text{ torr}}\right) = \frac{40656\text{ J/mol}}{8.314\text{ J/mol·K}}\left(\frac{1}{373.15\text{ K}} - \frac{1}{353.15\text{ K}}\right)$$
$$= (4890.)(0.0026799 - 0.0028317)$$
$$= -0.7424$$

$$\frac{P_2}{760.00\text{ torr}} = 0.4760$$

$$P_2 = \mathbf{361.8\ torr}$$

(2) for heavy water, D_2O: ΔH_{vap} = 41,606 J/mol
T_1 = 101.41°C + 273.15° = 374.56 K

Substituting,

$$\ln\left(\frac{P_2}{760.00\text{ torr}}\right) = \frac{41606\text{ J/mol}}{8.314\text{ J/mol·K}}\left(\frac{1}{374.56\text{ K}} - \frac{1}{353.15\text{ K}}\right)$$
$$= (5004)(0.0026698 - 0.0028317)$$
$$= -0.8101$$

$$\frac{P_2}{760.00\text{ torr}} = 0.4448$$

$$P_2 = \mathbf{338.0\ torr}$$

13-60. *Refer to Section 13-9.*

(a) The Clausius-Clapeyron equation is

$$\ln\left(\frac{P_2}{P_1}\right) = \frac{\Delta H_{vap}}{R}\left(\frac{1}{T_1} - \frac{1}{T_2}\right)$$

where
ΔH_{vap} = molar heat of vaporization (J/mol)
P_1 = vapor pressure at temperature, T_1 (K)
P_2 = vapor pressure at temperature, T_2 (K)
R = 8.314 J/mol·K

Expanding the equation we obtain:

$$\ln P_2 - \ln P_1 = \frac{\Delta H_{vap}}{R}\left(\frac{1}{T_1}\right) - \frac{\Delta H_{vap}}{R}\left(\frac{1}{T_2}\right)$$

$$\ln P_2 = \frac{-\Delta H_{vap}}{RT_2} + \left[\frac{\Delta H_{vap}}{R}\left(\frac{1}{T_1}\right) + \ln P_1\right]$$

If we let P_1 be a known vapor pressure of the substance at a particular temperature, T_1, and simplify by letting B stand for all the terms in the square brackets, we have

$$\ln P = \frac{-\Delta H_{vap}}{RT} + \text{B}$$

When $\ln P$ is plotted against $1/T$, we obtain a straight line with a slope of $-\Delta H_{vap}/R$ and y-intercept of B.

(b) Vapor pressure data for ethyl acetate, $CH_3COOC_2H_5$:

t (°C)	T (K)	$1/T$ (K^{-1})	P (torr)	$\ln P$
-43.4	229.8	4.35×10^{-3}	1	0.0
-23.5	249.7	4.00×10^{-3}	5	1.6
-13.5	259.7	3.85×10^{-3}	10.	2.30
-3.0	270.2	3.70×10^{-3}	20.	3.00
9.1	282.3	3.54×10^{-3}	40.	3.69
16.6	289.8	3.45×10^{-3}	60.	4.09
27.0	300.2	3.33×10^{-3}	100.	4.61
42.0	315.2	3.17×10^{-3}	200.	5.30
59.3	332.5	3.01×10^{-3}	400.	5.99
?	?	?	760.	6.63

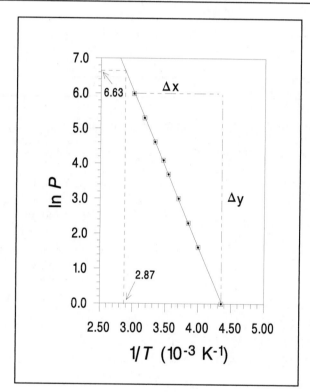

(c) The molar heat of vaporization of ethyl acetate is determined from the slope of the line. A linear regression fit to the data gives a slope of -4.471 with a coefficient of determination, $r^2 = 0.99958$, which indicates a very good fit. If you do not have access to a line-fitting program, the slope can be estimated by using two data points that are far apart as shown in the graph. This will work because the line fits the data well.

$$\text{slope} = \frac{-\Delta H_{vap}}{R} = \frac{\Delta y}{\Delta x} = \frac{\Delta \ln P}{\Delta 1/T} = \frac{(0.00 - 5.99)}{(4.35 \times 10^{-3} - 3.01 \times 10^{-3})} = -4.47 \times 10^3 \text{ K}$$

Therefore, $\Delta H_{vap} = -\text{slope} \times R = -(-4.47 \times 10^3 \text{ K})(8.314 \text{ J/mol·K}) = \textbf{3.72} \times \textbf{10}^4 \textbf{ J/mol}$

(d) The normal boiling point is the temperature at which the vapor pressure of ethyl acetate is 760 torr. From the graph, when $\ln P = \ln 760 = 6.63$,

$1/T = 2.87 \times 10^{-3}$ K^{-1} and solving, $T = \textbf{3.48} \times \textbf{10}^2 \textbf{ K}$ or $\textbf{75°C}$

13-62. *Refer to Section 13-9 and Example 13-4.*

Use the Clausius-Clapeyron equation to solve for the molar heat of vaporization of isopropyl alcohol, ΔH_{vap}.

We know: $P_1 = 100.$ torr $\quad\quad T_1 = 39.5°C + 273.2° = 312.7$ K

$\quad\quad\quad\quad\quad P_2 = 400.$ torr $\quad\quad T_2 = 67.8°C + 273.2° = 341.0$ K

$$\ln\left(\frac{P_2}{P_1}\right) = \frac{\Delta H_{vap}}{R}\left(\frac{1}{T_1} - \frac{1}{T_2}\right)$$

where $\quad \Delta H_{vap}$ = molar heat of vaporization (J/mol)
$\quad\quad\quad\quad P_1$ = vapor pressure at temperature, T_1 (K)
$\quad\quad\quad\quad P_2$ = vapor pressure at temperature, T_2 (K)
$\quad\quad\quad\quad R = 8.314$ J/mol·K

Plugging in, we have

$$\ln\left(\frac{400.\text{ torr}}{100.\text{ torr}}\right) = \frac{\Delta H_{vap}}{8.314\text{ J/mol·K}}\left(\frac{1}{312.7\text{ K}} - \frac{1}{341.0\text{ K}}\right)$$

$$1.386 = \frac{\Delta H_{vap}}{8.314\text{ J/mol·K}}(0.000265)$$

$$\Delta H_{vap} = \mathbf{4.34 \times 10^4 \text{ J/mol} \ \ or\ \ 43.4\ kJ/mol}$$

13-64. *Refer to Section 13-9 and Example 13-4.*

From Appendix E, the molar enthalpy of vaporization of mercury at the normal boiling point is 58.6 kJ/mol.

Using the Clausius-Clapeyron equation to find the vapor pressure at 25°C, we have

$$\ln\left(\frac{P_2}{P_1}\right) = \frac{\Delta H_{vap}}{R}\left(\frac{1}{T_1} - \frac{1}{T_2}\right)$$

where $\quad \Delta H_{vap}$ = molar heat of vaporization (J/mol)
$\quad\quad\quad\quad\quad\quad\quad = 58600$ J/mol
$\quad\quad\quad\quad P_1 = 760.$ torr
$\quad\quad\quad\quad P_2$ = vapor pressure at 25°C
$\quad\quad\quad\quad R = 8.314$ J/mol·K
$\quad\quad\quad\quad T_1 = 357°C + 273° = 630.$ K
$\quad\quad\quad\quad T_2 = 25°C + 273° = 298$ K

Substituting,

$$\ln\left(\frac{P_2}{760.\text{ torr}}\right) = \frac{58600\text{ J/mol}}{8.314\text{ J/mol·K}}\left(\frac{1}{630.\text{ K}} - \frac{1}{298\text{ K}}\right)$$

$$= -12.5$$

$$\frac{P_2}{760.\text{ torr}} = 3.86 \times 10^{-6}$$

$$P_2 = \mathbf{2.93 \times 10^{-3}\ torr}$$

13-66. *Refer to Section 13-13 and the Key Terms for Chapter 13.*

The critical point is the combination of critical temperature and critical pressure of a substance. The critical temperature is the temperature above which a gas cannot be liquefied, i.e., above this temperature, one cannot distinguish between a liquid and a gas. The critical pressure is the pressure required to liquefy a gas at its critical temperature. If the temperature is less than the critical temperature, the substance can be either a gas, liquid or solid, depending on the pressure.

(a) This point lies on the sublimation curve where the solid and gas phases are in equilibrium. So, both the solid and gaseous phases are present.

(b) This point is called the triple point where all three phases are in equilibrium with each other. So, the solid, liquid and gaseous phases are all present.

13-70. *Refer to Section 13-13 and Figure 13-17b.*

The melting point of carbon dioxide increases with increasing pressure, since the solid-liquid equilibrium line on its phase diagram slopes up and to the right. If the pressure on a sample of liquid carbon dioxide is increased at constant temperature, causing the molecules to get closer together, the liquid will solidify. This indicates that solid carbon dioxide has a higher density than the liquid phase. This is true for most substances. The notable exception is water.

13-72. *Refer to Section 13-13.*

(a) Phase diagram for butane: (not to scale)

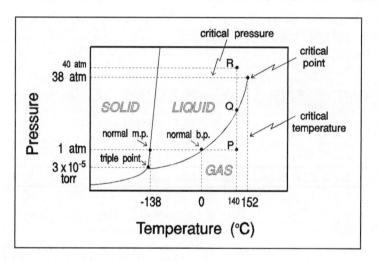

(b) As butane is compressed from 1 atm (point P) to 40 atm (point R) at 140°C, butane is converted from a gas to a liquid. Both phases are present simultaneously where the vapor pressure curve intersects the vertical isothermal line: $T = 140°C$ indicated by point Q. At point Q, both phases are in equilibrium.

(c) Since 200°C is a temperature greater than the critical temperature of 152°C, there is no pressure at which two phases exist. At all pressures the liquid phase cannot be distinguished from the gas phase.

13-74. *Refer to Section 13-13 and the phase diagram for sulfur.*

(a) solid (monoclinic)   (b) solid (rhombic)   (c) solid (rhombic)

(d) solid (rhombic)   (e) vapor   (f) liquid

13-76. *Refer to Section 13-13.*

Ice, i.e., solid water, floats in liquid water because the solid state is less dense than the liquid state. However, like most other substances, solid mercury is more dense than liquid mercury and therefore, solid mercury sinks when placed in liquid mercury.

MoF_6 molecular solid

BN   covalent (network) solid

Se_8 molecular solid

Pt   metallic solid

RbI   ionic solid

(a) SO_2F molecular solid

(b) MgF_2 ionic solid

(c) W   metallic solid

(d) Pb   metallic solid

(e) PF_5 molecular solid

Melting points of ionic compounds increase with increasing ion-ion interactions which are functions of d, the distance between the ions, and q, the charge on the ions:

$$F \propto \frac{q^+q^-}{d^2}$$

Due to the differences in the number of charges on the cations (Na^+, Mg^{2+} and Al^{3+}), the following order of increasing melting points is predicted:

$$NaF < MgF_2 < AlF_3$$

Simple cubic lattice:
   (Example: CsCl)

The eight corners of a cubic unit cell are occupied by one kind of atom, while the center of the unit cell is occupied by another kind of atom. Note that this is *not* the same as a body-centered cubic arrangement, because the atom at the center of the cell is *not* the same as those at the corners.

Body-centered cubic lattice (bcc):
   (Example: Na)

All eight corners *and* the point at the center of a cubic unit cell are occupied by a single kind of atom.

Face-centered cubic lattice (fcc):
   (Example: Ni)

The eight corners of a cubic unit cell as well as the central points on each of the six faces of the cube are occupied by the same kind of atom.

CsCl

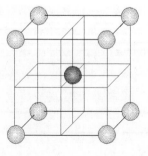

Unit cell 1: Cs^+ ◯ Cl^- ⬤
Unit cell 2: Cs^+ ⬤ Cl^- ◯

Na

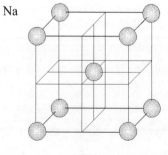

Na ◯

Ni

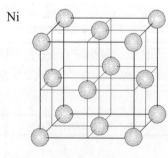

Ni ◯

(1) For CsCl (simple cubic): 1 Cs$^+$ cation in center **x** 1 = 1 cation in the unit cell

8 Cl$^-$ anions in corners **x** 1/8 = 1 anion in unit cell

(2) For NaCl (face-centered cubic):
(1 Na$^+$ cation in center **x** 1) + (12 Na$^+$ cations on edges **x** 1/4) = 4 cations in the unit cell
(6 Cl$^-$ anions on faces x 1/2) + (8 Cl$^-$ anions on corners x 1/8) = 4 anions in the unit cell

(3) For ZnS (face-entered cubic):
(4 Zn^{2+} cation in center **x** 1) = 4 cations in the unit cell
(6 S^{2-} anions on faces x 1/2) + (8 S^{2-} anions on corners x 1/8) = 4 anions in the unit cell

---

13-88. *Refer to Section 13-15, and Figures 13-28 and 13-29b.*

Consider the NaCl face-centered cubic structure with a unit cell edge represented as *a*, shown here.

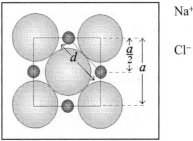
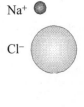

(a) The distance from Na$^+$ to its nearest neighbor is **$a/2$**.

(b) Each Na$^+$ ion has **6** equidistant nearest neighbors. They are **Cl$^-$** ions.

(c) The distance, *d*, from Na$^+$ to nearest Na$^+$ is the length of a hypotenuse of an isosceles right triangle, with sides equal to $a/2$.

$d^2 = (a/2)^2 + (a/2)^2$

$d = \sqrt{(a/2)^2 + (a/2)^2} = \sqrt{2(a/2)^2} = \sqrt{a^2/2} = \dfrac{a}{\sqrt{2}}$ or $\dfrac{a\sqrt{2}}{2}$

(d) Each Cl$^-$ ion has **6** equidistant nearest neighbors; they are **Na$^+$** ions.

13-90. *Refer to Section 13-16, Examples 13-9 and 13-10, and Exercise 13-84.*

Plan: (1) Calculate the volume of the unit cell, *V*.
(2) Calculate the mass, *m*, of Na atoms in the unit cell.
(3) Determine the density of Na:

$$\text{Density} = \frac{m_{\text{unit cell}}}{V_{\text{unit cell}}}$$

(1) let the length of the cube edge = a = 4.24 Å

$V_{\text{unit cell}} = a^3 = [4.24 \text{ Å} \times (1 \times 10^{-8} \text{ cm/Å})]^3 = 7.62 \times 10^{-23} \text{ cm}^3$

(2) A body-centered cubic unit cell contains 2 atoms in total:

in the corners: 8 Na atoms **x** 1/8 = 1 Na atom
in the center: 1 Na atom **x** 1   = 1 Na atom

? $m_{\text{unit cell}}$ = 2 Na atoms **x** $\dfrac{1 \text{ mol Na}}{6.02 \times 10^{23} \text{ atoms Na}}$ **x** $\dfrac{22.99 \text{ g Na}}{1 \text{ mol Na}}$ = 7.64 × 10^{-23} g

(3) Density of Na = $\dfrac{m}{V} = \dfrac{7.64 \times 10^{-23} \text{ g Na/unit cell}}{7.62 \times 10^{-23} \text{ cm}^3/\text{unit cell}}$ = **1.00 g/cm³**

The density of sodium as given by the *Handbook of Chemistry and Physics* is 0.97 g/cm³.

13-92. *Refer to Section 13-16, and Examples 13-9 and 13-10.*

Plan: (1) Calculate the mass of an Ir unit cell, m.
(2) Calculate the volume of an Ir unit cell, V.
(3) Calculate the density of Ir using

$$\text{Density} = \frac{\text{mass of unit cell } (m)}{\text{volume of unit cell } (V)}.$$

(1) $m_{\text{face-centered unit cell}}$ = mass of 4 Ir atoms

$$= 4 \text{ Ir atoms} \times \frac{1 \text{ mol Ir}}{6.022 \times 10^{23} \text{ Ir atoms}} \times \frac{192.2 \text{ g}}{1 \text{ mol Ir}}$$

$$= 1.277 \times 10^{-21} \text{ g}$$

(e) $V_{\text{cubic unit cell}} = (\text{length of unit cell edge}, a)^3$

For a face-centered cubic unit cell, the diagonal of a face is the hypotenuse of a right triangle and equals $4r$ where r is the radius of an Ir atom. From the Pythagorean Theorem,

$$(4r)^2 = a^2 + a^2$$
$$(4 \times 1.36 \text{ Å})^2 = 2a^2$$
$$a \text{ (Å)} = 3.85 \text{ Å}$$
$$a \text{ (cm)} = 3.85 \times 10^{-8} \text{ cm}$$

$$V_{\text{cubic unit cell}} = (a)^3 = (3.85 \times 10^{-8} \text{ cm})^3 = 5.71 \times 10^{-23} \text{ cm}^3$$

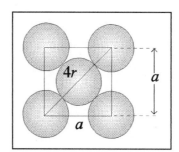

(3) Therefore,

$$\text{Density} = \frac{\text{mass of unit cell } (m)}{\text{volume of unit cell } (V)} = \frac{1.277 \times 10^{-21} \text{ g}}{5.71 \times 10^{-23} \text{ cm}^3} = \mathbf{22.4 \text{ g/cm}^3}$$

The Handbook of Chemistry and Physics (71st Ed.) gives the density of Ir as 22.421 g/cm^3.

13-94. *Refer to Section 13-16.*

(a) The cubic unit cell of diamond:

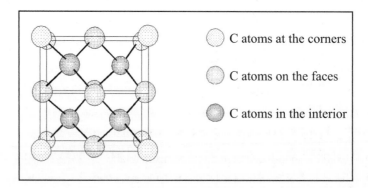

The unit cell contains 8 C atoms:
8 atoms at the corners x 1/8 = 1 atom
6 atoms on the faces x 1/2 = 3 atoms
4 atoms in the interior x 1 = 4 atoms

(b) Each C atom is at the center of a **tetrahedron** and hence has 4 nearest neighbors.

(c) Consider the lower right hand corner of diamond's unit cell:

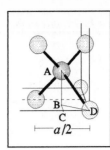

Triangle BCD is in the plane of the base of the unit cell.
Line BC = line CD = $a/4$.
So, (line BD)^2 = (line BC)^2 + (line CD)^2

$$\text{line BD} = \sqrt{(a/4)^2 + (a/4)^2} = \frac{a}{2\sqrt{2}}$$

Triangle ABD is perpendicular to the base of the unit cell. Line AD is the distance from any carbon atom to its nearest neighbor.

$$(\text{line AD})^2 = (\text{line AB})^2 + (\text{line BD})^2$$

$$\text{line AD} = \sqrt{\left(\frac{a}{4}\right)^2 + \left(\frac{a}{2\sqrt{2}}\right)^2} = \sqrt{3}\left(\frac{a}{4}\right)$$

(d) The unit cell edge is 3.567 Å. Therefore,

$$\text{C-C bond length} = \text{line AD} = \sqrt{3}(3.567 \text{ Å}/4) = \mathbf{1.545 \text{ Å}}$$

(e) $V_{\text{unit cell}} = a^3 = [3.567 \text{ Å} \times (1 \times 10^{-8} \text{ cm/Å})]^3 = 4.538 \times 10^{-23} \text{ cm}^3$

$$m_{\text{unit cell}} = 8 \text{ C atoms} \times \frac{1 \text{ mol}}{6.022 \times 10^{23} \text{ atoms}} \times \frac{12.01 \text{ g}}{1 \text{ mol}} = 1.595 \times 10^{-22} \text{ g}$$

$$\text{Density of C}_{\text{diamond}} = \frac{m}{V} = \frac{1.595 \times 10^{-22} \text{ g}}{4.538 \times 10^{-23} \text{ cm}^3} = \mathbf{3.515 \text{ g/cm}^3}$$

The density of C_{diamond} as given by the *Handbook of Chemistry and Physics* is 3.51 g/cm^3.

13-96. *Refer to Section 13-16 and Example 13-11.*

Plan: (1) Calculate the volume of the unit cell, V.
(2) Determine the mass of the unit cell, m, given the density.
(3) Knowing the number of atoms in a face-centered cube, find the moles of atoms in the unit cell.
(4) Determine the atomic weight of the unknown and its identity.

(1) let the length of the cube edge = a = 4.95 Å

$$V_{\text{unit cell}} = a^3 = [4.95 \text{ Å} \times (1 \times 10^{-8} \text{ cm/Å})]^3 = 1.21 \times 10^{-22} \text{ cm}^3$$

(2) $m_{\text{unit cell}} = D \times V_{\text{unit cell}} = (11.35 \text{ g/cm}^3)(1.21 \times 10^{-22} \text{ cm}^3) = 1.37 \times 10^{-21} \text{ g}$ since $D = \frac{m}{V}$

(3) A face-centered cubic unit cell contains 4 atoms:

in the corners   8 atoms $\times$ 1/8 = 1 atom
on the faces   6 atoms $\times$ 1/2 = 3 atoms

$$? \text{ mol element in unit cell} = 4 \text{ atoms} \times \frac{1 \text{ mol atoms}}{6.02 \times 10^{23} \text{ atoms}} = 6.64 \times 10^{-24} \text{ mol}$$

(4) $AW = \dfrac{\text{g element}}{\text{mol element}} = \dfrac{m_{\text{unit cell}}}{\text{mol}_{\text{unit cell}}} = \dfrac{1.37 \times 10^{-21} \text{ g}}{6.64 \times 10^{-24} \text{ mol}} = \mathbf{206 \text{ g/mol}}$

Therefore, the 4A element is **Pb (207.2 g/mol)**

13-98. *Refer to Section 13-14 and Figure 13-19.*

(a) Crystal diffraction studies use the x-ray region of the electromagnetic radiation.

(b) In the x-ray diffraction experiment, a monochromatic x-ray beam is attenuated by a system of slits and aimed at a crystal. The crystal is rotated to vary the angle of incidence. At certain angles which depend on the crystal's unit cell, x-rays are deflected and hit a photographic plate. After development the plate shows a set of symmetrically arranged spots. From the arrangement of the spots, the crystal structure can be determined.

(c) In order for diffraction to occur, the wavelength of the in-coming radiation must be about the same as the internuclear separations in the crystal.

13-100. *Refer to Section 13-14.*

In x-ray diffraction, the Bragg equation is used:

$$n\lambda = 2d\sin\theta$$

where
- n = 1 for the minimum diffraction angle
- λ = wavelength of Cu radiation
- θ = angle of incidence or diffraction angle, 19.98°
- d = spacing between parallel layers of Pt atoms, 2.256 Å

Solving for λ,

$$\lambda = \frac{2d\sin\theta}{n} = \frac{2 \times 2.256 \text{ Å} \times \sin 19.98°}{1} = \mathbf{1.542 \text{ Å}}$$

13-102. *Refer to Sections 13-16 and 13-17.*

When a metal is distorted (e.g., rolled into sheets or drawn into wire), new metallic bonds are formed and the environment around each atom is essentially unchanged. This can happen because the valence electrons of bonded metal atoms are only loosely associated with individual atoms, as though metal cations exist in a "cloud of electrons." In ionic solids, the lattice arrangements of cations and anions are more rigid. When an ionic solid is distorted, it is possible for cation-cation and anion-anion alignments to occur. But this will cause the solid to shatter due to electrostatic repulsions between ions of like charge.

13-104. *Refer to Section 13-17.*

According to band theory, the electrons in a substance must move within the conduction band in order to conduct electricity. The conduction band is a partially filled band or a band of vacant energy levels just higher in energy than a filled band. Metals conduct electricity because the highest energy electrons can easily move in a conduction band. For a metal, the electrical conductivity decreases as the temperature increases. The increase in temperature causes the metal ions to move more, which slows down the flow of electrons when an electric field is applied. A typical metalloid is a semiconductor. There is a small gap in energy between the highest-energy electrons and the conduction band. A semiconductor does not conduct electricity at low temperatures, but a small increase in temperature will excite some of the highest-energy electrons into the empty conduction band. So, electrical conductivity increases with increasing temperature for a metalloid.

13-106. *Refer to Sections 13-6 and 13-8.*

The first gas to evaporate would be the gas with the weakest intermolecular forces acting between the molecules. Since a lower boiling point is an indicator of weaker intermolecular forces of attraction at work, N_2 (b.p. −196°C) would be the first gas to evaporate, then Ar (b.p. −186°C), and lastly, O_2 (b.p. −183°C).

13-108. *Refer to Section 12-15 and Example 12-23.*

(1) Assuming $H_2O(g)$ obeys the ideal gas law: $PV = nRT$

$$P = \frac{nRT}{V} = \frac{(1.00\text{ mol})(0.0821\text{ L·atm/mol·K})(100.00°\text{C} + 273.15°)}{31.0\text{ L}} = \mathbf{0.988\ atm}$$

(2) Assuming $H_2O(g)$ obeys the van der Waals equation: $\left(P + \frac{n^2a}{V^2}\right)(V\text{-}nb) = nRT$

for H_2O, $a = 5.464\text{ L}^2\text{·atm/mol}^2$, $b = 0.03049\text{ L/mol}$

$$\left[P + \frac{(1.00\text{ mol})^2(5.464\text{ L}^2\text{·atm/mol}^2)}{(31.0\text{ L})^2}\right]\left[31.0\text{ L} - (1.00\text{ mol})\left(0.03049\frac{\text{L}}{\text{mol}}\right)\right]$$
$$= (1.00\text{ mol})(0.0821\text{ L·atm/mol·K})(100.00°\text{C} + 273.15°)$$

$$[P + 0.00569\text{ atm}][30.97\text{ L}] = 30.6\text{ L·atm}$$
$$P + 0.00569\text{ atm} = 0.988\text{ atm}$$
$$P = \mathbf{0.982\ atm}$$

(3) $\%\text{ difference} = \frac{P_{\text{ideal}} - P_{\text{real}}}{P_{\text{ideal}}} \times 100 = \frac{0.988 - 0.982}{0.988} \times 100 = \mathbf{0.6\%}$

Therefore, steam does not deviate significantly from ideality at these conditions.
This is expected because 100°C is the boiling point, and 1 atm is in the intermediate pressure range.

13-110. *Refer to Sections 13-2 and 13-8.*

Consider the following three molecules with formula $C_2H_2Cl_2$:

(1)

(2)

(3)

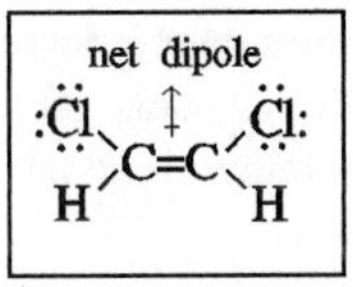

All of the compounds have dispersion forces of attraction operating between their molecules. Since the compounds are approximately the same size, their dispersion forces are about the same. Compounds (1) and (3) are polar and have permanent dipole-dipole interactions.

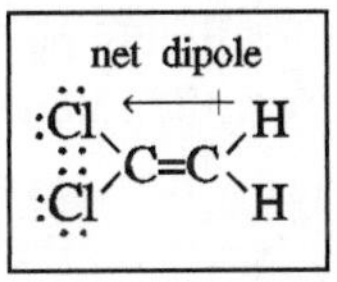

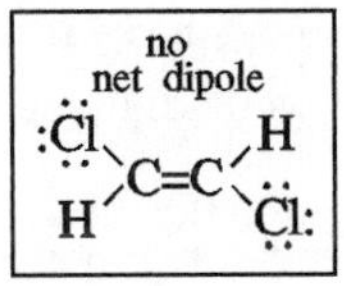

The compound with the lowest boiling point is the one with the weakest intermolecular forces of attraction. Compound (2) which is nonpolar and the same approximate size as the others should therefore have the lowest boiling point.

13-112. *Refer to Sections 13-7 and 13-8.*

(a) true

(b) false — The normal boiling point of a liquid is defined as the boiling point at 760 torr.

(c) false — As long as the temperature remains constant and some liquid remains, the vapor pressure of a liquid will remain the same. The escaping power of liquid molecules possessing sufficient kinetic energy to go into the gas phase stays the same regardless of the quantity of the liquid.

13-114. *Refer to Sections 13-2 and 13-9, and Exercise 13-113.*

The molar heat of vaporization, ΔH_{vap}, at the boiling point is a measure of the heat required to change 1 mole of a liquid into a gas. The stronger the intermolecular forces of attraction between the molecules, the greater will be the value of ΔH_{vap}. Since vapor pressure at a given temperature decreases with increasing intermolecular force, we expect the ΔH_{vap} values to be the reverse of the vapor pressure trend.

$$CS_2 < \text{acetone} < \text{ethanol}$$

This makes sense because CS_2 is a nonpolar molecule and has only dispersion forces at work holding the molecules together as a liquid. Acetone, CH_3COCH_3, is a larger polar molecule and has stronger dispersion forces (due to its larger size) as well as dipole-dipole forces present. Ethanol, CH_3CH_2OH, has dispersion forces and the very powerful hydrogen bonding present.

13-116. *Refer to Sections 13-13 and 13-14.*

(a) gas (b) condense, liquid (c) freeze, crystalline solid (d) sublimation, gas

13-118. *Refer to Sections 13-2, 13-8, 13-10 and 13-12.*

(a) liquid (T = 0°C at STP)

(b) liquid; freezes (or solidifies or crystallizes); crystalline solid

(c) sublimation; gas; 24°C

13-120. *Refer to Section 13-8.*

When a liquid boils, the bubbles are filled with the liquid's vapor.

13-122. *Refer to Section 13-13.*

At room temperature, I would expect to find the sealed container with the iodine crystals filled with a lightly purplish gas (iodine vapor) and tiny crystals of iodine on the inside walls and top of the container where the iodine gas redeposited. The equilibrium involved here is

$$I_2(s) \rightleftarrows I_2(g)$$

In the sealed container of liquid water, I would expect to find droplets of water on the inside of the container where the water vapor condensed. The equilibrium involved in this case is:

$$H_2O(\ell) \rightleftarrows H_2O(g)$$

13-124. *Refer to Section 13-2 and Figure 13-3.*

The molecule, HBr, is polar since the electronegativities of H and Br are 2.1 and 2.8, respectively.

The arrows between the larger, more electronegative Br atom of one HBr molecule and the smaller, less electronegative H atom of an adjacent HBr molecule represent dipole-dipole interactions.

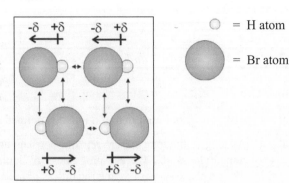

This question, "which freezes first, hot water or cold water," is a favorite of popular science magazines. This discussion is taken from the web site: http://math.ucr.edu/home/baez/physics/General/hot_water.html. Sir Francis Bacon, Descartes and even Aristotle are said to have remarked on it. There are five factors that can make the hot water freeze faster than expected.

(1) Evaporation: Evaporation occurs faster from warmer water. The mass of the hot water may decrease enough to make up for the greater temperature range it must cover to reach freezing. Also, there is a greater cooling effect from evaporation.

(2) Supercooling: The hot water sample may have a greater tendency to supercool, because it has less dissolved gas which can act as nucleation points for ice to form. Water that does not supercool may form a thin layer of ice at the surface which can insulate the rest of the water from the freezer and delay the freezing process.

(3) Convection: As water cools, convection currents within the sample occur as cooler, more dense water sinks, causing the water to circulate. However, water is most dense at 4°C, so at some point, colder liquid water will sink and a film of less dense ice will form at the top and insulate the sample from the cold. Some people speculate that the warmer sample will have more convection currents and will cool quicker.

(4) Dissolved Gases: Dissolved gases, such as oxygen and carbon dioxide, lower the freezing point of water. Cooler water has more dissolved gases and so the freezing point is slightly lower and the time that it would take to reach the slightly lower freezing point would be more.

(5) Conduction: When a hot water sample is put into a freezer, it has been observed that the sample container sometimes melts the ice-encrusted surface, which then allows for much better heat conduction than the frost on which the colder container rests. As a result, heat is drawn from the warmer container more rapidly.

Plan: (1) Find out the specific heat of the unknown metal in J/g·°C.
(2) Using the Law of Dulong and Petit, find out the atomic weight of the metal and then identify it.

(1)
$$|\text{heat lost by the metal}| = |\text{heat gained by the water}|$$
$$|\text{mass} \times \text{specific heat} \times \Delta t|_{\text{metal}} = |\text{mass} \times \text{specific heat}(\ell) \times \Delta t|_{\text{water}}$$
$$(100.2 \text{ g})(\text{specific heat of metal})(99.9°C - 36.6°C) = (50.6 \text{ g})(4.184 \text{ J/g·°C})(36.6°C - 24.8°C)$$

Solving, the specific heat of the metal = 0.394 J/g·°C

(2) $? \text{ AW} = \dfrac{25 \text{ J/mol·°C}}{0.394 \text{ J/g·°C}} = 63 \text{ g/mol}$

Therefore, the metal is likely to be **Cu**.

The vapor pressure of a liquid in equilibrium with its vapor cannot be treated like an ideal gas that obeys the gas laws; the equilibrium (liquid $\rightleftarrows$ vapor) controls the vapor pressure. As conditions are changed, the system adjusts itself until the system reaches equilibrium again; either the liquid which is present evaporates, or the vapor condenses.

In particular, if the temperature increases, more liquid evaporates becoming vapor to increase the vapor pressure. Mathematically, the saturated vapor pressure of a liquid increases exponentially instead of linearly with increasing temperature. Vapor pressure cannot be calculated from the ideal gas law: $P_1/T_1 = P_2/T_2$ since n, the number of moles of gas present, is not a constant but increases greatly with temperature.

(a) CH_3COOH

$$\begin{array}{c} H \quad :\!\overset{\displaystyle ..}{O}: \\ | \quad\quad \| \quad .. \\ H-\underset{|}{\overset{|}{C}}_1-\underset{..}{\overset{}{C}}_2-\overset{..}{\underset{..}{O}}_3-H \\ H \end{array}$$

HCOCH$_3$

$$\begin{array}{c} \cdot\overset{..}{\underset{}{O}}_2 H \\ \| \quad | \\ H-\underset{|}{\overset{|}{C}}_1-\underset{|}{\overset{|}{C}}_3-H \\ H \end{array}$$

CH_3COOH C_1 tetrahedral (sp^3)
  C_2 trigonal planar (sp^2)
  O_3 bent (sp^3 with 2 lone pairs)

$HCOCH_3$ C_1 trigonal planar (sp^2)
  O_2 trigonal planar (sp^2 with 2 lone pairs)
    (not really a central atom)
  C_3 tetrahedral (sp^3)

Both molecules are about the same size and are polar. Both molecules have dispersion and dipole-dipole intermolecular forces operating between their molecules. However, CH_3COOH has hydrogen bonds operating between its molecules, since CH_3COOH contains a hydrogen atom directly bonded to the small, highly electronegative atom, O; $HCOCH_3$ does not. Therefore, $HCOCH_3$ will have the lower boiling point.

(b) NHF_2

$$\begin{array}{c} :\overset{..}{\underset{..}{F}}-\overset{..}{\underset{}{N}}-\overset{..}{\underset{..}{F}}: \\ | \\ H \end{array}$$

BH_2Cl

$$\begin{array}{c} H-B-H \\ | \\ :\overset{..}{\underset{..}{Cl}}: \end{array}$$

NHF_2 is pyramidal (sp^3 with 1 lone pair) and BH_2Cl is trigonal planar (sp^2). Both molecules are polar and have dispersion and dipole-dipole intermolecular forces operating between their molecules. However, NHF_2 has hydrogen bonds operating between its molecules, and BH_2Cl does not. Therefore, BH_2Cl will have the lower boiling point.

(c) CH_3CH_2OH

$$\begin{array}{c} H \quad H \\ | \quad\quad | \\ H-\underset{|}{\overset{|}{C}}_1-\underset{|}{\overset{|}{C}}_2-\overset{..}{\underset{..}{O}}_3-H \\ H \quad H \end{array}$$

CH_3OCH_3

$$\begin{array}{c} H \quad\quad\quad H \\ | \quad\quad\quad\quad | \\ H-\underset{|}{\overset{|}{C}}_1-\overset{..}{\underset{..}{O}}_2-\underset{|}{\overset{|}{C}}_3-H \\ H \quad\quad\quad H \end{array}$$

CH_3CH_2OH C_1 tetrahedral (sp^3)
  C_2 tetrahedral (sp^3)
  O_3 bent (sp^3 with 2 lone pairs)

CH_3OCH_3 C_1 tetrahedral (sp^3)
  O_2 bent (sp^3 with 2 lone pairs)
  C_3 tetrahedral (sp^3)

Both molecules are about the same size and are polar. They have dispersion and dipole-dipole intermolecular forces operating between their molecules. However, CH_3CH_2OH has hydrogen bonds operating between its molecules, and CH_3OCH_3 does not. Therefore, CH_3OCH_3 will have the lower boiling point.

$\quad\quad\quad\quad$ (1) $\quad\quad$ (2) $\quad\quad$ (3)
(a)   Plan: g Si $\Rightarrow$ mol Si $\Rightarrow$ atoms Si $\Rightarrow$ atoms B

$$? \text{ atoms B} = 11.5 \text{ g Si} \times \frac{1 \text{ mol Si}}{28.09 \text{ g Si}} \times \frac{6.02 \times 10^{23} \text{ atoms Si}}{1 \text{ mol Si}} \times \frac{0.0010 \text{ boron atoms}}{100 \text{ Si atoms}} = \mathbf{2.5 \times 10^{18} \text{ atoms B}}$$

(b) $? \text{ g B} = 2.5 \times 10^{18} \text{ atoms B} \times \dfrac{1 \text{ mol B}}{6.02 \times 10^{23} \text{ atoms B}} \times \dfrac{10.81 \text{ g B}}{1 \text{ mol B}} = \mathbf{4.4 \times 10^{-5} \text{ g B}}$

Note: Even though the answer to Step (a) was rounded to the correct number of significant figures, the final answer was obtained by rounding only at the very end.

14 Solutions

14-2. *Refer to the Introduction to Chapter 14.*

	Type of Solution	Example	Solute	Solvent
(a)	solid dissolved in liquid	salt water	$NaCl(s)$	$H_2O(\ell)$
(b)	gas dissolved in gas	air (major components)	$O_2(g)$	$N_2(g)$
(c)	gas dissolved in liquid	$HCl(aq)$	$HCl(g)$	$H_2O(\ell)$
(d)	liquid dissolved in liquid	$CH_3COOH(aq)$	$CH_3COOH(\ell)$	$H_2O(\ell)$
(e)	solid dissolved in solid	brass	$Zn(s)$	$Cu(s)$

14-4. *Refer to Section 14-1 and Figure 14-1.*

Dissolution is favored when (a) solute-solute attractions and (b) solvent-solvent attractions are relatively small and (c) solvent-solute attractions are relatively large. Both processes (a) and (b) require energy, first to separate the solute particles from each other, then to separate the solvent molecules from each other. Process (c) releases energy as solute particles and solvent molecules interact. If the absolute value of heat absorbed in processes (a) and (b) is less than the absolute value of heat released in process (c), then the dissolving process is favored and releases heat.

14-6. *Refer to Section 14-1.*

There are two factors which control the spontaneity of a dissolution process: (1) the amount of heat absorbed or released and (2) the amount of increase in the disorder, or randomness of the system. All dissolution processes are accompanied by an increase in the disorder of both solvent and solute. Thus, their disorder factor is invariably favorable to solubility. Dissolution will always occur if the dissolution process is exothermic and the disorder term increases. Dissolution will occur when the dissolution process is endothermic if the disorder term is large enough to overcome the endothermicity, which opposes dissolution.

14-8. *Refer to Section 14-3.*

When two completely miscible, nonreactive liquids, A and B, are mixed in any proportions, the molecules of A and B will intermingle, and one phase only is always produced. When two completely immiscible liquids, C and D, are mixed in any proportions, two separate phases are always produced; one is pure C and the other is pure D. However, when liquid E is slowly added to liquid F, with which it is only partially miscible, in the beginning only one phase is present ($V_E \ll V_F$). When the solubility of liquid E in liquid F is exceeded, two phases result. After a very large quantity of liquid E is added to liquid F ($V_E \gg V_F$), one phase again is present.

14-10. *Refer to Sections 14-2, 14-3 and 14-4.*

(a) HCl in H_2O high solubility since HCl will ionize in water, which is very polar, to form ions.

(b) HF in H_2O high solubility since both are polar covalent molecules which are capable of forming hydrogen bonds.

(c) Al_2O_3 in H_2O low solubility since both Al^{3+} and O^{2-} in the ionic solid have high charge-to-size ratios and therefore have larger lattice energies. The dissolution of Al_2O_3 is very endothermic and is not favored.

(d) SiO_2 in H_2O low solubility since SiO_2 is a covalent solid called quartz and H_2O is very polar.

(e) Na_2SO_4 in C_6H_{14} low solubility since Na_2SO_4 is an ionic solid and C_6H_{14} is a nonpolar solvent.

14-12. *Refer to Sections 4-2, 14-2 and 14-3, and Exercise 14-10.*

(a) HCl in H_2O strong electrolyte

(b) HF in H_2O weak electrolyte

(c) Al_2O_3 in H_2O cannot be prepared in "reasonable" concentration

(d) SiO_2 in H_2O cannot be prepared in "reasonable" concentration

(e) Na_2SO_4 in C_6H_{14} cannot be prepared in "reasonable" concentration

---

14-14. *Refer to Sections 14-1, 14-2 and 14-3.*

(a) The solubility of a solid in a liquid does not depend appreciably on pressure.

(b) The solubility of a liquid in a liquid also is essentially independent of pressure, because the interactions (solute:solute, solvent:solvent, and solute:solvent) are not affected by pressure.

14-16. *Refer to Sections 14-2, 14-5 and 14-6.*

To determine if a sodium chloride solution is indeed saturated, one could add a known mass of solid sodium chloride, stir for an extended period of time, then quantitatively filter the remaining solid. If the weight of the dried NaCl is equal to what was originally added, the solution was saturated originally; if the weight is less, the solution was unsaturated. Another qualitative way would be to observe if a tiny added amount of NaCl would dissolve or not.

14-18. *Refer to Section 14-7 and Exercise 14-17.*

Henry's Law states: $C_{gas} = X_{gas} = kP_{gas}$ where $C_{gas} = X_{gas}$ = mole fraction of gas at a certain temperature
k = Henry's Law constant
P_{gas} = partial pressure of gas above the solution

For CH_4 at 25°C: For CH_4 at 50°C:

$X_{CH_4} = k P_{CH_4}$ $X_{CH_4} = k P_{CH_4}$
      $= (2.42 \times 10^{-5}$ atm$)(10.$ atm$)$ $= (1.73 \times 10^{-5}$ atm$^{-1})(10.$ atm$)$
$X_{CH_4} = \mathbf{2.4 \times 10^{-4}}$ $X_{CH_4} = \mathbf{1.7 \times 10^{-4}}$

Therefore, the solubility of $CH_4(g)$ **decreases** with increasing temperature.

14-20. *Refer to Section 14-2 and Table 14-1.*

The hydration energy of ions generally increases with increasing charge and decreasing size. The greater the hydration energy for an ion, the more strongly hydrated it is. In order of decreasing hydration energy:

(a) $Na^+ > Rb^+$ due to smaller size   (c) $Fe^{3+} > Fe^{2+}$ due to higher charge and smaller size

(b) $Cl^- > Br^-$ due to smaller size   (d) $Mg^{2+} > Na^+$ due to higher charge and smaller size

14-22. *Refer to Sections 14-1, 14-2 and 14-3.*

The dissolution of many solids in liquids is endothermic (requires heat) due to the large solute-solute attractions that must be overcome relative to the solvent-solute attractions.

The mixing of two miscible liquids is exothermic (releases heat) since solute-solute attractions are less than the solvent-solute attractions.

14-24. *Refer to Section 3-6.*

If s = solubility of A in $\left(\dfrac{\text{g A}}{100. \text{ g H}_2\text{O}}\right)$, then the maximum mass of A that will dissolve in 100. g of H_2O = s.

The solubility of A as a mass percent:

$$\text{solubility of A in} \left(\frac{\text{g A}}{100. \text{ g solution}}\right) = \frac{\text{g A}}{\text{g A} + 100. \text{ g H}_2\text{O}} \times 100\%$$

$$= \frac{s}{s + 100} \times 100\%$$

14-26. *Refer to Section 14-8 and Example 14-1.*

Plan: (1) Determine the mass of solution for 1 kg of water in 0.354 m NaCl solution.
(2) Calculate the mass of NaCl in 1 liter of solution.

(1) We know that a 0.354 m NaCl solution contains 0.354 moles of NaCl in 1000 g of water.

? g NaCl in 1000 g water = 0.354 mol $\times$ 58.44 g/mol = 20.7 g NaCl

? g solution = g solute + g solvent = 20.7 g NaCl + 1000 g water = 1020.7 g soln

(2) ? g NaCl in 1000 mL soln = 1000 mL soln $\times \dfrac{1.01 \text{ g soln}}{1 \text{ mL soln}} \times \dfrac{1000 \text{ g H}_2\text{O}}{1020.7 \text{ g soln}} \times \dfrac{0.354 \text{ mol NaCl}}{1000 \text{ g H}_2\text{O}} \times \dfrac{58.44 \text{ g NaCl}}{1 \text{ mol NaCl}}$

= 20.5 g NaCl

Therefore, to prepare 1.000 L of a 0.354 m NaCl soln, add 20.5 g of NaCl to a one-liter volumetric flask. Add enough distilled water to dissolve the NaCl(s), then continue to add water until the volume of the solution is exactly one liter and mix thoroughly.

14-28. *Refer to Section 14-8 and Example 14-1.*

We know: $m = \dfrac{\text{mol solute}}{\text{kg solvent}}$

? mol $C_6H_5COOH = \dfrac{56.5 \text{ g C}_6\text{H}_5\text{COOH}}{122 \text{ g/mol}} = 0.463$ mol C_6H_5COOH

? kg $C_2H_5OH = 350$ mL $C_2H_5OH \times \dfrac{0.789 \text{ g C}_2\text{H}_5\text{OH}}{1 \text{ mL C}_2\text{H}_5\text{OH}} \times \dfrac{1 \text{ kg}}{1000 \text{ g}} = 0.28$ kg C_2H_5OH

Therefore, $m = \dfrac{0.463 \text{ mol C}_6\text{H}_5\text{COOH}}{0.28 \text{ kg C}_2\text{H}_5\text{OH}}$

= **1.7 m C$_6$H$_5$COOH in C$_2$H$_5$OH**

14-30. *Refer to Section 14-8 and Example 14-3.*

Plan: (1) Calculate the moles of each component.
(2) Calculate the mole fraction.

(1) $? \text{ mol } C_2H_5OH = \dfrac{60.0 \text{ g}}{46.1 \text{ g/mol}} = 1.30 \text{ mol } C_2H_5OH$

$? \text{ mol } H_2O = \dfrac{40.0 \text{ g}}{18.0 \text{ g/mol}} = 2.22 \text{ mol } H_2O$

(2) $X_{C_2H_5OH} = \dfrac{\text{mol } C_2H_5OH}{\text{mol } C_2H_5OH + \text{mol } H_2O} = \dfrac{1.30 \text{ mol}}{1.30 \text{ mol} + 2.22 \text{ mol}} = \mathbf{0.369}$

$X_{H_2O} = \dfrac{\text{mol } H_2O}{\text{mol } H_2O + \text{mol } C_2H_5OH} = \dfrac{2.22 \text{ mol}}{2.22 \text{ mol} + 1.30 \text{ mol}} = \mathbf{0.631}$

Alternative method: $X_{H_2O} = 1 - X_{C_2H_5OH} = 1.000 - 0.369 = \mathbf{0.631}$

14-32. *Refer to Sections 3-6 and 14-8, and Examples 14-1 and 14-3.*

(1) $? \text{ mol } K_2SO_4 = \dfrac{12.50 \text{g}}{174.3 \text{ g/mol}} = 0.07172 \text{ mol } K_2SO_4$

$? \text{ mL soln} = \dfrac{\text{g soln}}{\text{density}} = \dfrac{100.00 \text{ g}}{1.083 \text{ g/mL}} = 92.34 \text{ mL}$

Therefore,

$M \text{ } K_2SO_4 = \dfrac{\text{mol } K_2SO_4}{\text{L soln}} = \dfrac{0.07172 \text{ mol}}{0.09234 \text{ L}} = \mathbf{0.7767 \textit{ M} \text{ } K_2SO_4}$

(2) $m \text{ } K_2SO_4 = \dfrac{\text{mol } K_2SO_4}{\text{kg } H_2O} = \dfrac{0.07172 \text{ mol}}{0.08750 \text{ kg}} = \mathbf{0.8197 \textit{ m} \text{ } K_2SO_4}$

since $\text{kg } H_2O = \text{kg soln} - \text{kg } K_2SO_4 = 0.10000 \text{ kg} - 0.01250 \text{ kg} = 0.08750 \text{ kg}$

(3) $\text{mass \% } K_2SO_4 = \dfrac{\text{g } K_2SO_4}{\text{g soln}} \times 100\% = \dfrac{12.50 \text{ g}}{100.00 \text{g}} \times 100\% = \mathbf{12.50\% \text{ } K_2SO_4}$

(4) $X_{H_2O} = \dfrac{\text{mol } H_2O}{\text{mol } H_2O + \text{mol } K_2SO_4} = \dfrac{\left(\dfrac{87.50 \text{ g}}{18.02 \text{ g/mol}}\right)}{\left(\dfrac{87.50 \text{ g}}{18.02 \text{ g/mol}} + \dfrac{12.50 \text{ g}}{174.3 \text{ g/mol}}\right)} = \dfrac{4.856 \text{ mol}}{4.856 \text{ mol} + 0.07172 \text{ mol}} = \mathbf{0.9854}$

14-34. *Refer to Section 14-8, and Examples 14-1 and 14-2.*

(a) Plan: (1) Assuming 100 g of solution, calculate the moles of $C_6H_{12}O_6$. Note that the solution density is not required for the calculation.
(2) Calculate the mass of water.
(3) Determine the molality of $C_6H_{12}O_6$.

(1) Recall: $\% \text{ by mass} = \dfrac{\text{g solute}}{\text{g solution}} \times 100\%$, so solving for the mass of $C_6H_{12}O_6$ in 100 g soln, we have

$? \text{ g } C_6H_{12}O_6 \text{ in 100 g soln} = \dfrac{21.0\% \text{ } C_6H_{12}O_6 \times 100 \text{ g soln}}{100\%} = 21.0 \text{ g } C_6H_{12}O_6$

$? \text{ mol } C_6H_{12}O_6 = \dfrac{21.0 \text{ g}}{180.2 \text{ g/mol}} = 0.117 \text{ mol } C_6H_{12}O_6$

(2) ? g H_2O = g soln - g $C_6H_{12}O_6$ = 100.0 g - 21.0 g = 79.0 g

(3) $m\ C_6H_{12}O_6 = \dfrac{\text{mol } C_6H_{12}O_6}{\text{kg } H_2O} = \dfrac{0.117 \text{ mol}}{0.0790 \text{ kg}} = \textbf{1.48 } \boldsymbol{m}\textbf{ } \boldsymbol{C_6H_{12}O_6}$

(b) Molality is independent of density and temperature since it is a measure of the number of moles of substance dissolved in 1 kilogram of solvent. Only concentration units that involve volume, e.g., molarity, are dependent on density and temperature. Therefore, the molality at a higher temperature would be the **same** as the molality at 20°C.

14-36. Refer to Section 14-9.

The vapor pressure of a liquid depends upon the ease with which the molecules are able to escape from the surface of the liquid. The vapor pressure of a liquid always decreases when nonvolatile solutes (ions or molecules) are dissolved in it, since after dissolution there are fewer solvent molecules at the surface to vaporize.

14-38. Refer to Section 14-9 and Example 14-4.

(a) From Raoult's Law,

$\Delta P_{\text{benzene}} = X_{\text{naphthalene}}\, P^{\circ}_{\text{benzene}}$ where $\Delta P_{\text{benzene}}$ = vapor pressure lowering of benzene, C_6H_6
$\qquad\qquad\qquad\qquad\qquad\qquad\qquad\qquad P^{\circ}_{\text{benzene}}$ = vapor pressure of pure benzene, 74.6 torr at 20°C
$\qquad\qquad\qquad\qquad\qquad\qquad\qquad\qquad X_{\text{naphthalene}}$ = mole fraction of naphthalene, $C_{10}H_8$

? mol benzene = $\dfrac{140.0 \text{ g } C_6H_6}{78.11 \text{ g/mol}}$ = 1.792 mol C_6H_6

? mol naphthalene = $\dfrac{22.5 \text{ g } C_{10}H_8}{128 \text{ g/mol}}$ = 0.176 mol $C_{10}H_8$

$X_{\text{naphthalene}} = \dfrac{\text{mol } C_{10}H_8}{\text{mol } C_{10}H_8 + \text{mol } C_6H_6} = \dfrac{0.176 \text{ mol}}{0.176 \text{ mol} + 1.792 \text{ mol}} = 0.0894$

Substituting into Raoult's Law, $\Delta P_{\text{benzene}}$ = 0.0894 x 74.6 torr = **6.67 torr**

(b) $P_{\text{benzene}} = P^{\circ}_{\text{benzene}} - \Delta P_{\text{benzene}}$ = 74.6 torr – 6.67 torr = **67.9 torr**

14-40. Refer to Section 14-9, Example 14-5 and the figure accompanying Exercise 14-43.

Plan: (1) Calculate the mole fraction of each component in the solution.
　　　 (2) By assuming the solution to be ideal, apply Raoult's Law to calculate the partial pressures of acetone and chloroform.

(1) Let X = mole fraction of the components

$X_{\text{acetone}} = \dfrac{0.550 \text{ mol acetone}}{(0.550 \text{ mol acetone}) + (0.550 \text{ mol chloroform})} = 0.500$

$X_{\text{chloroform}} = \dfrac{0.550 \text{ mol chloroform}}{(0.550 \text{ mol acetone}) + (0.350 \text{ mol chloroform})} = 0.500 \ (= 1 - X_{\text{acetone}})$

(2) From Raoult's Law:

$P_{\text{acetone}} = X_{\text{acetone}}\, P^{\circ}_{\text{acetone}} = (0.500)(345 \text{ torr}) = \textbf{172 torr}$

$P_{\text{chloroform}} = X_{\text{chloroform}}\, P^{\circ}_{\text{chloroform}} = (0.500)(295 \text{ torr}) = \textbf{148 torr}$

Since the acetone-chloroform system is expected to show negative deviation from an ideal solution, P_{acetone} should be less than 172 torr and $P_{\text{chloroform}}$ should be less than 148 torr.

14-42. *Refer to Sections 14-9 and 12-11, Examples 14-5 and 14-6, and Exercise 14-40 Solution.*

Plan: (1) Calculate the total pressure using Exercise 14-40 Solution: $P_{acetone} = 172$ torr, $P_{chloroform} = 148$ torr.
(2) Since the mole fraction of a component in a gaseous mixture equals the ratio of its partial pressure to the total pressure, the composition of the vapor above the solution can be calculated.

(1) $P_{total} = P_{acetone} + P_{chloroform} = 172$ torr $+ 148$ torr $= $ **320. torr**

(2) In the vapor, $X_{acetone} = \dfrac{P_{acetone}}{P_{total}} = \dfrac{172 \text{ torr}}{320. \text{ torr}} = \textbf{0.538}$

$X_{chloroform} = \dfrac{P_{chloroform}}{P_{total}} = \dfrac{148 \text{ torr}}{320. \text{ torr}} = \textbf{0.462}$

14-44. *Refer to Section 14-9 and Exercise 14-43.*

Assuming real behavior for a chloroform/acetone solution in which the mole fraction of chloroform, $CHCl_3$, is 0.3 and using the dashed (curved) lines on the diagram in Exercise 14-43,

(a) $P_{chloroform} = $ **65 torr**   (b) $P_{acetone} = $ **215 torr**   (c) $P_{total} = $ **280 torr** $(P_{total} = P_{chloroform} + P_{acetone})$

14-46. *Refer to Section 14-11, Example 14-7 and Table 14-2.*

Plan: (1) Find ΔT_b.
(2) Determine T_b for the ethylene glycol solution.

(1) From Table 14-2, K_b for $H_2O = 0.512$ °C/m; B.P. $= 100.00$°C.

$\Delta T_b = K_b\, m = (0.512 \text{ °C}/m)(3.05\, m) = 1.56$°C

(2) Boiling point of the ethylene glycol solution, $T_{b(soln)} = T_{b(solvent)} + \Delta T_b = 100.00$°C $+ 1.56$°C $= $ **101.56°C**

14-48. *Refer to Section 14-12, Examples 14-8 and 14-9, and Table 14-2.*

Plan: (1) Find ΔT_f.
(2) Determine T_f for the ethylene glycol solution.

(1) From Table 14-2, K_f for $H_2O = 1.86$ °C/m; F.P. $= 0.00$°C.

$\Delta T_f = K_f\, m = (1.86 \text{ °C}/m)(3.05\, m) = 5.67$°C

(2) Freezing point of the ethylene glycol solution, $T_{f(soln)} = T_{f(solvent)} - \Delta T_f = 0.00$°C $- 5.67$°C $= $ **–5.67°C**

14-50. *Refer to Sections 14-11 and 14-12, and Table 14-2.*

Consider a 0.175 m solution of a nonvolatile nonelectrolyte in the solvents listed in Table 14-2.

(a) The greatest freezing point depression, ΔT_f, occurs in a **camphor** solution since $\Delta T_f = K_f\, m$ and this solvent has the largest value of K_f.

(b) The lowest freezing point, $T_{f(solution)}$, occurs in a solution of the non-electrolyte in **water**. Pure water freezes at 0°C, the lowest freezing point of all the listed solvents. The effect of freezing point depression on the freezing point of a 0.175 m solution is minor compared to the actual freezing point of the pure solvents.

(c) The greatest boiling point elevation, ΔT_b, occurs in a solution of **camphor** since $\Delta T_b = K_b m$ and this solvent has the largest value of K_b.

(d) The highest boiling point, $T_{b(solution)}$, for all the listed solvents occurs in a solution of **nitrobenzene** since it has the highest boiling point.

14-52. *Refer to Section 14-12, Example 14-9 and Table 14-2.*

Plan: (1) Find ΔT_f.
(2) Determine T_f for the lemon juice.

(1) From Table 14-2, K_f for H_2O = 1.86 °C/m; F.P. = 0.00°C.

$\Delta T_f = K_f m = (1.86 \text{ °C}/m)(10.0 \ m) = 18.6°C$

(2) Freezing point of lemon juice, $T_{f(soln)} = T_{f(solvent)} - \Delta T_f = 0.00°C - 18.6°C = \mathbf{-18.6°C}$

14-54. *Refer to Section 14-11 and Example 14-7.*

Plan: (1) Calculate the molality of the nonelectrolyte solution.
(2) Find ΔT_b.
(3) Determine T_b for the nonelectrolyte solution.

(1) we know: molality, $m = \dfrac{\text{mol nonelectrolyte}}{\text{kg solvent}}$

? mol nonelectrolyte = 3.0 g $\times \dfrac{1 \text{ mol}}{137 \text{ g}}$ = 0.022 mol

? kg solvent = 250.0 mL ethanol $\times \dfrac{0.789 \text{ g}}{1 \text{ mL}} \times \dfrac{1 \text{ kg}}{1000 \text{ g}}$ = 0.197 kg ethanol

$m = \dfrac{0.022 \text{ mol nonelectrolyte}}{0.197 \text{ kg ethanol}} = 0.11 \ m$

(2) K_b for ethanol = 1.22 °C/m; B.P. = 78.41°C.

$\Delta T_b = K_b \ m = (1.22 \text{ °C}/m)(0.11 \ m) = 0.13°C$

(3) Boiling point of the ethyl alcohol solution, $T_{b(soln)} = T_{b(solvent)} + \Delta T_b = 78.41°C + 0.13°C = \mathbf{78.54°C}$

14-56. *Refer to Section 14-12 and Example 14-9.*

Plan: (1) Determine the molality of the solute, Zn, in the solid solution.
(2) Calculate the melting point (freezing point) of brass using $\Delta T_f = K_f m$.

(1) Assume 100. g of brass containing 12 g of Zn and 88 g of Cu.

$m \text{ Zn} = \dfrac{\text{mol solute}}{\text{kg solvent}} = \dfrac{\text{mol Zn}}{\text{kg Cu}} = \dfrac{12 \text{ g Zn}/65.4 \text{ g/mol}}{0.088 \text{ kg Cu}} = 2.1 \ m$

(2) $\Delta T_f = K_f m = 23 \text{ °C}/m \times 2.1 \ m = 48°C$

$T_{f(brass)} = T_{f(Cu)} - \Delta T_f = 1083°C - 48°C = \mathbf{1035°C}$

14-58. *Refer to Section 14-11 and Table 14-2.*

Plan: (1) Solve for the molality of the solution using $\Delta T_b = K_b m$.
           (2) Calculate the mass of $C_{10}H_8$.

From Table 14-2, for nitrobenzene: $T_b = 210.88°C$ $K_b = 5.24 \ °C/m$

(1) $m = \dfrac{\Delta T_b}{K_b} = \dfrac{214.20°C - 210.88°C}{5.24 \ °C/m} = 0.634 \ m \ C_{10}H_8$

(2)   Recall: $m = \dfrac{\text{mol solute}}{\text{kg solvent}} = \dfrac{\text{g solute/MW solute}}{\text{kg solvent}}$

     Solving,

     ? g $C_{10}H_8 = (m)(\text{MW } C_{10}H_8)(\text{kg } C_6H_5NO_2) = (0.634 \ m)(128 \ g/mol)(0.355 \ kg) = \mathbf{28.8 \ g \ C_{10}H_8}$

14-60. *Refer to Sections 14-13 and 2-9, Example 14-10, Exercise 2-52 Solution, and Table 14-2.*

Plan: (1) Solve for the molality and the approximate formula weight of the nonelectrolyte, using $\Delta T_f = K_f m$.
           (2) Determine the simplest (empirical) formula from the % composition data.
           (3) Using the approximate formula weight of the simplest formula, determine the true molecular formula and the exact molecular weight.

From Table 14-2, for benzene: $T_f = 5.48°C$ $K_f = 5.12 \ °C/m$

(1) $m = \dfrac{\Delta T_f}{K_f} = \dfrac{0.53°C}{5.12 \ °C/m} = 0.10 \ m$

     (Note: use $0.1035 \ m$ to do the molecular weight calculation, remembering there are only 2 significant figures.)

     Recall: $m = \dfrac{\text{mol solute}}{\text{kg solvent}} = \dfrac{\text{g solute/MW solute}}{\text{kg solvent}}$

     Solving,

     ? MW solute $= \dfrac{\text{g solute}}{m \times \text{kg benzene}} = \dfrac{0.500 \ g}{0.10 \ m \times 0.0750 \ kg} = 64 \ g/mol$ (to 2 significant figures)

(2)   Assume 100 g of sample containing 37.5 g C, 12.5 g H and 50.0 g O

     ? mol C $= \dfrac{37.5 \ g}{12.0 \ g/mol} = 3.12 \ mol$ Ratio $= \dfrac{3.12}{3.12} = 1$

     ? mol H $= \dfrac{12.5 \ g}{1.008 \ g/mol} = 12.4 \ mol$ Ratio $= \dfrac{12.4}{3.12} = 4$

     ? mol O $= \dfrac{50.0 \ g}{16.0 \ g/mol} = 3.12 \ mol$ Ratio $= \dfrac{3.12}{3.12} = 1$

     Therefore, the simplest formula is CH_4O (FW = 32.04 g/mol)

(3) $n = \dfrac{\text{molecular weight}}{\text{simplest formula weight}} = \dfrac{64 \ g/mol}{32 \ g/mol} = 2$

     The true molecular formula for the organic compound is $(CH_4O)_2 = \mathbf{C_2H_8O_2}$ with a molecular weight (to 4 significant figures) of **64.08 g/mol**.

14-62. *Refer to Section 14-12 and Fundamental Algebra.*

(a)   Plan: (1) Calculate the molality of the solution using $\Delta T_f = K_f m$
                 (2) Calculate the total number of moles of solutes.
                 (3) Calculate the masses of $C_{10}H_8$ and $C_{14}H_{10}$ and the % composition of each in the sample.

(1) From Table 14-2, for benzene: $T_b = 80.1°C$ $K_b = 2.53 °C/m$

$T_f = 5.48°C$ $K_f = 5.12 °C/m$

$$m = \frac{\Delta T_f}{K_f} = \frac{5.48°C - 4.85°C}{5.12 °C/m} = 0.12 \; m \text{ solutes}$$

(2) Recall: $m = \dfrac{\text{mol solute}}{\text{kg solvent}}$

? mol solutes = 0.12 m x 0.360 kg = 0.043 mol solutes

(3) Let   x = g $C_{10}H_8$ and $\dfrac{x}{128 \text{ g/mol}} = \text{mol } C_{10}H_8$

6.00 g - x = g $C_{14}H_{10}$ $\dfrac{6.00 \text{ g - x}}{178 \text{ g/mol}} = \text{mol } C_{14}H_{10}$

Therefore,

the total moles of solute = mol $C_{10}H_8$ + mol $C_{14}H_{10}$ = $\dfrac{x}{128} + \dfrac{6.00 - x}{178}$

So,   total moles = 0.043 mol = $\dfrac{x}{128} + \dfrac{6.00 - x}{178}$

When we multiply both sides of the equation by (128)(178), we obtain,

$$980 = (178x) + (768 - 128x)$$
$$50x = 212$$
$$x = 4.2 \text{ g } C_{10}H_8$$
$$6.00 - x = 1.8 \text{ g } C_{14}H_{10}$$

And,   % $C_{10}H_8$ = $\dfrac{4.2 \text{ g } C_{10}H_8}{6.00 \text{ g sample}}$ x 100% = **70%**

% $C_{14}H_{10}$ = $\dfrac{1.8 \text{ g } C_{14}H_{10}}{6.00 \text{ g sample}}$ x 100% = **30%**

(b) $\Delta T_b = K_b m_{total} = (2.53 °C/m)(0.12 \; m) = 0.30°C$

$T_{b(solution)} = T_{b(benzene)} + \Delta T_b = 80.1°C + 0.30°C = $ **80.4°C**

14-64.   *Refer to Section 14-14.*

The highest particle concentration is expected in 0.10 M $Al(NO_3)_3$. Theoretically, $Al(NO_3)_3$ dissociates completely into 4 ions whereas $Ca(NO_3)_2$ dissociates into 3 ions and $LiNO_3$ only dissociates into 2 ions, according to:

$$Al(NO_3)_3(aq) \rightarrow Al^{3+}(aq) + 3NO_3^-(aq)$$
$$Ca(NO_3)_2(aq) \rightarrow Ca^{2+}(aq) + 2NO_3^-(aq)$$
$$LiNO_3(aq) \rightarrow Li^+(aq) + NO_3^-(aq)$$

The solution with the greatest number of ions will conduct electricity the strongest. Therefore, **$Al(NO_3)_3$** is expected to be the best conductor of electricity among the three, even though there is some degree of association of ions.

14-66.   *Refer to Section 14-14 and Table 14-3.*

The ideal value for the van't Hoff factor, i, for strong electrolytes at infinite dilution is the total number of ions present in a formula unit.

(a) Na_2SO_4 $i_{ideal} = $ **3**   (b) NaOH $i_{ideal} = $ **2**   (c) $Al_2(SO_4)_3$ $i_{ideal} = $ **5**   (d) $CaSO_4$ $i_{ideal} = $ **2**

14-68. Refer to Section 14-14 and Table 14-3.

The value for the van't Hoff factor, i, for strong electrolytes in dilute solution approximates the total number of ions present in a formula unit. So, $i = 2$ for $KClO_3$ and $i = 3$ for $CaCl_2$. The value of i for the weak electrolyte such as CH_3COOH is between 1 and 2. The value of i for a nonelectrolyte such as CH_3OH is 1.

We know that $\Delta T_f = iK_f m$. For solutions of the same solvent at the same molality, the solute with the larger van't Hoff factor, i, has the lower freezing point. Therefore,

$$T_f \text{ for } CaCl_2 < T_f \text{ for } KClO_3 < T_f \text{ for } CH_3COOH < T_f \text{ for } CH_3OH$$

14-70. Refer to Sections 14-9 and 14-14.

Note: The van't Hoff factor, i, must be included in all calculations involving colligative properties, including vapor pressure lowering.

Plan: (1) Determine the total number of moles of ions in the solution.
(2) Determine the mole fraction of water in the solution.
(3) Calculate the vapor pressure above the solution.

(1) For NaCl, NaBr and NaI, the ideal van't Hoff factor, i_{ideal}, is 2.

? mol ions = (2 x mol NaCl) + (2 x mol NaBr) + (2 x mol NaI)

$$= \left(2 \times \frac{1.25 \text{ g NaCl}}{58.4 \text{ g/mol}}\right) + \left(2 \times \frac{1.25 \text{ g NaBr}}{103 \text{ g/mol}}\right) + \left(2 \times \frac{1.25 \text{ g NaI}}{150. \text{ g/mol}}\right)$$

= 0.0837 mol ions

(2) $X_{water} = \dfrac{\text{mol } H_2O}{\text{mol ions} + \text{mol } H_2O} = \dfrac{(150. \text{ g}/18.0 \text{ g/mol})}{0.0837 \text{ mol} + (150. \text{ g}/18.0 \text{ g/mol})} = 0.990$

(3) At 100°C, the boiling point of water, the vapor pressure of pure water, $P^\circ_{water} = 760$ torr.

$P_{water} = X_{water} P^\circ_{water} = 0.990 \times 760 \text{ torr} = \textbf{752 torr}$

14-72. Refer to Section 14-14.

We know that $\Delta T_f = iK_f m$. For solutions of the same solvent at the same molality, the solute with the larger van't Hoff factor, i, has the lower freezing point.

Since $Na_2SO_4(aq)$ will dissociate into 3 ions ($i_{ideal} = 3$) and $CaSO_4(aq)$ will dissociate into 2 ions ($i_{ideal} = 2$), **0.100 m Na_2SO_4** has the lower freezing point.

14-74. Refer to Section 14-14.

We know that $\Delta T_f = iK_f m$. For solutions of the same solvent at the same molality, the solute with the larger van't Hoff factor, i, has the higher boiling point. Since all are 1.1 m aqueous solutions:

(a) NaCl and LiCl solutions should boil at the same temperature, because both have $i_{ideal} = 2$.

(b) Na_2SO_4 ($i_{ideal} = 3$) solution boils at a higher temperature than that of NaCl ($i_{ideal} = 2$).

(c) Na_2SO_4 ($i_{ideal} = 3$) solution boils at a higher temperature than that of HCl ($i_{ideal} = 2$).

(d) HCl ($i_{ideal} = 2$) solution boils at a higher temperature than that of $C_6H_{12}O_6$ ($i_{ideal} = 1$, since it is a nonelectrolyte).

(e) $C_6H_{12}O_6$ (i_{ideal} = 1) and CH_3OH (i_{ideal} = 1) solutions should boil at the same temperature, since both are nonelectrolytes. However, CH_3OH is somewhat volatile, which would lower the boiling point.

(f) CH_3COOH solution ($1 < i < 2$ because acetic acid is a weak acid and only partially dissociates into its ions, H^+ and CH_3COO^-) boils at a higher temperature than that of CH_3OH (i_{ideal} = 1, since it is a nonelectrolyte). Note that both substances are somewhat volatile, which could affect their respective boiling points.

(g) KCl (i_{ideal} = 2) solution boils at a higher temperature than CH_3COOH solution ($1 < i < 2$, since acetic acid is a weak acid).

14-76. *Refer to Section 14-11, Example 14-7 and Table 14-2.*

Plan: (1) Calculate the molality of the NaCl solution.
          (2) Find ΔT_b.
          (3) Determine T_b for the aqueous solution - synthetic ocean water.

(1) we know: molality, $m = \dfrac{mol\ NaCl}{kg\ solvent}$

? mol NaCl = 36.0 g $\times \dfrac{1\ mol}{58.44\ g}$ = 0.616 mol

? kg H_2O = 1000. mL $H_2O \times \dfrac{0.998\ g}{1\ mL} \times \dfrac{1\ kg}{1000\ g}$ = 0.998 kg H_2O

$m = \dfrac{0.616\ mol\ NaCl}{0.998\ kg\ H_2O}$ = 0.617 m

(2) From Table 14-2, K_b for H_2O = 0.512 °C/m; B.P. = 100.00°C.

$\Delta T_b = iK_b\, m$ = (2)(0.512 °C/m)(0.617 m) = 0.632°C

(3) Boiling point of synthetic seawater, $T_{b(soln)} = T_{b(solvent)} + \Delta T_b$ = 100.00°C + 0.63°C = **100.63°C**

14-78. *Refer to Section 14-14 and Example 14-12.*

Plan: (1) Calculate $m_{effective}$.
          (2) Determine the % ionization of CH_3COOH.

(1) From Table 14-2, for water: T_f = 0°C, K_f = 1.86 °C/m

$m_{effective} = \dfrac{\Delta T_f}{K_f} = \dfrac{0.0000°C - (-0.188°C)}{1.86\ °C/m}$ = 0.101 m

(2) Let   x = molality of CH_3COOH that ionizes
    Then   x = molality of H^+ and CH_3COO^- that formed

Consider:	CH_3COOH	$\rightleftharpoons$	H^+	+	CH_3COO^-
Start	0.100 m		$\approx$ 0 m		0 m
Change	- x m		+ x m		+ x m
Final	(0.100 - x) m		x m		x m

$M_{effective} = m_{CH_3COOH} + m_{H^+} + m_{CH_3COO^-}$
  0.101 = (0.100 - x) + x + x
  0.101 = 0.100 + x
    x = 0.001 m

% ionization = $\dfrac{m_{ionized}}{m_{original}} \times 100\% = \dfrac{0.001\ m}{0.100\ m} \times 100\%$ = **1%**

14-80. *Refer to Section 14-14 and Example 14-12.*

Plan: (1) Find ΔT_f for the solution if CsCl had been a nonelectrolyte.
(2) Determine the van't Hoff factor, *i*.
(3) Calculate $m_{effective}$ for CsCl.
(4) Calculate the % dissociation.

(1) If CsCl were a nonelectrolyte

$\Delta T_f = K_f m = (1.86\ °C/m)(0.121\ m) = 0.225°C$ for water, $K_f = 1.86\ °C/m$

(2) The van't Hoff factor, $i = \dfrac{\Delta T_{f(actual)}}{\Delta T_{f(nonelectrolyte)}} = \dfrac{0°C - (-0.403°C)}{0.225°C} = \mathbf{1.79}$

(3) $m_{effective} = i \times m_{stated} = 1.79 \times 0.121\ m = 0.217\ m$

(4) Let x = molality of CsCl that apparently dissociated
Then x = molality of Cs^+ and Cl^- that formed

Consider:	CsCl	⇄	Cs^+	+	Cl^-
Start	0.121 *m*		0 *m*		0 *m*
Change	- x *m*		+ x *m*		+ x *m*
Final	(0.121 - x) *m*		x *m*		x *m*

$$m_{effective} = m_{CsCl} + m_{Cs^+} + m_{Cl^-}$$
$$0.217 = (0.121 - x) + x + x$$
$$0.217 = 0.121 + x$$
$$x = 0.096\ m$$

$$\%\ \text{ionization} = \frac{m_{dissociated}}{m_{original}} \times 100\% = \frac{0.096\ m}{0.121\ m} \times 100\% = \mathbf{79\%}$$

14-82. *Refer to Section 14-8.*

Assume 1 liter of solution.

? g NaCl in 1 L soln = 1.00×10^{-4} mol NaCl x 58.4 g/mol = 5.84×10^{-3} g NaCl

The contribution of NaCl to the mass of 1 liter of water is nearly negligible. Since the density of water (1.00 g/mL) is essentially the same as the density of the solution:

$$M = \frac{1.00 \times 10^{-4}\ \text{mol NaCl}}{1\ \text{L soln}} \cong \frac{1.00 \times 10^{-4}\ \text{mol NaCl}}{1\ \text{kg}\ H_2O} = m$$

However, if acetonitrile (density at 20.0°C = 0.786 g/mL) were the solvent, 1 liter of solution which would be essentially pure solvent would have a mass of only 786 g. We see that molarity of NaCl in acetonitrile would *not* be equivalent to the molality, but would be less.

14-84. *Refer to Section 14-15 and Example 14-13.*

osmotic pressure, $\pi = iMRT = (1)(0.0200\ M)(0.0821\ \text{L·atm/mol·K})(75°C + 273°) = \mathbf{0.571\ atm}$

14-86. *Refer to Sections 14-11, 14-12 and 14-15, and Exercise 14-85.*

Plan: (1) Determine the molality of the aqueous solution.
            (2) Calculate ΔT_f and ΔT_b.

(1) Recall, osmotic pressure, $\pi = iMRT$

$$M = \frac{\pi}{iRT} = \frac{1.25 \text{ atm}}{(1)(0.0821 \text{ L·atm/mol·K})(273 \text{ K})} = 0.0558 \ M$$

For dilute aqueous solution, $M \cong m$ (See Exercise 14-82 Solution).

Therefore, $? \ m = 0.0558 \ m$

(2) From Table 14-2, for water: $K_f = 1.86 \ °C/m$ $K_b = 0.512 \ °C/m$

$\Delta T_f = iK_f m = (1)(1.86 \ °C/m)(0.0558 \ m) = \textbf{0.104°C}$

$\Delta T_b = iK_b m = (1)(0.512 \ °C/m)(0.0558 \ m) = \textbf{0.0286°C}$

14-88. *Refer to Section 14-15 and Table 14-3.*

From Table 14-3, the van't Hoff factor, i, for $1.00 \ m$ K_2CrO_4 is 1.95. Assume that i for $1.20 \ m$ is the same.

$$m_{effective} = i \times m_{stated} = 1.95 \times 1.20 \ m = 2.34 \ m$$

At this fairly concentrated concentration, we cannot say that $M \cong m$, so we must calculate the molarity using density of the solution given as 1.25 g/mL. Note: if a $1.20 \ m$ K_2CrO_4 soln contained 1000 g of water, there was also 1.20 mol (233 g) of K_2CrO_4 in the solution as well, hence the unit factor: 1000 g H_2O/1233 g soln.

$$? \ M \ K_2CrO_4 \text{ soln} = \frac{1.20 \text{ mol}}{1000 \text{ g } H_2O} \times \frac{1000 \text{ g } H_2O}{1233 \text{ g soln}} \times \frac{1.25 \text{ g soln}}{1 \text{ mL soln}} \times \frac{1000 \text{ mL soln}}{1 \text{ L soln}} = 1.22 \ M \ \text{(use 1.2165 in next calculation)}$$

osmotic pressure, $\pi = iMRT = (1.95)(1.22 \ M)(0.0821 \text{ L·atm/mol·K})(25°C + 273°) = \textbf{58.0 atm}$ (3 significant figures)

14-90. *Refer to Sections 14-12 and 14-15, Table 14-2, and Example 14-14.*

From Table 14-2, for water: $K_f = 1.86 \ °C/m$

(a) $m = \dfrac{\text{mol substance}}{\text{kg solvent}} = \dfrac{0.0110 \text{ g}/2.00 \times 10^4 \text{ g/mol}}{0.0100 \text{ kg}} = 5.50 \times 10^{-5} \ m$

    $\Delta T_f = iK_f m = (1)(1.86 \ °C/m)(5.50 \times 10^{-5} \ m) = 1.02 \times 10^{-4} \ °C$

    $T_{f(soln)} = T_{f(water)} - \Delta T_f = 0°C - 1.02 \times 10^{-4} \ °C = \textbf{–1.02} \times \textbf{10}^{\textbf{–4}} \ \textbf{°C}$

(b) For dilute solutions, $M \cong m$,

    $\pi = MRT \cong mRT = (5.50 \times 10^{-5} \ m)(0.0821 \text{ L·atm/mol·K})(25°C + 273°) = \textbf{1.35} \times \textbf{10}^{\textbf{–3}} \textbf{ atm}$ or $\textbf{1.02 torr}$

(c) From the equation given in (a),

    % error in $T_{f(soln)}$ = % error in ΔT_f = % error in MW

    $? \ \%$ error in $T_{f(soln)} = \dfrac{\text{error in } T_f}{T_f} \times 100\% = \dfrac{0.001°C}{1.02 \times 10^{-4} \ °C} \times 100\% = \textbf{1000\%}$

Therefore, an error of only 0.001°C in the freezing point temperature corresponds to a 1000% error in the macromolecule's molecular weight.

(d) Since the osmotic pressure, $\pi = MRT = \dfrac{nRT}{V} = (\text{mol/MW})\dfrac{RT}{V}$, % error in π = % error in MW

    $? \ $ error in $\pi = \dfrac{\text{error in } \pi}{\pi} \times 100\% = \dfrac{0.1 \text{ torr}}{1.02 \text{ torr}} \times 100\% = \textbf{10\% error}$

Therefore, an error of 0.1 torr in osmotic pressure gives only a 10% error in molecular weight.

14-92. *Refer to Table 14-4 and the Key Terms for Chapter 14.*

(a) sol — a colloidal suspension of a solid dispersed in a liquid, e.g., detergents in water

(b) gel — a colloidal suspension of a solid dispersed in a liquid; a semirigid sol, e.g., jelly

(c) emulsion — a colloidal suspension of a liquid in a liquid, e.g., some cough medicines

(d) foam — a colloidal suspension of a gas in a liquid, e.g., bubbles in a bubble bath

(e) solid sol — a colloidal suspension of a solid in a solid, e.g., dirty ice

(f) solid emulsion — a colloidal suspension of a liquid dispersed in a solid, e.g., some kinds of sea ice containing pockets of brine

(g) solid foam — a colloidal suspension of a gas dispersed in a solid, e.g., marshmallows

(h) solid aerosol — a colloidal suspension of a solid in a gas, e.g., fine dust

(i) liquid aerosol — a colloidal suspension of a liquid in gas, e.g., insect spray

---

14-94. *Refer to Section 14-18 and the Key Terms for Chapter 14.*

Hydrophilic colloids are colloidal particles that attract water molecules, whereas hydrophobic colloids are colloidal particles that repel water molecules.

14-96. *Refer to Section 14-18.*

Soaps and detergents are both emulsifying agents. Solid soaps are usually sodium salts of long chain organic acids called fatty acids with the general formula, $R\text{-}COO^-Na^+$. On the other hand, synthetic detergents contain sulfonate, $-SO_3^-$, sulfate, $-SO_4^-$, or phosphate groups instead of carboxylate groups, $-COO^-$.

"Hard" water contains Fe^{3+}, Ca^{2+} and/or Mg^{2+} ions, all of which displace Na^+ from soap molecules and give an undesirable precipitate coating. However, detergents do not form precipitates with the ions of "hard" water.

A typical equation between soap and hard water that contains Ca^{2+} ions is shown below:

$$Ca^{2+}(aq) + 2RCOO^-Na^+(aq) \rightarrow (RCOO^-)_2Ca^{2+}(s) + 2Na^+(aq)$$

14-98. *Refer to Section 14-14 and Exercise 14-88 Solution.*

Plan: (1) Evaluate the van't Hoff factor, i, from the freezing point data.
        (2) Calculate the boiling point of the solution.
        (3) Calculate the osmotic pressure of the solution.

(1) $? \; m$ of $AlCl_3$ soln $= \dfrac{\text{mol } AlCl_3}{\text{kg solvent}} = \dfrac{1.56 \text{ g}/133.3 \text{ g/mol}}{0.0500 \text{ kg } H_2O} = 0.234 \; m \; AlCl_3$

we know: $\Delta T_f = iK_f m$, so

$i = \dfrac{\Delta T_f}{K_f m} = \dfrac{0°C - (-1.61°C)}{(1.86°C/m)(0.234 \; m \; AlCl_3)} = 3.70$

(2) $\Delta T_b = iK_b m = (3.70)(0.512 \; °C/m)(0.234 \; m) = 0.443 \; °C$

$T_{b(\text{soln})} = T_{b(\text{solvent})} + \Delta T_b = 100.00°C + 0.443°C = \mathbf{100.44°C}$ (to 2 decimal places)

(3) $? \, M \, AlCl_3 \, soln = \dfrac{0.234 \, mol \, AlCl_3}{1000 \, g \, H_2O} \times \dfrac{50.0 \, g \, H_2O}{51.56 \, g \, soln} \times \dfrac{1.002 \, g \, soln}{1 \, mL \, soln} \times \dfrac{1000 \, mL \, soln}{1 \, L \, soln} = 0.227 \, M$

osmotic pressure, $\pi = iMRT = (3.70)(0.227 \, M)(0.0821 \, L \cdot atm/mol \cdot K)(25°C + 273°) = \mathbf{20.5 \, atm}$

14-100. *Refer to Sections 14-12 and 14-13, and Exercise 13-20 Solution.*

From Table 14-2,   for water: $T_f = 0°C$ $K_f = 1.86 \, °C/m$
                       for benzene: $T_f = 5.48°C$ $K_f = 5.12 \, °C/m$

Plan: (1) Solve for the molality of CH_3COOH in water or benzene using $\Delta T_f = K_f m$ (assume $i = 1$).
          (2) Calculate the apparent molecular weight of CH_3COOH in each solvent.

(a) in aqueous solution:

   (1) $m_{apparent} = \dfrac{\Delta T_f}{K_f} = \dfrac{0.310°C}{1.86 \, °C/m} = 0.167 \, m \, CH_3COOH \, in \, H_2O$

   (2) Recall: $m = \dfrac{mol \, solute}{kg \, solvent} = \dfrac{g \, solute/MW}{kg \, solvent}$

      Assume 100.00 g of solution containing 1.00 g of CH_3COOH in 99.00 g of water.

      $? \, MW_{apparent} = \dfrac{g \, CH_3COOH}{m_{apparent} \times kg \, H_2O} = \dfrac{1.00 \, g}{0.167 \, m \times 0.09900 \, kg} = \mathbf{60.5 \, g/mol}$

The true molecular weight of acetic acid is 60.05 g/mol. Acetic acid is a weak acid and dissociates very slightly in water; the van't Hoff factor, i, is then only slightly larger than 1. The behavior of acetic acid approximates that of a nonelectrolyte in water.

(b) in benzene solution:

   (1) $m_{apparent} = \dfrac{\Delta T_f}{K_f} = \dfrac{0.441°C}{5.12 \, °C/m} = 0.0861 \, m \, CH_3COOH \, in \, benzene$

   (2) Assume 100.00 g of solution containing 1.00 g of CH_3COOH in 99.00 g of benzene.

      $? \, MW_{apparent} = \dfrac{g \, CH_3COOH}{m_{apparent} \times kg \, benzene} = \dfrac{1.00 \, g}{0.0861 \, m \times 0.09900 \, kg} = \mathbf{117 \, g/mol}$

The apparent molecular weight, determined in benzene is almost twice as large as the true molecular weight. The reason is that in benzene, the acetic acid dimerizes due to hydrogen bonding (refer to the figure in Exercise 13-20 Solution). Most of the acetic acid exists as one particle called a dimer, consisting of 2 acetic acid molecules held together by hydrogen bonds.

14-102. *Refer to Section 14-18.*

(a) hydrophobic - hydrocarbons   (b) hydrophilic - starch

(c) The hydrophilic dispersion of starch in water is much easier to make and maintain. Using hydrogen bonding, the very polar water molecules surround and isolate each starch molecule from one another. The starch molecules cannot coalesce and therefore remain dispersed.

14-104. *Refer to Section 14-15.*

Plan: (1) Determine the total molarity of the drug-sugar mixture dissolved in water, using $\pi = MRT$ ($i = 1$).
          (2) Determine the mass of lactose in the solution.
          (3) Calculate the % lactose present.

(1) $M_{lactose} + M_{drug} = M_{total} = \dfrac{\pi}{RT} = \dfrac{(519/760)\ atm}{(0.0821\ L\cdot atm/mol\cdot K)(25°C + 273)} = 0.0279\ M_{total}$

(2) Let $\quad x$ = g lactose, $C_{12}H_{22}O_{11}$ (MW: 342 g/mol)

$\qquad\quad 1 - x$ = g drug, $C_{21}H_{23}O_5N$ (MW: 369 g/mol)

$M_{total} = \dfrac{mol\ lactose + mol\ drug}{L\ solution}$

Substituting,

$$0.0279\ M = \left(\dfrac{\dfrac{x}{342\ g/mol} + \dfrac{1.00 - x}{369\ g/mol}}{0.100\ L} \right)$$

$$0.00279 = \dfrac{x}{342} + \dfrac{1.00 - x}{369}$$

Multiplying both sides by the product, (342)(369), we obtain

$$352 = 369x + 342(1.00 - x)$$

$$10. = 27x$$

$$x = 0.37\ g\ lactose$$

(3) ? lactose by mass $= \dfrac{0.37\ g\ lactose}{1.00\ g\ mixture} \times 100\% = \textbf{37\% lactose by mass}$

14-106. *Refer to Sections 14-1 and 14-9.*

To have an ideal solution, the solvent-solvent, solute-solute and solvent-solute intermolecular forces should be as nearly identical as possible. Solution (c) consisting of $CH_4(\ell)$ dissolved in $CH_3CH_3(\ell)$ would be the most ideal.

14-108. *Refer to Chapter 14.*

In chemistry, the word "dissolution" is used very specifically to describe the process by which a solute is dispersed by a solvent to form a solution. In popular or common usage, it is defined as the decomposition of a whole into fragments or parts.

When we speak of the dissolution of an estate, we are applying the common definition and we know that the estate will be broken up into parts. In chemistry, when we discuss the ease of solute dissolution, we are implying a full range of chemical and physical processes that take place when a solute dissolves.

14-110. *Refer to Section 14-14 and Table 14-3.*

The actual value for the van't Hoff factor can be larger than the ideal value if more particles are produced than expected, e.g., a molecular compound might be considered to have $i_{ideal} = 1$, but if it slightly ionizes, the i_{actual} would be larger than 1.

In another instance, suppose we thought we were making a solution of the molecule S_8 by weighing out a known mass. Sulfur also exists as S_2, S_4 and S_6. If these other allotropes were accidentally present in the solution, the observed i would be greater than expected, since there would be more particles present than expected.

Another case might involve dissolving an unstable compound in solution. If it decomposed into two or more soluble parts, the observed i would again be greater than expected.

Plan: (A) Calculate $m_{effective}$ at each concentration.
          (B) Determine the % ionization of the monoprotic acid.

(1) For 1.0 *m* acid:

(A) From Table 14-2, for water: $T_f = 0°C$, $K_f = 1.86$ °C/*m*

$$m_{effective} = \frac{\Delta T_f}{K_f} = \frac{0.0000°C - (-3.35°C)}{1.86 \ °C/m} = 1.80 \ m$$

(B) Let   x = molality of HA that ionizes
     Then   x = molality of H^+ and A^- that formed

Consider:	HA	$\rightleftarrows$	H^+	+	A^-
Start	1.0 *m*		$\approx 0 \ m$		0 *m*
Change	- x *m*		+ x *m*		+ x *m*
Final	(1.0 - x) *m*		x *m*		x *m*

$m_{effective} = m_{HA} + m_{H^+} + m_{A^-}$
$1.80 = (1.0 - x) + x + x$
$1.80 = 1.0 + x$
$x = 0.8 \ m$

$$\% \text{ ionization} = \frac{m_{ionized}}{m_{original}} \times 100\% = \frac{0.8 \ m}{1.0 \ m} \times 100\% = \textbf{80\%}$$

(2) For 2.0 *m* acid:

(A) From Table 14-2, for water: $T_f = 0°C$, $K_f = 1.86$ °C/*m*

$$m_{effective} = \frac{\Delta T_f}{K_f} = \frac{0.0000°C - (-6.10°C)}{1.86 \ °C/m} = 3.28 \ m$$

(B) Let   x = molality of HA that ionizes
     Then   x = molality of H^+ and A^- that formed

Consider:	HA	$\rightleftarrows$	H^+	+	A^-
Start	2.0 *m*		$\approx 0 \ m$		0 *m*
Change	- x *m*		+ x *m*		+ x *m*
Final	(2.0 - x) *m*		x *m*		x *m*

$m_{effective} = m_{HA} + m_{H^+} + m_{A^-}$
$3.28 = (2.0 - x) + x + x$
$3.28 = 2.0 + x$
$x = 1.3 \ m$

$$\% \text{ ionization} = \frac{m_{ionized}}{m_{original}} \times 100\% = \frac{1.3 \ m}{2.0 \ m} \times 100\% = \textbf{65\%}$$

(3) For 4.0 *m* acid:

(A) From Table 14-2, for water: $T_f = 0°C$, $K_f = 1.86$ °C/*m*

$$m_{effective} = \frac{\Delta T_f}{K_f} = \frac{0.0000°C - (-8.70°C)}{1.86 \ °C/m} = 4.68 \ m$$

(B) Let   x = molality of HA that ionizes
     Then   x = molality of H^+ and A^- that formed

Consider:	HA	$\rightleftarrows$	H^+	+	A^-
Start	4.0 *m*		$\approx 0 \ m$		0 *m*
Change	- x *m*		+ x *m*		+ x *m*
Final	(4.0 - x) *m*		x *m*		x *m*

$$m_{\text{effective}} = m_{HA} + m_{H^+} + m_{A^-}$$
$$4.68 = (4.0 - x) + x + x$$
$$4.68 = 4.0 + x$$
$$x = 0.68 \, m$$

$$\% \text{ ionization} = \frac{m_{\text{ionized}}}{m_{\text{original}}} \times 100\% = \frac{0.68 \, m}{4.0 \, m} \times 100\% = \mathbf{17\%}$$

As you can see, the percent ionization decreases as the solution concentration increases.

Acid Concentration	Percent Ionization
1.0 m	80%
2.0 m	65%
4.0 m	17%

14-114. *Refer to Section 14-8.*

(a) The solution on the right is more concentrated. Concentration can be expressed as a ratio of the amount of solute to the amount of solution. The solution on the right has the same amount of solute as the solution on the left, but it is dissolved in less solution, so it is more concentrated.

(b) The 1.00 m solution is the one on the left, because it was prepared by dissolving 1.00 mol of K_2CrO_4 in 1.00 kg of water (1.00 liter = 1.00 kg since the density of water is 1.00 g/mL).

(c) The 1.00 M solution is the one on the right, because it was prepared by dissolving 1.00 mol of K_2CrO_4 to make 1.00 L of solution, as measured by the volumetric flask.

14-116. *Refer to Section 14-5.*

Rock candy consists of crystals of sugar on a stick or string. It could be made by first preparing a supersaturated solution of sugar.

(1) Heat some water to boiling in a glass or metal container with smooth sides and dissolve as much sugar as possible.

(2) Remove the hot saturated sugar solution carefully from the heat source and allow it to cool slowly. A supersaturated solution should result. A smooth container is critical so that the sugar doesn't start to crystallize before you're ready.

(3) Then put a stick or string with rough sides into the solution, perhaps with some sugar crystals, called seed crystals, imbedded in it. This will cause crystallization to begin and large sugar crystals should form on the stick or string in time.

14-118. *Refer to Sections 13-2 and 14-2.*

From the data given, DDT, a toxic pesticide now banned in the United States, is not very soluble in water, but is soluble in fat and fatty tissue. Therefore, it probably is not ionic, so therefore it does not have ion-ion interactions. It probably does not have hydrogen bonding in operation to any extent, because it is not very soluble in water, but is soluble in fat. So, most likely DDT has only dipole-dipole and dispersion forces in operation between its molecules.

DDT

In fact, DDT is a polar molecule and so there are dipole-dipole and dispersion forces present. Its official IUPAC name is 1,1,1-trichloro-2,2-bis(p-chlorophenyl)ethane and its structure is given here.

Plan: (1) Determine the osmotic pressure (in torr) required to force the sap to the top of the tree.
         (2) Calculate the molarity of the sugar solution (sap).

(1) ? height of the tree (mm) = 40. ft $\times \dfrac{12 \text{ in}}{1 \text{ ft}} \times \dfrac{2.54 \text{ cm}}{1 \text{ in}} \times \dfrac{10 \text{ mm}}{1 \text{ cm}} = 1.2 \times 10^4$ mm

The pressure exerted by a column of sugar solution in the tree can be stated as 1.2×10^4 mm sugar solution. However, pressure is more often given in units of the height of an equivalent column of mercury in millimeters, called torr, which can be easily converted to atmospheres. The unit factor relating a column of mercury to a column of a liquid exerting the same pressure involves their densities. In this case since the density of the solution is only 1.10/13.6 that of mercury as determined from the relative densities, a 13.6 mm column of solution exerts the same pressure as a 1.10 mm column of mercury.

Thus,

? pressure (atm) = 1.2×10^4 mm sugar soln $\times \dfrac{1.10 \text{ mm Hg}}{13.6 \text{ mm sugar soln}} \times \dfrac{1 \text{ torr}}{1 \text{ mm Hg}} \times \dfrac{1 \text{ atm}}{760 \text{ torr}} = \mathbf{1.3 \text{ atm}}$

(2) Recall, osmotic pressure, $\pi = iMRT$

$M = \dfrac{\pi}{iRT} = \dfrac{1.3 \text{ atm}}{(1)(0.0821 \text{ L·atm/mol·K})(273 \text{ K} + 15°\text{C})} = \mathbf{0.054 \text{ } M}$

Note: If the numbers are not rounded off to 2 significant figures until the end, the answer is 0.055 *M*.

15 Chemical Thermodynamics

15-2. *Refer to the Key Terms for Chapters 1 and 15.*

(a) Heat is a form of energy that flows between two samples of matter due to their differences in temperature.

(b) Temperature is a measure of the intensity of heat, i.e., the hotness or coldness of an object.

(c) The system refers to the substances of interest in a process, i.e., it is the part of the universe that is under investigation.

(d) The surroundings refer to everything in the environment of the system of interest.

(e) The thermodynamic state of a system refers to a set of conditions that completely specifies all of the thermodynamic properties of the system.

(f) Work is the application of a force through a distance. For physical or chemical changes that occur at constant pressure, the work done on the system is $-P\Delta V$.

15-4. *Refer to Sections 1-1, 15-1 and 15-10, and Figure 15-1.*

(a) When heat is given off by a system, that system is labeled an exothermic process. Figure 15-1 describes such a process: the combustion of 1 mole of methane at 25°C:

$$CH_4(g) + 2O_2(g) \rightarrow CO_2(g) + 2H_2O(\ell) + 890 \text{ kJ}$$

(b) In this example, the system's volume at constant pressure decreases since the system initially has 3 moles of gas, but produces only 1 mole of gas, since volume is directly proportional to moles of gas at constant temperature and pressure. When gas is consumed in a process (e.g. in this process: 3 moles gaseous reactants $\rightarrow$ 1 mole gaseous product), the surroundings are doing work on the system.

Note: When we look at the volume of a system, we generally consider only the gases, since the volume of 1 mole of gas is so much greater than the volume of 1 mole of a liquid or solid, which are much more dense.

15-6. *Refer to Sections 1-1 and 15-1.*

An endothermic process absorbs heat energy from its surroundings; an exothermic process releases heat energy to its surroundings. If a reaction is endothermic in one direction, it is exothermic in the opposite direction. For example, the melting of 1 mole of ice water is an endothermic process requiring 6.02 kJ of heat:

$$H_2O(s) + 6.02 \text{ kJ} \rightarrow H_2O(\ell)$$

The reverse process, the freezing of 1 mole of liquid water, releasing 6.02 kJ of heat, is an exothermic process:

$$H_2O(\ell) \rightarrow H_2O(s) + 6.02 \text{ kJ}$$

15-8. *Refer to Section 15-1.*

According to the First Law of Thermodynamics, the total amount of energy in the universe is constant. When an incandescent light is turned on, electrical energy is converted mainly into light and heat energy. A small fraction of the energy is converted to chemical energy, which is why the filament eventually burns out.

A state function is a variable that defines the state of a system; it is a function that is independent of the pathway by which a process occurs. Therefore, the change in a state function depends only on the initial and the final value, not on how that change occurred.

(a) Your bank balance is a state function, because it depends only on the difference between your deposits and withdrawals.

(b) The mass of a candy bar is a state function.

(c) However, your weight is not a state function. Weight depends on the gravitational attraction of your body to the center of the earth, which changes depending on where you are.

(d) The heat lost by perspiration during a climb up a mountain along a fixed path is not a state function, because it depends on the person - her/his size, build, metabolism and degree of fitness.

15-12.   *Refer to Sections 15-5, 15-6 and 15-17.*

(a) ΔH, the enthalpy change or heat of reaction, is the heat change of a reaction occurring at some constant pressure and temperature.

$\Delta H°$ is the standard enthalpy change of a reaction that occurs at 1 atm pressure. Unless otherwise stated, the reaction temperature is 25°C.

(b) As stated in (a), $\Delta H°_{rxn}$ is the standard enthalpy change of a reaction occurring at 1 atm pressure.

$\Delta H°_f$, the standard molar enthalpy of formation of a substance, is the enthalpy change for a reaction in which 1 mole of the substance in a specific state is formed from its elements in their standard states.

15-14.   *Refer to Sections 15-4 and 15-5.*

(i) Since the reaction is endothermic, (a) enthalpy increases, (b) $H_{product} > H_{reactant}$ and (c) ΔH is positive.

(ii) Since the reaction is exothermic, (a) enthalpy decreases, (b) $H_{reactant} > H_{product}$ and (c) ΔH is negative.

15-16.   *Refer to Section 15-5 and Example 15-5.*

Balanced equation: $CH_3OH(g) + \frac{3}{2}O_2(g) \rightarrow CO_2(g) + 2H_2O(\ell)$ $\Delta H = -764$ kJ/mol rxn

(a) Plan: heat evolved/mol rxn $\overset{(1)}{\Rightarrow}$ heat evolved/mol CH_3OH $\overset{(2)}{\Rightarrow}$ heat evolved/g CH_3OH $\overset{(3)}{\Rightarrow}$ heat evolved

? heat evolved (kJ) $= \dfrac{764 \text{ kJ}}{\text{mol rxn}} \times \dfrac{1 \text{ mol rxn}}{1 \text{ mol } CH_3OH} \times \dfrac{1 \text{ mol } CH_3OH}{32.0 \text{ g } CH_3OH} \times 115.0 \text{ g } CH_3OH = \textbf{2750 kJ evolved}$

(b) Plan: heat evolved $\overset{(1)}{\Rightarrow}$ mol reaction $\overset{(2)}{\Rightarrow}$ mol O_2 $\overset{(3)}{\Rightarrow}$ g O_2

? g $O_2 = 925$ kJ $\times \dfrac{1 \text{ mol rxn}}{764 \text{ kJ}} \times \dfrac{1.5 \text{ mol } O_2}{1 \text{ mol rxn}} \times \dfrac{32.0 \text{ g } O_2}{1 \text{ mol } O_2} = \textbf{58.1 g } \mathbf{O_2}$

15-18.   *Refer to Section 15-5 and Example 15-5.*

Balanced equation: $PbO(s) + C(s) \rightarrow Pb(s) + CO(g)$

Since the equation involves one mole of PbO, ΔH can be expressed in the units of kJ/mol PbO.

? heat supplied to the reaction $= \dfrac{5.95 \text{ kJ}}{13.43 \text{ g PbO}} \times \dfrac{223.2 \text{ g PbO}}{1 \text{ mol PbO}} = 98.9$ kJ/mol PbO

Therefore, since the heat is being added to the reaction, $\Delta H = \textbf{+98.9 kJ/mol rxn}$

15-20. *Refer to Section 15-3 and Appendix K.*

The standard molar enthalpy of formation, ΔH_f°, is the amount of heat absorbed when 1 mole of the substance is produced from its elements in their standard states. At 25°C, ΔH_f° of liquid water is −285.8 kJ/mol and ΔH_f° of water vapor is −241.8 kJ/mol. This means that more heat is released when liquid water is formed from its elements, then when gaseous water is formed from its elements. So, the formation reaction of liquid water is more exothermic, which means that $H_2O(\ell)$ has a lower enthalpy than $H_2O(g)$. See the Solution to Exercise 15-22.

15-22. *Refer to Sections 15-1 and 15-5.*

Consider the balanced reactions: (1) $CH_4(g) + 2O_2(g) \rightarrow CO_2(g) + 2H_2O(\ell)$ $\Delta H_1 = (-)$

(2) $CH_4(g) + 2O_2(g) \rightarrow CO_2(g) + 2H_2O(g)$ $\Delta H_2 = (-)$

The only difference between them is that Reaction (1) involves water in the liquid phase and Reaction (2) involves water as water vapor. Since more heat is released when $H_2O(g) \rightarrow H_2O(\ell)$, as shown in the adjacent diagram, Reaction (1) is more exothermic than Reaction (2).

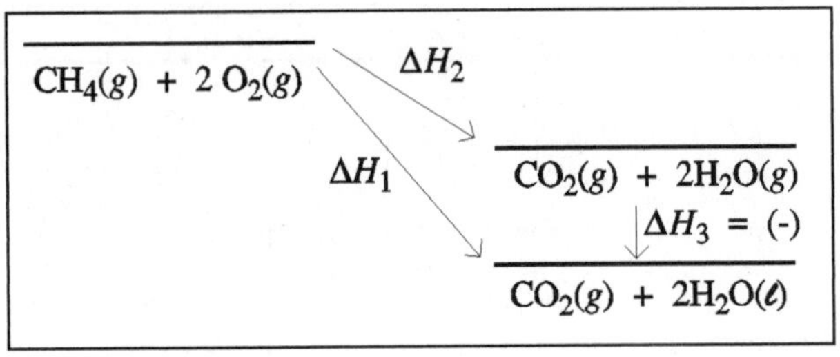

$$CH_4(g) + 2\,O_2(g)$$

$$\Delta H_2$$

$$\Delta H_1$$

$$CO_2(g) + 2H_2O(g)$$

$$\Delta H_3 = (-)$$

$$CO_2(g) + 2H_2O(\ell)$$

15-24. *Refer to Section 15-6 and the Key Terms for Chapter 15.*

The thermochemical standard state of a substance is its most stable state under standard pressure (1 atm) and at some specific temperature (usually 25°C). "Thermochemical" refers to the observation, measurement and prediction of energy changes that accompany physical changes or chemical reaction. "Standard" refers to the set conditions of 1 atm pressure and 25°C. The "state" of a substance is its phase: gas, liquid or solid. "Substance" is any kind of matter all specimens of which have the same chemical composition and physical properties.

15-26. *Refer to Section 15-7 and Appendix K.*

The standard molar enthalpy of formation, ΔH_f°, of elements in their standard states is zero. From the tabulated values of standard molar enthalpies in Appendix K, we can identify the standard states of elements.

(a) chlorine $Cl_2(g)$

(b) chromium $Cr(s)$

(c) bromine $Br_2(\ell)$

(d) iodine $I_2(s)$

(e) sulfur $S(s,\text{rhombic})$

(f) nitrogen $N_2(g)$

15-28. *Refer to Section 15-7, Example 15-6 and Appendix K.*

Hint: Use Appendix K to identify an element's standard state since its ΔH_f° value is equal to zero.

(a) $Ca(s) + O_2(g) + H_2(g) \rightarrow Ca(OH)_2(s)$

(b) $6C(s, \text{graphite}) + 3H_2(g) \rightarrow C_6H_6(\ell)$

(c) $Na(s) + \frac{1}{2}H_2(g) + C(s,\text{graphite}) + \frac{3}{2}O_2(g) \rightarrow NaHCO_3(s)$

(d) $Ca(s) + F_2(g) \rightarrow CaF_2(s)$

(e) $\frac{1}{4}P_4(s,\text{white}) + \frac{3}{2}H_2(g) \rightarrow PH_3(g)$

(f) $3C(s,\text{graphite}) + 4H_2(g) \rightarrow C_3H_8(g)$

(g) $S(s,\text{rhombic}) \rightarrow S(g)$

(h) $H_2(g) + \frac{1}{2}O_2(g) \rightarrow H_2O(\ell)$

15-30. *Refer to Section 15-5.*

The balanced equation for the standard molar enthalpy of formation of $Li_2O(s)$ is: $2Li(s) + \frac{1}{2}O_2(g) \rightarrow Li_2O(s)$

$? \text{ kJ/mol } Li_2O(s) = \dfrac{146 \text{ kJ}}{3.47 \text{ g Li}} \times \dfrac{6.94 \text{ g Li}}{1 \text{ mol Li}} \times \dfrac{2 \text{ mol Li}}{1 \text{ mol } Li_2O} = 584 \text{ kJ/mol}$

And so, $\Delta H^{\circ}_{f\, Li_2O(s)} = \mathbf{-584 \text{ kJ/mol}}$ since the reaction is exothermic.

15-32. *Refer to Section 15-8 and Examples 15-7 and 15-8.*

To obtain the desired equation,

(1) divide the first equation by 2 to give 2 moles of HCl on the reactant side,

(2) multiply the second equation by 2, giving 2 moles of HF on the product side. Then,

(3) reverse the third equation, so that H_2O, H_2 and $\frac{1}{2}O_2$ are eliminated when the modified equations are added together.

$$\begin{array}{lr} & \Delta H^{\circ} \\ 2HCl(g) + \frac{1}{2}O_2(g) \rightarrow H_2O(\ell) + Cl_2(g) & -101.2 \text{ kJ/mol rxn} \\ H_2(g) + F_2(g) \rightarrow 2HF(\ell) & -1200.0 \text{ kJ/mol rxn} \\ \underline{H_2O(\ell) \rightarrow H_2(g) + \frac{1}{2}O_2(g)} & \underline{+285.8 \text{ kJ/mol rxn}} \\ 2HCl(g) + F_2(g) \rightarrow 2HF(\ell) + Cl_2(g) & \mathbf{-1015.4 \text{ kJ/mol rxn}} \end{array}$$

15-34. *Refer to Section 15-8 and Examples 15-7 and 15-8.*

To obtain the desired equation,

(1) multiply the first equation by 2 to give 2 moles of SO_2 on the product side, then

(2) reverse the second equation and multiply by 2, giving 2 moles of SO_3 on the reactant side.

$$\begin{array}{lr} & \Delta H^{\circ} \\ 2S(s) + 2O_2(g) \rightarrow 2SO_2(g) & -593.6 \text{ kJ/mol rxn} \\ \underline{2SO_3(g) \rightarrow 2S(s) + 3O_2(g)} & \underline{+791.2 \text{ kJ/mol rxn}} \\ 2SO_3(g) \rightarrow 2SO_2(g) + O_2(g) & \mathbf{+197.6 \text{ kJ/mol rxn}} \end{array}$$

15-36. *Refer to Section 15-8 and Examples 15-7 and 15-8.*

To obtain the desired hydrogenation equation,

(1) use the first equation as it is to give 2 moles of H_2 on the reactant side,

(2) use the second equation as it is to give 1 mole of C_3H_4 on the reactant side, then

(3) reverse the third equation to give 1 mole of C_3H_8 on the product side.

$$\begin{array}{lr} & \Delta H^{\circ} \\ 2H_2(g) + O_2(g) \rightarrow 2H_2O(\ell) & -571.6 \text{ kJ/mol rxn} \\ C_3H_4(g) + 4O_2(g) \rightarrow 3\,CO_2(g) + 2H_2O(\ell) & -1937 \text{ kJ/mol rxn} \\ \underline{3CO_2(g) + 4H_2O(\ell) \rightarrow C_3H_8(g) + 5O_2(g)} & \underline{+2220. \text{ kJ/mol rxn}} \\ C_3H_4(g) + 2H_2(g) \rightarrow C_3H_8(g) & \mathbf{-289 \text{ kJ/mol rxn}} \end{array}$$

15-38. *Refer to Section 15-8, Example 15-9 and Appendix K.*

(a) Balanced equation: $NH_4NO_3(s) \rightarrow N_2O(g) + 2H_2O(\ell)$

$$\Delta H^{\circ}_{rxn} = [\Delta H^{\circ}_{f\ N_2O(g)} + 2\Delta H^{\circ}_{f\ H_2O(\ell)}] - [\Delta H^{\circ}_{f\ NH_4NO_3(s)}]$$
$$= [(1\ mol)(82.05\ kJ/mol) + (2\ mol)(-285.8\ kJ/mol)] - [(1\ mol)(-365.6\ kJ/mol)]$$
$$= \mathbf{-124.0\ kJ/mol\ rxn}$$

(b) Balanced equation: $2FeS_2(s) + \frac{11}{2}O_2(g) \rightarrow Fe_2O_3(s) + 4SO_2(g)$

$$\Delta H^{\circ}_{rxn} = [\Delta H^{\circ}_{f\ Fe_2O_3(s)} + 4\Delta H^{\circ}_{f\ SO_2(g)}] - [2\Delta H^{\circ}_{f\ FeS_2(s)} + \frac{11}{2}\Delta H^{\circ}_{f\ O_2(g)}]$$
$$= [(1\ mol)(-824.2\ kJ/mol) + (4\ mol)(-296.8\ kJ/mol)] - [(2\ mol)(-177.5\ kJ/mol) + (\tfrac{11}{2}\ mol)(0\ kJ/mol)]$$
$$= \mathbf{-1656\ kJ/mol\ rxn}$$

(c) Balanced equation: $SiO_2(s) + 3C(s,graphite) \rightarrow SiC(s) + 2CO(g)$

$$\Delta H^{\circ}_{rxn} = [\Delta H^{\circ}_{f\ SiC(s)} + 2\Delta H^{\circ}_{f\ CO(g)}] - [\Delta H^{\circ}_{f\ SiO_2(s)} + 3\Delta H^{\circ}_{f\ C(s,graphite)}]$$
$$= [(1\ mol)(-65.3\ kJ/mol) + (2\ mol)(-110.5\ kJ/mol)] - [(1\ mol)(-910.9\ kJ/mol) + (3\ mol)(0\ kJ/mol)]$$
$$= \mathbf{+624.6\ kJ/mol\ rxn}$$

15-40. *Refer to Section 15-5 and Examples 15-4 and 15-5.*

(1) Balanced equation for combustion of propane: $C_3H_8(g) + 5O_2(g) \rightarrow 3CO_2(g) + 4H_2O(g)$

$$\Delta H^{\circ}_{combustion} = [3\Delta H^{\circ}_{f\ CO_2(g)} + 4\Delta H^{\circ}_{f\ H_2O(g)}] - [\Delta H^{\circ}_{f\ C_3H_8(g)} + 5\Delta H^{\circ}_{f\ O_2(g)}]$$
$$= [(3\ mol)(-393.5\ kJ/mol) + (4\ mol)(-241.8\ kJ/mol)]$$
$$- [(1\ mol)(-103.8\ kJ/mol) + (5\ mol)(0\ kJ/mol)]$$
$$= -2043.9\ kJ/mol\ C_3H_8$$

$$\text{heat released (kJ/g)} = \frac{2043.9\ kJ}{1\ mol\ C_3H_8} \times \frac{1\ mol\ C_3H_8}{44.09\ g\ C_3H_8} = \mathbf{46.36\ kJ/g\ C_3H_8}$$

(2) Balanced equation for combustion of octane: $C_8H_{18}(\ell) + \frac{25}{2}O_2(g) \rightarrow 8CO_2(g) + 9H_2O(g)$

$$\Delta H^{\circ}_{combustion} = [8\Delta H^{\circ}_{f\ CO_2(g)} + 9\Delta H^{\circ}_{f\ H_2O(g)}] - [\Delta H^{\circ}_{f\ C_8H_{18}(\ell)} + \frac{25}{2}\Delta H^{\circ}_{f\ O_2(g)}]$$
$$= [(8\ mol)(-393.5\ kJ/mol) + (9\ mol)(-241.8\ kJ/mol)]$$
$$- [(1\ mol)(-268.8\ kJ/mol) + (\tfrac{25}{2}\ mol)(0\ kJ/mol)]$$
$$= -5055.4\ kJ/mol\ C_8H_{18}$$

$$\text{heat released (kJ/g)} = \frac{5055.4\ kJ}{1\ mol\ C_8H_{18}} \times \frac{1\ mol\ C_8H_{18}}{114.2\ g\ C_8H_{18}} = \mathbf{44.27\ kJ/g\ C_8H_{18}}$$

Note: The sign convention for ΔH° tells the reader whether heat is being released or absorbed. However, when the question asks for "heat released" or "heat absorbed," the value of heat is a positive number. When the words "released" or "absorbed" are used, the sign convention is *not* used.

15-42. *Refer to Section 15-5 and Example 15-4.*

Balanced equation: $8Al(s) + 3Fe_3O_4(s) \rightarrow 4Al_2O_3(s) + 9Fe(s)$ $\Delta H^{\circ} = -3350.\ kJ/mol\ rxn$

Plan: (1) Determine the limiting reactant.
        (2) Calculate the heat released based on the limiting reactant.

(1) $?\ mol\ Al = \dfrac{27.6\ g\ Al}{27.0\ g/mol} = 1.02\ mol\ Al$ $?\ mol\ Fe_3O_4 = \dfrac{69.12\ g\ Fe_3O_4}{231.6\ g/mol} = 0.2984\ mol\ Fe_3O_4$

   $\text{Required ratio} = \dfrac{8\ mol\ Al}{3\ mol\ Fe_3O_4} = 2.67$ $\text{Available ratio} = \dfrac{1.02\ mol\ Al}{0.2984\ mol\ Fe_3O_4} = 3.42$

Available ratio > Required ratio; Fe_3O_4 is the limiting reactant.

(2) $\Delta H° = 69.12 \text{ g } Fe_3O_4 \times \dfrac{1 \text{ mol } Fe_3O_4}{231.6 \text{ g } Fe_3O_4} \times \dfrac{-3350. \text{ kJ}}{3 \text{ mol } Fe_3O_4} = -333.3 \text{ kJ}$

Therefore, there are **+333 kJ** of heat released.

Note: The sign convention for $\Delta H°$ tells the reader whether heat is being released or absorbed. However, when the question asks for "heat released" or "heat absorbed," the value of heat is a positive number. When the words "released" or "absorbed" are used, the sign convention is *not* used.

15-44.   *Refer to Sections 15-5 and 15-8, Example 15-10, and Exercise 15-38(c).*

Plan: Use Hess's Law to solve for $\Delta H_f°$ of silicon carbide.

Balanced equation: $SiO_2(s) + 3C(\text{graphite}) \rightarrow SiC(s) + 2CO(g)$

$\Delta H°_{rxn} = [\Delta H_f° \text{ } SiC(s) + 2\Delta H_f° \text{ } CO(g)] - [\Delta H_f° \text{ } SiO_2(s) + 3\Delta H_f° \text{ } C(\text{graphite})]$

$624.6 \text{ kJ} = [(1 \text{ mol})(\Delta H_f° \text{ } SiC(s)) + (2 \text{ mol})(-110.5 \text{ kJ/mol})] - [(1 \text{ mol})(-910.9 \text{ kJ/mol}) + (3 \text{ mol})(0 \text{ kJ/mol})]$

$624.6 \text{ kJ} = (1 \text{ mol})(\Delta H_f° \text{ } SiC(s)) + 689.9 \text{ kJ}$

$\Delta H_f° \text{ } SiC(s) = $ **−65.3 kJ/mol SiC(s)**

15-46.   *Refer to Sections 15-7 and 15-9, and Table 15-3.*

(a) For a reaction occurring in the gaseous phase, the net enthalpy change, $\Delta H°_{rxn}$, equals the sum of the bond energies in the reactants minus the sum of the bond energies in the products:

$$\Delta H°_{rxn} = \Sigma \text{ B.E.}_{\text{reactants}} - \Sigma \text{ B.E.}_{\text{products}}$$

If the products have higher bond energy and are therefore more stable than the reactants, the reaction is exothermic. If the opposite is true, the reaction is endothermic.

(b) Consider $O_2(g)$: $\Delta H_f° \text{ } O_2(g) = 0$ kJ/mol since the standard state of oxygen is $O_2(g)$
$-\Sigma \text{ B.E.}_{O_2(g)} = -\text{B.E.}_{O=O} = -498$ kJ/mol

Therefore, one cannot say that $\Delta H_f° \text{ substance} = -\Sigma \text{ B.E.}_{\text{substance}}$. Bond energies are a measure of the energy involved in breaking of one mole of bonds in a gaseous substance to form gaseous atoms of the elements. The value of $\Delta H_f°$ is a measure of the energy involved in making one mole of the substance from its elements in their standard states. They differ in two major aspects: (1) In bond energy considerations, all the bonds are broken to give free atoms, while in $\Delta H_f°$ determinations, some bonds may still be maintained as diatomic or polyatomic free elements (e.g., $O_2(g)$ or $P_4(s)$.) (2) The standard states of the elements are not necessarily the gaseous state. Moreover, the $\Delta H_f°$ equation is an exact calculation, but the bond energy equation is only an estimation of $\Delta H_f°$.

15-48.   *Refer to Section 15-9, Tables 15-2 and 15-3, and Examples 15-11 and 15-12.*

(a) Balanced equation in terms of Lewis structures of the reactants and products:

$\Delta H°_{rxn} = \Sigma \text{ B.E.}_{\text{reactants}} - \Sigma \text{ B.E.}_{\text{products}}$ in the gas phase

$= [\text{B.E.}_{C=C} + 4\text{B.E.}_{C-H} + \text{B.E.}_{Br-Br}] - [\text{B.E.}_{C-C} + 4\text{B.E.}_{C-H} + 2\text{B.E.}_{C-Br}]$

$= [(1 \text{ mol})(602 \text{ kJ/mol}) + (4 \text{ mol})(413 \text{ kJ/mol}) + (1 \text{ mol})(193 \text{ kJ/mol})]$

$\qquad - [(1 \text{ mol})(346 \text{ kJ/mol}) + (4 \text{ mol})(413 \text{ kJ/mol}) + (2 \text{ mol})(285 \text{ kJ/mol})]$

$= $ **−121 kJ/mol rxn**

(b) Balanced equation in terms of Lewis structures of the reactants and products:

$$\text{H-\ddot{\underset{..}{O}}-\ddot{\underset{..}{O}}-H} \ (g) \quad \rightarrow \quad \text{H-\ddot{\underset{..}{O}}-H} \ (g) \quad + \quad \tfrac{1}{2} \ :\ddot{O}=\ddot{O}: \ (g)$$

$\Delta H^\circ_{rxn} = \Sigma \ B.E._{reactants} - \Sigma \ B.E._{products}$ in the gas phase

$= [2B.E._{O-H} + B.E._{O-O}] - [2B.E._{O-H} + 1/2 \ B.E._{O=O}]$

$= [(2 \ mol)(463 \ kJ/mol) + (1 \ mol)(146 \ kJ/mol)] - [(2 \ mol)(463 \ kJ/mol) + (1/2 \ mol)(498 \ kJ/mol)]$

$= \mathbf{-103 \ kJ/mol \ rxn}$

15-50.   *Refer to Section 15-9, Table 15-2, and Examples 15-11 and 15-12.*

Balanced equation: $CCl_2F_2(g) + F_2(g) \rightarrow CF_4(g) + Cl_2(g)$

$\Delta H^\circ_{rxn} = \Sigma \ B.E._{reactants} - \Sigma \ B.E._{products}$ in the gas phase

$= [2B.E._{C-Cl} + 2B.E._{C-F} + B.E._{F-F}] - [4B.E._{C-F} + B.E._{Cl-Cl}]$

$= [(2 \ mol)(339 \ kJ/mol) + (2 \ mol)(485 \ kJ/mol) + (1 \ mol)(155 \ kJ/mol)]$

$\qquad\qquad\qquad - [(4 \ mol)(485 \ kJ/mol) + (1 \ mol)(242 \ kJ/mol)]$

$= \mathbf{-379 \ kJ/mol \ rxn}$

15-52.   *Refer to Section 15-9, Table 15-2, Examples 15-11 and 15-12, and Appendix K.*

(1) Balanced equation for standard heat of formation of HCl: $\frac{1}{2} H_2(g) + \frac{1}{2} Cl_2(g) \rightarrow HCl(g)$

$\Delta H^\circ_{rxn} = \Sigma \ B.E._{reactants} - \Sigma \ B.E._{products}$ in the gas phase

$= [\frac{1}{2} B.E._{H-H} + \frac{1}{2} B.E._{Cl-Cl}] - [B.E._{H-Cl}]$

$= [(0.5 \ mol)(436 \ kJ/mol) + (0.5 \ mol)(242 \ kJ/mol)] - [(1 \ mol)(432 \ kJ/mol)]$

$= \mathbf{-93 \ kJ/mol \ HCl}$

For HCl(g), $\Delta H^\circ_f = -92.31$ kJ/mol

(2) Balanced equation for standard heat of formation of HF: $\frac{1}{2} H_2(g) + \frac{1}{2} F_2(g) \rightarrow HF(g)$

$\Delta H^\circ_{rxn} = \Sigma \ B.E._{reactants} - \Sigma \ B.E._{products}$ in the gas phase

$= [\frac{1}{2} B.E._{H-H} + \frac{1}{2} B.E._{F-F}] - [B.E._{H-F}]$

$= [(0.5 \ mol)(436 \ kJ/mol) + (0.5 \ mol)(155 \ kJ/mol)] - [(1 \ mol)(565 \ kJ/mol)]$

$= \mathbf{-270 \ kJ/mol \ HF}$

For HF(g), $\Delta H^\circ_f = -271$ kJ/mol

15-54.   *Refer to Section 15-9 and Appendix K.*

The ΔH°_{rxn} of this reaction: $PCl_3(g) \rightarrow P(g) + 3Cl(g)$
is equal to 3 times the average P-Cl bond energy in $PCl_3(g)$ since this reaction involves the breaking of 3 P-Cl bonds.

$\Delta H^\circ_{rxn} = [\Delta H^\circ_{f \ P(g)} + 3\Delta H^\circ_{f \ Cl(g)}] - [\Delta H^\circ_{f \ PCl_3(g)}]$

$= [(1 \ mol)(314.6 \ kJ/mol) + (3 \ mol)(121.7 \ kJ/mol)] - [(1 \ mol)(-306.4 \ kJ/mol)]$

$= 986 \ kJ/mol \ rxn$

Therefore, the average bond energy of an P-Cl bond in $PCl_3(g)$ is (986/3) kJ or **329 kJ**.

15-56. *Refer to Section 15-9, Appendix K, and Exercise 15-54.*

The ΔH°_{rxn} of the reaction: $PCl_5(g) \rightarrow P(g) + 5Cl(g)$
is equal to 5 times the average P-Cl bond energy in PCl_5, since this reaction involves the breaking of 5 P-Cl bonds.

$\Delta H^{\circ}_{rxn} = [\Delta H^{\circ}_{f\ P(g)} + 5\Delta H^{\circ}_{f\ Cl(g)}] - [\Delta H^{\circ}_{f\ PCl_5(g)}]$

$\quad = [(1\ mol)(314.6\ kJ/mol) + (5\ mol)(121.7\ kJ/mol)] - [(1\ mol)(-398.9\ kJ/mol)]$

$\quad = 1322\ kJ/mol\ rxn$

Therefore, the average bond energy of a P-Cl bond in $PCl_5(g)$ is (1322/5) kJ or **264 kJ**.

It takes less energy to break an average P-Cl bond in PCl_5 than one in PCl_3 because P is a relatively small atom and Cl is relatively large. It is more difficult to squeeze 5 atoms of Cl around a P than 3 atoms of Cl. Therefore, those 5 atoms of Cl in PCl_5 are not held as tightly and have weaker P-Cl bonds.

15-58. *Refer to Section 15-9, Table 15-2 and Example 15-12.*

$\Delta H^{\circ}_{rxn} = [5B.E._{C-H} + B.E._{C-C} + B.E._{C-N} + 2B.E._{N-H}] - [4B.E._{C-H} + B.E._{C=C} + 3B.E._{N-H}]$

Substituting,

$53.6\ kJ = [(5\ mol)(413\ kJ/mol) + (1\ mol)(346\ kJ/mol) + (1\ mol)(B.E._{C-N}) + (2\ mol)(391\ kJ/mol)]$
$\quad\quad\quad\quad - [(4\ mol)(413\ kJ/mol) + (1\ mol)(602\ kJ/mol) + (3\ mol)(391\ kJ/mol)]$

$53.6\ kJ = (1\ mol)(B.E._{C-N}) - 234\ kJ$

$B.E._{C-N} = $ **288 kJ/mol**

Table 15-2 gives the bond energy for an average C-N bond as 305 kJ/mol.

15-60. *Refer to Sections 1-13 and 15-4, Example 15-1 and Exercise 1-60 Solution.*

Plan: (1) Determine the heat gained by the calorimeter.
          (2) Find the heat capacity of the calorimeter (calorimeter constant).

(1) |heat lost|$_{iron}$ = |heat gained|$_{water}$ + |heat gained|$_{calorimeter}$

|specific heat x mass x Δt|$_{iron}$ = |specific heat x mass x Δt|$_{water}$ + |heat gained|$_{calorimeter}$

$(0.444\ J/g\cdot°C)(93.3\ g)(65.58°C - 19.68°C) = (4.184\ J/g\cdot°C)(75.0\ g)(19.68°C - 16.95°C) + $ |heat gained|$_{calorimeter}$

$1.90\ x\ 10^3\ J = 8.57\ x\ 10^2\ J + $ |heat gained|$_{calorimeter}$

Therefore, |heat gained|$_{calorimeter} = 1.90\ x\ 10^3\ J - 857\ J = 1.04\ x\ 10^3\ J$

(2) heat capacity of calorimeter (J/°C) $= \dfrac{|heat\ gained|_{calorimeter}}{\Delta T} = \dfrac{1.04\ x\ 10^3\ J}{19.68°C - 16.95°C} = $ **381 J/°C**

15-62. *Refer to Sections 1-13 and 15-4, Example 15-1 and Exercise 1-60 Solution.*

|heat lost|$_{metal}$ = |heat gained|$_{water}$ + |heat gained|$_{calorimeter}$

|specific heat x mass x Δt|$_{metal}$ = |specific heat x mass x Δt|$_{water}$ + |calorimeter constant x Δt|$_{cal}$

$(Sp.\ Ht)(36.5\ g)(100.0°C - 32.5°C) = (4.184\ J/g\cdot°C)(50.0\ mL\ x\ 0.997\ g/mL)(32.5°C - 25.0°C)$
$\quad\quad\quad\quad\quad\quad\quad\quad\quad\quad\quad\quad + (1.87\ J/°C)(32.5°C - 25.0°C)$

$(Sp.\ Ht.)(2.46\ x\ 10^3\ J) = 1.6\ x\ 10^3\ J + 14\ J$

Specific heat of the metal = **0.66 J/g·°C**

Balanced equation: $Pb(NO_3)_2(aq) + 2NaI(aq) \rightarrow PbI_2(s) + 2NaNO_3(aq)$

(a) |heat released| = |heat gained|$_{soln}$ + |heat gained|$_{calorimeter}$

$\qquad$ = |specific heat × mass × Δt|$_{soln}$ + |heat capacity × Δt|$_{calorimeter}$

$\qquad$ = (4.184 J/g·°C)(200. g)(24.2°C - 22.6°C)

$\qquad\qquad$ + (472 J/°C)(24.2°C - 22.6°C)

$\qquad$ = 1.3 × 10^3 J + 7.6 × 10^2 J

$\qquad$ **= 2.1 × 10^3 J**

(b) This is a possible limiting reactant problem because amounts of both reactants are given. In this case, we are given stoichiometric amounts of both reactants.

$\text{mol Pb(NO}_3)_2 = \dfrac{6.62 \text{ g}}{331 \text{g/mol}} = 0.0200 \text{ mol}$ $\qquad\qquad$ $\text{mol NaI} = \dfrac{6.00 \text{ g}}{149.9 \text{g/mol}} = 0.0400 \text{ mol}$

$\Delta H_{rxn} = \dfrac{-2.1 \times 10^3 \text{ J}}{0.0200 \text{ mol Pb(NO}_3)_2} \times \dfrac{1 \text{ mol Pb(NO}_3)_2}{1 \text{ mol rxn}} = \textbf{-1.0 × 10}^5 \textbf{ J/mol rxn or -1.0 × 10}^2 \textbf{ kJ/mol rxn}$

(a) $2C_6H_6(\ell) + 15 O_2(g) \rightarrow 12CO_2(g) + 6H_2O(\ell)$

(b) |heat released| = |heat gained|$_{water}$ + |heat gained|$_{calorimeter}$

$\qquad$ = |specific heat × mass × Δt|$_{water}$ + |heat capacity × Δt|$_{calorimeter}$

$\qquad$ = (4.184 J/g·°C)(945 g)(32.692°C - 23.640°C)

$\qquad\qquad$ + (891 J/°C)(32.692°C - 23.640°C)

$\qquad$ = 3.58 × 10^4 J + 8.07 × 10^3 J

$\qquad$ = 4.39 × 10^4 J or 43.9 kJ

Since heat is released in this reaction (the temperature of the water increased), ΔE is a negative quantity.

$\Delta E = -\dfrac{43.9 \text{ kJ}}{1.048 \text{ g C}_6\text{H}_6(\ell)} = \textbf{-41.9 kJ/g C}_6\textbf{H}_6(\ell)$

$\Delta E = -\dfrac{43.9 \text{ kJ}}{1.048 \text{ g C}_6\text{H}_6(\ell)} \times \dfrac{78.11 \text{ g}}{1 \text{ mol}} = \textbf{-3270 kJ/mol C}_6\textbf{H}_6(\ell)$ $\qquad$ (to 3 significant figures)

Balanced equation: $Mg(s) + 2HCl(aq) \rightarrow MgCl_2(aq) + H_2(g)$

(1) |heat released| = |heat gained|$_{soln}$ + |heat gained|$_{calorimeter}$

$\qquad$ = |specific heat × mass × Δt|$_{soln}$ + |heat capacity × Δt|$_{calorimeter}$

$\qquad$ = (4.184 J/g·°C)[(100. mL × 1.10 g/mL) + 1.22 g](45.5°C - 23.0°C)

$\qquad\qquad$ + (562 J/°C)(45.5°C - 23.0°C)

$\qquad$ = 1.05 × 10^4 J + 1.26 × 10^4 J

$\qquad$ **= 2.31 × 10^4 J**

Note: The mass of the solution equals the mass of the HCl solution plus the mass of the magnesium strip.

(2) This is a possible limiting reactant problem because amounts of both reactants are given. In this case, it is clear that Mg is the limiting reactant, since

$$\text{mol Mg} = \frac{1.22 \text{ g}}{24.3 \text{ g/mol}} = 0.0502 \text{ mol} \qquad\qquad \text{mol HCl} = 6.02 \text{ } M \times 0.100 \text{ L} = 0.612 \text{ mol}$$

$$\Delta H_{rxn} = \frac{-2.31 \times 10^4 \text{ J}}{0.0502 \text{ mol Mg}} \times \frac{1 \text{ mol Mg}}{1 \text{ mol rxn}} = \textbf{–4.60} \times \textbf{10}^5 \textbf{ kJ or –460. kJ/mol rxn}$$

15-70. *Refer to Sections 15-4 and 15-10, and Examples 15-2 and 15-14.*

Balanced equation: $2C_{10}H_{22}(\ell) + 31 \text{ } O_2(g) \rightarrow 20CO_2(g) + 22H_2O(\ell)$

$$|\text{heat released}| = |\text{heat gained}|_{water} + |\text{heat gained}|_{calorimeter}$$
$$= |\text{specific heat} \times \text{mass} \times \Delta t|_{water} + |\text{heat capacity} \times \Delta t|_{calorimeter}$$
$$= (4.184 \text{ J/g·°C})(1250.0 \text{ g})(26.4°C - 24.6°C) + (2450 \text{ J/°C})(26.4°C - 24.6°C)$$
$$= 9400 \text{ J} + 4400 \text{ J} \quad \text{(each value has 2 significant figures)}$$
$$= 13800 \text{ J} \quad \text{(3 significant figures - see rules for adding numbers)}$$

Since heat is released in this reaction (the temperature of the water increased), ΔE is a negative quantity.

$$\Delta E = -\frac{13800 \text{ J}}{6.620 \text{ g } C_{10}H_{22}(\ell)} = \textbf{–2.08} \times \textbf{10}^3 \textbf{ J/g } \textbf{C}_{10}\textbf{H}_{22}(\ell)$$

$$\Delta E = -\frac{13800 \text{ J}}{6.620 \text{ g } C_{10}H_{22}(\ell)} \times \frac{142.3 \text{ g}}{1 \text{ mol}} \times \frac{1 \text{ kJ}}{1000 \text{ J}} = \textbf{–297 kJ/mol } \textbf{C}_{10}\textbf{H}_{22}(\ell)$$

15-72. *Refer to Sections 15-10.*

(a) When heat is absorbed by a system or added to a system, q is "+." When heat is released or removed from a system, q is "–."

(b) When work is done on a system, w is "+." When work is done by a system, w is "–."

15-74. *Refer to Section 15-10.*

We know: $\Delta E = q + w$ where ΔE = change in internal energy
q = heat absorbed by the system from the surroundings
w = work done on the system by the surroundings

(a) $q > 0$ and $w > 0$: $\Delta E = +$ (internal energy increases)

(b) $q = w = 0$: $\Delta E = 0$ (no change in internal energy)

(c) $q < 0$ and $w > 0$: $\Delta E = +$ or $-$, depending on the relative values of q and w

---

15-76. *Refer to Section 15-10.*

For the system: $q = -175 \text{ J}$, $w_{electrical} = +96 \text{ J}$ and $w_{PV} = -257 \text{ J}$

$$\Delta E = q + w_{total} = q + (w_{electrical} + w_{PV}) = -175 \text{ J} + [+96 \text{ J} + (-257 \text{ J})] = \textbf{–336 J}$$

15-78. *Refer to Section 15-10, Example 15-13 and Exercise 15-77.*

Plan: Evaluate $\Delta n_{gas} = n_{gaseous\ products} - n_{gaseous\ reactants}$. The sign of the work term is opposite that of Δn_{gas} since $w = -P\Delta V = -\Delta n RT$ at constant P and T.

(a) $2SO_2(g) + O_2(g) \rightarrow 2SO_3(g)$

$\Delta n_{gas} = 2$ mol - 3 mol = -1 mol.   Therefore, $w > 0$ and the work is done by the surroundings on the system.

(b) $CaCO_3(s) \rightarrow CaO(s) + CO_2(g)$

$\Delta n_{gas} = 1$ mol - 0 mol = +1 mol.   Therefore, work < 0, and the work is done by the system on the surroundings.

(c) $CO_2(g) + H_2O(\ell) + CaCO_3(s) \rightarrow Ca^{2+}(aq) + 2HCO_3^{-}(aq)$

$\Delta n_{gas} = 0$ mol - 1 mol = -1 mol.   Therefore, work > 0 and work is done on the system by the surroundings.

15-80. *Refer to Sections 15-10 and 15-11.*

(a) The balanced equation for the oxidation of 1 mole of HCl: $HCl(g) + 1/4\ O_2(g) \rightarrow 1/2\ Cl_2(g) + 1/2\ H_2O(g)$

$$work = -P\Delta V = -\Delta n_{gas}RT = -(n_{gaseous\ products} - n_{gaseous\ reactants})RT$$
$$= -(1\ mol - 5/4\ mol)(8.314\ J/mol \cdot K)(200°C + 273°)$$
$$= +983\ \textbf{J}$$

Work is a positive number, therefore, work is done on the system by the surroundings. As the system "shrinks" from 5/4 mole of gas to 1 mole of gas, work is done on the system by the surroundings to decrease the volume. (Recall that $V \propto n$ at constant T and P.)

(b) The balanced reaction for the decomposition of 1 mole of NO: $NO(g) \rightarrow 1/2\ N_2(g) + 1/2\ O_2(g)$

$$work = -P\Delta V = -\Delta n_{gas}RT = -(1\ mol - 1\ mol)RT = \textbf{0 J}$$

There is no work done since the number of moles of gas, and hence the volume of the system, remains constant.

15-82. *Refer to the Introduction to Section 15-12, Sections 15-12 and 15-15.*

When fuel, e.g., gasoline, is burned, it first undergoes a physical change as it is converted from a liquid to the gaseous state. In the carburetor, the fuel is mixed with oxygen, and a spark ignites the mixture. The fuel then undergoes a chemical change as it reacts with oxygen gas to produce carbon dioxide and water. This reaction happens spontaneously. Let us consider gasoline as being primarily octane; the reaction in the engine is:

$$2C_8H_{18}(g) + 25\ O_2(g) \rightarrow 16CO_2(g) + 18H_2O(g) + heat$$

and it is exothermic, producing a great deal of heat. The Second Law of Thermodynamics states that in spontaneous changes, the universe tends toward a state of increasing entropy, $\Delta S_{universe} > 0$. Does this make sense in this case? Absolutely. We are first going from a system containing 2 moles of liquid fuel to 2 moles of gaseous fuel - a big increase in entropy. Then before reaction we have 27 moles of gas, and after reaction we have a system containing 34 moles of gas. Entropy involves an increase in the relative positions of the molecules with respect to each other and the energies they can have. The entropy of this system has definitely increased after the combustion reaction has occurred.

15-84. *Refer to Section 15-14.*

The Third Law of Thermodynamics states that the entropy of a pure, perfect crystalline substance is zero at 0 K. This means that all substances have some entropy (dispersal of energy and/or matter, i.e. disorder) except when the substance is a pure, perfect, motionless, vibrationless crystal at absolute zero Kelvin. This also implies that the entropy of a substance can be expressed on an absolute basis.

15-86. *Refer to Section 15-13.*

(a) The probability that a coin will come up heads in one flip is ½ = **0.5**.

(b) The probability that a coin comes up heads two times in a row is ½ x ½ = ¼ = **0.25**.

(c) The probability that the coin comes up heads 10 times in a row is $(½)^{10}$ = **1/1024 = 0.000977**.

15-88. *Refer to Section 15-13, Exercise 15-87 and Figure 15-13.*

Consider the following arrangements of molecules:

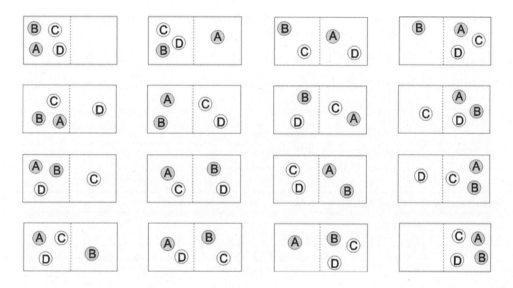

(a) A total of **16** different arrangements are possible.

(b) A mixture of unlike molecules in at least one of the flasks can be found in **14 out of 16** arrangements.

(c) The probability that at least one of the flasks contains a mixture of unlike molecules is 14/16 or **7/8**.

(d) The probability that the gases are not mixed is 2/16 or **1/8**.

15-90. *Refer to Section 15-13.*

(a) heating glass to its softening temperature   entropy is increasing for the glass, so ΔS is positive

(b) sugar dissolving in coffee   entropy is increasing for both sugar and coffee, so ΔS is positive

(c) $CaCO_3$ is precipitating   entropy is decreasing for $CaCO_3$, so ΔS is negative

15-92. *Refer to Section 15-14 and Table 15-4.*

(a) $H_2O(s) \rightarrow H_2O(\ell)$ entropy (dispersal of energy) increased as the solid melted to a liquid

(b), (c), (d) $H_2O(\ell) \rightarrow H_2O(s)$ entropy (dispersal of energy) decreased as the liquid froze to a solid. In (d), as the solid was cooled to −10.°C, the entropy continued to decrease.

When the volume occupied by one mole of Ar at 0°C is halved, there is a *decrease* in entropy (dispersal of energy), as signified by the negative sign of the entropy change, −5.76 J/(mol rxn)·K. In the smaller volume there are fewer energy levels available for the argon molecules to occupy and so, there is a decrease in entropy in the smaller volume.

15-96. *Refer to Sections 15-13 and 15-14, and Table 15-4.*

(a) increase in entropy

When the NaCl dissolves, the ions disperse throughout the water. This allows the ions and the water molecules to transfer energy to each other. Dispersal of matter allows for more dispersal of energy.

(b) decrease in entropy

When some of the NaCl precipitates out as the saturated solution cools, there are fewer number of ways to distribute the same total energy.

(c) decrease in entropy

The solid phase is always more ordered than the liquid phase of a substance.

(d) increase in entropy

The gas phase is always more disordered than the liquid phase of a substance.

(e) increase in entropy

The reaction is producing 2 moles of gas from 1 mole of gas. Energy is more dispersed in a system with 2 moles of gas than in a system with 1 mole of gas.

(f) decrease in entropy

The reaction is the opposite of (e).

15-98. *Refer to Section 15-14 and Example 15-18.*

(a) ΔS of MgO(s) < ΔS of NaF(s)

The higher ion charges in MgO (2+ and 2−) as compared to the ion charges in NaF (1+ and 1−) hold the MgO ionic solid together more tightly so the ions vibrate less, leading to lower absolute entropy.

(b) ΔS of Au(s) < ΔS of Hg(ℓ)

Solids generally have lower entropy than liquids.

(c) ΔS of H$_2$O(g) < ΔS of H$_2$S(g)

For similar molecules, absolute entropy generally increases with increasing size.

(d) ΔS of CH$_3$OH(ℓ) < ΔS of C$_2$H$_5$OH(ℓ)

See (c).

(e) ΔS of NaOH(s) < ΔS of NaOH(aq)

When substances are mixed, in this case, dissolved in water, the absolute entropy is always higher than either substance by itself.

15-100. *Refer to Section 15-14 and Example 15-17.*

Entropy increases and the change in entropy is positive when a reaction occurs
(1) when there are more gaseous products than gaseous reactants and
(2) when there are more aqueous products than aqueous reactants if no gases are present.

(a) entropy change is negative   2 mol gaseous products → 1 mol gaseous product

(b) entropy change is negative   4 mol gaseous products → 2 mol gaseous product

(c) entropy change is positive   0 mol gaseous products → 1 mol gaseous product

(d) entropy change is negative   1/2 mol gaseous products → 0 mol gaseous product

(e) entropy change is negative   2 mol aqueous products → 0 mol gaseous product

Consider the boiling of a pure liquid at constant pressure.

(a) $\Delta S_{\text{system}} > 0$ (b) $\Delta H_{\text{system}} > 0$ (c) $\Delta T_{\text{system}} = 0$

(a) Balanced equation: $4HCl(g) + O_2(g) \rightarrow 2Cl_2(g) + 2H_2O(g)$

$$\Delta S^{\circ}_{\text{rxn}} = [2S^{\circ}_{Cl_2(g)} + 2S^{\circ}_{H_2O(g)}] - [4S^{\circ}_{HCl(g)} + S^{\circ}_{O_2(g)}]$$
$$= [(2 \text{ mol})(+223.0 \text{ J/mol·K}) + (2 \text{ mol})(+188.7 \text{ J/mol·K})]$$
$$- [(4 \text{ mol})(+186.8 \text{ J/mol·K}) + (1 \text{ mol})(+205.0 \text{ J/mol·K})]$$
$$= \mathbf{-128.8 \text{ J/(mol rxn)·K}}$$

The reaction is producing 4 moles of gas from 5 moles of gas. The energy and mass in the system is becoming less dispersed as the number of moles of gas decreases; entropy is decreasing and the change in entropy is expected to be negative.

(b) Balanced equation: $PCl_3(g) + Cl_2(g) \rightarrow PCl_5(g)$

$$\Delta S^{\circ}_{\text{rxn}} = [S^{\circ}_{PCl_5(g)}] - [S^{\circ}_{PCl_3(g)} + S^{\circ}_{Cl_2(g)}]$$
$$= [(1 \text{ mol})(+353 \text{ J/mol·K})] - [(1 \text{ mol})(+311.7 \text{ J/mol·K}) + (1 \text{ mol})(+223.0 \text{ J/mol·K})]$$
$$= \mathbf{-182 \text{ J/(mol rxn)·K}}$$

The reaction is producing 1 mole of gas from 2 moles of gas. For the same reasoning as shown in (a), the entropy is decreasing and the change in entropy is expected to be negative.

(c) Balanced equation: $2N_2O(g) \rightarrow 2N_2(g) + O_2(g)$

$$\Delta S^{\circ}_{\text{rxn}} = [2S^{\circ}_{N_2(g)} + S^{\circ}_{O_2(g)}] - [2S^{\circ}_{N_2O(g)}]$$
$$= [(2 \text{ mol})(+191.5 \text{ J/mol·K}) + (1 \text{ mol})(+205.0 \text{ J/mol·K})] - [(2 \text{ mol})(+219.7 \text{ J/mol·K})]$$
$$= \mathbf{+148.6 \text{ J/(mol rxn)·K}}$$

The reaction is producing 3 mole of gas from 2 moles of gas. The entropy is increasing and the change in entropy is expected to be positive.

(a) always spontaneous: (iii) $\Delta H < 0, \Delta S > 0$

(b) always nonspontaneous: (ii) $\Delta H > 0, \Delta S < 0$

(c) spontaneous or nonspontaneous, depending on T and the magnitudes of ΔH and ΔS: (i) $\Delta H > 0, \Delta S > 0$
     (iv) $\Delta H < 0, \Delta S < 0$

Plan: Use the Gibbs free energy change equation, $\Delta G = \Delta H - T\Delta S$, to calculate ΔG_{rxn} from ΔH_{rxn} and ΔS_{rxn}. Remember to change the units of ΔS to kJ/K.

$$\Delta G^{\circ}_{\text{rxn}} = \Delta H^{\circ}_{\text{rxn}} - T\Delta S^{\circ}_{\text{rxn}}$$
$$= -57.20 \text{ kJ} - (25.00°C + 273.15°)(-0.17583 \text{ kJ/K})$$
$$= \mathbf{-4.78 \text{ kJ}}$$

Since $\Delta G^{\circ}_{\text{rxn}} < 0$, the reaction is spontaneous. Its exothermicity ($\Delta H < 0$) is the driving force for spontaneity.

15-110. *Refer to Sections 15-8 and 15-16.*

Since $\Delta G°$ is a state function like $\Delta H°$, we can use Hess's Law type of manipulations to determine the $\Delta G_f°$. The balanced equation representing the $\Delta G_f°$ of $HBr(g)$ is: $\frac{1}{2}H_2(g) + \frac{1}{2}Br_2(\ell) \rightarrow HBr(g)$

	$\Delta G°$
$\frac{1}{2}Br_2(\ell) \rightarrow \frac{1}{2}Br_2(g)$	1.57 kJ
$H(g) + Br(g) \rightarrow HBr(g)$	−339.09 kJ
$\frac{1}{2}Br_2(g) \rightarrow Br(g)$	80.85 kJ
$\frac{1}{2}H_2(g) \rightarrow H(g)$	203.247 kJ
$\frac{1}{2}H_2(g) + \frac{1}{2}Br_2(\ell) \rightarrow HBr(g)$	**−53.42 kJ/mol rxn**

15-112. *Refer to Section 15-16, Example 15-20 and Appendix K.*

Plan: Calculate $\Delta H_{rxn}°$ and $\Delta S_{rxn}°$, then use the Gibbs free energy change equation , $\Delta G = \Delta H - T\Delta S$, to determine $\Delta G_{rxn}°$.

(a) Balanced equation: $3NO_2(g) + H_2O(\ell) \rightarrow 2HNO_3(\ell) + NO(g)$

$\Delta H_{rxn}° = [2\Delta H_{f\,HNO_3(\ell)}° + \Delta H_{f\,NO(g)}°] - [3\Delta H_{f\,NO_2(g)}° + \Delta H_{f\,H_2O(\ell)}°]$

$\qquad = [(2\ mol)(-174.1\ kJ/mol) + (1\ mol)(+90.25\ kJ/mol)] - [(3\ mol)(+33.2\ kJ/mol) + (1\ mol)(-285.8\ kJ/mol)]$

$\qquad = \textbf{−71.75 kJ/mol rxn}$

$\Delta S_{rxn}° = [2S_{HNO_3(\ell)}° + S_{NO(g)}°] - [3S_{NO_2(g)}° + S_{H_2O(\ell)}°]$

$\qquad = [(2\ mol)(+155.6\ J/mol·K) + (1\ mol)(+210.7\ J/mol·K)]$

$\qquad\qquad - [(3\ mol)(+240.0\ J/mol·K) + (1\ mol)(+69.91\ J/mol·K)]$

$\qquad = \textbf{−268.0 J/(mol rxn)·K}$

$\Delta G_{rxn}° = \Delta H_{rxn}° - T\Delta S_{rxn}° = -71.75\ kJ - (298.15\ K)(-0.268\ kJ/K) = \textbf{+8.15 kJ/mol rxn}$

(b) Balanced equation: $SnO_2(s) + 2CO(g) \rightarrow 2CO_2(g) + Sn(s,white)$

$\Delta H_{rxn}° = [2\Delta H_{f\,CO_2(g)}° + \Delta H_{f\,Sn(s)}°] - [\Delta H_{f\,SnO_2(s)}° + 2\Delta H_{f\,CO(g)}°]$

$\qquad = [(2\ mol)(-393.5\ J/mol) + (1\ mol)(0\ kJ/mol)] - [(1\ mol)(-580.7\ kJ/mol) + (2\ mol)(-110.5\ kJ/mol)]$

$\qquad = \textbf{+14.7 kJ/mol rxn}$

$\Delta S_{rxn}° = [2S_{CO_2(g)}° + S_{Sn(s)}°] - [S_{SnO_2(s)}° + 2S_{CO(g)}°]$

$\qquad = [(2\ mol)(+213.6\ J/mol·K) + (1\ mol)(+51.55\ J/mol·K)]$

$\qquad\qquad - [(1\ mol)(+52.3\ J/mol·K) + (2\ mol)(+197.6\ J/mol·K)]$

$\qquad = \textbf{+31.2 J/(mol rxn)·K}$

$\Delta G_{rxn}° = \Delta H_{rxn}° - T\Delta S_{rxn}° = +14.7\ kJ - (298.15\ K)(+0.0312\ kJ/K) = \textbf{+5.4 kJ/mol rxn}$

(c) Balanced equation: $2Na(s) + 2H_2O(\ell) \rightarrow 2NaOH(aq) + H_2(g)$

$\Delta H_{rxn}° = [2\Delta H_{f\,NaOH(aq)}° + \Delta H_{f\,H_2(g)}°] - [2\Delta H_{f\,Na(s)}° + 2\Delta H_{f\,H_2O(\ell)}°]$

$\qquad = [(2\ mol)(-469.6\ kJ/mol) + (1\ mol)(0\ kJ/mol)] - [(2\ mol)(0\ kJ/mol) + (2\ mol)(-285.8\ kJ/mol)]$

$\qquad = \textbf{−367.6 kJ/mol rxn}$

$\Delta S_{rxn}° = [2S_{NaOH(aq)}° + S_{H_2(g)}°] - [2S_{Na(s)}° + 2S_{H_2O(\ell)}°]$

$\qquad = [(2\ mol)(+49.8\ J/mol·K) + (1\ mol)(+130.6\ J/mol·K)]$

$\qquad\qquad - [(2\ mol)(+51.0\ J/mol·K) + (2\ mol)(+69.91\ J/mol·K)]$

$\qquad = \textbf{−11.62 J/(mol rxn)·K}$

$\Delta G_{rxn}° = \Delta H_{rxn}° - T\Delta S_{rxn}° = -367.6\ kJ - (298.15\ K)(-0.01162\ kJ/K) = \textbf{−364.1 kJ/mol rxn}$

15-114. *Refer to Sections 15-16 and 15-17.*

Recall: Gibbs free energy change equation: $\Delta G = \Delta H - T\Delta S$

(a) false An exothermic reaction ($\Delta H < 0$) will be spontaneous ($\Delta G < 0$) only if either ΔS is positive, or, in the event ΔS is negative, the absolute value of $T\Delta S$ is smaller than that of ΔH.

(b) true From the Gibbs free energy change equation; the $T\Delta S$ term has a negative sign in front.

(c) false A reaction with $\Delta S_{sys} > 0$ will be spontaneous ($\Delta G < 0$) only if either ΔH is negative, or, in the event ΔH is positive, its absolute value is smaller than that of $T\Delta S$.

15-116. *Refer to Sections 15-16 and 15-17, and Appendix K.*

Balanced equation: $2H_2O_2(\ell) \rightarrow 2H_2O(\ell) + O_2(g)$

(a) $\Delta H^\circ_{rxn} = [2\Delta H^\circ_{f\ H_2O(\ell)} + \Delta H^\circ_{f\ O_2(g)}] - [2\Delta H^\circ_{f\ H_2O_2(\ell)}]$
$= [(2\text{ mol})(-285.8\text{ kJ/mol}) + (1\text{ mol})(0\text{ kJ/mol})] - [(2\text{ mol})(-187.8\text{ kJ/mol})]$
$= \mathbf{-196.0\ kJ/mol\ rxn}$

$\Delta G^\circ_{rxn} = [2\Delta G^\circ_{f\ H_2O(\ell)} + \Delta G^\circ_{f\ O_2(g)}] - [2\Delta G^\circ_{f\ H_2O_2(\ell)}]$
$= [(2\text{ mol})(-237.2\text{ kJ/mol}) + (1\text{ mol})(0\text{ kJ/mol})] - [(2\text{ mol})(-120.4\text{ kJ/mol})]$
$= \mathbf{-233.6\ kJ/mol\ rxn}$

$\Delta S^\circ_{rxn} = [2S^\circ_{H_2O(\ell)} + S^\circ_{O_2(g)}] - [2S^\circ_{H_2O_2(\ell)}]$
$= [(2\text{ mol})(+69.91\text{ J/mol·K}) + (1\text{ mol})(+205.0\text{ J/mol·K})] - [(2\text{ mol})(+109.6\text{ J/mol·K})]$
$= \mathbf{+125.6\ J/(mol\ rxn)\cdot K}$

(b) Hydrogen peroxide, $H_2O_2(\ell)$, will be stable if $\Delta G^\circ > 0$ for the above balanced reaction at some temperature, i.e., if the above reaction is non-spontaneous. However, $\Delta H^\circ_{rxn} < 0$ and $\Delta S^\circ_{rxn} > 0$ for the decomposition of $H_2O_2(\ell)$ and the reaction is spontaneous ($\Delta G^\circ < 0$) for all temperatures. Hence, there is no temperature at which $H_2O_2(\ell)$ is stable at 1 atm.

15-118. *Refer to Section 15-17.*

Dissociation reactions, such as $HCl(g) \rightarrow H(g) + Cl(g)$, require energy to break bonds and therefore are endothermic with positive ΔH values. The ΔS values for such reactions are positive since 2 or more particles are being formed from 1 molecule, causing the system to become more energetically dispersed. Under the circumstances when ΔH and ΔS are both positive, the spontaneity of the reaction is favored at higher temperatures.

15-120. *Refer to Section 15-17, Examples 15-22 and 15-23, and Appendix K.*

Plan: Evaluate ΔH_{rxn} and ΔS_{rxn}. To assess the temperature range over which the reaction is spontaneous, use the signs of ΔH and ΔS and the Gibbs free energy change equation, $\Delta G = \Delta H - T\Delta S$. Assume that ΔH and ΔS are independent of temperature.

(a) Balanced equation: $CaCO_3(s) + H_2SO_4(\ell) \rightarrow CaSO_4(s) + H_2O(\ell) + CO_2(g)$

$\Delta H^\circ_{rxn} = [\Delta H^\circ_{f\ CaSO_4(s)} + \Delta H^\circ_{f\ H_2O(\ell)} + \Delta H^\circ_{f\ CO_2(g)}] - [\Delta H^\circ_{f\ CaCO_3(s)} + \Delta H^\circ_{f\ H_2SO_4(\ell)}]$
$= [(1\text{ mol})(-1433\text{ kJ/mol}) + (1\text{ mol})(-285.8\text{ kJ/mol}) + (1\text{ mol})(-393.5\text{ kJ/mol})]$
$- [(1\text{ mol})(-1207\text{ kJ/mol}) + (1\text{ mol})(-814.0\text{ kJ/mol})]$
$= -91.3\text{ kJ/mol rxn}$

$$\Delta S^\circ_{rxn} = [S^\circ_f\,CaSO_4(s) + S^\circ_f\,H_2O(\ell) + S^\circ_f\,CO_2(g)] - [S^\circ_f\,CaCO_3(s) + S^\circ_f\,H_2SO_4(\ell)]$$

$$= [(1\ mol)(+107\ J/mol{\cdot}K) + (1\ mol)(+69.91\ J/mol{\cdot}K) + (1\ mol)(+213.6\ J/mol{\cdot}K)]$$
$$- [(1\ mol)(+92.9\ J/mol{\cdot}K) + (1\ mol)(+156.9\ J/mol{\cdot}K)]$$

$$= +141\ J/(mol\ rxn){\cdot}K$$

Since ΔH is negative and ΔS is positive, the reaction is **spontaneous at all temperatures**.

(b) Balanced equation: $2HgO(s) \rightarrow 2Hg(\ell) + O_2(g)$

$$\Delta H^\circ_{rxn} = [2\Delta H^\circ_f\,Hg(\ell) + \Delta H^\circ_f\,O_2(g)] - [2\Delta H^\circ_f\,HgO(s)]$$
$$= [(2\ mol)(0\ kJ/mol) + (1\ mol)(0\ kJ/mol)] - [(2\ mol)(-90.83\ kJ/mol)]$$
$$= +181.7\ kJ/mol\ rxn$$

$$\Delta S^\circ_{rxn} = [2S^\circ_f\,Hg(\ell) + S^\circ_f\,O_2(g)] - [2S^\circ_f\,HgO(s)]$$
$$= [(2\ mol)(+76.02\ J/mol{\cdot}K) + (1\ mol)(+205.0\ J/mol{\cdot}K)] - [(2\ mol)(+70.29\ J/mol{\cdot}K)]$$
$$= +216.5\ J/(mol\ rxn){\cdot}K$$

At equilibrium, $\Delta G^\circ_{rxn} = 0 = \Delta H^\circ_{rxn} - T\Delta S^\circ_{rxn}$, and solving for T_{eq}

$$T_{eq} = \frac{\Delta H_{rxn}}{\Delta S_{rxn}} = \frac{181.7\ kJ}{0.2165\ kJ/K} = 839.3\ K$$

Since ΔH and ΔS are positive, the reaction is **spontaneous at $T > 839.3$ K**.

(c) Balanced equation: $CO_2(g) + C(s) \rightarrow 2CO(g)$

$$\Delta H^\circ_{rxn} = [2\Delta H^\circ_f\,CO(g)] - [\Delta H^\circ_f\,CO_2(g) + \Delta H^\circ_f\,C(s)]$$
$$= [(2\ mol)(-110.5\ kJ/mol)] - [(1\ mol)(-393.5\ kJ/mol) + (1\ mol)(0\ kJ/mol)]$$
$$= +172.5\ kJ/mol\ rxn$$

$$\Delta S^\circ_{rxn} = [2S^\circ_f\,CO(g)] - [S^\circ_f\,CO_2(g) + S^\circ_f\,C(s)]$$
$$= [(2\ mol)(+197.6\ J/mol{\cdot}K)] - [(1\ mol)(+213.6\ J/mol{\cdot}K) + (1\ mol)(+5.740\ J/mol{\cdot}K)]$$
$$= +175.9\ J/(mol\ rxn){\cdot}K$$

At equilibrium, $\Delta G^\circ_{rxn} = 0 = \Delta H^\circ_{rxn} - T\Delta S^\circ_{rxn}$, and solving for T_{eq}

$$T_{eq} = \frac{\Delta H_{rxn}}{\Delta S_{rxn}} = \frac{172.5\ kJ}{0.1759\ kJ/K} = 980.7\ K$$

Since ΔH and ΔS are positive, the reaction is **spontaneous at $T > 980.7$ K**.

(d) Balanced equation: $2Fe_2O_3(s) \rightarrow 4Fe(s) + 3O_2(g)$

$$\Delta H^\circ_{rxn} = [4\Delta H^\circ_f\,Fe(s) + 3\Delta H^\circ_f\,O_2(g)] - [2\Delta H^\circ_f\,Fe_2O_3(s)]$$
$$= [(4\ mol)(0\ kJ/mol) + (3\ mol)(0\ kJ/mol)] - [(2\ mol)(-824.2\ kJ/mol)]$$
$$= +1648\ kJ/mol\ rxn$$

$$\Delta S^\circ_{rxn} = [4S^\circ_f\,Fe(s) + 3S^\circ_f\,O_2(g)] - [2S^\circ_f\,Fe_2O_3(s)]$$
$$= [(4\ mol)(+27.3\ J/mol{\cdot}K) + (3\ mol)(+205.0\ J/mol{\cdot}K)] - [(2\ mol)(+87.40\ J/mol{\cdot}K)]$$
$$= +549.4\ J/(mol\ rxn){\cdot}K$$

$$T_{eq} = \frac{\Delta H_{rxn}}{\Delta S_{rxn}} = \frac{1648\ kJ}{0.5494\ kJ/K} = 3000.\ K$$

Since ΔH and ΔS are positive, the reaction is **spontaneous at $T > 3000.$ K**.

15-122. *Refer to Section 15-17, Appendix K and Example 15-21.*

(a) The process is: $H_2O(\ell) \rightarrow H_2O(g)$

$\Delta H^\circ_{rxn} = \Delta H^\circ_{f\ H_2O(g)} - \Delta H^\circ_{f\ H_2O(\ell)} = (1\ mol)(-241.8\ kJ/mol) - (1\ mol)(-285.8\ kJ/mol) = +44.0\ kJ$

$\Delta S^\circ_{rxn} = S^\circ_{H_2O(g)} - S^\circ_{H_2O(\ell)} = (1\ mol)(+188.7\ J/mol\cdot K) - (1\ mol)(+69.91\ J/mol\cdot K) = +118.8\ J/K$

$T_{eq} = \dfrac{\Delta H_{rxn}}{\Delta S_{rxn}} = \dfrac{44.0\ kJ}{0.1188\ kJ/K} = 370\ K\ or\ \mathbf{97°C}$

(b) The known boiling point of water is, of course, 100°C. The discrepancy is because we assumed that the standard values of enthalpy of formation and entropy in Appendix K are independent of temperature. However, these tabulated values were determined at 25°C; we are using them to solve a problem at 100°C. Nevertheless, this assumption allows us to estimate the boiling point of water with reasonable accuracy.

15-124. *Refer to Section 15-17 and Example 15-19.*

Balanced equation: $C(s, diamond) \rightarrow C(s, graphite)$

(a) $\Delta G^\circ_{rxn} = \Delta G^\circ_{f\ C(s,\ graphite)} - \Delta G^\circ_{f\ C(s,\ diamond)} = (1\ mol)(0\ kJ/mol) - (1\ mol)(+2.900\ kJ/mol) = -2.900\ kJ$

Since $\Delta G^\circ_{rxn} < 0$, the reaction is spontaneous at standard conditions.

(b) From common experience, we know that diamonds do not readily change to graphite. Thermodynamics tells us if a reaction will occur but it says nothing about the *rate* at which a reaction will occur. The rate at which diamonds change to graphite is very, very slow.

(c) $\Delta H^\circ_{rxn} = \Delta H^\circ_{f\ C(s,\ graphite)} - \Delta H^\circ_{f\ C(s,\ diamond)} = (1\ mol)(0\ kJ/mol) - (1\ mol)(+1.897\ kJ/mol) = -1.897\ kJ$

$\Delta S^\circ_{rxn} = S^\circ_{C(s,\ graphite)} - S^\circ_{C(s,\ diamond)} = (1\ mol)(+5.740\ J/mol\cdot K) - (1\ mol)(+2.38\ J/mol\cdot K) = +3.36\ J/K$

Since ΔH_{rxn} is negative and ΔS_{rxn} is positive, the reaction is spontaneous at all temperatures, i.e., there is no temperature at which diamond and graphite are in equilibrium.

(d) At higher pressures, atoms are pressed closer together and the density increases. At a certain pressure, carbon as graphite transforms itself into the higher density crystalline form called diamond, which is more stable at the higher pressure..

15-126. *Refer to Sections 15-7 and 15-8, Example 15-10, and Appendix K.*

Plan: Use Hess's Law and solve for ΔH°_f of the organic compound.

(a) Balanced equation: $C_6H_{12}(\ell) + 9O_2(g) \rightarrow 6CO_2(g) + 6H_2O(\ell)$

$\Delta H^\circ_{combustion} = [6\Delta H^\circ_{f\ CO_2(g)} + 6\Delta H^\circ_{f\ H_2O(\ell)}] - [\Delta H^\circ_{f\ C_6H_{12}(\ell)} + 9\Delta H^\circ_{f\ O_2(g)}]$

$-3920\ kJ = [(6\ mol)(-393.5\ kJ/mol) + (6\ mol)(-285.8\ kJ/mol)]$
$\qquad\qquad\qquad - [(1\ mol)\Delta H^\circ_{f\ C_6H_{12}(\ell)} + (9\ mol)(0\ kJ/mol)]$

$-3920\ kJ = -4075.8\ kJ - (1\ mol)\Delta H^\circ_{f\ C_6H_{12}(\ell)}$

$\Delta H^\circ_{f\ C_6H_{12}(\ell)} = \mathbf{-156\ kJ/mol\ C_6H_{12}(\ell)}$

(b) Balanced equation: $C_6H_5OH(s) + 7O_2(g) \rightarrow 6CO_2(g) + 3H_2O(\ell)$

$\Delta H^\circ_{combustion} = [6\Delta H^\circ_{f\ CO_2(g)} + 3\Delta H^\circ_{f\ H_2O(\ell)}] - [\Delta H^\circ_{f\ C_6H_5OH(s)} + 7\Delta H^\circ_{f\ O_2(g)}]$

$-3053\ kJ = [(6\ mol)(-393.5\ kJ/mol) + (3\ mol)(-285.8\ kJ/mol)]$
$\qquad\qquad\qquad - [(1\ mol)\Delta H^\circ_{f\ C_6H_5OH(s)} + (7\ mol)(0\ kJ/mol)]$

$-3053\ kJ = -3218.4\ kJ - (1\ mol)\Delta H^\circ_{f\ C_6H_5OH(s)}$

$\Delta H^\circ_{f\ C_6H_5OH(s)} = \mathbf{-165\ kJ/mol\ C_6H_5OH(s)}$

The vaporization process is: ethanol(ℓ) $\rightarrow$ ethanol(g)

$$\Delta E = q + w \qquad \text{where} \qquad \begin{aligned} \Delta E &= \text{change in internal energy} \\ q &= \text{heat absorbed by the system} \\ w &= \text{work done on the system} \end{aligned}$$

(1) The heat absorbed by the system,

$q = \Delta H_{vap} \times$ g ethanol $= +855$ J/g $\times$ 12.5 g $= \mathbf{+10700\ J}$

(2) The work done on the system in going from a liquid to a gas,

$w = -P\Delta V = -P(V_{gas} - V_{liquid})$

where $\quad V_{gas} = \dfrac{nRT}{P} = \dfrac{(12.5\ \text{g}/46.1\ \text{g/mol})(0.0821\ \text{L·atm/mol·K})(78.0°C + 273.15°)}{1.00\ \text{atm}} = 7.82\ \text{L}$

$V_{liquid} = 12.5$ g ethanol $\times \dfrac{1.00\ \text{mL ethanol}}{0.789\ \text{g ethanol}} = 15.8$ mL or 0.0158 L

Therefore,

$w = -P\Delta V = -(1\ \text{atm})(7.82\ \text{L} - 0.02\ \text{L}) = -7.80$ L·atm (the negative value means the system is doing the work)

To find a factor to convert L·atm to J, we can equate two values of the molar gas constant, R

$$0.0821\ \text{L·atm/mol·K} = 8.314\ \text{J/mol·K}$$
$$1\ \text{L·atm} = 101\ \text{J}$$

And so, $w = -7.80$ L·atm $\times \dfrac{101\ \text{J}}{1\ \text{L·atm}} = \mathbf{-788\ J}$

(3) Finally, $\Delta E = q + w = 10700$ J $+ (-788$ J$) = \mathbf{9900\ J}$

(a) heat gained by calorimeter $= 0.01520$ g $C_{10}H_8 \times \dfrac{1\ \text{mol } C_{10}H_8}{128.16\ \text{g } C_{10}H_8} \times \dfrac{5156.8\ \text{kJ}}{1\ \text{mol } C_{10}H_8} = 0.6116$ kJ

We know: |heat gained by calorimeter| = |heat capacity $\times \Delta t$| where t is temperature in °C

Therefore, heat capacity $= \dfrac{|\text{heat gained by calorimeter}|}{|\Delta t|} = \dfrac{0.6116\ \text{kJ}}{0.212°C} = \mathbf{2.88\ kJ/°C}$

(b) |heat released in the reaction| $= 0.1040$ g $C_8H_{18} \times \dfrac{1\ \text{mol } C_8H_{18}}{114.22\ \text{g/mol}} \times \dfrac{5451.4\ \text{kJ}}{1\ \text{mol } C_8H_{18}} = 4.964$ kJ

We also know:

|heat released in the reaction| = |heat gained by calorimeter|

Substituting, $\qquad\qquad$ 4.964 kJ = |heat capacity $\times \Delta t$|

$= |2.88\ \text{kJ/°C} \times \Delta t|$

$\Delta t = 1.72°C$

Therefore, $t_{final} = t_{initial} + \Delta t = 22.102°C + 1.72°C = \mathbf{23.82°C}$.

When a rubber band is stretched: $\Delta H < 0$, since heat is released

$\Delta S < 0$, since the rubber band is becoming more ordered (more linear); therefore

$\Delta G > 0$, since the process does not occur spontaneously

When the stretched rubber band is relaxed, the signs of the thermodynamic state functions change:

$\Delta H > 0$, since heat is absorbed (that's why your hand feels colder)

$\Delta S > 0$, since the rubber band is becoming more disordered; therefore

$\Delta G < 0$, since the process occurs spontaneously

The spontaneous process that occurs when the stretched rubber band is allowed to return to its original, random arrangement of polymer molecules, must be driven by the increase in the mass and energy dispersal of the system, since the reaction is endothermic ($\Delta H > 0$).

(a)

$$|\text{heat lost}|_{\text{metal}} = |\text{heat gained}|_{\text{water}}$$
$$|\text{specific heat} \times \text{mass} \times \Delta t|_{\text{metal}} = |\text{specific heat} \times \text{mass} \times \Delta t|_{\text{water}}$$
$$(\text{specific heat of metal})(32.6 \text{ g})(99.83°C - 24.41°C) = (4.184 \text{ J/g·°C})(100.0 \text{ g})(24.41°C - 23.62°C)$$
$$(\text{specific heat of metal})(2.46 \times 10^3) = 330$$
$$\text{specific heat of metal} = \textbf{0.13 J/g·°C}$$

Therefore, according to this calculation, the metal is **tungsten, W** (specific heat = 0.135 J/g·°C).

(b)

$$|\text{heat lost}|_{\text{metal}} = |\text{heat gained}|_{\text{water}} + |\text{heat gained}|_{\text{calorimeter}}$$
$$|\text{specific heat} \times \text{mass} \times \Delta t|_{\text{metal}} = |\text{specific heat} \times \text{mass} \times \Delta t|_{\text{water}}$$
$$+ |\text{heat capacity} \times \Delta t|_{\text{calorimeter}}$$
$$(\text{specific heat of metal})(32.6 \text{ g})(99.83°C - 24.41°C) = (4.184 \text{ J/g·°C})(100.0 \text{ g})(24.41°C - 23.62°C)$$
$$+ (410 \text{ J/°C})(24.41°C - 23.62°C)$$
$$(\text{specific heat of metal})(2.46 \times 10^3) = 330 + 320$$
$$\text{specific heat of metal} = 0.26 \text{ J/g·°C}$$

Yes, the identification of the metal was different. When the heat capacity of the calorimeter is taken into account, the specific heat of the metal is 0.26 J/g·°C and the metal is identified as molybdenum, Mo (specific heat = 0.250 J/g·°C).

Calculation: $\text{Activity Time Equivalent (min)} = \dfrac{\text{Food Fuel Value (kcal)}}{\text{Energy Output (kcal/min)}}$

Food	Fuel Value (kcal)	Activity Time Equivalent (min)				
		Sitting (1.7 kcal/min)	Walking (5.5 kcal/min)	Cycling (10 kcal/min)	Swimming (8.4 kcal/min)	Running (19 kcal/min)
Apple	100	59	18	10	12	5.3
Cola	105	62	19	11	13	5.5
Malted milk	500	290	91	50	60	26
Pasta	195	110	35	20	23	10
Hamburger	350	210	64	35	42	18
Steak	1000	590	180	100	120	53

15-138. *Refer to Section 15-3 and Fundamental Algebra.*

? kJ of energy found in 100. g protein = 100. g protein x $\frac{17 \text{ kJ}}{1 \text{ g protein}}$ = 1700 kJ

? kJ of energy found in 100. g fat = 100. g fat x $\frac{39 \text{ kJ}}{1 \text{ g fat}}$ = 3900 kJ

The difference in energy content is the amount of energy that must be burned up by walking instead of resting, so that the person doesn't gain weight:

difference in energy content = 3900 - 1700 = 2200 kJ

$$\text{? time required to walk instead of rest to burn off 2200 kJ} = \frac{\text{difference in energy content}}{\text{difference in utilization rate}}$$
$$= \frac{(3900 - 1700)\text{ kJ}}{(1250 - 335)\text{ kJ/hr}}$$
$$= \frac{2200 \text{ kJ}}{915 \text{ kJ/hr}}$$
$$= \mathbf{2.4\ hr}$$

15-140. *Refer to Sections 1-13 and 15-4.*

$$|\text{heat lost}|_{\text{lead}} = |\text{heat gained}|_{\text{water}} + |\text{heat gained}|_{\text{calorimeter}}$$
$$|\text{specific heat x mass x } \Delta t|_{\text{lead}} = |\text{specific heat x mass x } \Delta t|_{\text{water}} + |\text{heat capacity x } \Delta t|_{\text{calorimeter}}$$
$$(\text{Sp. Ht. of Pb})(43.6\text{ g})(100.0°\text{C} - 26.8°\text{C}) = (4.184\text{ J/g·°C})(50.0\text{ g})(26.8°\text{C} - 25.0°\text{C}) + (18.6\text{ J/°C})(26.8°\text{C} - 25.0°\text{C})$$
$$(\text{Specific heat of Pb})(3190) = 380 + 33$$
$$\text{Specific heat of Pb} = \mathbf{0.13\ J/g·°C}$$
$$\text{Molar heat capacity of Pb} = 0.13\text{ J/g·°C x } 207.2\text{ g/mol} = \mathbf{27\ J/mol·°C}$$

15-142. *Refer to Section 15-4.*

(a) Heat gain by calorimeter = (4572 J/°C)(27.93°C - 24.76°C) = 1.449 x 10^4 J or 14.49 kJ

Fuel value of butter = $\frac{14.49 \text{ kJ}}{0.483 \text{ g}}$ = **30.0 kJ/g**

(b) Nutritional Calories/g butter = $\frac{30.0 \text{ kJ/g}}{4.184 \text{ kJ/kilocalorie}}$ = **7.17 kilocalorie/g**

(c) Nutritional Calories/5.00 g pat of butter = (7.17 kilocalorie/g) x 5.00 g = **35.9 kilocalorie**

16 Chemical Kinetics

16-2. *Refer to Sections 16-5 and 16-6.*

The collision theory of reaction rates states that molecules, atoms or ions must collide effectively in order to react. For an effective collision to occur, the reacting species must have (1) at least a minimum amount of energy in order to break old bonds and make new ones, and (2) the proper orientation toward each other.

Transition state theory complements collision theory. When particles collide with enough energy to react, called the activation energy, E_a, the reactants form a short-lived, high energy activated complex, or transition state, before forming the products. The transition state also could revert back to the reactants.

16-4. *Refer to the Introduction to Chapter 16.*

In Chapter 15, we learned that reactions which are thermodynamically favorable have negative ΔG values and occur spontaneously as written. However, thermodynamics cannot be used to determine the rate of a reaction. Kinetically favorable reactions must be thermodynamically favorable *and* have a low enough activation energy to occur at a reasonable rate at a certain temperature.

16-6. *Refer to Section 16-3.*

The coefficients of the balanced overall equation bear no necessary relationship to the exponents to which the concentrations are raised in the rate law expression. The exponents are determined experimentally and describe how the concentrations of each reactant affect the reaction rate. The exponents are related to the rate-determining (slow) step in a sequence of mainly unimolecular and bimolecular reactions called the mechanism of the reaction. It is the mechanism which lays out exactly the order in which bonds are broken and made as the reactants are transformed into the products of the reaction.

16-8. *Refer to Section 16-1 and Example 16-1.*

(a) $3ClO^-(aq) \rightarrow ClO_3^-(aq) + 2Cl^-(aq)$ rate of reaction $= -\dfrac{\Delta[ClO^-]}{3\Delta t} = \dfrac{\Delta[ClO^-]}{\Delta t} = \dfrac{\Delta[Cl^-]}{2\Delta t}$

(b) $2SO_2(g) + O_2(g) \rightarrow 2SO_3(g)$ rate of reaction $= -\dfrac{\Delta[SO_2]}{2\Delta t} = -\dfrac{\Delta[O_2]}{\Delta t} = \dfrac{\Delta[SO_3]}{2\Delta t}$

(c) $C_2H_4(g) + Br_2(g) \rightarrow C_2H_4Br_2(g)$ rate of reaction $= -\dfrac{\Delta[C_2H_4]}{\Delta t} = -\dfrac{\Delta[Br_2]}{\Delta t} = \dfrac{\Delta[C_2H_4Br_2]}{\Delta t}$

(d) $(C_2H_5)_2(NH)_2 + I_2 \rightarrow (C_2H_5)_2N_2 + 2HI$ rate of reaction $= -\dfrac{\Delta[(C_2H_5)_2(NH)_2]}{\Delta t} = -\dfrac{\Delta[I_2]}{\Delta t} = \dfrac{\Delta[(C_2H_5)_2N_2]}{\Delta t} = \dfrac{\Delta[HI]}{2\Delta t}$

16-10. *Refer to Section 16-1 and Example 16-1.*

Balanced reaction: $4NH_3 + 5O_2 \rightarrow 4NO + 6H_2O$ rate of reaction $= -\dfrac{\Delta[NH_3]}{4\Delta t} = -\dfrac{\Delta[O_2]}{5\Delta t} = \dfrac{\Delta[NO]}{4\Delta t} = \dfrac{\Delta[H_2O]}{6\Delta t}$

Substituting,

rate of reaction $= -\dfrac{\Delta[NH_3]}{4\Delta t} = \dfrac{1.20\ M\ NH_3}{4 \times 1\ min} = 0.300\ M/min$

Therefore,

$$\text{rate of disappearance of } O_2 = -\frac{\Delta[O_2]}{\Delta t} = 5 \times \text{rate of reaction} = 5 \times 0.300 \; M/\text{min} = \mathbf{1.50 \; M/min}$$

$$\text{rate of appearance of NO} = \frac{\Delta[NO]}{\Delta t} = 4 \times \text{rate of reaction} = 4 \times 0.300 \; M/\text{min} = \mathbf{1.20 \; M/min}$$

$$\text{rate of appearance of } H_2O = \frac{\Delta[H_2O]}{\Delta t} = 6 \times \text{rate of reaction} = 6 \times 0.300 \; M/\text{min} = \mathbf{1.80 \; M/min}$$

16-12.　**Refer to Section 16-2.**

Some fireworks are bright because of the burning of magnesium:

$$2Mg + O_2 \rightarrow 2MgO.$$

This reaction gives off much energy as heat and light. The magnesium metal pieces could be interspersed in the body of the firework near the fuse. To get the best visual display, this oxidation reaction cannot go too fast or too slow. The speed of this reaction can be controlled by the size of magnesium pieces. If the pieces were large, the reaction would take place at a slower rate if it occurred at all due to the smaller surface area of magnesium exposed to the air. If the pieces were too small, the reaction would occur too quickly; there would be one large burst of light and the beauty of the sparks would be lost.

16-14.　**Refer to Section 16-3 and Example 16-2.**

The simplest approach to this problem is to assume that the initial concentrations of NO and O_2 for the first experiment are each 1 M. Then for the second experiment, the initial concentration of NO is 1/2 M and that of O_2 is 2 M. Let us substitute these values into the rate-law expression,

$$\text{rate} = k[NO]^2[O_2]$$

First experiment: $\quad\quad\quad\quad$ rate $= k(1 \; M)^2(1 \; M) = k$

Second experiment: $\quad\quad\quad\quad$ rate $= k(\frac{1}{2} \; M)^2(2 \; M) = \frac{1}{2}k$

Therefore, the rate of reaction in the second experiment would be **1/2** times that of the first experiment.

16-16.　**Refer to Section 16-4.**

Plan:　Use dimensional analysis and the rate-law expression to determine the units of k, the rate constant, in the following general equation:

$$\text{rate } (M/s) = k[A]^x \quad\quad\quad \text{where} \quad\quad x = \text{the overall order of the reaction}$$
$$[A] = \text{the reactant concentration } (M)$$

	Overall Reaction Order	Example	Units of k
(a)	1	rate $= k[A]$	$(M/s)/M = s^{-1}$
(b)	2	rate $= k[A]^2$	$(M/s)/M^2 = M^{-1} \cdot s^{-1}$
(c)	3	rate $= k[A]^3$	$(M/s)/M^3 = M^{-2} \cdot s^{-1}$
(d)	1.5	rate $= k[A]^{1.5}$	$(M/s)/M^{1.5} = M^{-0.5} \cdot s^{-1}$

The form of the rate-law expression: rate = $k[A]^x[B]^y[C]^z$

Step 1: rate dependence on [A]. Consider Experiments 1 and 3:

Method 1: By observation, [B] and [C] do not change; [A] increases by a factor of 3. However, the reaction rate does not change. Therefore, changing [A] does not affect reaction rate and the reaction is zero order with respect to A. In all subsequent determinations, the effect of A can be ignored.

Method 2: A mathematical solution is obtained by substituting the experimental values of Experiments 1 and 3 into rate-law expressions and dividing the latter by the former. Note: the calculations are easier when the experiment with the larger rate is in the numerator.

$$\frac{\text{Expt 3}}{\text{Expt 1}} \qquad \frac{5.0 \times 10^{-4} \ M/min}{5.0 \times 10^{-4} \ M/min} = \frac{k(0.30 \ M)^x(0.10 \ M)^y(0.20 \ M)^z}{k(0.10 \ M)^x(0.10 \ M)^y(0.20 \ M)^z}$$

$$1 = 3^x$$

$$x = 0$$

Step 2: rate dependence on [B]. Consider Experiments 1 and 2:

Method 1: [B] changes by a factor of 3; [C] does not change; the reaction rate also changes by a factor of 3 (= $1.5 \times 10^{-3}/5.0 \times 10^{-4}$). The reaction rate is directly proportional to [B] and y must be equal to 1. The reaction is first order with respect to B.

Method 2:

$$\frac{\text{Expt 2}}{\text{Expt 1}} \qquad \frac{1.5 \times 10^{-3} \ M/min}{5.0 \times 10^{-4} \ M/min} = \frac{k(0.20 \ M)^0(0.30 \ M)^y(0.20 \ M)^z}{k(0.10 \ M)^0(0.10 \ M)^y(0.20 \ M)^z}$$

$$3 = 3^y$$

$$y = 1$$

Step 3: rate dependence on [C]. Consider Experiments 2 and 4:

Method 1: [B] does not change; [C] changes by a factor of 3; the reaction rate changes by a factor of 3 (= $4.5 \times 10^{-3}/1.5 \times 10^{-3}$). The reaction rate is directly proportional to [C] and z must be equal to 1. The reaction is first order with respect to C.

Method 2:

$$\frac{\text{Expt 4}}{\text{Expt 2}} \qquad \frac{4.5 \times 10^{-3} \ M/min}{1.5 \times 10^{-3} \ M/min} = \frac{k(0.40 \ M)^0(0.30 \ M)^1(0.60 \ M)^z}{k(0.20 \ M)^0(0.30 \ M)^1(0.20 \ M)^z}$$

$$3 = 3^z$$

$$z = 1$$

The rate-law expression is: rate = $k[A]^0[B]^1[C]^1 = k[B][C]$. To calculate the value of k, substitute the values from any one of the experiments into the rate-law expression and solve for k. If we use the data from Experiment 1,

$$5.0 \times 10^{-4} \ M/min = k(0.10 \ M)(0.20 \ M)$$

$$k = 2.5 \times 10^{-2} \ M^{-1} \cdot min^{-1}$$

The rate-law expression is now: **rate = $(2.5 \times 10^{-2} \ M^{-1} \cdot min^{-1})[B][C]$**

The simplest approach to this problem is to assume that the initial concentrations of A and B_2 are each 1 M. Then after the change, the concentrations of A and B_2 are each 1/2 M. Let us substitute these values into the rate-law expression,

rate = $k[A]^2[B_2]$

Initial: rate = $k(1 \ M)^2(1 \ M) = k$

After the change: rate = $k(\frac{1}{2} \ M)^2(\frac{1}{2} \ M) = (1/8) \ k$

Therefore, the rate of reaction would **decrease** by a factor of **8**.

Balanced equation: $2ClO_2(aq) + 2OH^-(aq) \rightarrow ClO_3^-(aq) + ClO_2^-(aq) + H_2O(\ell)$

(a) The form of the rate-law expression: rate $= k[ClO_2]^x[OH^-]^y$

Step 1: rate dependence on $[ClO_2]$. Consider Experiments 1 and 3.

Method 1: By observation, $[OH^-]$ is constant and $[ClO_2]$ increases by a factor of 2 ($= 0.024/0.012$). The rate of reaction increases by a factor of 4 ($= 8.28 \times 10^{-4}/2.07 \times 10^{-4}$). The reaction rate increases as the square of $[ClO_2]$ and x equals 2. The reaction is second order with respect to ClO_2.

Method 2:

Expt 2
Expt 1
$$\frac{8.28 \times 10^{-4}\ M/s}{2.07 \times 10^{-4}\ M/s} = \frac{k(0.024\ M)^x(0.012\ M)^y}{k(0.012\ M)^x(0.012\ M)^y}$$

$$4 = 2^x$$
$$x = 2$$

Step 2: rate dependence on $[OH^-]$. Consider Experiments 1 and 2:

Method 1: By observation, $[ClO_2]$ is constant and $[OH^-]$ increases by a factor of 2 ($= 0.024/0.012$). The rate of reaction increases by a factor of 2 ($= 4.14 \times 10^{-4}/2.07 \times 10^{-4}$). The reaction rate is directly proportional to $[OH^-]$ and y equals 1. The reaction is first order with respect to OH^-.

Method 2:

Expt 2
Expt 1
$$\frac{4.14 \times 10^{-4}\ M/s}{2.07 \times 10^{-4}\ M/s} = \frac{k(0.012\ M)^1(0.024\ M)^y}{k(0.012\ M)^1(0.012\ M)^y}$$

$$2 = 2^y$$
$$y = 1$$

The rate-law expression is: rate $= k[ClO_2]^2[OH^-]^1 = k[ClO_2]^2[OH^-]$

(b) The reaction is second order with respect to ClO_2, first order with respect to OH^- and third order overall.

(c) Using the data from Experiment 1 to calculate k, we have
$$2.07 \times 10^{-4}\ M/s = k(0.012\ M)^2(0.012\ M)$$
$$k = 1.2 \times 10^2\ M^{-2} \cdot s^{-1}$$

The rate-law expression is now: **rate $= (1.2 \times 10^2\ M^{-2} \cdot s^{-1})[ClO_2]^2[OH^-]$**

Balanced equation: $A + B \rightarrow C$

The form of the rate-law expression: rate $= k[A]^x[B]^y$

Step 1: rate dependence on $[B]$. Consider Experiments 1 and 2:

Method 1: $[A]$ does not change; $[B]$ changes by a factor of 1.5 ($= 0.30/0.20$); reaction rate changes also by a factor of 1.5 ($= 7.5 \times 10^{-6}/5.0 \times 10^{-6}$). The reaction rate is directly proportional to $[B]$ and y equals 1. The reaction is first order with respect to B.

Method 2:

Expt 2
Expt 1
$$\frac{7.5 \times 10^{-6}\ M/s}{5.0 \times 10^{-6}\ M/s} = \frac{k(0.10\ M)^x(0.30\ M)^y}{k(0.10\ M)^x(0.20\ M)^y}$$

$$1.5 = 1.5^y$$
$$y = 1$$

Step 2: rate dependence on [A].

There is no pair of experiments in which [A] is changing and [B] is constant. Therefore, one may choose any 2 experiments in which [A] is varying and use Method 2. If we choose Experiments 1 and 3:

Expt 3
Expt 1
$$\frac{4.0 \times 10^{-5} \ M/s}{5.0 \times 10^{-6} \ M/s} = \frac{k(0.20 \ M)^x(0.40 \ M)^1}{k(0.10 \ M)^x(0.20 \ M)^1}$$

$$8 = (2)^x(2)^1$$
$$4 = 2^x$$
$$x = 2$$

Therefore, the reaction is second order with respect to A.

The rate-law expression is: rate $= k[A]^2[B]^1 = k[A]^2[B]$

Using the data from Experiment 1 to calculate k, we have

$$5.0 \times 10^{-6} \ M/s = k(0.10 \ M)^2(0.20 \ M)$$
$$k = 2.5 \times 10^{-3} \ M^{-2}{\cdot}s^{-1}$$

The rate-law expression is now: **rate $= (2.5 \times 10^{-3} \ M^{-2}{\cdot}s^{-1})[A]^2[B]$**

16-26   *Refer to Section 16-3, Examples 16-3 and 16-4, and Exercise 16-18 Solution.*

Balanced equation: $A + B \rightarrow C$

(a) The form of the rate-law expression: rate $= k[A]^x[B]^y$

Step 1: rate dependence on [B]. Consider Experiments 1 and 2:

Method 1: [A] does not change; [B] changes by a factor of 2 ($= 0.20/0.10$); reaction rate changes by a factor of 4 ($= 8.0 \times 10^{-4}/2.0 \times 10^{-4}$). The reaction rate quadruples when the concentration doubles. Therefore, y equals 2. The reaction is second order with respect to B.

Method 2:

Expt 2
Expt 1
$$\frac{8.0 \times 10^{-4} \ M/s}{2.0 \times 10^{-4} \ M/s} = \frac{k(0.10 \ M)^x(0.20 \ M)^y}{k(0.10 \ M)^x(0.10 \ M)^y}$$

$$4.0 = 2.0^y$$
$$y = 2$$

Step 2: rate dependence on [A].

There is no pair of experiments in which [A] is changing and [B] is constant. Therefore, one may choose any 2 experiments in which [A] is varying and use Method 2. If we choose Experiments 1 and 3:

Expt 3
Expt 1
$$\frac{2.56 \times 10^{-2} \ M/s}{2.0 \times 10^{-4} \ M/s} = \frac{k(0.20 \ M)^x(0.40 \ M)^2}{k(0.10 \ M)^x(0.10 \ M)^2}$$

$$128 = (2)^x(4)^2$$
$$8 = 2^x$$
$$x = 3$$

Therefore, the reaction is third order with respect to A.

The rate-law expression is: rate $= k[A]^3[B]^2$

(b) Using the data from Experiment 1 to calculate k, we have

$$2.0 \times 10^{-4} \ M/s = k(0.10 \ M)^3(0.10 \ M)^2$$
$$k = 20. \ M^{-4}{\cdot}s^{-1}$$

The rate-law expression is now: **rate $= (20. \ M^{-4}{\cdot}s^{-1})[A]^3[B]^2$**

16-28. *Refer to Section 16-3.*

The rate-law expression: rate = $k[A][B]^2$

Plan: (1) Use the data for Experiment 1 and the rate-law expression to calculate the rate constant, k.
(2) Substitute the given values into the complete rate-law expression to determine the reaction rate.

(1) Substituting,

$$0.150 \ M/s = k(1.00 \ M)(0.200 \ M)^2$$
$$k = 3.75 \ M^{-2}\cdot s^{-1}$$

(2) Expt 2: rate = $(3.75 \ M^{-2}\cdot s^{-1})(2.00 \ M)(0.200 \ M)^2 =$ **0.300 M/s**
Expt 3: rate = $(3.75 \ M^{-2}\cdot s^{-1})(2.00 \ M)(0.400 \ M)^2 =$ **1.20 M/s**

16-30. *Refer to Section 16-4.*

The half-life of a reactant is the time required for half of that reactant to be converted into products. For a first order reaction, the half-life is independent of concentration so that the same time is required to consume half of any starting amount or concentration of the reactant. On the other hand, the half-life of a second-order reaction *does* depend on the starting amount of the reactant.

16-32. *Refer to Section 16-4 and Example 16-8.*

For the reaction, $2NO_2 \rightarrow 2NO + O_2$, the units of the rate constant, $M^{-1}\cdot min^{-1}$, tell us that the reaction is second order overall. Since the reaction has only one reactant, NO_2, the reaction is second order with respect to NO_2.

The integrated rate equation for a reaction that is second order with respect to NO_2 as the only reactant:

$$\frac{1}{[NO_2]} - \frac{1}{[NO_2]_0} = akt \qquad\qquad \text{where} \quad a = \text{stoichiometric coefficient of } NO_2$$

Substituting,

$$\frac{1}{1.25 \ M} - \frac{1}{2.00 \ M} = (2)(1.70 \ L\cdot mol^{-1}\cdot min^{-1})t$$

$$t = \frac{0.800 \ M^{-1} - 0.500 \ M^{-1}}{(2)(1.70 \ L\cdot mol^{-1}\cdot min^{-1})}$$

$$t = \textbf{0.0882 min or 5.29 s}$$

16-34. *Refer to Section 16-4, and Examples 16-5 and 16-6.*

Balanced equation: $CS_2 \rightarrow CS + S$ first order reaction with $k = 2.8 \times 10^{-7} \ s^{-1}$ at 1000°C

(a) For the first order reaction:

$$t_{1/2} = \frac{0.693}{ak} = \frac{0.693}{(1)(2.8 \times 10^{-7} \ s^{-1})} = \textbf{2.5} \times \textbf{10}^6 \ \textbf{s} \qquad\qquad \text{where } a = \text{stoichiometric coefficient of } CS_2 = 1$$

(b) The integrated first order rate equation: $\ln\left(\dfrac{[A]_0}{[A]}\right) = akt$

Solving for t, we have

$$t = \frac{1}{ak} \ln\left(\frac{[A]_0}{[A]}\right) = \frac{1}{(1)(2.8 \times 10^{-7} \ s^{-1})} \ln\left(\frac{2.00 \ g}{0.75 \ g}\right) = 3.5 \times 10^6 \ s \times \frac{1 \ min}{60 \ s} \times \frac{1 \ hr}{60 \ min} \times \frac{1 \ day}{24 \ hr} = \textbf{41 days}$$

(c) If 0.75 g of CS_2 remains, then 1.25 g of CS_2 (= 2.00 g - 0.75 g) had been converted to the products, CS and S.

$$? \ g \ CS \ \text{formed} = 1.25 \ g \ CS_2 \times \frac{1 \ mol \ CS_2}{76.1 \ g \ CS_2} \times \frac{1 \ mol \ CS}{1 \ mol \ CS_2} \times \frac{44.1 \ g \ CS}{1 \ mol \ CS} = \textbf{0.724 g CS}$$

(d) Substituting into the integrated first order rate equation - see (b),

$$\ln\left(\frac{2.00\ g}{?\ g\ A}\right) = (1)(2.8 \times 10^{-7}\ s^{-1})(45\ days \times \frac{24\ hr}{1\ day} \times \frac{60\ min}{1\ hr} \times \frac{60\ s}{1\ min})$$

$$\ln\left(\frac{2.00\ g}{?\ g\ A}\right) = 1.1$$

taking the inverse ln of both sides,

$$\frac{2.00\ g}{?\ g\ A} = 3.0$$

? g A remaining after 45.0 days = **0.67 g**

Note: To minimize rounding errors, keep all your numbers in your calculator until the end, then round to the appropriate number of significant figures.

16-36. *Refer to Section 16-4, and Examples 16-7, 16-8 and 16-9.*

The rate equation, rate = $(1.4 \times 10^{-10}\ M^{-1} \cdot s^{-1})([NO_2]^2$, tells us that the reaction is second order with respect to NO_2 as the only reactant.

(a) For the second order reaction, $2NO_2 \rightarrow 2NO + O_2$, where a = stoichiometric coefficient of NO_2:

$$t_{1/2} = \frac{1}{ak[NO_2]_0} = \frac{1}{(2)(1.4 \times 10^{-10}\ M^{-1} \cdot s^{-1})(3.00\ mol/2.00\ L)} = \textbf{2.4 x 10}^9\ \textbf{s\ or\ 76\ yrs}$$

(b) The integrated second order rate equation: $\frac{1}{[NO_2]} - \frac{1}{[NO_2]_0} = akt$

Substituting,

$$\frac{1}{[NO_2]} - \frac{1}{(1.50\ M)} = (2)(1.4 \times 10^{-10}\ M^{-1} \cdot s^{-1})(115\ yr \times 3.15 \times 10^7\ s/yr)$$

$$\frac{1}{[NO_2]} = 1.0\ M^{-1} + 0.667\ M^{-1} = 1.7\ M^{-1}$$

$$[NO_2] = \textbf{0.59}\ \textbf{M}$$

$$?\ g\ NO_2\ remaining = 2.00\ L \times \frac{0.59\ mol}{1.00\ L} \times \frac{46.0\ g}{1.0\ mol} = \textbf{54 g NO}_2$$

(c) $[NO_2]_{reacted} = [NO_2]_0 - [NO_2] = 1.50\ M - 0.59\ M = \textbf{0.91}\ \textbf{M}$

$[NO]_{produced} = [NO_2]_{reacted} = \textbf{0.91}\ \textbf{M}$

16-38. *Refer to Section 16-4 and Example 16-6.*

The integrated first order rate equation: $\ln\left(\frac{A_0}{A_t}\right) = akt$

If 99.0% of the cyclopropane disappeared, then 1.0% of it remains. It is not necessary to know the actual starting concentration to do this calculation.

Substituting, $\qquad \ln\left(\frac{100.0\%}{1.0\%}\right) = (1)(2.74 \times 10^{-3}\ s^{-1})t$

$$4.61 = (2.74 \times 10^{-3}\ s^{-1})t$$

$$t = \textbf{1680 s\ or\ 28.0 min}$$

The integrated first order rate equation: $\ln\left(\frac{[A]_0}{[A]_t}\right) = akt$

We have: $[A]_0 = 0.75\ M$ $\quad$ $[A]_t = 0.20\ M$ $\quad$ $t = 47.0$ min

Substituting, $\quad\ln\left(\frac{0.75\ M}{0.25\ M}\right) = k(47.0\ \text{min})$ $\quad$ where $a = 1$

Solving, $\quad k = \textbf{0.023 min}^{-1}$

Consider the decomposition reaction: $2HI \rightarrow H_2 + I_2$

(a)

t (s)	[HI] (mmol/L)	ln[HI]	1/[HI]
0.	5.46	1.697	0.183
250.	4.10	1.411	0.244
500.	2.73	1.004	0.366
750.	1.37	0.3148	0.730

(1) For a zero order reaction, a plot of [HI] vs. t gives a straight line.

(2) For a first order reaction, a plot of ln[HI] vs. t gives a straight line.

(3) For a second order reaction, a plot of 1/[HI] vs. t gives a straight line.

(1) 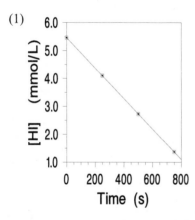 (2) 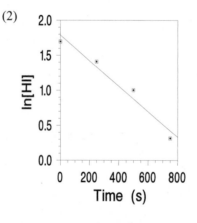 (3)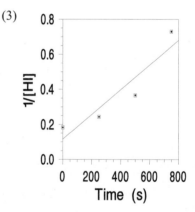

The data lay on a straight line only for Plot (1), the graph of [HI] vs. t. Therefore, the reaction is **zero order** with respect to HI. The slope of the line $= -0.00546\ mM \cdot s^{-1}$, using a least mean square regression fitting program. However, the slope can be estimated from any two points on the line. If we use the first and last points:

$$\text{slope} = \frac{\Delta y}{\Delta x} = \frac{\Delta[HI]}{\Delta t} = \frac{1.37\ mM - 5.46\ mM}{750\ s - 0\ s} = -5.45 \times 10^{-3}\ mM \cdot s^{-1}.$$

Table 16-2 summarizes: for a zero order reaction,

$$\text{slope} = -ak \quad\quad \text{where} \quad\quad a = \text{stoichiometric coefficient}$$
$$k = \text{rate constant}$$

Rearranging, $\quad k = \dfrac{\text{slope}}{-a} = \dfrac{-0.00546\ mM \cdot s^{-1}}{-2} = 0.00273\ mM \cdot s^{-1}$

The rate equation is: $\textbf{rate} = k[HI]^0 = \textbf{\textit{k}} = \textbf{0.00273}\ \textbf{\textit{mM}} \cdot \textbf{s}^{-1}$

The integrated rate equation is: $[HI] = [HI]_0 - akt = \textbf{5.46}\ \textbf{\textit{mM}} - \textbf{(2)(2.73} \times \textbf{10}^{-3}\ \textbf{\textit{mM}} \cdot \textbf{s}^{-1}\textbf{)}\textbf{\textit{t}}$

(b) At 600 s, $[HI] = 5.46\ mM - (2)(2.73 \times 10^{-3}\ mM \cdot s^{-1})(600\ s) = 2.18\ mM$ or **2.18 mmoles/L**

Balanced equation: $2HI(g) \rightarrow H_2(g) + I_2(g)$

(a) The units of the rate constant, $0.080 \ M \cdot s^{-1}$, relate that the reaction is **zeroth order**.

(b) For zeroth order:

$$[HI]_0 - [HI] = akt$$
$$1.50 \ M - 0.15 \ M = (2)(0.080 \ M \cdot s^{-1})t$$
$$t = \mathbf{8.4 \ s}$$

The integrated first order rate equation: $\ln\left(\dfrac{A_0}{A_t}\right) = akt$

Substituting, $\qquad \ln\left(\dfrac{100\%}{25\%}\right) = (1)(3.0 \times 10^{-3} \ d^{-1})t \qquad$ since if 75% decomposed, 25% remains

$$1.39 = (3.0 \times 10^{-3} \ d^{-1})t$$
$$t = \mathbf{462 \ d}$$

Balanced equation: $C_2H_5Cl \rightarrow C_2H_4 + HCl \qquad$ This is a first order equation.

(a) The Arrhenius equation:

$k = Ae^{-E_a/RT} \qquad$ where $\qquad$ k = specific rate constant (s^{-1})
$\qquad\qquad\qquad\qquad\qquad\qquad\quad A$ = constant = $1.58 \times 10^{13} \ s^{-1}$
$\qquad\qquad\qquad\qquad\qquad\qquad\quad E_a$ = activation energy (J/mol) = 237,000 J/mol
$\qquad\qquad\qquad\qquad\qquad\qquad\quad R$ = 8.314 J/mol·K
$\qquad\qquad\qquad\qquad\qquad\qquad\quad T$ = 298.0 K

Substituting,

$k = (1.58 \times 10^{13} \ s^{-1})e^{-(237000 \ \text{J/mol})/(8.314 \ \text{J/mol·K} \times 298 \ \text{K})} = (1.58 \times 10^{13} \ s^{-1})e^{-95.66} = \mathbf{4.52 \times 10^{-29} \ s^{-1}}$

(b) Substituting into the Arrhenius equation,

$k = (1.58 \times 10^{13} \ s^{-1})e^{-(237000 \ \text{J/mol})/(8.314 \ \text{J/mol·K} \times 548 \ \text{K})} = (1.58 \times 10^{13} \ s^{-1})e^{-52.0} = \mathbf{4.04 \times 10^{-10} \ s^{-1}}$

Note: To find $e^{-95.66}$, take the inverse ln of -95.66 on your calculator. inv ln of $-95.66 = 2.85 \times 10^{-42}$. Keep one more significant figure and round off to three significant figures at the end, particularly when working with logarithms.

Catalysts are substances which increase the reaction rate when added to a system by providing an alternative mechanism with a lower activation energy. Although a catalyst may enter into a reaction, it does not appear in the overall balanced equation; it is a reactant in one step and a product in another. It is not consumed during a reaction.

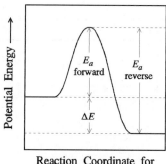

Reaction Coordinate for
Uncatalyzed Reaction

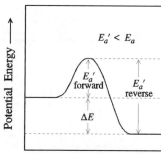

Reaction Coordinate for
Catalyzed Reaction

(a) An increase in temperature does increase the initial rate of reaction. In order for a reaction to occur, the reactants must collide effectively. As the temperature increases, the velocity of the reactants increases and there are more collisions. Also, the reactant species must have a certain energy, called the activation energy, to produce products upon collision. When the temperature increases, more of the reactant species have that necessary amount of kinetic energy, so the reaction proceeds at a faster rate. Recall that the average kinetic energy of a container of molecules in the gas phase is directly proportional to the absolute temperature.

(b) A gas phase reaction is faster than the same reaction in the solid phase because the reacting species in the gaseous phase can move more quickly, have more collisions, resulting in a faster reaction rate.

16-54. *Refer to Section 16-8 and Example 16-12.*

The Arrhenius equation can be presented as:

$$\ln\left(\frac{k_2}{k_1}\right) = \frac{E_a}{R}\left(\frac{T_2 - T_1}{T_1 T_2}\right)$$

where

k_2/k_1 = ratio of rate constants = 3.000
E_a = activation energy (J/mol)
R = 8.314 J/mol·K
T_1 = 600.0 K
T_2 = 610.0 K

Substituting,

$$\ln 3.000 = \frac{E_a}{8.314 \text{ J/mol·K}}\left(\frac{610. \text{ K} - 600. \text{ K}}{(600. \text{ K})(610. \text{ K})}\right)$$

$$1.0986 = \frac{E_a}{8.314 \text{ J/mol·K}}(2.7 \times 10^{-5} \text{ K}^{-1})$$

$$E_a = \textbf{3.4} \times \textbf{10}^\textbf{5} \textbf{ J/mol rxn or 340 kJ/mol rxn}$$

16-56. *Refer to Section 16-8 and Example 16-12.*

The Arrhenius equation can be presented as:

$$\ln\left(\frac{k_2}{k_1}\right) = \frac{E_a}{R}\left(\frac{T_2 - T_1}{T_1 T_2}\right)$$

where

k_2/k_1 = ratio of rate constants = 3.000
E_a = activation energy (J/mol)
R = 8.314 J/mol·K
T_1 = 298 K
T_2 = 308 K

Substituting,

$$\ln 3.000 = \frac{E_a}{8.314 \text{ J/mol·K}}\left(\frac{308 \text{ K} - 298 \text{ K}}{(298 \text{ K})(308 \text{ K})}\right)$$

$$1.0986 = \frac{E_a}{8.314 \text{ J/mol·K}}(1.1 \times 10^{-4} \text{ K}^{-1})$$

$$E_a = \textbf{8.4} \times \textbf{10}^\textbf{4} \textbf{ J/mol rxn or 84 kJ/mol rxn}$$

16-58. *Refer to Sections 16-6 and 16-8, Figure 16-10 and Exercise 16-56 Solution.*

For a particular reaction: $\Delta E° = 51.51$ kJ/mol reaction

(1) From the Arrhenius equation:

$$\ln\left(\frac{k_2}{k_1}\right) = \frac{E_a}{8.314 \text{ J/mol·K}}\left(\frac{323.15 \text{ K} - 273.15 \text{ K}}{273.15 \text{ K} \times 323.15 \text{ K}}\right)$$

$$\ln\left(\frac{8.9 \times 10^{-4} \text{ s}^{-1}}{8.0 \times 10^{-7} \text{ s}^{-1}}\right) = \frac{E_a}{8.314 \text{ J/mol·K}}(5.664 \times 10^{-4} \text{ K}^{-1})$$

$$E_a = \textbf{1.03} \times \textbf{10}^\textbf{5} \textbf{ J/mol rxn or 103 kJ/mol rxn}$$

(2) The reaction coordinate diagram for the reaction is:

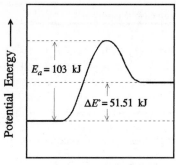

Reaction Coordinate

16-60. *Refer to Section 16-8.*

Plan: The Arrhenius equation can be rearranged: $\ln k = -\left(\dfrac{E_a}{R}\right)\left(\dfrac{1}{T}\right) + \ln A$.

Plot $\ln k$ against $1/T$.

The slope of the line $= -E_a/R$ and solve for E_a.

T (K)	$1/T$ (K^{-1})	k (s^{-1})	$\ln k$
600	1.67×10^{-3}	3.30×10^{-9}	-19.53
650	1.54×10^{-3}	2.19×10^{-7}	-15.33
700	1.43×10^{-3}	7.96×10^{-6}	-11.74
750	1.33×10^{-3}	1.80×10^{-4}	-8.623
800	1.25×10^{-3}	2.74×10^{-3}	-5.900
850	1.18×10^{-3}	3.04×10^{-2}	-3.493
900	1.11×10^{-3}	2.58×10^{-1}	-1.355

(a) By plotting the data as shown, we obtained
    slope $= -E_a/R = -3.25 \times 10^4$ K

Therefore,

$E_a = -$ (slope) $\times R$
$= -(-3.25 \times 10^4$ K$) \times 8.314$ J/mol·K
$= \mathbf{2.70 \times 10^5}$ **J/mol or 270. kJ/mol**

(b) On the x axis, $1/T = 1/500$ K $= 0.00200$ K^{-1}

From the graph, we can estimate:
    y $= \ln k = -30.2$
Therefore, $k = \mathbf{7.7 \times 10^{-14}}$ **s^{-1}** at 500 K

(c) On the y axis, $\ln k = \ln (5.00 \times 10^{-5}) = -9.9$
From the graph, we can estimate:
    x $= 1/T = 0.00137$ K^{-1}
Therefore, $T = \mathbf{730}$ **K** when $k = 5.00 \times 10^{-5}$ s^{-1}

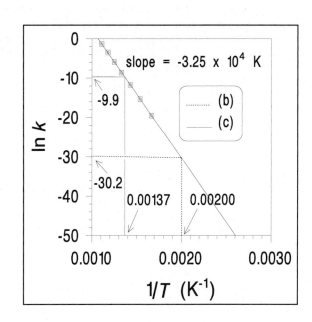

The hydration reaction of CO_2, $CO_2 + H_2O \rightarrow H_2CO_3$, is enzyme catalyzed. The rate of reaction does not depend on $[CO_2]$ or $[H_2O]$. This is deduced from the fact that it only takes 1 molecule of enzyme to react with 10^6 molecules of CO_2. Therefore, the reaction is zero-order with respect to CO_2.: rate $= k[CO_2]^0 = k$.

Plan: mol enzyme/L $\overset{(1)}{\Rightarrow}$ molecules enzyme/L $\overset{(2)}{\Rightarrow}$ hydration rate (molecules CO_2/L·s)
$\overset{(3)}{\Rightarrow}$ hydration rate (molecules CO_2/L·min) $\overset{(4)}{\Rightarrow}$ hydration rate (mol CO_2/L·min)
$\overset{(5)}{\Rightarrow}$ hydration rate (g CO_2/L·min)

$? \text{ g } CO_2 \text{ hydrated/L·min} = \dfrac{1.0 \times 10^{-6} \text{ mol enzyme}}{1 \text{ liter}} \times \dfrac{6.02 \times 10^{23} \text{ molecules enzyme}}{1 \text{ mol enzyme}} \times \dfrac{10^6 \text{ molecules } CO_2}{1 \text{ molecule enzyme} \times 1\text{s}}$

$\times \dfrac{60 \text{ s}}{1 \text{ min}} \times \dfrac{1 \text{ mol } CO_2}{6.02 \times 10^{23} \text{ molecules } CO_2} \times \dfrac{44 \text{ g } CO_2}{1 \text{ mol } CO_2}$

$= \textbf{2600 g } CO_2\textbf{/L·min}$

(a) From the Arrhenius equation: $\ln\left(\dfrac{k_2}{k_1}\right) = \dfrac{E_a}{R}\left(\dfrac{T_2 - T_1}{T_1 T_2}\right)$

Substituting,

$\ln\dfrac{k_2}{9.16 \times 10^{-3} \text{ s}^{-1}} = \dfrac{88 \times 10^3 \text{ J/mol}}{8.314 \text{ J/mol·K}}\left(\dfrac{298 \text{ K} - 273 \text{ K}}{(273 \text{ K})(298 \text{ K})}\right) = 3.3 \text{ (or } 3.2526)$

$\dfrac{k_2}{9.16 \times 10^{-3} \text{ s}^{-1}} = 26$

$k_2 = \textbf{0.24 s}^{-1}$

(b) Substituting into the following version of the Arrhenius equation and solving for T_2,

$\ln\dfrac{k_2}{k_1} = \dfrac{E_a}{R}\left(\dfrac{1}{T_1} - \dfrac{1}{T_2}\right)$

$\ln\left(\dfrac{3.00 \times 10^{-2} \text{ s}^{-1}}{9.16 \times 10^{-3} \text{ s}^{-1}}\right) = \dfrac{88 \times 10^3 \text{ J/mol}}{8.314 \text{ J/mol·K}}\left(\dfrac{1}{273 \text{ K}} - \dfrac{1}{T_2}\right)$

$1.186 = (1.06 \times 10^4 \text{ K})(3.66 \times 10^{-3} \text{ K}^{-1} - 1/T_2)$

$1.12 \times 10^{-4} = 3.66 \times 10^{-3} \text{ K}^{-1} - 1/T_2$

$1/T_2 = 3.66 \times 10^{-3} - 1.12 \times 10^{-4}$

$1/T_2 = 3.55 \times 10^{-3}$

$T_2 = \textbf{282 K or 9°C}$

The rate-law expression for the reaction is: rate $= k[Cl_2][H_2S]$. If the rate-law expression derived from a proposed mechanism is different from this expression, the mechanism cannot be the correct one.

(a) The rate-law expression consistent with the slow step of this mechanism is: rate $= k[Cl_2]$ and this cannot be a mechanism for the reaction.

(b) The rate-law expression consistent with the slow step of this mechanism is: rate $= k[Cl_2][H_2S]$ and this is a possible mechanism for the reaction.

(c) The rate-law expression consistent with this mechanism is more complicated. It can be determined by following the procedure presented in Section 16-7:

From Step 3 (the slow step), $\qquad$ rate $= k_3[HS][Cl]$ $\qquad$ where HS and Cl are intermediates

Since HS and Cl are intermediates, their concentrations must be expressed in terms of the reactants, Cl_2 and H_2S. For a fast, equilibrium step, we know:

$$rate_{1f} = rate_{1r}$$

From Step 1, $\qquad k_{1f}[Cl_2] = k_{1r}[Cl][Cl] = k_{1r}[Cl]^2$

$$[Cl] = \left(\frac{k_{1f}}{k_{1r}}\right)^{1/2}[Cl_2]^{1/2}$$

From Step 2, $\qquad k_{2f}[Cl][H_2S] = k_{2r}[HCl][HS]$

$$[HS] = \left(\frac{k_{2f}}{k_{2r}}\right)\frac{[Cl][H_2S]}{[HCl]} = \left(\frac{k_{2f}}{k_{2r}}\right)\left(\frac{k_{1f}}{k_{1r}}\right)^{1/2}\frac{[Cl_2]^{1/2}[H_2S]}{[HCl]}$$

Substituting $\qquad$ rate $= k_3\left(\frac{k_{2f}}{k_{2r}}\right)\left(\frac{k_{1f}}{k_{1r}}\right)^{1/2}\frac{[Cl_2]^{1/2}[H_2S]}{[HCl]} \times \left(\frac{k_{1f}}{k_{1r}}\right)^{1/2}[Cl_2]^{1/2}$

$$= k\frac{[Cl_2][H_2S]}{[HCl]}$$

And so, this **cannot** be a mechanism for this reaction.

16-68. *Refer to Section 16-7.*

(a) Overall reaction: **$2A + 2B \rightarrow E + G$**

Using the second step which is the slow, rate-determining step, we can write: rate $= k_2[D][B]$.
However, D is an intermediate; its concentration must be expressed in terms of the reactants, A and B.

For a fast, equilibrium step, we know:

$$rate_{1f} = rate_{1r}$$

From Step 1, $\qquad k_{1f}[A]^2[B] = k_{1r}[D]$

$$[D] = \left(\frac{k_{1f}}{k_{1r}}\right)[A]^2[B]$$

Substituting into the rate law expression obtained in the slow, rate-determining step:

$$rate = k_2\left(\frac{k_{1f}}{k_{1r}}\right)[A]^2[B][B].$$

$$\textbf{rate} = \textbf{\textit{k}}[\textbf{A}]^2[\textbf{B}]^2$$

(b) **Overall reaction: $A + B + D \rightarrow G$** $\qquad$ where C and F are intermediates

The rate-law expression consistent with this mechanism is more complicated. It can be determined by following the procedure presented in Section 16-7:

From Step 3 (the slow step), $\qquad$ rate $= k_3[F]$ $\qquad$ where F is an intermediate

Since F is an intermediate, its concentration must ultimately be expressed as a function of the reactants, A, B and D.

For a fast, equilibrium step, we know:

Starting from the top in Step 1, $\quad$ rate$_{1f}$ = rate$_{1r}$

$$k_{1f}[A][B] = k_{1r}[C] \qquad \text{where C is an intermediate}$$

$$[C] = \left(\frac{k_{1f}}{k_{1r}}\right)[A][B]$$

From Step 2, $\qquad k_{2f}[C][D] = k_{2r}[F]$

$$[F] = \left(\frac{k_{2f}}{k_{2r}}\right)\left(\frac{k_{1f}}{k_{1r}}\right)[A][B][D]$$

Substituting for [F] $\qquad$ rate $= k_3[F] = k_3\left(\frac{k_{2f}}{k_{2r}}\right)\left(\frac{k_{1f}}{k_{1r}}\right)[A][B][D]$

Therefore $\qquad\qquad$ **rate = k[A][B][D]**

16-70. *Refer to Section 16-7.*

Overall reaction: $2O_3(g) \rightarrow 3O_2(g)$

The reaction mechanism: $\qquad O_3 \rightleftarrows O_2 + O \qquad$ (fast, equilibrium)

$\qquad\qquad\qquad\qquad\qquad\quad O + O_3 \rightarrow 2O_2 \qquad$ (slow)

From Step 2 (the slow step), $\qquad$ rate $= k_2[O][O_3] \qquad$ where O is an intermediate

From the slow step, rate $= k_2[O][O_3]$. However, O is an intermediate and its concentration must be expressed in terms of the reactant, O_3. For a fast, equilibrium step, we know:

$$\text{rate}_{1f} = \text{rate}_{1r}$$

$$k_{1f}[O_3] = k_{1r}[O_2][O]$$

$$[O] = \frac{k_{1f}}{k_{1r}}\frac{[O_3]}{[O_2]}$$

Substituting for [O] in the original rate equation, we have

$$\text{rate} = k_2[O][O_3] = k_2\left(\frac{k_{1f}}{k_{1r}}\frac{[O_3]}{[O_2]}\right)[O_3] \quad \text{or} \quad \textbf{rate} = k\,\frac{[O_3]^2}{[O_2]}$$

16-72. *Refer to Section 16-7.*

The reaction mechanism: $\qquad N_2 + Cl \rightleftarrows N_2Cl \qquad$ (fast, equilibrium)

$$N_2Cl + Cl \xrightarrow{k_2} Cl_2 + N_2 \qquad \text{(slow)}$$

(a) The intermediate species is **N_2Cl**.

(b) From the slow step, rate $= k_2[N_2Cl][Cl]$. However, N_2Cl is an intermediate and its concentration must be expressed in terms of the reactants, N_2 and Cl. For a fast, equilibrium step, we know:

$$\text{rate}_{1f} = \text{rate}_{1r}$$

$$k_{1f}[N_2][Cl] = k_{1r}[N_2Cl]$$

$$[N_2Cl] = \frac{k_{1f}}{k_{1r}}[N_2][Cl]$$

Substituting for [N_2Cl] in the original rate equation, we have

$$\text{rate} = k_2[N_2Cl][Cl] = k_2\left(\frac{k_{1f}}{k_{1r}}[N_2][Cl]\right)[Cl] \quad \text{or} \quad \text{rate} = k[N_2][Cl]^2$$

Yes, the mechanism is consistent with the experimental rate law, rate $= k[N_2][Cl]^2$.

The reaction mechanism:

$$H_2 \rightleftarrows 2H \qquad \text{(fast, equilibrium)}$$

$$H + CO \xrightarrow{k_2} HCO \qquad \text{(slow)}$$

$$H + HCO \xrightarrow{k_3} H_2CO \qquad \text{(fast)}$$

(a) Balanced equation: $H_2 + CO \rightarrow H_2CO$

(b) From the slow step, rate = $k_2[H][CO]$. However, H is an intermediate and its concentration must be expressed in terms of the reactant, H_2. For a fast, equilibrium step, we know:

$$\text{rate}_{1f} = \text{rate}_{1r}$$

$$k_{1f}[H_2] = k_{1r}[H]^2$$

$$[H] = \left(\frac{k_{1f}}{k_{1r}}\right)^{1/2}[H_2]^{1/2}$$

Substituting for [H] in the original rate equation, we have

$$\text{rate} = k_2[H][CO] = k_2\left(\frac{k_{1f}}{k_{1r}}\right)^{1/2}[H_2]^{1/2}[CO] = k[H_2]^{1/2}[CO]$$

Yes, the mechanism is consistent with the observed rate dependence: rate = $k[H_2]^{1/2}[CO]$.

(a) The transition state is a short-lived, high-energy intermediate state that the reactants must convert into before the products are formed.

(b) Yes, the activation energy and the transition state are related concepts. The activation energy, E_a, is the additional energy that must be absorbed by the reactants in their ground states to allow them to reach the transition state.

(c) The higher the activation energy, the more energy that is required by the reactants to form products and the slower is the overall reaction.

From Exercise 16-64, we know that the reaction, $N_2O_5 \rightarrow NO_2 + NO_3$, is first order with specific rate constant, $k = 0.24 \text{ s}^{-1}$ at 25°C.

(a) The integrated rate equation: $\ln \dfrac{A_0}{A} = akt$

where A_0 = 2.80 mol in 3.00 L (the volume is irrelevant in this question)
a = stoichiometric coefficient of N_2O_5

Substituting, $\ln\left(\dfrac{2.80 \text{ mol}}{? \text{ mol } N_2O_5}\right) = (0.24 \text{ s}^{-1})(2.00 \text{ min} \times 60 \text{ s/min}) = 29 \text{ (28.8 to 3 sig. figs.)}$

$$\dfrac{2.80 \text{ mol}}{? \text{ mol } N_2O_5} = 3.2 \times 10^{12} \quad \text{(using 28.8 then rounding to 2 sig. figs.)}$$

? mol N_2O_5 = **8.7×10^{-13} mol N_2O_5 remaining after 2.00 min**

Note: Any small change in the value of k can greatly alter the answer.

(b) If 99.0% of N_2O_5 have decomposed, then 1.0% of N_2O_5 remains.

Substituting into the integrated rate equation, we have $\ln\left(\dfrac{100.0\%}{1.0\%}\right) = (0.24 \text{ s}^{-1})t$ Solving, $t = $ **19 s**

Balanced equation: $4Hb + 3CO \rightarrow Hb_4(CO)_3$ where Hb is hemoglobin

(a) The form of the rate-law expression: rate $= k[Hb]^x[CO]^y$

    Step 1: rate dependence on [Hb]. Consider Experiments 1 and 2:

        Method 1: By observation, [CO] does not change. [Hb] increases by a factor of 2 (= 6.72/3.36). The rate of disappearance of Hb also increases by a factor of 2 (= 1.88/0.941). The rate is directly proportional to [Hb] and x is equal to 1. The reaction is first order with respect to Hb.

        Method 2:

$$\frac{\text{Expt 2}}{\text{Expt 1}} \quad \frac{1.88 \text{ μmol/L·s}}{0.941 \text{ μmol/L·s}} = \frac{k(6.72 \text{ μmol/L})^x(1.00 \text{ μmol/L})^y}{k(3.36 \text{ μmol/L})^x(1.00 \text{ μmol/L})^y}$$

$$2 = (2)^x$$

$$x = 1$$

    Step 2: rate dependence on [CO]. Consider Experiments 2 and 3.

        Method 1: By observation, [Hb] does not change. [CO] increases by a factor of 3 (= 3.00/1.00). The rate of disappearance of Hb also increases by a factor of 3 (= 5.64/1.88). The rate is directly proportional to [CO] and y is equal to 1. The reaction is first order with respect to CO.

        Method 2:

$$\frac{\text{Expt 3}}{\text{Expt 2}} \quad \frac{5.64 \text{ μmol/L·s}}{1.88 \text{ μmol/L·s}} = \frac{k(6.72 \text{ μmol/L})^x(3.00 \text{ μmol/L})^y}{k(6.72 \text{ μmol/L})^x(1.00 \text{ μmol/L})^y}$$

$$3 = (3)^y$$

$$y = 1$$

    The rate-law expression is: **rate** $= k[Hb]^1[CO]^1 = \mathbf{k[Hb][CO]}$

(b) Substituting, the data from Experiment 1 into the rate-law expression to calculate k, we have,

$$0.941 \text{ μmol/L·s} = k(3.36 \text{ μmol/L})(1.00 \text{ μmol/L})$$

$$k = \mathbf{0.280 \text{ L/μmol·s}}$$

(c) Substituting into the complete rate-law expression, gives

$$\text{rate} = (0.280 \text{ L/μmol·s})[Hb][CO]$$

$$= (0.280 \text{ L/μmol·s})(1.50 \text{ μmol/L})(0.600 \text{ μmol/L})$$

$$= \mathbf{0.252 \text{ μmol/L·s}}$$

The shape of a reactant can affect the rate of a reaction. Collision theory tells us that in order for a reaction to occur there must be an effective collision. For a collision to be effective, the reacting species must (1) possess at least a certain minimum energy, called the activation energy, and (2) have the proper orientations toward each other at the time of collision. Because orientation is important, the shape of the molecule must also be important.

formula unit: $HCl(aq) + NaOH(aq) \rightarrow NaCl(aq) + H_2O(\ell)$

total ionic: $H^+(aq) + Cl^-(aq) + Na^+(aq) + OH^-(aq) \rightarrow Na^+(aq) + Cl^-(aq) + H_2O(\ell)$

net ionic: $H^+(aq) + OH^-(aq) \rightarrow H_2O(\ell)$

To find the ΔE of the reaction:

(1) $\Delta H^\circ_{rxn} = [\Delta H^\circ_{f\ NaCl(aq)} + \Delta H^\circ_{f\ H_2O(\ell)}] - [\Delta H^\circ_{f\ HCl(aq)} + \Delta H^\circ_{f\ NaOH(aq)}]$

   $= [(1\ mol)(-407.1\ kJ/mol) + (1\ mol)(-285.8\ kJ/mol)]$

   $- [(1\ mol)(-167.4\ kJ/mol) + (1\ mol)(-469.6\ kJ/mol)]$

   $= -55.9\ kJ/mol\ rxn$

(2) Because there are no gases involved in this reaction, $\Delta E = \Delta H = -55.9$ kJ/mol rxn.
     This means that energy is released in this reaction.

The potential energy diagram for the reaction is:

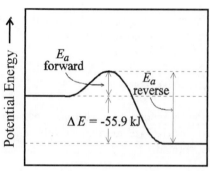

Given the equation: $3O_2 \rightarrow 2O_3$, we definitely **cannot** say that O_2 reacts with itself to form O_3. This equation is the overall reaction, and does not give any information about the mechanism of the reaction, i.e. the order in which bonds are broken and formed to create the products from the reactant. It is only telling us that for every 3 moles of O_2 that react, 2 moles of O_3 are formed. There is no information in the equation as to how that happens. However, using thermodynamic tables and Hess's Law, we find that $\Delta H_{rxn} = +286$ kJ and $\Delta S_{rxn} = -137.4$ J/mol K, so ΔG_{rxn} is positive at all temperatures and the reaction must be nonspontaneous at all temperatures.

The activation energy, E_a, is the additional energy that must be absorbed by the reactants in their ground states to allow them to reach the transition state. Consider the following three endothermic reactions; they are identical except that they occur in different phases:

$$A(g) + B(g) \rightarrow C(g)$$
$$A(\ell) + B(\ell) \rightarrow C(\ell)$$
$$A(s) + B(s) \rightarrow C(s)$$

We know that the potential energy of the reactants as gases is greater than that of the liquid reactants, and the potential energy of the solid reactants is the lowest. Let us assume that (1) the potential energy of the high-energy transition state for a reaction is independent of the phase of the reactants and products, and (2) the mechanism of the reaction doesn't change. Therefore, the activation energy of the gaseous reaction is expected to be less than the liquid-phase reaction, which is less than the solid-phase reaction.

So, we expect:

E_a for gas-phase rxn $< E_a$ for liquid-phase rxn $< E_a$ for solid-phase rxn

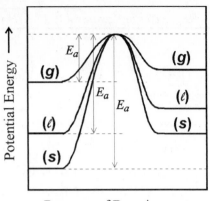

16-90.   *Refer to Section 16-6 and Figure 16-10b.*

The potential energy diagram for the generic reaction shown in Figure 16-10b is shown at the right.

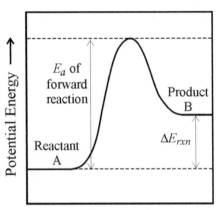

(a) The reaction can be expressed as: A → B.

(b) The reaction is endothermic because ΔE for the reaction is positive; the reaction absorbs energy as product B is formed.

(c) The activation energy of the forward reaction is shown as the difference between the potential energy of reactant, A, and the energy of the transition (excited) state.

16-92.   *Refer to Sections 13-9 and 16-8.*

Plan: (1) Use the Clausius-Clapeyron equation to calculate the steam temperature in the pressure cooker. Assume that ΔH_{vap} for H_2O is independent of temperature.

(2) Use the Arrhenius equation to calculate the activation energy for the process of steaming vegetables.

(1) From the Clausius-Clapeyron equation: $\ln\left(\dfrac{P_2}{P_1}\right) = \dfrac{\Delta H_{vap}}{R}\left(\dfrac{1}{T_1} - \dfrac{1}{T_2}\right)$

where ΔH_{vap} = molar heat of vaporization for H_2O, 40.7 kJ/mol
P_1 = atmospheric pressure, 15 psi
P_2 = cooker pressure = P_1 + gauge pressure = (15 + 15) psi
T_1 = boiling point of water at 1 atm, 100.0°C
T_2 = steam temperature in the pressure cooker

Substituting,

$$\ln\left(\frac{30 \text{ psi}}{15 \text{ psi}}\right) = \frac{40.7 \times 10^3 \text{ J/mol}}{8.314 \text{ J/mol·K}}\left(\frac{1}{373 \text{ K}} - \frac{1}{T_2}\right)$$

$$0.69 = (4.90 \times 10^3 \text{ K})(2.68 \times 10^{-3} \text{ K}^{-1} - 1/T_2)$$

$$1.4 \times 10^{-4} \text{ K}^{-1} = (2.68 \times 10^{-3} \text{ K}^{-1} - 1/T_2)$$

$$T_2 = 394 \text{ K}$$

(2) From the Arrhenius equation: $\ln\left(\dfrac{k_2}{k_1}\right) = \dfrac{E_a}{R}\left(\dfrac{T_2 - T_1}{T_1 T_2}\right)$

where $T_1 = 373$ K
$T_2 = 394$ K
k_1 = rate constant for cooking vegetables at atmospheric pressure
k_2 = rate constant for cooking vegetables in the pressure cooker
$k_2/k_1 = 3$ since the cooking process is 3 times faster in the pressure cooker

Substituting,

$$\ln 3 = \dfrac{E_a}{8.314 \text{ J/mol·K}}\left(\dfrac{394 \text{ K} - 373 \text{ K}}{(373 \text{ K})(394 \text{ K})}\right)$$

$E_a = \textbf{6.4 x 10}^4$ **J/mol or 64 kJ/mol**

16-94. *Refer to Section 16-7 and Figure 16-12.*

The overall reaction: $H_2(g) + I_2(g) \rightarrow 2HI(g)$ $\qquad$ rate = $k[H_2][I_2]$

The reaction mechanism: (1) $\qquad I_2 \rightleftarrows 2I$ $\qquad$ (fast, equilibrium)
$\qquad$ (2) $\qquad I + H_2 \rightleftarrows H_2I$ $\qquad$ (fast, equilibrium)
$\qquad$ (3) $\qquad H_2I + I \rightarrow 2HI$ $\qquad$ (slow)
$\qquad\qquad\qquad H_2 + I_2 \rightarrow 2HI$ $\qquad$ (overall)

Step 1 $\qquad$ **Reactant** $\qquad$ **Excited State (first peak)** $\qquad$ **Products (first trough)**

$\qquad$:I̤-I̤: (H-H present) $\qquad$:I̤—I̤: (H-H present) $\qquad$ 2 :I̤· (H-H present)

Step 2 $\qquad$ **Reactants** $\qquad$ **Excited State (second peak)** $\qquad$ **Product (second trough)**

$\qquad$ H-H and :I̤· (:I̤· present) $\qquad$ H–H---:I̤: (:I̤· present) $\qquad$ H-H-I̤: (:I̤· present)
$\qquad\qquad\qquad\qquad\qquad\qquad$ weaker bond $\qquad\qquad\qquad\qquad$ (9 e^- present) *

Step 3 $\qquad$ **Reactants** $\qquad$ **Excited State (third peak)** $\qquad$ **Products (final)**

$\qquad$:I̤· and H-H-I̤: * $\qquad$:I̤---H---H---:I̤: $\qquad$ 2 H-I̤:

* The extra weak H-H bond contains only one electron.

16-96. *Refer to Sections 15-3, 15-8 and 16-6, Appendix K and Figure 16-10.*

Balanced equation: $O_3(g) + NO(g) \rightarrow NO_2(g) + O_2(g)$ $\qquad\qquad E_a = 9.6$ kJ/mol reaction

(a) $\Delta H^\circ_{rxn} = [\Delta H^\circ_{f\,NO_2(g)} + \Delta H^\circ_{f\,O_2(g)}] - [\Delta H^\circ_{f\,O_3(g)} + \Delta H^\circ_{f\,NO(g)}]$
$\qquad\qquad = [(1 \text{ mol})(33.2 \text{ kJ/mol}) + (1 \text{ mol})(0 \text{ kJ/mol})] - [(1 \text{ mol})(143 \text{ kJ/mol}) + (1 \text{ mol})(90.25 \text{ kJ/mol})]$
$\qquad\qquad = \textbf{−200. kJ/mol reaction}$

(b) For the reaction, $\Delta E^\circ = \Delta H^\circ$ since $\Delta n_{gas} = 0$.

Therefore, the activation energy plot for this reaction is:

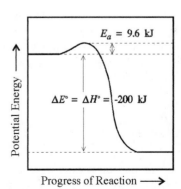

17 Chemical Equilibrium

17-2. *Refer to Section 17-2.*

Equilibrium constants do not have units because in the strict thermodynamic definition of the equilibrium constant, the *activity* of a component is used, not its concentration. The activity of a species in an ideal mixture is the ratio of its concentration or partial pressure to a standard concentration (1 *M*) or pressure (1 atm). Because activity is a ratio, it is unitless and the equilibrium constant involving activity is also unitless.

17-4. *Refer to Sections 17-1 and 17-2.*

Equilibrium can be defined as a state in which no observable changes occur as time goes by.

(a) Static equilibrium is an important concept in physics. A body is said to be "in static equilibrium" if a body at rest will stay at rest. (Note: this happens when its acceleration and angular acceleration are zero.)

Examples of static equilibrium are (1) a ladder leaning against a wall, (2) a block of wood resting on a table and (3) a picture hanging on a wall.

(b) Dynamic equilibrium is a state in which no net change takes place because two opposing processes are occurring at the same time.

For example (1) consider the movement of skiers on a busy day on the ski slopes when the number of skiers going up the hill on the chair lift equals the number of skiers skiing down the hill. If you took pictures during the day, the number of skiers at the top of the slope and the number at the bottom remain unchanged. (2) Another similar example would be children on a slide in the park. (3) A more scientific example is the evaporation of water in a closed container. Once the system is at equilibrium, the number of water molecules in the gaseous phase and the number of water molecules in the liquid phase remain unchanged. However, liquid molecules are constantly entering the gaseous phase while gaseous molecules are constantly condensing to form liquid molecules of water. We write this as:

$$H_2O(\ell) \rightleftarrows H_2O(g)$$

17-6. *Refer to Section 17-2.*

Consider the equilibrium: $A(g) \rightleftarrows B(g)$ $\qquad\qquad K_c = \dfrac{[B]}{[A]}$

The magnitude of the equilibrium constant, K_c, is a measure of the extent to which a reaction occurs. If the equilibrium lies far to the right, then this means that at equilibrium most of the reactants would be converted into products and the value of K_c would be much greater than one. If the equilibrium lies far to the left, then at equilibrium, most of the reactants remain unreacted and there are very little products formed. The value of K_c would be a very small fraction.

17-8. *Refer to Section 17-2.*

The magnitude of an equilibrium constant tells us *nothing* about how fast the system will reach equilibrium. Equilibrium constants are thermodynamic quantities, whereas the speed of a reaction is a kinetic quantity. The two are not related. Rather, an equilibrium constant is a measure of the extent to which a reaction occurs.

(a) The equilibrium constant is related to the specific rate constants of the forward and reverse reactions. Consider the following 1 step reaction: A $\rightleftarrows$ B, where k_f is the specific rate constant of the forward reaction and k_r is the specific rate constant of the reverse reaction. At equilibrium, the rates of the forward reaction and the reverse reaction are equal.

$$\text{Rate}_f = \text{Rate}_r$$

$$k_f[A] = k_r[B]$$

$$K_c = \frac{k_f}{k_r} = \frac{[B]}{[A]}$$

K_c, the conventional equilibrium constant, is equal to the ratio of the forward rate constant divided by the reverse rate constant.

(b) The rate expressions for the forward and reverse reactions can NOT be written from the balanced chemical equation. You need the experimentally-derived rate law expression or the mechanism for the reaction. The only exception to this is if you know the reaction occurs in only 1 step - a 1-step mechanism, as in the example given in (a). In that case, the balanced chemical reaction is the rate-determining step (it's the only step) and you can determine the rate law expressions.

(c) However, the equilibrium constant expression (also called the mass action expression) can be written from the balanced chemical equation. Regardless of the mechanism by which a reaction occurs, the concentrations of reaction intermediates always cancel out and the equilibrium constant expression has the same form.

17-12. *Refer to Section 17-2 and Figure 17-2.*

(a) Consider the equilibrium: $3A(g) + B(g) \rightleftarrows 2C(g)$. Assuming that the concentrations of A and B are both 1 M initially, the *changes* that the concentrations undergo are related to the stoichiometric coefficients. When the reaction is at equilibrium at t_e, the system has little A remaining. The change in [B] is 1/3 that of the [A], and the increase in the [C] is equal to twice the loss of [B]. In other words:

Let   x = mol/L of B that react. Then
      3x = mol/L of A that react, and
      2x = mol/L of C that are produced.

	3A	+	B	$\rightleftarrows$	2C
initial	1.0 mol/L		1.0 mol/L		0 mol/L
change	- 3x mol/L		- x mol/L		+ 2x mol/L
at equilibrium	(1.0 - 3x) mol/L		(1.0 - x) mol/L		2x mol/L

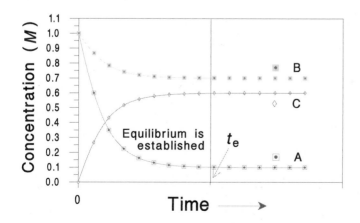

In this diagram illustrating the situation when $K_c > 1$, let us assume that x = 0.3 at equilibrium and work through this problem. There will tend to be more products and less reactants at equilibrium. From the equilibrium concentrations, we can calculate K_c for this example.

$$K_c = \frac{[C]^2}{[A]^3[B]} = \frac{(0.60)^2}{(0.10)^3(0.70)} = 510$$

(to 2 significant figures)

Note the overall look of the graph. The concentrations of A and B decrease greatly and the concentration of C increases greatly as the system approaches equilibrium.

(b) For the same equilibrium with $K_c < 1$, there will be more reactants and little products when the system reaches equilibrium.

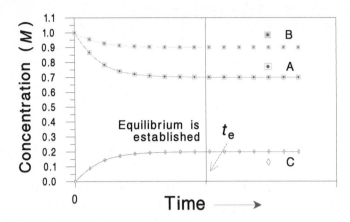

In this diagram, let us assume that $x = 0.1$ at equilibrium. From the equilibrium concentrations, we can calculate K_c for this example.

$$K_c = \frac{[C]^2}{[A]^3[B]} = \frac{(0.20)^2}{(0.70)^3(0.90)} = 0.13$$

Note the overall look of the graph. The concentrations of A and B decrease slightly and the concentration of C increases slightly as the system approaches equilibrium.

17-14.   *Refer to Section 17-2 and Exercise 17-2 Solution.*

We omit concentrations of pure solids and pure liquids from equilibrium constant expressions because their activity is taken to be 1.

17-16.   *Refer to Section 17-2 and Exercise 17-2 Solution.*

The concentrations of pure solids and pure liquids are omitted from the equilibrium constant expression because their activity is taken to be 1. Therefore, when writing equilibrium constant expressions, we would omit:

pure solids: $CaCO_3(s)$, $NaOH(s)$, $I_2(s)$ and $C(graphite)$
pure liquids: $H_2SO_4(\ell)$ and $CH_3COOH(\ell)$

17-18.   *Refer to Section 17-2 and 17-11.*

(a) $K_c = \dfrac{[H_2S]^2[O_2]^3}{[H_2O]^2[SO_2]^2}$

(b) $K_c = \dfrac{[NO]^4[H_2O]^6}{[NH_3]^4[O_2]^5}$

(c) $K_c = \dfrac{[PCl_5]}{[PCl_3][Cl_2]}$

(d) $K_c = [HF]$

(e) $K_c = \dfrac{[SO_3]^2}{[SO_2]^2[O_2]}$

17-20.   *Refer to Sections 17-2 and 17-11.*

(a) $K_c = \dfrac{[H_2O]^2[O_2]}{[H_2O_2]^2}$

(b) $K_c = \dfrac{[SO_2]^2}{[O_2]^3}$

(c) $K_c = \dfrac{1}{[NH_3][HCl]}$

(d) $K_c = \dfrac{[NO_2]^2}{[N_2O_4]}$

(e) $K_c = \dfrac{[HCl]^4[O_2]}{[Cl_2]^2[H_2O]^2}$

17-22.   *Refer to Section 17-2.*

The products are favored in those reactions in which K > 1: (b), (c) and (d).

17-24. *Refer to Section 17-2 and Example 17-1.*

Balanced equation: $N_2(g) + O_2(g) \rightleftarrows 2NO(g)$

$$K_c = \frac{[NO]^2}{[N_2][O_2]} = \frac{(1.1 \times 10^{-5})^2}{(6.4 \times 10^{-3})(1.7 \times 10^{-3})} = \mathbf{1.1 \times 10^{-5}}$$

17-26. *Refer to Section 17-2 and Example 17-1.*

Balanced equation: $PCl_3(g) + Cl_2(g) \rightleftarrows PCl_5(g)$

$$K_c = \frac{[PCl_5]}{[PCl_3][Cl_2]} = \frac{(12)}{(10.)(9.0)} = \mathbf{0.13}$$

17-28. *Refer to Section 17-2 and Example 17-2.*

Balanced equation: $A(g) + B(g) \rightleftarrows C(g) + 2D(g)$

Plan: (1) Determine the concentrations of the species of interest.
(2) Determine the concentrations of all species after equilibrium is reached.
(3) Calculate K_c.

(1) $[A]_{initial} = [B]_{initial} = \dfrac{1.00 \text{ mol}}{0.400 \text{ L}} = 2.50 \ M$

$[C]_{equil} = \dfrac{0.20 \text{ mol}}{0.400 \text{ L}} = 0.50 \ M$

(2)

	A	+	B	$\rightleftarrows$	C	+	2D
initial	2.50 M		2.50 M		0 M		0 M
change	- 0.50 M		- 0.50 M		+ 0.50 M		+ 1.00 M
at equilibrium	2.00 M		2.00 M		0.50 M		1.00 M

(3) $K_c = \dfrac{[C][D]^2}{[A][B]} = \dfrac{(0.50)(1.00)^2}{(2.00)(2.00)} = \mathbf{0.12}$

17-30. *Refer to Section 17-3 and Example 17-3.*

Balanced equation: $H_2(g) + Br_2(g) \rightleftarrows 2HBr(g)$

$$K_c = \frac{[HBr]^2}{[H_2][Br_2]} = 7.9 \times 10^{11}$$

(a) $1/2 \ H_2(g) + 1/2 \ Br_2(g) \rightleftarrows HBr(g)$

$$K_c' = \frac{[HBr]}{[H_2]^{1/2}[Br_2]^{1/2}} = \sqrt{K_c} = \mathbf{8.9 \times 10^5}$$

(b) $2HBr(g) \rightleftarrows H_2(g) + Br_2(g)$

$$K_c'' = \frac{[H_2][Br_2]}{[HBr]^2} = \frac{1}{K_c} = \mathbf{1.3 \times 10^{-12}}$$

(c) $4HBr(g) \rightleftarrows 2H_2(g) + 2Br_2(g)$

$$K_c''' = \frac{[H_2]^2[Br_2]^2}{[HBr]^4} = \frac{1}{K_c^2} = \mathbf{1.6 \times 10^{-24}}$$

17-32. *Refer to Sections 17-2, and Examples 17-1 and 17-2.*

Balanced equation: $H_2(g) + I_2(g) \rightleftarrows 2HI(g)$

(a) Let $\quad$ x = moles of H_2 or I_2 that react. Then
$\qquad$ 2x = moles of HI that are produced.

	H_2	+	I_2	$\rightleftarrows$	$2HI$
initial	9.84×10^{-4} mol		1.38×10^{-3} mol		0 mol
change	$- x$ mol		$- x$ mol		$+ 2x$ mol
at equilibrium	$(9.84 \times 10^{-4} - x)$ mol		$(1.38 \times 10^{-3} - x)$ mol		$2x$ mol

At equilibrium, mol $I_2 = (1.38 \times 10^{-3} - x)$ mol $= 4.73 \times 10^{-4}$ mol I_2

$$x = 9.1 \times 10^{-4} \text{ mol}$$

Therefore, mol $H_2 = 9.84 \times 10^{-4} - x = 9.84 \times 10^{-4}$ mol $- 9.1 \times 10^{-4}$ mol $= \mathbf{7.4 \times 10^{-5} \text{ mol } H_2}$

mol $HI = 2x = 2(9.1 \times 10^{-4}$ mol$) = \mathbf{1.8 \times 10^{-3} \text{ mol } HI}$

(b) Let V = volume of container in liters. Then, at equilibrium,

$$[H_2] = \frac{7.4 \times 10^{-5} \text{ mol}}{V} \qquad [I_2] = \frac{4.73 \times 10^{-4} \text{ mol}}{V} \qquad [HI] = \frac{1.8 \times 10^{-3} \text{ mol}}{V}$$

$$K_c = \frac{[HI]^2}{[H_2][I_2]} = \frac{\left(\frac{1.8 \times 10^{-3}}{V}\right)^2}{\left(\frac{7.4 \times 10^{-5}}{V}\right)\left(\frac{4.73 \times 10^{-4}}{V}\right)} = 93$$

Note: In this exercise, the volume of the container cancels out in the equilibrium expression. This does not often happen, so you must work in units of molarity, not moles, in the equilibrium expression.

17-34. *Refer to Section 17-2 and Example 17-1.*

Balanced equation: $SbCl_5(g) \rightleftarrows SbCl_3(g) + Cl_2(g)$

(a) Plan: (1) Determine the equilibrium concentrations of $SbCl_5$, $SbCl_3$ and Cl_2 at some high temperature.

(2) Evaluate K_c at that temperature by substituting the equilibrium concentrations into the K_c expression.

(a) $[SbCl_5] = \dfrac{(6.91 \text{ g}/299 \text{ g/mol})}{5.00 \text{ L}} = 4.62 \times 10^{-3} \ M$

$[SbCl_3] = \dfrac{(16.45 \text{ g}/228.15 \text{ g/mol})}{5.00 \text{ L}} = 1.44 \times 10^{-2} \ M$

$[Cl_2] = \dfrac{(5.11 \text{ g}/70.9 \text{ g/mol})}{5.00 \text{ L}} = 1.44 \times 10^{-2} \ M$

(2) $K_c = \dfrac{[SbCl_3][Cl_2]}{[SbCl_5]} = \dfrac{(1.44 \times 10^{-2})(1.44 \times 10^{-2})}{(4.62 \times 10^{-3})} = \mathbf{4.49 \times 10^{-2}}$

(b) Plan: (1) Determine the initial concentration of $SbCl_5$.

(2) Determine the equilibrium concentrations of $SbCl_5$, $SbCl_3$ and Cl_2 using the K_c expression.

(1) $[SbCl_5] = \dfrac{(25.0 \text{ g}/299 \text{ g/mol})}{5.00 \text{ L}} = 0.0167 \ M$

(2) Since $[Cl_2] = [SbCl_3] = 0$, the forward reaction will predominate. Some $SbCl_5$ will react and equal moles of $SbCl_3$ and Cl_2 will be produced.

Let x = moles per liter of $SbCl_5$ that react. Then

x = moles per liter of $SbCl_3$ produced = moles per liter of Cl_2 produced.

	SbCl$_5$	$\rightleftarrows$	SbCl$_3$	+	Cl$_2$
initial	0.0167 M		0 M		0 M
change	- x M		+ x M		+ x M
at equilibrium	(0.0167 - x) M		x M		x M

$$K_c = \frac{[SbCl_3][Cl_2]}{[SbCl_5]} = \frac{(x)(x)}{(0.0167 - x)} = 4.49 \times 10^{-2}$$

The quadratic equation: $x^2 + (4.49 \times 10^{-2})x - 7.50 \times 10^{-4} = 0$

Solving,

$$x = \frac{-4.49 \times 10^{-2} \pm \sqrt{(4.49 \times 10^{-2})^2 - 4(1)(-7.51 \times 10^{-4})}}{2(1)} = \frac{-4.49 \times 10^{-2} \pm 7.08 \times 10^{-2}}{2}$$

$$= 1.30 \times 10^{-2} \text{ or } -5.78 \times 10^{-2} \text{ (discard)}$$

Note: There are always two solutions when solving quadratic equations, but only one is meaningful in this type of chemical problem. The other solution, -5.79×10^{-2}, is discarded because a negative value for concentration has no physical meaning in this problem.

Therefore, at equilibrium: $[SbCl_5] = 0.0167 - x = \mathbf{3.7 \times 10^{-3}\ M}$

$[SbCl_3] = [Cl_2] = x = \mathbf{1.30 \times 10^{-2}\ M}$

17-36. *Refer to Section 17-3 and Example 17-3.*

Balanced equation: $2NO(g) + O_2(g) \rightleftarrows 2NO_2(g)$ $K_c = \dfrac{[NO_2]^2}{[NO]^2[O_2]} = 1538$

$2NO_2(g) \rightleftarrows 2NO(g) + O_2(g)$ $K_c' = \dfrac{[NO]^2[O_2]}{[NO_2]^2} = \dfrac{1}{K_c} = \mathbf{6.502 \times 10^{-4}}$

17-38. *Refer to Section 17-4.*

Many systems are not at equilibrium. The mass action expression, also called the reaction quotient, Q, is a measure of how far a system is from equilibrium and in what direction the system must go to get to equilibrium. The reaction quotient has the same form as the equilibrium constant, K, but the concentration values put into Q are the actual values found in the system at that given moment.

(a) If $Q = K$, the system is at equilibrium.

(b) If $Q < K$, the system has greater concentrations of reactants than it would have if it were at equilibrium. The forward reaction will dominate until equilibrium is established.

(c) If $Q > K$, the system has greater concentrations of products than it would have if it were at equilibrium. The reverse reaction will dominate until equilibrium is reached.

17-40. *Refer to Section 17-4.*

If $Q > K$ for a reversible reaction, the reverse reaction occurs to a greater extent than the forward reaction until equilibrium is reached.

If $Q < K$, the forward reaction occurs to a greater extent than the reverse reaction until equilibrium is reached.

17-42. *Refer to Section 17-4 and Example 17-4.*

Balanced equation: $H_2CO \rightleftarrows H_2 + CO$ $K_c = 0.50$

The reaction quotient, Q, for this reaction at the given moment is: $Q = \dfrac{[H_2][CO]}{[H_2CO]} = \dfrac{(0.80)(0.25)}{(0.50)} = 0.40$

Under the given conditions, $Q < K_c$.

(a) false The reaction is not at equilibrium since $Q \neq K_c$.

(b) false The reaction is not at equilibrium. However, the reaction will continue to proceed until equilibrium is reached.

(c) true When $Q < K_c$, the system has more reactants and less products than it would have at equilibrium. Equilibrium will be reached by forming more H_2 and CO and using up more H_2CO.

(d) false The forward rate of the reaction is more than that of the reverse reaction since $Q < K_c$. The forward and reverse rates are equal only at equilibrium.

17-44. *Refer to Section 17-5 and Example 17-7.*

Balanced equation: $N_2(g) + C_2H_2(g) \rightleftarrows 2HCN(g)$ $K_c = 2.3 \times 10^{-4}$

Plan: Determine the equilibrium concentration of HCN using the K_c expression.

Let x = moles per liter of N_2 that react. Then
x = moles per liter of C_2H_2 that react, and
2x = moles per liter of HCN that are formed.

	N_2	+	C_2H_2	$\rightleftarrows$	2HCN
initial	3.5 *M*		2.5 *M*		0 *M*
change	- x *M*		- x *M*		+ 2x *M*
at equilibrium	(3.5 - x) *M*		(2.5 - x) *M*		2x *M*

$$K_c = \frac{[HCN]^2}{[N_2][C_2H_2]} = \frac{(2x)^2}{(3.5 - x)(2.5 - x)} = 2.3 \times 10^{-4}$$

Note: In the following calculations, the answers that are in parentheses contain too many significant figures, but are the ones used in the progressive calculation in order to minimize rounding errors.

Rearranging into a quadratic equation: $4.0x^2 + (1.38 \times 10^{-3})x - 2.01 \times 10^{-3} = 0$
Solving,

$$x = \frac{-1.38 \times 10^{-3} \pm \sqrt{(1.38 \times 10^{-3})^2 - 4(4.0)(-2.01 \times 10^{-3})}}{(2)(4.0)} = \frac{-1.38 \times 10^{-3} \pm 0.179}{8.00} = 0.0222 \text{ or } -0.0225 \text{ (discard)}$$

Note: There are always two solutions when solving quadratic equations, but only one is meaningful in this type of chemical problem. The other solution, -0.0225, is discarded because a negative value for concentration of HCN has no physical meaning in this problem.

Therefore, at equilibrium: [HCN] = 2x = **0.044 *M*** (two significant figures)

17-46. *Refer to Section 17-5 and Example 17-7.*

Balanced equation: $PCl_3(g) + Cl_2(g) \rightleftarrows PCl_5(g)$ $K_c = 96.2$

Plan: Determine the equilibrium concentration of Cl_2 using the equilibrium expression.

Let x = moles per liter of PCl_3 that react = moles per liter of Cl_2 that react. Then
x = moles per liter of PCl_5 that are formed.

	PCl_3	+	Cl_2	$\rightleftarrows$	PCl_5
initial	0.22 M		5.5 M		0 M
change	- x M		- x M		+ x M
at equilibrium	(0.22 - x) M		(5.5 - x) M		x M

$$K_c = \frac{[PCl_5]}{[PCl_3][Cl_2]} = \frac{(x)}{(0.22 - x)(5.5 - x)} = 96.2$$

Rearranging into a quadratic equation: $96.2x^2 - 551x + 116 = 0$

Solving, $x = \dfrac{550 \pm \sqrt{(550)^2 - 4(96.2)(116)}}{(2)(96.2)} = \dfrac{550 \pm 508}{192} = 0.22$ or 5.5 (discard)

Note: The solution, x = 5.5 is meaningless since it would result in a negative equilibrium concentration for PCl_3.

Therefore, at equilibrium: $[Cl_2] = 5.5 - x = $ **5.3 M**

Note: The equilibrium concentration of PCl_3 is not zero, but a small positive number. Solve for $[PCl_3]$ by plugging the calculated equilibrium concentrations of Cl_2 and PCl_5 into the equilibrium expression.

$$K_c = \frac{[PCl_5]}{[PCl_3][Cl_2]} = \frac{x}{[PCl_3](5.5 - x)} = \frac{0.22}{[PCl_3](5.3)} = 96.2 \qquad \text{Solving, } [PCl_3] = 0.00043 \ M$$

17-48. *Refer to Section 17-5 and Example 17-6.*

Balanced equation: $3Fe(s) + 4H_2O(g) \rightleftarrows Fe_3O_4(s) + 4H_2(g)$ $\qquad K_c = 4.6$ at 850°C

Plan: (1) Determine the initial concentration of H_2O.

        (2) Determine the equilibrium concentration of H_2 using the K_c expression.

Note: In the following calculations, the answers that are in parentheses contain too many significant figures, but are the ones used in the progressive calculation in order to minimize rounding errors.

(1) $[H_2O] = \dfrac{(26 \ g/18.0 \ g/mol)}{10.0 \ L} = 0.14 \ M$ (0.144 M - use this to minimize rounding errors, then round at the end)

(2) Since initially, $[H_2] = 0$, the forward reaction will predominate. Note that this is a heterogeneous equilibrium, so we can ignore the solids.

Let   4x = moles per liter of H_2O that react. Then

        4x = moles per liter of H_2 produced.

	$3Fe(s)$	+	$4H_2O(g)$	$\rightleftarrows$	$Fe_3O_4(s)$	+	$4H_2(g)$
initial			0.144 M				0 M
change			- 4x M				+ 4x M
at equilibrium			(0.144 - 4x) M				4x M

$$K_c = \frac{[H_2]^4}{[H_2O]^4} = \frac{(4x)^4}{(0.144 - 4x)^4} = 4.6$$

If we take the 4th root of both sides, we have: $\dfrac{4x}{0.144 - 4x} = \sqrt[4]{4.6} = 1.5$ (1.465)

Solving for x, we have: $4x = 0.211 - 5.86x$ (knowing our numbers are only good to 2 significant figures)

                    $9.86x = 0.212$

                      $x = 0.021$ (0.02150)

Therefore, at equilibrium: $[H_2O] = 0.144 - 4x = 0.058 \ M$

                          $[H_2] = 4x = $ **0.086 M**

Balanced equation: $CO_2(g) + C(\text{graphite}) \rightleftarrows 2CO(g)$ $K_c = 10.0$ at 850°C

Plan: (1) Determine the initial concentration of CO.
            (2) Determine the equilibrium concentration of CO_2 using the K_c expression.

(1) $[CO] = \dfrac{(22.5 \text{ g}/28.0 \text{ g/mol})}{2.50 \text{ L}} = 0.321 \ M$

(2) Since initially $[CO_2] = 0$, the reverse reaction will predominate. Note that this is a heterogeneous equilibrium, so we can ignore the solid graphite.

Let $2x$ = moles per liter of CO that react. Then
        x = moles per liter of CO_2 produced.

	$CO_2(g)$	+	$C(\text{graphite})$	$\rightleftarrows$	$2CO(g)$
initial	0 M				0.321 M
change	+ x M				- 2x M
at equilibrium	x M				(0.321 - 2x) M

$K_c = \dfrac{[CO]^2}{[CO_2]} = \dfrac{(0.321 - 2x)^2}{(x)} = 10.0$

Note: Even though the reverse reaction is favored, we still write the equilibrium expression for the reaction as originally written. This is a major point!

Rearranging into a quadratic equation: $4x^2 - 11.3x + 0.103 = 0$
Solving,

$x = \dfrac{+11.3 \pm \sqrt{(11.3)^2 - 4(4)(0.103)}}{(2)(4)} = \dfrac{+11.3 \pm 11.2}{8.00} = 0.009$ or 2.81 (discard)

Note: There are always two solutions when solving quadratic equations, but only one is meaningful in this type of problem. The other solution, $x = 2.81$, is discarded because a negative value for the equilibrium concentration of CO would result and this has no physical meaning.

Therefore, at equilibrium: $[CO_2] = 0.009 \ M$

$? \text{ g } CO_2 = 2.50 \text{ L} \times \dfrac{0.009 \text{ mol } CO_2}{1 \text{ L}} \times \dfrac{44 \text{ g } CO_2}{1 \text{ mol } CO_2} = \mathbf{1 \ g}$ (1 sig. fig.)

Balanced equation: $2HI(g) \rightleftarrows H_2(g) + I_2(g)$ $K_c = 0.830$

Plan: (1) Determine the initial concentration of HI.
            (2) Determine the equilibrium concentrations of HI, H_2 and I_2 using the K_c expression.

(1) $[HI] = \dfrac{(72.5 \text{ g}/127.9 \text{ g/mol})}{1.50 \text{ L}} = 0.378 \ M$ (0.3779)

(2) Since initially, $[H_2] = [I_2] = 0$, the forward reaction will predominate.

Let $2x$ = moles per liter of HI that react. Then
        x = moles per liter of H_2 and I_2 produced.

	$2HI(g)$	$\rightleftarrows$	$H_2(g)$	+	$I_2(g)$
initial	0.378 M		0 M		0 M
change	- 2x M		+ x M		+ x M
at equilibrium	(0.378 - 2x) M		x M		x M

$$K_c = \frac{[H_2][I_2]}{[HI]^2} = \frac{(x)^2}{(0.3779 - 2x)^2} = 0.830 \quad \text{(We'll round at the end of the problem to 3 significant figures.)}$$

If we take the square root of both sides, we have: $\dfrac{x}{0.3779 - 2x} = \sqrt{0.830} = 0.9110$

Solving for x, we have:
$$x = 0.3443 - 1.822x$$
$$2.822x = 0.3443$$
$$x = 0.1220$$

Therefore, at equilibrium:
$$[HI] = 0.3779 - 2x = \textbf{0.134 } \textit{\textbf{M}}$$
$$[H_2] = [I_2] = x = \textbf{0.122 } \textit{\textbf{M}}$$

17-54. *Refer to Section 17-6 and Example 17-9.*

When an equilibrium system involving gases is subjected to an increase in pressure resulting from a decrease in volume, the concentrations of the gases increase and there may or may not be a shift in the equilibrium. If there is the same number of moles of gas on each side of the equation, equilibrium is not affected. If the number of moles of gas on each side of the equation is different, the general rule is that such an increase in pressure shifts a system in the direction that produces the smaller number of moles of gas.

(a) shift to left

(b) shift to right

(c) equilibrium is unaffected

(d) equilibrium is unaffected

17-56. *Refer to Section 17-6, Examples 17-9 and 17-10, and Exercise 17-54 Solution.*

Balanced equation: $A(g) + 3B(g) \rightleftarrows 2C(g) + 3D(g) + \text{heat}$ (the reaction is exothermic)

(a) Whenever the temperature changes, K_c changes as well. In fact, the only variable that affects K_c is temperature. When the reaction is exothermic, adding heat causes (ii) K_c to decrease. (At that point, $Q_c > K_c$ and the reaction shifts to the left.)

(b) K_c (iii) stays the same when more A is added since the temperature is constant.
(The equilibrium will shift to the right.)

(c) K_c (iii) stays the same when more C is added since T is constant. (The equilibrium will shift to the left.)

(d) K_c (iii) stays the same when a little D is removed since T is constant. (The equilibrium will shift to the right.)

(e) K_c (iii) stays the same when the pressure is decreased by increasing the volume of the container since T is constant. (The equilibrium will shift to the right.)

17-58. *Refer to Section 17-6, Example 17-9 and Exercise 17-54 Solution.*

Balanced equation: $2C(s) + O_2(g) \rightleftarrows 2CO(g)$

If the total pressure were decreased, the equilibrium would **shift to the right** to create more gas molecules.

17-60. *Refer to Section 17-6 and Examples 17-9 and 17-10.*

Balanced equation: $6CO_2(g) + 6H_2O(\ell) \rightleftarrows C_6H_{12}O_6(s) + 6O_2(g)$ $\qquad \Delta H° = 2801.69 \text{ kJ/mol rxn}$

(a) If $[CO_2]$ is decreased, (i) the equilibrium will shift to the left.

(b) If P_{O_2} is increased, (ii) the equilibrium will shift to the left.

(c) If one-half of $C_6H_{12}O_6(s)$ is removed, (iii) the equilibrium is unaffected since $C_6H_{12}O_6$ is a solid and does not appear in the equilibrium expression.

(d) If the total pressure is decreased, the equilibrium is unaffected since the total number of moles of gas is the same on both sides of the equation.

(e) If the temperature is increased, (i) the equilibrium will shift to the right since the forward reaction is endothermic ($\Delta H° > 0$) and the forward reaction will absorb heat to minimize the effect of raising the temperature and adding heat.

(f) If a catalyst is added, (iii) the equilibrium is unaffected. The reaction would simply reach equilibrium at a faster rate.

17-62. *Refer to Section 17-6, Examples 17-9 and 17-10, and Exercise 17-54 Solution.*

When the pressure is increased by decreasing the volume, the equilibrium in question :

(a) shifts to right

(b) is not affected

(c) shifts to left

(d) shifts to right

(e) shifts to right

17-64. *Refer to Sections 17-8, and Example 17-11.*

Balanced equation: $A(g) + B(g) \rightleftarrows C(g) + D(g)$

(a) $K_c = \dfrac{[C][D]}{[A][B]} = \dfrac{(1.60 \text{ mol}/1.00 \text{ L})^2}{(0.40 \text{ mol}/1.00 \text{ L})^2} = \mathbf{16}$

(b) Since we are adding both reactant and product to the system, the value of Q_c must be evaluated to determine the direction of the reaction.

New $[B] = 0.40\ M + 0.20\ M = 0.60\ M$
New $[C] = 1.60\ M + 0.20\ M = 1.80\ M$

$Q_c = \dfrac{[C][D]}{[A][B]} = \dfrac{(1.80\ M)(1.60\ M)}{(0.40\ M)(0.60\ M)} = 12$

Since $Q_c < K_c$, the forward reaction proceeds.

Let   x = moles per liter of A or B that react *after* the addition of 0.20 moles per liter of A and C. Then
x = moles per liter of C produced = moles per liter of D produced.

	A	+	B	$\rightleftarrows$	C	+	D
initial	0.40 *M*		0.40 *M*		1.60 *M*		1.60 *M*
mol/L added	0 *M*		+ 0.20 *M*		+ 0.20 *M*		0 *M*
new system	0.40 *M*		0.60 *M*		1.80 *M*		1.60 *M*
change	- x *M*		- x *M*		+ x *M*		+ x *M*
at equil	(0.40 - x) *M*		(0.60 - x) *M*		(1.80 + x) *M*		(1.60 + x) *M*

$K_c = \dfrac{[C][D]}{[A][B]} = \dfrac{(1.80 + x)(1.60 + x)}{(0.40 - x)(0.60 - x)} = 16$

The quadratic equation: $15x^2 - 19.4x + 0.96 = 0$

Solving, $x = \dfrac{19.4 \pm \sqrt{(19.4)^2 - 4(15)(0.96)}}{2(15)} = \dfrac{19.4 \pm 17.9}{30} = 0.050$ or 1.24 (discard)

Therefore, the new equilibrium concentration of A is (0.40 - x) or **0.35 *M***

268

17-66. *Refer to Sections 17-5 and 17-8, and Example 17-12.*

Balanced equation: $A(g) \rightleftarrows B(g) + C(g)$

(a) $K_c = \dfrac{[B][C]}{[A]} = \dfrac{(0.25)^2}{0.30} = \mathbf{0.21}$

(b) If the volume is suddenly doubled, the initial concentrations will be halved and the system is no longer at equilibrium. We learned in Section 17-5 that the equilibrium will then shift to the side with the greater number of moles of gas, i.e., the right side.

Let x = number of moles per liter of A that react *after* the volume is doubled.
     x = number of moles per liter of B produced = number of moles per liter of C produced.

	A	$\rightleftarrows$	B	+	C
initial	0.30 *M*		0.25 *M*		0.25 *M*
new system	0.15 *M*		0.125* *M*		0.125* *M*
change	- x *M*		+ x *M*		+ x *M*
at equilibrium	(0.15 - x) *M*		(0.125* + x) *M*		(0.125* + x) *M*

*this number really should have 2 significant figures. The numbers will be rounded off at the end.

$K_c = \dfrac{[B][C]}{[A]} = \dfrac{(0.125 + x)^2}{(0.15 - x)} = 0.21$

The quadratic equation: $x^2 + 0.46x - 0.016 = 0$

Solving, $x = \dfrac{-0.46 \pm \sqrt{(0.46)^2 - 4(1)(-0.016)}}{2(1)} = \dfrac{-0.46 \pm 0.52}{2} = 0.03$ or -0.49 (discard)

Therefore, [A] = 0.15 - x = **0.12 *M***
         [B] = [C] = 0.12 + 0.03 = **0.15 *M***

(c) If the volume is suddenly halved, the initial equilibrium concentrations will be doubled and this system is no longer at equilibrium. The equilibrium will shift to the left side, the side with the lesser number of moles of gas.

Let x = number of moles per liter of B that react *after* the volume is halved, and
     x = number of moles per liter of C that react after the volume is halved. Then
     x = number of moles of A that are produced.

	A	$\rightleftarrows$	B	+	C
initial	0.30 *M*		0.25 *M*		0.25 *M*
new system	0.60 *M*		0.50 *M*		0.50 *M*
change	+ x *M*		- x *M*		- x *M*
at equilibrium	(0.60 + x) *M*		(0.50 - x) *M*		(0.50 - x) *M*

$K_c = \dfrac{[B][C]}{[A]} = \dfrac{(0.50 - x)^2}{(0.60 + x)} = 0.21$

The quadratic equation: $x^2 - 1.21x + 0.12 = 0$

Solving, $x = \dfrac{1.21 \pm \sqrt{(-1.21)^2 - 4(1)(0.12)}}{2(1)} = \dfrac{1.21 \pm 0.99}{2} = 0.11$ or 1.1 (discard)

Therefore, [A] = 0.60 + x = **0.71 *M***
         [B] = [C] = 0.50 - x = **0.39 *M***

17-68. *Refer to Sections 17-6 and 17-8, Example 17-12, and Exercise 17-66 Solution.*

Balanced equation: $N_2O_4(g) \rightleftarrows 2NO_2(g)$ $\qquad\qquad\qquad\qquad$ $K_c = 5.84 \times 10^{-3}$

(a) $[N_2O_4]_{initial} = \dfrac{4.00 \text{ g}/92.0 \text{ g/mol}}{2.00 \text{ L}} = 0.0217 \ M$

Let $\ x$ = number of moles per liter of N_2O_4 that react. Then
$\quad 2x$ = number of moles per liter of NO_2 that are produced.

	N_2O_4	$\rightleftarrows$	$2NO_2$
initial	0.0217 M		0 M
change	- x M		+ 2x M
at equilibrium	(0.0217 - x) M		2x M

$K_c = \dfrac{[NO_2]^2}{[N_2O_4]} = \dfrac{(2x)^2}{(0.0217 - x)} = 5.84 \times 10^{-3}$

The quadratic equation: $4x^2 + (5.84 \times 10^{-3})x - 1.27 \times 10^{-4} = 0$

Solving, $\ x = \dfrac{-5.84 \times 10^{-3} \pm \sqrt{(5.84 \times 10^{-3})^2 - 4(4)(-1.27 \times 10^{-4})}}{2(4)}$

$= \dfrac{-5.84 \times 10^{-3} \pm 4.55 \times 10^{-2}}{8} = 4.95 \times 10^{-3} \ $ or $\ -6.41 \times 10^{-3}$ (discard)

Therefore, at equilibrium $\qquad [N_2O_4] = 0.0217 - x = \mathbf{0.0167 \ M}$
$\qquad\qquad\qquad\qquad\qquad\qquad [NO_2] = 2x = \mathbf{9.90 \times 10^{-3} \ M}$

(b) When the volume is suddenly increased (2.00 L $\rightarrow$ 3.00 L), the concentrations of N_2O_4 and NO_2 are decreased by a factor of 2/3 and the equilibrium shifts to the right.

Let $\ x$ = number of moles per liter of N_2O_4 that react *after* the volume is increased. Then
$\quad 2x$ = number of moles of NO_2 that are produced.
$[N_2O_4]_{new} = 1.67 \times 10^{-2} \ M \times 2/3 = 1.11 \times 10^{-2} \ M$
$[NO_2]_{new} = 9.90 \times 10^{-3} \ M \times 2/3 = 6.60 \times 10^{-3} \ M$

	N_2O_4	$\rightleftarrows$	$2 \ NO_2$
initial	1.67 x 10^{-2} M		9.90 x 10^{-3} M
new system	1.11 x 10^{-2} M		6.60 x 10^{-3} M
change	- x M		+ 2x M
at equilibrium	(1.11 x 10^{-2} - x) M		(6.60 x 10^{-3} + 2x) M

$K_c = \dfrac{[NO_2]^2}{[N_2O_4]} = \dfrac{(6.60 \times 10^{-3} + 2x)^2}{(1.11 \times 10^{-2} - x)} = 5.84 \times 10^{-3}$

The quadratic equation: $4x^2 + 0.0322x - 2.12 \times 10^{-5} = 0$

Solving, $\ x = \dfrac{-0.0322 \pm \sqrt{(0.0322)^2 - 4(4)(-2.12 \times 10^{-5})}}{2(4)}$

$= \dfrac{-0.0322 \pm 0.0371}{8} = 6.1 \times 10^{-4} \ $ or $\ -8.66 \times 10^{-3}$ (discard)

Therefore, at equilibrium $\qquad [N_2O_4] = 1.11 \times 10^{-2} - x = \mathbf{1.05 \times 10^{-2} \ M}$
$\qquad\qquad\qquad\qquad\qquad\qquad [NO_2] = 6.60 \times 10^{-3} + 2x = \mathbf{7.82 \times 10^{-3} \ M}$

(c) When the volume in (a) is suddenly halved (2.00 L → 1.00 L), the concentrations of N_2O_4 and NO_2 are doubled and the equilibrium shifts to the left side.

Let x = number of moles per liter of N_2O_4 that are produced *after* the volume is halved. Then
$2x$ = number of moles of NO_2 that are consumed.

	N_2O_4	$\rightleftharpoons$	$2\,NO_2$
initial	$1.67 \times 10^{-2}\ M$		$9.90 \times 10^{-3}\ M$
new system	$0.0334\ M$		$0.0198\ M$
change	$+\ x\ M$		$-\ 2x\ M$
at equilibrium	$(0.0334 + x)\ M$		$(0.0198 - 2x)\ M$

$$K_c = \frac{[NO_2]^2}{[N_2O_4]} = \frac{(0.0198 - 2x)^2}{(0.0334 + x)} = 5.84 \times 10^{-3}$$

The quadratic equation: $4x^2 - 0.0850x + 1.97 \times 10^{-4} = 0$

Solving, $x = \dfrac{0.0850 \pm \sqrt{(-0.0850)^2 - 4(4)(1.97 \times 10^{-4})}}{2(4)}$

$= \dfrac{0.0850 \pm 0.0638}{8} = 2.65 \times 10^{-3}$ or 0.0186 (discard)

Therefore, at equilibrium $[N_2O_4] = 0.0334 + x = \mathbf{0.0360\ M}$
$[NO_2] = 0.0198 - 2x = \mathbf{0.0145\ M}$

17-70.   *Refer to Sections 17-9 and 17-10.*

The values of K_p and K_c are numerically equal for reactions in which there are equal numbers of moles of gases on both sides of the equation, i.e., $\Delta n_{gas} = 0$. K_p and K_c are numerically equal for reactions 17(a), 17(c), 17(d) only.

17-72.   *Refer to Section 17-9 and Example 17-13.*

Balanced equation: $C(graphite) + CO_2(g) \rightleftharpoons 2CO(g)$

Plan: (1) Calculate the partial pressures of CO and CO_2.
(2) Determine K_p.

(1) Since the CO_2 gas stream contains 4.0×10^{-3} mol percent CO, the mole fraction of CO is 4.0×10^{-5}.

P_{CO} = mole fraction CO x P_{total} = $(4.0 \times 10^{-5})(1.00\ atm)$ = 4.0×10^{-5} atm

P_{CO_2} = mole fraction CO_2 x P_{total} = 1.00 (to 3 significant figures) x 1 atm = 1.00 atm

(2) $K_p = \dfrac{(P_{CO})^2}{P_{CO_2}} = \dfrac{(4.0 \times 10^{-5}\ atm)^2}{1.00\ atm} = \mathbf{1.6 \times 10^{-9}}$ (Remember that equilibrium constants have no units.)

17-74.   *Refer to Section 17-9 and Example 17-13.*

Balanced equation: $N_2O_4(g) \rightleftharpoons 2NO_2(g)$

Plan: (1) Calculate the initial partial pressure of N_2O_4, $K_p = 0.715$ at $T = 47°C$ (320 K).
(2) Determine the partial pressures of N_2O_4 and NO_2 at equilibrium.

(1) $P_{N_2O_4} = M(RT) = (3.3 \text{ mol}/5.0 \text{ L})(0.0821 \text{ L·atm/mol·K})(320 \text{ K}) = 17 \text{ atm}$ (17.3 atm - used in calculations)

(2) Let x = partial pressure of N_2O_4 that that reacted. Then
$2x$ = partial pressure of NO_2 at equilibrium.

	N_2O_4	$\rightleftarrows$	$2 NO_2$
initial	17.3 atm		0 atm
change	- x atm		+ 2x atm
at equilibrium	(17.3 - x) atm		2x atm

$K_p = \dfrac{[NO_2]^2}{[N_2O_4]} = \dfrac{(2x)^2}{(17.3 - x)} = 0.715$ 	The quadratic equation: $4x^2 + 0.715x - 12.4 = 0$

Solving, $x = \dfrac{-0.715 \pm \sqrt{(0.715)^2 - 4(4)(-12.4)}}{2(4)}$

$= \dfrac{-0.715 \pm 14.1}{8} = 1.67$ or -1.85 (discard)

Therefore, at equilibrium 	$P_{N_2O_4} = 17.3 - x = \textbf{16 atm}$ (answer limited to 2 significant figures by the data)
$P_{NO_2} = 2x = \textbf{3.3 atm}$

17-76. *Refer to Section 17-9 and Example 17-13.*

Balanced equation: $H_2(g) + CO_2(g) \rightleftarrows CO(g) + H_2O(g)$

$K_p = \dfrac{P_{CO}P_{H_2O}}{P_{H_2}P_{CO_2}} = \dfrac{(0.180 \text{ atm})(0.252 \text{ atm})}{(0.387 \text{ atm})(0.152 \text{ atm})} = \textbf{0.771}$

17-78. *Refer to Section 17-10.*

Balanced equation: $Br_2(g) \rightleftarrows 2Br(g)$ 	$K_p = 2550$ at 4000 K

$K_c = K_p(RT)^{-\Delta n} = 2550(0.0821 \times 4000)^{-(2-1)} = \textbf{7.76}$

17-80. *Refer to Sections 17-5 and 17-6.*

Balanced equation: $Fe_2O_3(s) + 3H_2(g) \rightleftarrows 2Fe(s) + 3H_2O(g)$ 	$K_c = 8.11$ at 1000 K 	$\Delta H = 96$ kJ/mol rxn

(a) Let 	x = moles per liter of H_2 present initially
y = moles per liter of H_2 that react

	$Fe_2O_3(s)$	+	$3H_2$	$\rightleftarrows$	$2Fe(s)$	+	$3H_2O$
initial	-		x M		-		0 M
change			- y M				+ y M
at equilibrium			(x - y) M				y M

$K_c = \dfrac{[H_2O]^3}{[H_2]^3} = \dfrac{y^3}{(x-y)^3} = 8.11$ 	Taking the cube root of both sides: 	$\dfrac{y}{x-y} = 2.01$

$y = (x - y)2.01$
$y = 2.01x - 2.01y$
$3.01y = 2.01x$
$y = 0.668x$

272

Therefore,   the percentage of H_2 that reacts = 66.8%

the percentage of H_2 that remains unreacted = (100 - 66.8)% = **33.2%**

(b) At lower temperatures, the equilibrium will shift to the left since the reaction is endothermic ($\Delta H > 0$), and the percentage of H_2 that remains unreacted will be **greater**.

17-82.   *Refer to Section 17-12.*

(a) If $K >> 1$, the forward reaction is likely to be spontaneous, and ΔG°_{rxn} must be negative.

(b) If $K = 1$, then ΔG°_{rxn} is 0, and the reaction is at equilibrium when the concentrations of aqueous species are 1 M and the partial pressures of gaseous species are 1 atm, or when the numerator and denominator cancel out.

(c) If $K << 1$, the forward reaction is likely to be nonspontaneous and ΔG°_{rxn} must be positive.

17-84.   *Refer to Sections 17-12 and 15-15, Examples 17-17 and 17-18, and Appendix K.*

Balanced equation: $CO(g) + H_2O(g) \rightleftarrows CO_2(g) + H_2(g)$

(a) Plan: (1) Calculate ΔH°_{rxn} and ΔS°_{rxn} from data in Appendix K and substitute into the Gibbs free energy change equation to determine ΔG°_{rxn}.
(2) Calculate K_P, using $\Delta G^{\circ} = -RT \ln K_p$.

(1) $\Delta H^{\circ}_{rxn} = [\Delta H^{\circ}_{f\ CO_2(g)} + \Delta H^{\circ}_{f\ H_2(g)}] - [\Delta H^{\circ}_{f\ CO(g)} + \Delta H^{\circ}_{f\ H_2O(g)}]$

$= [(1\ mol)(-393.5\ kJ/mol) + (1\ mol)(0\ kJ/mol)]$

$- [(1\ mol)(-110.5\ kJ/mol) + (1\ mol)(-241.8\ kJ/mol)]$

$= -41.2\ kJ/mol\ rxn$

$\Delta S^{\circ}_{rxn} = [S^{\circ}_{CO_2(g)} + S^{\circ}_{H_2(g)}] - [S^{\circ}_{CO(g)} + S^{\circ}_{H_2O(g)}]$

$= [(1\ mol)(213.6\ J/mol \cdot K) + (1\ mol)(130.6\ J/mol \cdot K)]$

$- [(1\ mol)(197.6\ J/mol \cdot K) + (1\ mol)(188.7\ J/mol \cdot K)]$

$= -42.1\ J/K$

$\Delta G^{\circ} = \Delta H^{\circ} - T\Delta S^{\circ} = -41.2 \times 10^3\ J - (298\ K)(-42.1\ J/K) = -2.87 \times 10^4\ J/mol\ rxn$

(2) Substituting into $\Delta G^{\circ} = -RT \ln K_p$, we have

$-2.87 \times 10^4\ J = -(8.314\ J/K)(298\ K)\ln K_p$

$\ln K_p = 11.6$

$K_p = \mathbf{1.1 \times 10^5}$

(b) If we assume that ΔH° and ΔS° for the reaction are independent of temperature, we can determine ΔG° at 200°C (473 K) by substituting into the Gibbs free energy change equation.

$\Delta G^{\circ} = \Delta H^{\circ} - T\Delta S^{\circ} = -41.2 \times 10^3\ J - (473\ K)(-42.1\ J/K) = -2.13 \times 10^4\ J/mol\ rxn$

Then,

$\Delta G^{\circ} = -RT \ln K_p$

$-2.13 \times 10^4\ J = -(8.314\ J/K)(473\ K)\ln K_p$

$\ln K_p = 5.42$

$K_p = \mathbf{2.3 \times 10^2}$

(c) $\Delta G^{\circ}_{rxn} = [\Delta G^{\circ}_{f\ CO_2(g)} + \Delta G^{\circ}_{f\ H_2(g)}] - [\Delta G^{\circ}_{f\ CO(g)} + \Delta G^{\circ}_{f\ H_2O(g)}]$

$= [(1\ mol)(-394.4\ kJ/mol) + (1\ mol)(0\ kJ/mol)] - [(1\ mol)(-137.2\ kJ/mol) + (1\ mol)(-228.6\ kJ/mol)]$

$= -28.6\ kJ/mol\ rxn\ \ or\ \ -2.86 \times 10^4\ J/mol\ rxn$

Substituting into $\Delta G° = -RT \ln K_p$, we have

$$-2.86 \times 10^4 \text{ J} = -(8.314 \text{ J/K})(298 \text{ K})\ln K_p$$
$$\ln K_p = 11.5$$
$$K_p = \mathbf{1.0 \times 10^5}$$

17-86. *Refer to Section 17-12, Examples 17-17 and 17-18, and Appendix K.*

Balanced equation: $2SO_2(g) + O_2(g) \rightleftarrows 2SO_3(g)$

$$\Delta G°_{rxn} = [2\Delta G°_f \, _{SO_3(g)}] - [2\Delta G°_f \, _{SO_2(g)} + \Delta G°_f \, _{O_2(g)}]$$
$$= [(2 \text{ mol})(-371.1 \text{ kJ/mol})] - [(2 \text{ mol})(-300.2 \text{ kJ/mol}) + (1 \text{ mol})(0 \text{ kJ/mol})]$$
$$= -141.8 \text{ kJ/mol rxn}$$

We know $\qquad\qquad \Delta G°_{rxn} = -RT \ln K_p \qquad$ for a gas phase reaction

Substituting, $\qquad -141.8 \times 10^3 \text{ J} = -(8.314 \text{ J/K})(298.15 \text{ K})\ln K_p$
$$\ln K_p = 57.20$$
$$K_p = \mathbf{7.0 \times 10^{24}}$$

17-88. *Refer to Sections 17-12 and 17-13, and Examples 17-19 and 17-20.*

(a) $\quad \Delta G°_{rxn} = -RT \ln K_p$

$$= -(8.314 \text{ J/K})(298.15 \text{ K})(\ln 4.3 \times 10^6)$$
$$= \mathbf{-3.79 \times 10^4 \text{ J/mol rxn} \ or \ -37.9 \text{ kJ/mol rxn at } 25°C}$$

(b) Using the van't Hoff equation,

$$\ln \frac{K_{p\,T_2}}{K_{p\,T_1}} = \frac{\Delta H°(T_2 - T_1)}{RT_1 T_2}$$

$$\ln \frac{K_{p\,1073\,K}}{4.3 \times 10^6} = \frac{(-78.58 \times 10^3 \text{ J})(1073 \text{ K} - 298 \text{ K})}{(8.314 \text{ J/K})(298 \text{ K})(1073 \text{ K})} = -22.9$$

$$\frac{K_{p\,1073\,K}}{4.3 \times 10^6} = e^{-22.9} = 1.13 \times 10^{-10}$$

$$K_{p\,1073\,K} = \mathbf{4.9 \times 10^{-4}}$$

(c) $\Delta G°_{rxn} = -(8.314 \text{ J/K})(1073 \text{ K})(\ln 4.9 \times 10^{-4}) = \mathbf{+6.80 \times 10^4 \text{ J/mol rxn}}$
$$\mathbf{or \ +68.0 \text{ kJ/mol rxn at } 800°C}$$

(d) The forward reaction at 800°C is nonspontaneous (ΔG is positive) whereas at 25°C the forward reaction is spontaneous (ΔG is negative).

(e) The reaction mixture is heated to speed up the rate at which equilibrium is reached, not to shift the equilibrium toward more product. Heating actually decreases the amount of product present at equilibrium since the reaction is exothermic (ΔH is negative).

(f) The basic purpose of a catalyst is to speed up the reaction of interest. Recall from Chapter 16 that the presence of a catalyst increases the rates of both forward and reverse reactions to the same extent without affecting equilibrium. However, the catalyst increases the rate at which the reaction goes to equilibrium. Hopefully, the yield of desired product is adequate at equilibrium.

Balanced equation: $2CH_3COOH(g) \rightleftarrows (CH_3COOH)_2(g)$

Since the equilibrium constant at 25°C, 3.2×10^4, is much greater than 1, the **dimer** is favored at equilibrium.

17-92. *Refer to Section 17-9.*

Balanced equation: $C(s) + CO_2(g) \rightleftarrows 2CO(g)$

Given: $P_{CO} + P_{CO_2} = 1.00$ atm; $K_p = 1.50$

Let   x = partial pressure of CO = P_{CO}. Then,

      1.00 - x = partial pressure of $CO_2 = P_{CO_2}$

$$K_p = \frac{(P_{CO})^2}{P_{CO_2}} = \frac{x^2}{1.00 - x} = \mathbf{1.50}$$

The quadratic equation: $x^2 + 1.50x - 1.50 = 0$

Solving, $x = \dfrac{-1.50 \pm \sqrt{(1.50)^2 - 4(1)(-1.50)}}{2(1)} = \dfrac{-1.50 \pm 2.87}{2} = 0.685$ or -2.19 (discard)

Therefore, at equilibrium $P_{CO} = x = \mathbf{0.685\ atm}$

                                $P_{CO_2} = 1.00 - x = \mathbf{0.315\ atm}$

17-94. *Refer to Sections 17-6, 17-8 and 17-11, and Example 17-10.*

Balanced equation: $NH_4Cl(s) \rightleftarrows NH_3(g) + HCl(g)$ $\Delta H = 176$ kJ/mol rxn

(a) As the temperature decreases, the mass of NH_3 decreases because the reaction is endothermic ($\Delta H > 0$).

(b) When more NH_3 is added, the total mass of NH_3 is initially the sum of the original equilibrium amount plus the additional NH_3. After equilibrium is reestablished by shifting to the left, the final mass of NH_3 is greater than the original equilibrium amount but less than the total mass present immediately after the additional NH_3 was added.

(c) When more HCl is added, the equilibrium shifts to the left and the mass of NH_3 decreases.

(d) When more solid NH_4Cl is added, the equilibrium is unaffected if the total gas volume is unchanged.

(e) When more solid NH_4Cl is added and the total gas volume decreases, the concentration of the gases will increase and the equilibrium will then shift to the left, thereby decreasing the concentrations of NH_3 and HCl back to their original values. However, although there is no change in the concentration of NH_3, the mass of NH_3 decreases due to the volume shrinkage.

17-96. *Refer to Sections 17-13 and 17-10, and Example 17-20.*

Balanced equation: $2Cl_2(g) + 2H_2O(g) \rightleftarrows 4HCl(g) + O_2(g)$ $\Delta H° = +115$ kJ/mol rxn

                                                           $K_p = 4.6 \times 10^{-14}$ at 25°C

At 400°C (673 K): Substituting into the van't Hoff equation, we have

$$\ln \frac{K_{p\,T_2}}{K_{p\,T_1}} = \frac{\Delta H°(T_2 - T_1)}{RT_1T_2}$$

$$\ln \frac{K_{p\,673\,K}}{4.6 \times 10^{-14}} = \frac{(115 \times 10^3\ J)(673\ K - 298\ K)}{(8.314\ J/K)(298\ K)(673\ K)} = 25.9$$

$$\frac{K_{p\,673\,K}}{4.6 \times 10^{-14}} = e^{25.9} = 1.8 \times 10^{11}$$

$$K_{p\,673\,K} = \mathbf{8.3 \times 10^{-3}}$$

$K_c = K_p(RT)^{-\Delta n} = (8.3 \times 10^{-3})(0.0821 \times 673)^{-(5-4)} = \mathbf{1.5 \times 10^{-4}}$

At 800°C (1073 K): Substituting into the van't Hoff equation, we have

$$\ln \frac{K_{p\ 1073\ K}}{4.6 \times 10^{-14}} = \frac{(115 \times 10^3\ J)(1073\ K - 298\ K)}{(8.314\ J/K)(298\ K)(1073\ K)} = 33.5$$

$$K_{p\ 1073\ K} = \mathbf{16}$$

$$K_c = K_p (RT)^{-\Delta n} = (16)(0.0821 \times 1073)^{-(5 - 4)} = \mathbf{0.18}$$

17-98.   *Refer to Sections 17-10.*

Balanced equation: $CO(g) + 3H_2(g) \rightleftarrows CH_4(g) + H_2O(g)$

Plan: (1) Determine K_c at 1133 K.
   (2) Calculate K_p.

(1) $K_c = \dfrac{[CH_4][H_2O]}{[CO][H_2]^3} = \dfrac{(1.21 \times 10^{-4}\ mol/0.100\ L)(5.63 \times 10^{-8}\ mol/0.100\ L)}{(1.21 \times 10^{-4}\ mol/0.100\ L)(2.47 \times 10^{-4}\ mol/0.100\ L)^3} = 37.4$

(2) $K_p = K_c (RT)^{\Delta n} = (37.4)(0.0821 \times 1133)^{(2 - 4)} = \mathbf{0.00432}$

17-100. *Refer to Section 17-1 and the solution to Exercise 17-4.*

When the beam of a triple balance stops swinging, the balance has reached a static equilibrium. A body is said to be "in static equilibrium" if a body at rest will stay at rest. That statement is definitely true in this case.

17-102. *Refer to Section 17-1.*

The relationship between the forward and reverse reactions in an equilibrium must be occurring at the same time and at the same rate, by definition.

17-104. *Refer to Section 17-1.*

The actual masses of reactant species do not equal the masses of product species at equilibrium.

For example, consider the equilibrium:

$$2A \rightleftarrows A_2.$$

If K_c were very large, at equilibrium we would have essentially all A_2 and very little A. The Law of Conservation of Matter is not violated because wherever the system is, the total mass of the system remains constant:

$$\text{mass of A} + \text{mass of } A_2 = \text{constant}$$

Recall that for other reactions we have studied which we know obey the Law of Conservation of Matter, we assumed they went totally to completion. At the beginning of the reaction, there were only reactants and at the end, there were only products.

Balanced equation: $[Co(H_2O)_6]^{2+}(aq) + 4Cl^-(aq) \rightleftarrows CoCl_4^{2-}(aq) + 6H_2O(\ell)$ $K_c = \dfrac{[CoCl_4^{2-}]}{[[Co(H_2O)_6]^{2+}][Cl^-]^4}$

pink   blue

Assuming $K > 1$, initially in the beaker, I would "see" mostly the blue $CoCl_4^{2-}$ ions and fewer pink $[Co(H_2O)_6]^{2+}$ ions, along with colorless Cl^- ions. Overall, the solution would look blue.

When water is added, the reaction quotient, Q_c, would initially become greater than K_c since the concentrations of all the species would decrease. The system would then shift to the left toward the reactants to reach equilibrium, in much the same way as when working with gases and the volume of the container is increased, i.e. the concentrations of all species are decreased. In the gas case, the system shifts toward the side with the greater number of moles of gas. Here, the system shifts toward the side with the greater number of species. Water is not included as a relevant species since it does not appear in the equilibrium expression. The solution would become pinker.

Balanced equation: $2SO_2(g) + O_2(g) \rightleftarrows 2SO_3(g)$ $K_p = 6.98 \times 10^{24}$ at 298 K

(1) $\Delta G_{rxn}^\circ = -RT \ln K_p = -(8.314 \text{ J/K})(298 \text{ K})\ln(6.98 \times 10^{24}) = -1.42 \times 10^5$ J or **−142 kJ**

(2) Given: $\Delta G_{f\ SO_2(g)}^\circ = -300.194$ kJ/mol

$\Delta G_{rxn}^\circ = [2\Delta G_{f\ SO_3(g)}^\circ] - [2\Delta G_{f\ SO_2(g)}^\circ + \Delta G_{f\ O_2(g)}^\circ]$

−142 kJ $= [2\Delta G_{f\ SO_3(g)}^\circ] - [(2 \text{ mol})(-300.194 \text{ kJ/mol}) + (1 \text{ mol})(0 \text{ kJ/mol})]$

Solving, $\Delta G_{f\ SO_3(g)}^\circ =$ **−371 kJ/mol**   (In Appendix K, $\Delta G_{f\ SO_3(g)}^\circ = -371.1$ kJ/mol)

Plan: (1) Treat the vaporization process as a chemical reaction, then establish the relationship between K_{eq} and P_{H_2O}.

(2) Use the van't Hoff equation to evaluate P_{H_2O} at 75°C.

(1) For vaporization: $H_2O(\ell) \rightleftarrows H_2O(g)$ $K_{eq} = P_{H_2O}$

At temperature $T_1 = 100°C = 373$ K $K_1 = P_{H_2O}$ at 373 K = 1 atm

At temperature $T_2 = 75°C = 348$ K $K_2 = P_{H_2O}$ at 348 K = unknown

(2) van't Hoff equation:

$$\ln \frac{K_2}{K_1} = \frac{\Delta H°(T_2 - T_1)}{RT_1T_2} \quad \text{where } \Delta H° = 40.66 \text{ kJ/mol}$$

$$\ln \frac{K_2}{1 \text{ atm}} = \frac{(40660 \text{ J})(348 \text{ K} - 373 \text{ K})}{(8.314 \text{ J/K})(348 \text{ K})(373 \text{ K})}$$

$$\ln \frac{K_2}{1 \text{ atm}} = -0.94$$

$$K_2 = P_{H_2O} \text{ at 348 K} = \textbf{0.39 atm}$$

Note: In Appendix E, the vapor pressure of water, P_{H_2O}, at 348 K (75°C) = 289.1 torr = 0.380 atm. The answers agree very well (< 3% error).

18 Ionic Equilibria I: Acids and Bases

(a) Strong and weak acids are electrolytes and ionize to some degree in water, producing H^+, according to Arrhenius and Brønsted-Lowry definitions. They have a sour taste, change the color of many indicators, react with metal oxides and metal hydroxides to form salts and water. They differ in the degree to which they ionize in aqueous solution. Strong acids ionize almost completely, whereas weak acids generally ionize less than 5%.

(b) Strong and weak bases are also electrolytes. Both produce OH^- ions, according to Brønsted-Lowry definitions. They generally have a bitter taste, have a slippery feeling, change the color of many indicators, and react with protic acids to form salts and sometimes water. Strong bases are soluble metal hydroxides (ionic compounds), like NaOH and dissociate completely into their ions. When strong bases react with protic acids, salt and water are produced. Two categories of weak bases are (1) the soluble organic base (covalent compounds), like NH_3 or CH_3NH_2, and (2) the conjugate base of a weak acid, such as CN^- or F^-. Both produce OH^- ions in solution by reacting with water in an equilibrium (see below). When weak organic bases react with protic acids, only salt is produced.

$$(1) \ \ NH_3(aq) + H_2O(\ell) \rightleftarrows NH_4^+(aq) + OH^-(aq) \qquad\qquad (2) \ \ CN^-(aq) + H_2O(\ell) \rightleftarrows HCN(aq) + OH^-(aq)$$

18-4. *Refer to Sections 18-1 and 4-2, and Example 18-1.*

(a) $? \ M \ NaCl = \dfrac{\text{mol NaCl}}{\text{L soln}} = \dfrac{25.85 \ \text{g}/58.44 \ \text{g/mol}}{0.250 \ \text{L}} = \mathbf{1.77 \ M \ NaCl}$

(b) $? \ M \ H_2SO_4 = \dfrac{\text{mol } H_2SO_4}{\text{L soln}} = \dfrac{75.5 \ \text{g}/98.1 \ \text{g/mol}}{1.00 \ \text{L}} = \mathbf{0.770 \ M \ H_2SO_4}$

(c) $? \ M \ C_6H_5OH = \dfrac{\text{mol } C_6H_5OH}{\text{L soln}} = \dfrac{0.126 \ \text{g}/94.1 \ \text{g/mol}}{1.00 \ \text{L}} = \mathbf{1.34 \times 10^{-3} \ M \ C_6H_5OH}$

18-6. *Refer to Sections 18-1, 18-2 and 4-2, and Examples 18-1 and 18-2.*

These compounds are strong electrolytes.

(a) $0.45 \ M \ HBr(aq) \rightarrow 0.45 \ M \ H^+(aq) + 0.45 \ M \ Br^-(aq)$ $[H^+] = [Br^-] = \mathbf{0.45 \ M}$

(b) $0.045 \ M \ KOH(aq) \rightarrow 0.045 \ M \ K^+(aq) + 0.045 \ M \ OH^-(aq)$ $[K^+] = [OH^-] = \mathbf{0.045 \ M}$

(c) $0.0112 \ M \ CaCl_2(aq) \rightarrow 0.0112 \ Ca^{2+}(aq) + 0.0224 \ M \ Cl^-(aq)$ $[Ca^{2+}] = \mathbf{0.0112 \ M}; \ [Cl^-] = \mathbf{0.0224 \ M}$

18-8. *Refer to Sections 18-1 and 18-2, and Example 18-2.*

These compounds are strong electrolytes.

(a) $? \ M \ KOH = \dfrac{(1.15 \ \text{g}/56.1 \ \text{g/mol})}{1.50 \ \text{L}} = 0.0137 \ M$

	KOH	$\rightarrow$	K^+	+	OH^-
initial	0.0137 M		0 M		0 M
change	- 0.0137 M		+ 0.0137 M		+ 0.0137 M
final	0 M		0.0137 M		0.0137 M

Therefore, $[K^+] = \mathbf{0.0137 \ M}$ and $[OH^-] = \mathbf{0.0137 \ M}$

(b) $? M \, Ba(OH)_2 = \dfrac{(0.2505 \text{ g}/171.3 \text{ g/mol})}{0.250 \text{ L}} = 0.00585 \ M$ Therefore, $[Ba^{2+}] = \mathbf{0.00585 \ M}$ and $[OH^-] = \mathbf{0.0117 \ M}$

(c) $? M \, Ca(NO_3)_2 = \dfrac{(1.26 \text{ g}/164 \text{ g/mol})}{0.100 \text{ L}} = 0.0768 \ M$ Therefore, $[Ca^{2+}] = \mathbf{0.0768 \ M}$ and $[NO_3{}^-] = \mathbf{0.154 \ M}$

18-10. *Refer to Section 18-2.*

(a) $H_2O(\ell) + H_2O(\ell) \rightleftarrows H_3O^+(aq) + OH^-(aq)$

(b) $K_c = K_w = [H_3O^+][OH^-]$

(c) The equilibrium constant, K_w, is known as the ion product for water.

(d) In pure water, $[H_3O^+] = [OH^-]$. At 25°C, $[OH^-] = [H_3O^+] = 1.0 \times 10^{-7} \ M$.

(e) In acidic solutions, $[H_3O^+] > [OH^-]$. In basic solutions, $[OH^-] > [H_3O^+]$.

18-12. *Refer to Section 18-2.*

(a) A 0.060 M solution of NaOH does have 2 sources of OH^- ion: (1) OH^- from the complete dissociation of NaOH and (2) OH^- from the ionization of water. Since Source 1 dominates Source 2, the concentration of OH^- produced by the ionization of water is therefore neglected.

(b) From Source 1, $NaOH(aq) \rightarrow Na^+(aq) + OH^-(aq)$, $[OH^-] = 0.060 \ M$. To calculate $[OH^-]$ from Source 2, consider the ionization of water in a 0.060 M solution of NaOH.

Let x = moles per liter of H_3O^+ and OH^- produced by the ionization of water.

	$2H_2O(\ell)$	$\rightleftarrows$	$H_3O^+(aq)$	$+$	$OH^-(aq)$
initial			0 M		0.060 M
change			+ x M		+ x M
at equilibrium			x M		(0.060 + x) M

$K_w = [H_3O^+][OH^-] = x(0.060 + x) = 1.0 \times 10^{-14}$

We know that x << 0.060. Let us make the approximation that $0.060 + x \approx 0.060$.

Then $x \times 0.060 = 1.0 \times 10^{-14}$ and $x = 1.7 \times 10^{-13}$. So, our approximation that x << 0.060 is a good one.

Therefore, $[OH^-]$ from 0.060 M NaOH = 0.060 M
 $[OH^-]$ from the ionization of water = $1.7 \times 10^{-13} \ M$, and can be neglected.

18-14. *Refer to Section 18-2, Example 18-3, and Exercises 18-6, 18-7 and 18-8 Solutions.*

In pure water at 25°C, $[H_3O^+] = 1.00 \times 10^{-7} \ M$.

To determine $[H_3O^+]$ in basic solution, use the K_w expression: $K_w = [H_3O^+][OH^-] = 1.00 \times 10^{-14}$.

Solution	[OH⁻]	$[H_3O^+] = K_w/[OH^-]$
0.045 M KOH	0.045 M	$2.2 \times 10^{-13} \ M$
0.0105 M Sr(OH)₂	0.0210 M	$4.76 \times 10^{-13} \ M$
0.0385 M Ba(OH)₂	0.0117 M	$8.55 \times 10^{-13} \ M$
pure water	$1.00 \times 10^{-7} \ M$	$1.00 \times 10^{-7} \ M$

The $[H_3O^+]$ in basic solutions is much lower than that in pure water.

18-16. *Refer to Section 18-2 and Example 18-2.*

$$[H_3O^+] = \frac{K_w}{[OH^-]} = \frac{1.00 \times 10^{-14}}{8.78 \times 10^{-6}} = \mathbf{1.14 \times 10^{-9}\ \textit{M}} \qquad \text{at } 25°C$$

18-18. *Refer to Section 18-3 and Appendix A-2.*

When working with base 10 logarithms, the number of significant figures in the number (the antilogarithm) sets the number of significant digits in the mantissa (the decimal part of the logarithm). For example in (a), the number 0.00052 has two significant figures, so there are 2 decimal places in the log of that number.

(a) $\log 0.00052 = \mathbf{-3.28}$ (b) $\log 150.2 = \mathbf{2.1767}$ (c) $\log (5.8 \times 10^{-11}) = \mathbf{-10.24}$ (d) $\log (2.9 \times 10^{-7}) = \mathbf{-6.54}$

18-20. *Refer to Section 18-3, Examples 18-4 and 18-5, andTable18-2.*

We know: $pH = -\log [H_3O^+] = 7.45$.
Therefore, $[H_3O^+] = \text{antilog} (-7.45) = \mathbf{3.5 \times 10^{-8}\ \textit{M}}$

If we assume the blood is at 25°C, $[OH^-] = \dfrac{K_w}{[H_3O^+]} = \dfrac{1.00 \times 10^{-14}}{3.5 \times 10^{-8}} = \mathbf{2.9 \times 10^{-7}\ \textit{M}}$

If we assume the blood is at 37°C (body temperature), $[OH^-] = \dfrac{K_w}{[H_3O^+]} = \dfrac{2.4 \times 10^{-14}}{3.5 \times 10^{-8}} = \mathbf{6.9 \times 10^{-7}\ \textit{M}}$

18-22. *Refer to Sections 18-2 and 18-3, and Example 18-5.*

(a) $[H_3O^+] = [HCl] = 0.20\ M$
     $pH = -\log[H^+] = -\log (0.20) = \mathbf{0.70}$

(b) $[H_3O^+] = [HNO_3] = 0.050\ M$
     $pH = -\log[H^+] = -\log (0.050) = \mathbf{1.30}$

(c) $[H_3O^+] = [HClO_4] = \dfrac{0.65\ \text{g HClO}_4}{1\ L} \times \dfrac{1\ \text{mol HClO}_4}{100.45\ \text{g HClO}_4} = 6.5 \times 10^{-3}\ M$

     $pH = -\log (6.5 \times 10^{-3}\ M) = \mathbf{2.19}$

(d) $[OH^-] = [NaOH] = 9.8 \times 10^{-4}\ M$
     $pOH = -\log[OH^-] = 3.01; pH = 14.00 - 3.01 = \mathbf{10.99}$

18-24. *Refer to Section 18-3 and Example 18-4.*

We know: $pH = -\log [H_3O^+] = 3.52$.
Therefore, $[H_3O^+] = \text{antilog} (-3.52) = 3.0 \times 10^{-4}\ M$

Since HNO_3 is a strong acid, $[HNO_3] = [H_3O^+] = \mathbf{3.0 \times 10^{-4}\ \textit{M}}$

18-26. *Refer to Section 18-3, and Examples 18-5 and 18-6.*

	Solution	$[H_3O^+]$	$[OH^-]$	pH	pOH
(a)	0.075 M NaOH	$1.3 \times 10^{-13}\ M$	$0.075\ M$	12.88	1.12
(b)	0.075 M HCl	$0.075\ M$	$1.3 \times 10^{-13}\ M$	1.12	12.88
(c)	0.075 M $Ca(OH)_2$	$6.7 \times 10^{-14}\ M$	$0.15\ M$	13.18	0.82

(a) H_3PO_4 is probably a slightly stronger acid than H_3AsO_4. P is probably slightly more electronegative than As, since electronegativity increases upward in a group, even though both are given an electronegativity of 2.1. The P atom would pull the electron density in the O-H bond away from the O-H bond toward itself a little more than the As atom would, making the O-H bond slightly weaker in H_3PO_4 and producing a slightly stronger acid. This deduction is verified by the K_a values in Appendix F: K_a for H_3PO_4 is 7.5×10^{-3} and K_a for H_3AsO_4 is 2.5×10^{-4}.

(b) H_3AsO_4 is a stronger acid than H_3AsO_3, because H_3AsO_4 has one more O atom. The extra oxygen atom with its high electronegativity, helps pull the electron density away from the H-O bond in H_3AsO_4, thereby weakening the H-O bond. The weaker the H-O bond, the easier it will break in solution and the stronger the acid will be. This deduction is verified by the data in Appendix F: K_a for H_3AsO_4 is 2.5×10^{-4} and K_a for H_3AsO_3 is 6.0×10^{-10}.

18-30.   *Refer to Section 18-4.*

Balanced equation for the ionization of a weak acid: $HA(aq) + H_2O(\ell) \rightleftarrows H_3O^+(aq) + A^-(aq)$

In Chapter 17 (Section 17-11), we stated that for heterogeneous equilibria, terms for pure liquids and pure solids do not appear in K expressions because their activity values are essentially 1. In the discussion in Section 18-4, these terms are temporarily included. In dilute solutions, activities are equal to concentrations.

$$K = \frac{(H_3O^+)(A^-)}{(HA)(H_2O)} \quad \text{where (x) represents the activity of x} \qquad K_a = \frac{[H_3O^+][A^-]}{[HA](1)}$$

The symbol for the equilibrium constant is K_a. It is called the acid ionization constant.

18-32.   *Refer to Section 18-4 and Examples 18-15 and 18-16.*

(1) For triethylamine, $(C_2H_5)_3N$

Balanced equation: $(C_2H_5)_3N + H_2O \rightleftarrows (C_2H_5)_3NH^+ + OH^-$ $\qquad\qquad K_b = 5.2 \times 10^{-4}$

Let $x = [(C_2H_5)_3N]$ that ionizes. Then $x = [(C_2H_5)_3NH^+]$ produced $= [OH^-]$ produced.

$$K_b = \frac{[(C_2H_5)_3NH^+][OH^-]}{[(C_2H_5)_3N]} = 5.2 \times 10^{-4} = \frac{x^2}{0.018 - x} \approx \frac{x^2}{0.018} \qquad \text{Solving, } x = 3.1 \times 10^{-3}$$

In this case, 3.1×10^{-3} is greater than 5% of 0.018 ($= 9.0 \times 10^{-4}$). A simplifying assumption cannot be made and the original quadratic equation must be solved: $x^2 + (5.2 \times 10^{-4})x - 9.4 \times 10^{-6} = 0$

$$x = \frac{-(5.2 \times 10^{-4}) \pm \sqrt{(5.2 \times 10^{-4})^2 - 4(1)(-9.4 \times 10^{-6})}}{2(1)} = \frac{-5.2 \times 10^{-4} \pm 6.2 \times 10^{-3}}{2}$$

$$= 2.8 \times 10^{-3} \text{ or } -3.4 \times 10^{-3} \text{ (discard)}$$

Therefore, $[OH^-] = x = \mathbf{2.8 \times 10^{-3}}\ \mathbf{\mathit{M}}$

(2) For trimethylamine, $(CH_3)_3N$

Balanced equation: $(CH_3)_3N + H_2O \rightleftarrows (CH_3)_3NH^+ + OH^-$ $\qquad\qquad K_b = 7.4 \times 10^{-5}$

Let $x = [(CH_3)_3N]$ that ionizes. Then $x = [(CH_3)_3NH^+]$ produced $= [OH^-]$ produced.

$$K_b = \frac{[(CH_3)_3NH^+][OH^-]}{[(CH_3)_3N]} = 7.4 \times 10^{-5} = \frac{x^2}{0.018 - x} \approx \frac{x^2}{0.018} \qquad \text{Solving, } x = 1.2 \times 10^{-3}$$

In this case, 1.2×10^{-3} is greater than 5% of 0.018 ($= 9.0 \times 10^{-4}$). A simplifying assumption cannot be made and the original quadratic equation must be solved: $x^2 + (7.4 \times 10^{-5})x - 1.3 \times 10^{-6} = 0$

$$x = \frac{-(7.4 \times 10^{-5}) \pm \sqrt{(7.4 \times 10^{-5})^2 - 4(1)(-1.3 \times 10^{-6})}}{2(1)}$$

$$= \frac{-7.4 \times 10^{-5} \pm 2.3 \times 10^{-3}}{2}$$

$$= 1.1 \times 10^{-3} \text{ or } -1.2 \times 10^{-3} \text{ (discard)}$$

Therefore, $[OH^-] = x = \mathbf{1.1 \times 10^{-3} \ M}$

Yes, $[OH^-]$ in triethylamine(aq) is greater than $[OH^-]$ in trimethylamine(aq) for the same concentration.

Note: Even if the simplifying assumption may not always be applicable, it is a good idea to always use it in the beginning for several reasons. (1) It is quick and easy to use, and (2) it usually works. (3) Even when it does not yield the correct answer, it *usually* gives a reasonable estimate of the true value.

18-34. *Refer to Sections 18-2 and 18-3, and Table 18-2.*

(a) K_w at 37°C $= 2.4 \times 10^{-14} = [H_3O^+][OH^-]$.   For pure water, $[H_3O^+] = [OH^-]$

    $[H_3O^+] = \sqrt{K_w} = \sqrt{2.4 \times 10^{-14}} = 1.5 \times 10^{-7}$ at 37°C

    pH $= -\log [H_3O^+] = -\log (1.5 \times 10^{-7}) = \mathbf{6.82 \ at \ 37°C}$

(b) Pure water at 37°C is **neutral** since $[H_3O^+] = [OH^-]$. The pH of a neutral aqueous solution is 7.0 only at 25°C.

18-36. *Refer to Section 18-4 and Exercise 18-30 Solution.*

Reaction of proton-accepting weak base, B, with water: $B(aq) + H_2O(\ell) \rightleftarrows BH^+(aq) + OH^-(aq)$

In Chapter 17 (Section 17-11), we stated that for heterogeneous equilibria, terms for pure liquids and pure solids do not appear in K expressions because their activity values are essentially 1. In the discussion in Section 18-4, these terms are temporarily included. In dilute solutions, activities are equal to concentrations.

$$K = \frac{(BH^+)(OH^-)}{(B)(H_2O)}$$ where (x) represents the activity of x $$K_b = \frac{[BH^+][OH^-]}{[B](1)}$$

K_b, the ionization constant for bases, is used for calculations involving the equilibria of weak bases.

18-38. *Refer to Section 18-4 and Example 18-8.*

Balanced equation: $HX + H_2O \rightleftarrows H_3O^+ + X^-$

Since HX is 1.07% ionized, $[HX]_{reacted} = 0.0750 \ M \times 0.0107 = 8.02 \times 10^{-4} \ M$

	HX	+	H₂O	$\rightleftarrows$	H₃O⁺	+	X⁻
initial	0.0750 M				≈ 0 M		0 M
change	- 8.02 × 10⁻⁴ M				+ 8.02 × 10⁻⁴ M		+ 8.02 × 10⁻⁴ M
at equilibrium	0.0742 M				8.02 × 10⁻⁴ M		8.02 × 10⁻⁴ M

pH $= -\log (8.02 \times 10^{-4}) = \mathbf{3.096}$

$$K_a = \frac{[H_3O^+][X^-]}{[HX]} = \frac{(8.02 \times 10^{-4})^2}{0.0742} = \mathbf{8.67 \times 10^{-6}}$$

Balanced equation: $C_3H_7COOH + H_2O \rightleftarrows H_3O^+ + C_3H_7COO^-$

Let x = mol/L of C_3H_7COOH that reacts. Then
  x = mol/L of H_3O^+ produced = mol/L of $C_3H_7COO^-$ produced.

Since pH = 3.21, $[H_3O^+] = x$ = antilog $(-3.21) = 6.2 \times 10^{-4}$ M

	C_3H_7COOH	+	H_2O	$\rightleftarrows$	H_3O^+	+	$C_3H_7COO^-$
initial	0.025 M				≈ 0 M		0 M
change	- x M				+ x M		+ x M
at equilibrium	(0.025 - x) M				x M		x M

$$K_a = \frac{[H_3O^+][C_3H_7COO^-]}{[C_3H_7COOH]} = \frac{x^2}{0.025 - x} = \frac{(6.2 \times 10^{-4})^2}{0.025 - 6.2 \times 10^{-4}} = \mathbf{1.6 \times 10^{-5}}$$

Balanced equation: $C_6H_5COOH + H_2O \rightleftarrows H_3O^+ + C_6H_5COO^-$ $K_a = 6.3 \times 10^{-5}$

Let x = mol/L of C_6H_5COOH that reacts. Then
  x = mol/L of H_3O^+ produced = mol/L of $C_6H_5COO^-$ produced.

	C_6H_5COOH	+	H_2O	$\rightleftarrows$	H_3O^+	+	$C_6H_5COO^-$
initial	0.52 M				≈ 0 M		0 M
change	- x M				+ x M		+ x M
at equilibrium	(0.52 - x) M				x M		x M

$$K_a = \frac{[H_3O^+][C_6H_5COO^-]}{[C_6H_5COOH]} = \frac{x^2}{0.52 - x} = 6.3 \times 10^{-5}$$

Assume that 0.52 - x $\approx$ 0.52. Then $x^2/0.52 = 6.3 \times 10^{-5}$ and $x = 5.7 \times 10^{-3}$. The simplifying assumption is justified since 5.7×10^{-3} is much less than 5% of 0.52 (= 0.03). However, in this example, it does make a slight difference in the concentration of C_6H_5COOH.

Therefore at equilibrium: $[C_6H_5COOH] = \mathbf{0.51}$ $\boldsymbol{M}$
            $[H_3O^+] = [C_6H_5COO^-] = x = \mathbf{5.7 \times 10^{-3}}$ $\boldsymbol{M}$
            $[OH^-] = K_w/[H_3O^+] = \mathbf{1.8 \times 10^{-12}}$ $\boldsymbol{M}$

Balanced equation: $HF + H_2O \rightleftarrows H_3O^+ + F^-$ $K_a = 7.2 \times 10^{-4}$

Let x = mol/L of HF that reacts. Then
  x = mol/L of H_3O^+ produced = mol/L of F^- produced.

	HF	+	H_2O	$\rightleftarrows$	H_3O^+	+	F^-
initial	0.33 M				≈ 0 M		0 M
change	- x M				+ x M		+ x M
at equilibrium	(0.33 - x) M				x M		x M

$$K_a = \frac{[H_3O^+][F^-]}{[HF]} = \frac{x^2}{0.33 - x} = 7.2 \times 10^{-4}$$

Assume that 0.33 - x $\approx$ 0.33. Then $x^2/0.33 = 7.2 \times 10^{-4}$ and x = 0.015. The simplifying assumption is justified since 0.015 is less than 5% of 0.33 (= 0.016).

Therefore at equilibrium: $[H_3O^+] = 0.015$ M pH = **1.82**

18-46. *Refer to Section 18-4, Example 18-15, Table 18-6 and Appendix G.*

Balanced equation: $NH_3 + H_2O \rightleftarrows NH_4^+ + OH^-$ $\qquad\qquad K_b = 1.8 \times 10^{-5}$

Let $x = [NH_3]$ that ionizes. Then
$\qquad x = [NH_4^+]$ produced $= [OH^-]$ produced.

	NH_3	$+$	H_2O	$\rightleftarrows$	NH_4^+	$+$	OH^-
initial	2.05 M				0 M		≈ 0 M
change	$- x$ M				$+ x$ M		$+ x$ M
at equilibrium	(2.05 - x) M				x M		x M

$K_b = \dfrac{[NH_4^+][OH^-]}{[NH_3]} = 1.8 \times 10^{-5} = \dfrac{x^2}{2.05 - x} \approx \dfrac{x^2}{2.05}$ $\qquad$ Solving, $x = 6.1 \times 10^{-3}$ (6.07×10^{-3})

Since 6.1×10^{-3} is less than 5% of 2.05, the approximation is justified.

Therefore, $[OH^-] = 6.1 \times 10^{-3}$ M $\qquad\qquad$ $pH = -\log \dfrac{K_w}{[OH^-]} = -\log\left(\dfrac{1.0 \times 10^{-14}}{6.1 \times 10^{-3}}\right) = \mathbf{11.79}$

$\qquad pOH = -\log (6.07 \times 10^{-3}\ M) = \mathbf{2.21}$ $\qquad\qquad$ or $pH = 14.00 - pOH = 14.00 - 2.21 = \mathbf{11.79}$

Note: Keep all the significant figures and round at the end. Remember the number of decimal places in pH or pOH values are set by the number of significant figures in the $[H^+]$ or $[OH^-]$; this is a result of working with logarithms.

18-48. *Refer to Section 18-4, Examples 18-10 and 18-11, and Appendix F.*

Balanced equation: $HCOOH + H_2O \rightleftarrows H_3O^+ + HCOO^-$ $\qquad\qquad K_a = 1.8 \times 10^{-4}$

Let $x = [HCOOH]$ that ionizes. Then
$\qquad x = [H_3O^+]$ produced $= [HCOO^-]$ produced.

	$HCOOH$	$+$	H_2O	$\rightleftarrows$	H_3O^+	$+$	$HCOO^-$
initial	0.0425 M				≈ 0 M		0 M
change	$- x$ M				$+ x$ M		$+ x$ M
at equilibrium	(0.0425 - x) M				x M		x M

$K_a = \dfrac{[H_3O^+][HCOO^-]}{[HCOOH]} = \dfrac{x^2}{0.0425 - x} = 1.8 \times 10^{-4}$

Assuming $0.0425 - x \approx 0.0425$, then $x^2/0.0425 = 1.8 \times 10^{-4}$ and $x = 2.8 \times 10^{-3}$. However, the simplifying assumption may not be justified since 2.8×10^{-3} is 7% of 0.0425.

Let us check ourselves by solving the original quadratic equation: $x^2 + (1.8 \times 10^{-4})x - 7.6 \times 10^{-6} = 0$

$$x = \dfrac{-(1.8 \times 10^{-4}) \pm \sqrt{(1.8 \times 10^{-4})^2 - 4(1)(-7.6 \times 10^{-6})}}{2(1)} = \dfrac{-1.8 \times 10^{-4} \pm 5.5 \times 10^{-3}}{2}$$
$$= 2.7 \times 10^{-3} \text{ or } -2.8 \times 10^{-3} \text{ (discard)}$$

The two solutions for x are nearly the same to two significant digits. Therefore, in this case, if an approximate answer is desired, the solution using the simplifying assumption is probably adequate.

% ionization $= \dfrac{[HCOOH]_{ionized}}{[HCOOH]_{initial}} \times 100\% = \dfrac{2.7 \times 10^{-3}\ M}{0.0425\ M} \times 100\% = \mathbf{6.4\%}$

18-50. *Refer to Section 18-4 and Example 18-12.*

For Weak Acid #1 $\qquad pK_a = -\log K_a = -\log (7.2 \times 10^{-5}) = \mathbf{4.14}$

For Weak Acid #2 $\qquad pK_a = -\log K_a = -\log (4.2 \times 10^{-10}) = \mathbf{9.38}$

18-52. *Refer to Section 18-4.*

Balanced equation: $C_5H_5N(aq) + H_2O(\ell) \rightleftarrows C_5H_5NH^+(aq) + OH^-(aq)$

Since a $0.00500\ M\ C_5H_5N$ solution is 0.053% ionized, $[C_5H_5N]_{reacted} = 0.00500\ M \times 0.00053 = 2.7 \times 10^{-6}\ M$

	C_5H_5N	+	H_2O	$\rightleftarrows$	$C_5H_5NH^+$	+	OH^-
initial	0.00500 M				0 M		$\approx 0\ M$
change	- 2.7 x 10^{-6} M				+ 2.7 x 10^{-6} M		+ 2.7 x 10^{-6} M
at equilibrium	0.00500 M				2.7 x 10^{-6} M		2.7 x 10^{-6} M

$$K_b = \frac{[C_5H_5NH^+][OH^-]}{[C_5H_5N]} = \frac{(2.7 \times 10^{-6})^2}{0.00500} = 1.5 \times 10^{-9}$$

$pK_b = -\log(1.5 \times 10^{-9}) = \mathbf{8.82}$

This agrees with the K_b value given in Appendix G and Table 18-6.

18-54. *Refer to Section 18-4.*

Balanced equation: $CH_3NH_2(aq) + H_2O(\ell) \rightleftarrows CH_3NH_3^+(aq) + OH^-(aq)$

$$K_b = \frac{[CH_3NH_3^+][OH^-]}{[CH_3NH_2]} = \frac{(2.0 \times 10^{-3})^2}{0.0080} = \mathbf{5.0 \times 10^{-4}}$$

This agrees with the K_b value given in Appendix G and Table 18-6.

18-56. *Refer to Section 18-4, Example 18-15, Table 18-6 and Appendix G.*

(a) Balanced equation: $NH_3 + H_2O \rightleftarrows NH_4^+ + OH^-$ $\qquad K_b = 1.8 \times 10^{-5}$

Let x = $[NH_3]$ that ionizes. Then
x = $[NH_4^+]$ produced = $[OH^-]$ produced.

	NH_3	+	H_2O	$\rightleftarrows$	NH_4^+	+	OH^-
initial	0.15 M				0 M		$\approx 0\ M$
change	- x M				+ x M		+ x M
at equilibrium	(0.15 - x) M				x M		x M

$$K_b = \frac{[NH_4^+][OH^-]}{[NH_3]} = 1.8 \times 10^{-5} = \frac{x^2}{0.15 - x} \approx \frac{x^2}{0.15} \qquad \text{Solving, } x = 1.6 \times 10^{-3}$$

Since 1.6×10^{-3} is less than 5% of 0.15, the approximation is justified.

Therefore, $[OH^-] = \mathbf{1.6 \times 10^{-3}\ M}$

$$\% \text{ ionization} = \frac{[NH_3]_{ionized}}{[NH_3]_{initial}} \times 100\% = \frac{1.6 \times 10^{-3}\ M}{0.15\ M} \times 100\% = \mathbf{1.1\%}$$

$$pH = -\log\frac{K_w}{[OH^-]} = -\log\left(\frac{1.0 \times 10^{-14}}{1.6 \times 10^{-3}}\right) = \mathbf{11.20}$$

(b) Balanced equation: $CH_3NH_2 + H_2O \rightleftarrows CH_3NH_3^+ + OH^-$ $\qquad K_b = 5.0 \times 10^{-4}$

Let x = $[CH_3NH_2]$ that ionizes. Then
x = $[CH_3NH_3^+]$ produced = $[OH^-]$ produced.

	CH_3NH_2	+	H_2O	$\rightleftarrows$	$CH_3NH_3^+$	+	OH^-
initial	0.15 M				0 M		$\approx 0\ M$
change	- x M				+ x M		+ x M
at equilibrium	(0.15 - x) M				x M		x M

$$K_b = \frac{[CH_3NH_3^+][OH^-]}{[CH_3NH_2]} = 5.0 \times 10^{-4} = \frac{x^2}{0.15 - x} \approx \frac{x^2}{0.15} \qquad \text{Solving, } x = 8.7 \times 10^{-3}$$

Since 8.7×10^{-3} is greater than 5% of 0.15 (= 7.5×10^{-3}), we should not use the simplifying assumption; the correct answer is obtained by solving the original quadratic equation: $x^2 + (5.0 \times 10^{-4})x - 7.5 \times 10^{-5} = 0$

$$x = \frac{-(5.0 \times 10^{-4}) \pm \sqrt{(5.0 \times 10^{-4})^2 - 4(1)(-7.5 \times 10^{-5})}}{2(1)} = \frac{-5.0 \times 10^{-4} \pm 1.73 \times 10^{-2}}{2}$$

$$= 8.4 \times 10^{-3} \text{ or } -8.9 \times 10^{-3} \text{ (discard)}$$

Therefore, $[OH^-] = \mathbf{8.4 \times 10^{-3}\ M}$

$$\% \text{ ionization} = \frac{[CH_3NH_2]_{ionized}}{[CH_3NH_2]_{initial}} \times 100\% = \frac{8.4 \times 10^{-3}\ M}{0.15\ M - 8.4 \times 10^{-3}\ M} \times 100\% = \mathbf{5.9\%}$$

$$pH = -\log \frac{K_w}{[OH^-]} = -\log\left(\frac{1.0 \times 10^{-14}}{8.4 \times 10^{-3}}\right) = \mathbf{11.92}$$

Note: Even if the simplifying assumption may not always be applicable, it is a good idea to always use it in the beginning for several reasons. (1) It is quick and easy to use, and (2) it usually works. (3) Even when it does not yield the correct answer, it *usually* gives a reasonable estimate of the true value. For example, in this exercise, using the simplifying assumption gave the answer, $x = 8.7 \times 10^{-3}$, which was fairly close to the correct answer, $x = 8.4 \times 10^{-3}$.

18-58. *Refer to Section 18-5, Example 18-17, Table 18-7 and Appendix F.*

Balanced equations:
$H_3AsO_4 + H_2O \rightleftarrows H_3O^+ + H_2AsO_4^-$ $\qquad K_{a1} = 2.5 \times 10^{-4}$
$H_2AsO_4^- + H_2O \rightleftarrows H_3O^+ + HAsO_4^{2-}$ $\qquad K_{a2} = 5.6 \times 10^{-8}$
$HAsO_4^{2-} + H_2O \rightleftarrows H_3O^+ + AsO_4^{3-}$ $\qquad K_{a3} = 3.0 \times 10^{-13}$

First Step:

Let $x = [H_3AsO_4]_{ionized}$. $\qquad$ Then $[H_3AsO_4] = (0.100 - x)\ M$
$\qquad\qquad\qquad\qquad\qquad\qquad\qquad [H_3O^+] = [H_2AsO_4^-] = x\ M$

$$K_{a1} = \frac{[H_3O^+][H_2AsO_4^-]}{[H_3AsO_4]} = \frac{x^2}{(0.100 - x)} = 2.5 \times 10^{-4} \approx \frac{x^2}{0.100} \qquad \text{Solving, } x = 5.0 \times 10^{-3}$$

Since 5.0×10^{-3} is 5% of 0.100, the simplifying assumption is valid.

Therefore, $[H_3O^+] = [H_2AsO_4^-] = x = 5.0 \times 10^{-3}\ M$
$\qquad\qquad [H_3AsO_4] = 0.100 - x = 0.095\ M$

Second Step:

Let $y = [H_2AsO_4^-]_{ionized}$. $\qquad$ Then $[H_2AsO_4^-] = (5.0 \times 10^{-3} - y)\ M$
$\qquad\qquad\qquad\qquad\qquad\qquad\qquad\ [H_3O^+] = (5.0 \times 10^{-3} + y)\ M$
$\qquad\qquad\qquad\qquad\qquad\qquad\qquad\ [HAsO_4^{2-}] = y\ M$

$$K_{a2} = \frac{[H_3O^+][HAsO_4^{2-}]}{[H_2AsO_4^-]} = \frac{(5.0 \times 10^{-3} + y)(y)}{(5.0 \times 10^{-3} - y)} = 5.6 \times 10^{-8} \approx \frac{(5.0 \times 10^{-3})y}{(5.0 \times 10^{-3})}$$

Solving, $y = 5.6 \times 10^{-8}$

Therefore, because the simplifying assumptions are valid, $\qquad [H_3O^+] = [H_2AsO_4^-] = 5.0 \times 10^{-3}\ M$
$\qquad\qquad\qquad\qquad\qquad\qquad\qquad\qquad\qquad\qquad\qquad\qquad\qquad\ [HAsO_4^{2-}] = 5.6 \times 10^{-8}\ M$

Third Step:

Let z = $[\text{HAsO}_4{}^{2-}]_{\text{ionized}}$.   Then $[\text{HAsO}_4{}^{2-}] = (5.6 \times 10^{-8} - z)\ M$

$[\text{H}_3\text{O}^+] = (5.0 \times 10^{-3} + z)\ M$

$[\text{AsO}_4{}^{3-}] = z\ M$

$$K_{a3} = \frac{[\text{H}_3\text{O}^+][\text{AsO}_4{}^{3-}]}{[\text{HAsO}_4{}^{-}]} = \frac{(5.0 \times 10^{-3} + z)(z)}{(5.6 \times 10^{-8} - z)} = 3.0 \times 10^{-13} = \frac{(5.0 \times 10^{-3})z}{(5.6 \times 10^{-8})} \qquad \text{Solving, } z = 3.4 \times 10^{-18}$$

Therefore, $[\text{AsO}_4{}^{3-}] = 3.4 \times 10^{-18}\ M$ because the simplifying assumptions are valid.

0.100 M H₃AsO₄ Solution		0.100 M H₃PO₄ Solution	
Species	Concentration (M)	Species	Concentration (M)
H_3AsO_4	0.095	H_3PO_4	0.076
H_3O^+	0.0050	H_3O^+	0.024
$\text{H}_2\text{AsO}_4{}^-$	0.0050	$\text{H}_2\text{PO}_4{}^-$	0.024
$\text{HAsO}_4{}^{2-}$	5.6×10^{-8}	$\text{HPO}_4{}^{2-}$	6.2×10^{-8}
OH^-	2.0×10^{-12}	OH^-	4.2×10^{-13}
$\text{AsO}_4{}^{3-}$	3.4×10^{-18}	$\text{PO}_4{}^{3-}$	9.3×10^{-19}

18-60.   *Refer to Section 18-5, Example 18-18 and Appendix F.*

Balanced equations: $\text{H}_2\text{SeO}_4 + \text{H}_2\text{O} \rightleftarrows \text{H}_3\text{O}^+ + \text{HSeO}_4{}^-$ K_{a1} = very large

$\text{HSeO}_4{}^- + \text{H}_2\text{O} \rightleftarrows \text{H}_3\text{O}^+ + \text{SeO}_4{}^{2-}$ $K_{a2} = 1.2 \times 10^{-2}$

First Step:

Since K_{a1} is very large, H_2SeO_4 is a strong electrolyte and totally dissociates into H_3O^+ and $\text{HSeO}_4{}^-$. Therefore,

$[\text{H}_2\text{SeO}_4] \approx 0\ M$

$[\text{HSeO}_4{}^-] = 0.15\ M$

$[\text{H}_3\text{O}^+] = 0.15\ M$

Second Step:

Let x = $[\text{HSeO}_4{}^-]_{\text{ionized}}$.   Then $[\text{HSeO}_4{}^-] = (0.15 - x)\ M$

$[\text{H}_3\text{O}^+] = (0.15 + x)\ M$

$[\text{SeO}_4{}^{2-}] = x\ M$

$$K_{a2} = \frac{[\text{H}_3\text{O}^+][\text{SeO}_4{}^{2-}]}{[\text{HSeO}_4{}^-]} = \frac{(0.15 + x)(x)}{(0.15 - x)} = 1.2 \times 10^{-2} \approx \frac{(0.15)x}{(0.15)} \qquad \text{Solving, } x = 1.2 \times 10^{-2}$$

In this case, 1.2×10^{-2} is greater than 5% of 0.15 (= 7.5×10^{-3}). A simplifying assumption cannot be made and the original quadratic equation must be solved: $x^2 + 0.16x - 1.8 \times 10^{-3} = 0$

$$x = \frac{-0.16 \pm \sqrt{(0.16)^2 - 4(1)(-1.8 \times 10^{-3})}}{2(1)} = \frac{-0.16 \pm 0.18}{2} = 0.01 \ \text{ or } -0.17 \text{ (discard)}$$

Therefore, $[\text{H}_3\text{O}^+] = 0.15 + x = \mathbf{0.16}\ \boldsymbol{M}$

$[\text{OH}^-] = K_w/[\text{H}_3\text{O}^+] = \mathbf{6.2 \times 10^{-14}}\ \boldsymbol{M}$

$[\text{HSeO}_4{}^-] = 0.15 - x = \mathbf{0.14}\ \boldsymbol{M}$

$[\text{SeO}_4{}^{2-}] = x = \mathbf{0.01}\ \boldsymbol{M}$

18-62. Refer to Section 18-5, Example 18-17 and Appendix F.

Balanced equations: $(COOH)_2 + H_2O \rightleftarrows H_3O^+ + COOCOOH^-$ $K_{a1} = 5.9 \times 10^{-2}$

                      $COOCOOH^- + H_2O \rightleftarrows H_3O^+ + (COO)_2^{2-}$ $K_{a2} = 6.4 \times 10^{-5}$

Let x = $[(COOH)_2]_{ionized}$.   Then $[(COOH)_2] = (0.045 - x)\ M$

                            $[H_3O^+] = [COOCOOH^-] = x\ M$

$$K_1 = \frac{[H_3O^+][COOCOOH^-]}{[(COOH)_2]} = \frac{x^2}{(0.045 - x)} = 5.9 \times 10^{-2} \approx \frac{x^2}{0.045} \qquad \text{Solving, } x = 0.052$$

Since 0.052 is more than 100% of 0.045, the simplifying assumptions do not hold and we must solve the original quadratic equation: $x^2 + (5.9 \times 10^{-2})x - 2.7 \times 10^{-3} = 0$

$$x = \frac{-5.9 \times 10^{-2} \pm \sqrt{(5.9 \times 10^{-2})^2 - 4(1)(-2.7 \times 10^{-3})}}{2(1)} = \frac{-5.9 \times 10^{-2} \pm 0.12}{2} = 0.030 \text{ or } -0.090 \text{(discard)}$$

Therefore, $[(COOH)_2] = 0.045 - x = 0.045 - 0.030 = 0.015\ M$

             $[H_3O^+] = [COOCOOH^-] = x = 0.030\ M$

             pH = $-\log (0.030) = $ **1.52**

Note: $[H_3O^+]$ from the second ionization is equal to the value of $K_{a2} = 6.4 \times 10^{-5}$. Therefore, the $[H_3O^+]$ furnished by the second ionization is negligible compared to that from the first ionization and was ignored.

18-64. Refer to Section 18-6.

(a) Solvolysis is the reaction of a substance with the solvent in which it is dissolved. Common solvents used include $H_2O(\ell)$, $NH_3(\ell)$, $H_2SO_4(\ell)$ and $CH_3COOH(\ell)$. There are many others. For example, glacial acetic acid, $CH_3COOH(\ell)$, is commonly used in non-aqueous titrations with weak acids:

$$\underset{\text{solute}}{C_6H_5NH_3^+} + \underset{\text{solvent}}{CH_3COOH} \rightleftarrows CH_3COOH_2^+ + \underset{\text{aniline}}{C_6H_5NH_2}$$

(b) Hydrolysis is the reaction of a substance with the solvent, water, or its ions, OH^- and H_3O^+, e.g., the hydrolysis of the ion, CH_3COO^- to give a basic solution:

$$\underset{\text{solute}}{CH_3COO^-} + \underset{\text{solvent}}{H_2O} \rightleftarrows CH_3COOH + OH^-$$

18-66. Refer to Section 18-7.

Some cations in aqueous solution, such as Na^+, K^+, and Ba^{2+}, undergo no real significant reaction with the water molecules in a solution. This is because their acid strengths are much less than the acid strength of water. In other words, these cations are such weak acids that they do not react with water. Because of this, when these cations are dissolved in water, the pH is unchanged.

18-68. Refer to Sections 18-7, 18-8, 18-9 and 18-10.

The pH of aqueous salt solutions depends on whether or not the ions produced by the dissociation of the salt will hydrolyze (react with water). If the cation of the salt hydrolyzes more than the anion, the solution is acidic. If the anion hydrolyzes more than the cation, the solution is basic. If the cation and the anion hydrolyze to the same extent or if neither hydrolyzes appreciably, the resulting solution is neutral.

Salts produced by a strong base reacting with a strong acid are

> (d) $BaSO_4$ ($Ba(OH)_2$ with H_2SO_4) and
> (e) $NaClO_3$ ($NaOH$ with $HClO_3$).

Note: in dilute solution, H_2SO_4 is completely ionized and is considered to be a strong diprotic acid.

18-72. *Refer to Section 18-8.*

The solution of a salt derived from a strong base and weak acid is basic because the anion of a weak acid reacts with water (hydrolysis) to form hydroxide ions. Consider the soluble salt $NaClO$ prepared by reacting $NaOH$, a strong base, and $HClO$, a weak acid. The salt dissociates completely in water and the conjugate base of the weak acid, ClO^-, hydrolyzes, producing OH^- ions.

$$NaClO(s) \xrightarrow[100\%]{H_2O} Na^+(aq) + ClO^-(aq)$$

$$ClO^-(aq) + H_2O(\ell) \rightleftarrows HClO(aq) + OH^-(aq)$$

18-74. *Refer to Section 18-8, Example 18-19 and Appendix F.*

Balanced equation: $N_3^-(aq) + H_2O(\ell) \rightleftarrows HN_3(aq) + OH^-(aq)$ (K_a for hydrazoic acid, $HN_3 = 1.9 \times 10^{-5}$)

$$K = K_b = \frac{[HN_3][OH^-]}{[N_3^-]} = \frac{K_w}{K_{a(HN_3)}} = \frac{1.0 \times 10^{-14}}{1.9 \times 10^{-5}} = \mathbf{5.3 \times 10^{-10}}$$

18-76. *Refer to Section 18-8, Example 18-19 and Appendix F.*

(a) for NO_2^-, $K_b = \dfrac{K_w}{K_{a(HNO_2)}} = \dfrac{1.0 \times 10^{-14}}{4.5 \times 10^{-4}} = \mathbf{2.2 \times 10^{-11}}$

(b) for BrO^-, $K_b = \dfrac{K_w}{K_{a(HBrO)}} = \dfrac{1.0 \times 10^{-14}}{2.5 \times 10^{-9}} = \mathbf{4.0 \times 10^{-6}}$

(c) for $HCOO^-$, $K_b = \dfrac{K_w}{K_{a(HCOOH)}} = \dfrac{1.0 \times 10^{-14}}{1.8 \times 10^{-4}} = \mathbf{5.6 \times 10^{-11}}$

The mathematical relationship between K_a, the ionization constant for a weak acid, and K_b, the base hydrolysis constant for the anion of the weak acid is $K_w = K_a \times K_b$.

The weaker the acid, the smaller is its K_a, the more its anion will hydrolyze, and the larger is K_b for the anion.

18-78. *Refer to Section 18-8, Example 18-20, and Appendix F.*

(a) Balanced equations: $NaCH_3COO \rightarrow Na^+ + CH_3COO^-$ (to completion)
                         $CH_3COO^- + H_2O \rightleftarrows CH_3COOH + OH^-$ (reversible)

Let $x = [CH_3COO^-]_{hydrolyzed}$ Then, $1.2 - x = [CH_3COO^-]$; $x = [CH_3COOH] = [OH^-]$

$$K_b = \frac{K_w}{K_a} = \frac{1.0 \times 10^{-14}}{1.8 \times 10^{-5}} = 5.6 \times 10^{-10} = \frac{[CH_3COOH][OH^-]}{[CH_3COO^-]} = \frac{x^2}{1.2 - x} \approx \frac{x^2}{1.2}$$ Solving, $x = 2.6 \times 10^{-5}$

$[OH^-] = 2.6 \times 10^{-5}$ M; $pOH = 4.59$; $pH = 14 - 4.59 = \mathbf{9.41}$

(b) Balanced equations: $KOBr \rightarrow K^+ + OBr^-$ (to completion)

                          $OBr^- + H_2O \rightleftarrows HOBr + OH^-$ (reversible)

Let $x = [OBr^-]_{hydrolyzed}$ Then, $1.2 - x = [OBr^-]$; $x = [HOBr] = [OH^-]$

$$K_b = \frac{K_w}{K_a} = \frac{1.0 \times 10^{-14}}{2.5 \times 10^{-9}} = 4.0 \times 10^{-6} = \frac{[HOBr][OH^-]}{[OBr^-]} = \frac{x^2}{1.2 - x} \approx \frac{x^2}{1.2}$$ Solving, $x = 2.2 \times 10^{-3}$

$[OH^-] = 2.2 \times 10^{-3}$ M; pOH = 2.66; pH = 14 - 2.66 = **11.34**

(c) Balanced equations: $LiCN \rightarrow Li^+ + CN^-$ (to completion)

                          $CN^- + H_2O \rightleftarrows HCN + OH^-$ (reversible)

Let $x = [CN^-]_{hydrolyzed}$ Then, $1.2 - x = [CN^-]$; $x = [HCN] = [OH^-]$

$$K_b = \frac{K_w}{K_a} = \frac{1.0 \times 10^{-14}}{4.0 \times 10^{-10}} = 2.5 \times 10^{-5} = \frac{[HCN][OH^-]}{[CN^-]} = \frac{x^2}{1.2 - x} \approx \frac{x^2}{1.2}$$ Solving, $x = 5.5 \times 10^{-3}$

$[OH^-] = 5.5 \times 10^{-3}$ M; pOH = 2.26; pH = 14 - 2.26 = **11.74**

18-80.   *Refer to Section 18-8, Example 18-20, and Appendix F.*

(a) Balanced equations: $KOI \rightarrow K^+ + OI^-$ (to completion)

                          $OI^- + H_2O \rightleftarrows HOI + OH^-$ (reversible)

Let $x = [OI^-]_{hydrolyzed}$ Then, $0.15 - x = [OI^-]$; $x = [HOI] = [OH^-]$

$$K_b = \frac{K_w}{K_a} = \frac{1.0 \times 10^{-14}}{2.3 \times 10^{-11}} = 4.3 \times 10^{-4} = \frac{[HOI][OH^-]}{[OI^-]} = \frac{x^2}{0.15 - x} \approx \frac{x^2}{0.15}$$ Solving, $x = 8.1 \times 10^{-3}$

In this case, 8.1×10^{-3} is greater than 5% of 0.15 (= 7.5×10^{-3}). A simplifying assumption cannot be made and the original quadratic equation must be solved: $x^2 + (4.3 \times 10^{-4})x - 6.4 \times 10^{-5} = 0$.

$$x = \frac{-(4.3 \times 10^{-4}) \pm \sqrt{(4.3 \times 10^{-4})^2 - 4(1)(-6.4 \times 10^{-5})}}{2(1)} = \frac{-4.3 \times 10^{-4} \pm 0.016}{2}$$

$$= 7.8 \times 10^{-3} \quad \text{or} \quad -8.2 \times 10^{-3} \text{ (discard)}$$

Therefore, $[OH^-] = 7.8 \times 10^{-3}$ M; pOH = 2.11; pH = 14 - 2.11 = **11.89**

(b) Balanced equations: $KF \rightarrow K^+ + F^-$ (to completion)

                          $F^- + H_2O \rightleftarrows HF + OH^-$ (reversible)

Let $x = [F^-]_{hydrolyzed}$ Then, $0.15 - x = [F^-]$; $x = [HF] = [OH^-]$

$$K_b = \frac{K_w}{K_a} = \frac{1.0 \times 10^{-14}}{7.2 \times 10^{-4}} = 1.4 \times 10^{-11} = \frac{[HF][OH^-]}{[F^-]} = \frac{x^2}{0.15 - x} \approx \frac{x^2}{0.15}$$ Solving, $x = 1.4 \times 10^{-6}$

In this case, 1.4×10^{-6} is much less than 5% of 0.15 (= 7.5×10^{-3}). The simplifying assumption can be made.

$[OH^-] = 1.4 \times 10^{-6}$ M; pOH = 5.85; pH = 14 - 5.85 = **8.15**

18-82.   *Refer to Section 18-9.*

NH_4Cl ammonium chloride

C_5H_5NHBr pyridinium bromide

$CH_3NH_3NO_3$ methylammonium nitrate

$[(CH_3)_3NH]_2SO_4$ trimethylammonium sulfate

(a) for $(CH_3)_2NH_2^+$, $\quad K_a = \dfrac{K_w}{K_{b((CH_3)_2NH)}} = \dfrac{1.0 \times 10^{-14}}{7.4 \times 10^{-4}} = \mathbf{1.4 \times 10^{-11}}$

(b) for $C_5H_5NH^+$, $\quad K_a = \dfrac{K_w}{K_{b(C_5H_5N)}} = \dfrac{1.0 \times 10^{-14}}{1.5 \times 10^{-9}} = \mathbf{6.7 \times 10^{-6}}$

(c) for $(CH_3)_3NH^+$, $\quad K_a = \dfrac{K_w}{K_{b((CH_3)_3N)}} = \dfrac{1.0 \times 10^{-14}}{7.4 \times 10^{-5}} = \mathbf{1.4 \times 10^{-10}}$

18-86. *Refer to Section 18-9, Example 18-21 and Appendix G.*

(a) Balanced equations: $\quad NH_4NO_3 \rightarrow NH_4^+ + NO_3^-$ $\qquad$ (to completion)

$\qquad\qquad\qquad\qquad\quad NH_4^+ + H_2O \rightleftarrows NH_3 + H_3O^+$ $\qquad$ (reversible)

Let x = $[NH_4^+]_{hydrolyzed}$ $\qquad$ Then, $0.26 - x = [NH_4^+]$; x = $[NH_3] = [H_3O^+]$

$K_a = \dfrac{K_w}{K_{b(NH_3)}} = \dfrac{1.0 \times 10^{-14}}{1.8 \times 10^{-5}} = 5.6 \times 10^{-10} = \dfrac{[NH_3][H_3O^+]}{[NH_4^+]} = \dfrac{x^2}{0.26 - x} \approx \dfrac{x^2}{0.26}$

Solving, x = 1.2×10^{-5} $\qquad$ Therefore, $[H_3O^+] = 1.2 \times 10^{-5}$ *M*; pH = **4.92**

(b) Balanced equations: $\quad CH_3NH_3NO_3 \rightarrow CH_3NH_3^+ + NO_3^-$ $\qquad$ (to completion)

$\qquad\qquad\qquad\qquad\quad CH_3NH_3^+ + H_2O \rightleftarrows CH_3NH_2 + H_3O^+$ $\qquad$ (reversible)

Let x = $[CH_3NH_3^+]_{hydrolyzed}$ $\qquad$ Then, $0.26 - x = [CH_3NH_3^+]$; x = $[CH_3NH_2] = [H_3O^+]$

$K_a = \dfrac{K_w}{K_{b(CH_3NH_2)}} = \dfrac{1.0 \times 10^{-14}}{5.0 \times 10^{-4}} = 2.0 \times 10^{-11} = \dfrac{[CH_3NH_2][H_3O^+]}{[CH_3NH_3^+]} = \dfrac{x^2}{0.26 - x} \approx \dfrac{x^2}{0.26}$

Solving, x = 2.3×10^{-6} $\qquad$ Therefore, $[H_3O^+] = 2.3 \times 10^{-6}$ *M*; pH = **5.64**

(c) Balanced equations: $\quad C_6H_5NH_3NO_3 \rightarrow C_6H_5NH_3^+ + NO_3^-$ $\qquad$ (to completion)

$\qquad\qquad\qquad\qquad\quad C_6H_5NH_3^+ + H_2O \rightleftarrows C_6H_5NH_2 + H_3O^+$ $\qquad$ (reversible)

Let x = $[C_6H_5NH_3^+]_{hydrolyzed}$ $\qquad$ Then, $0.26 - x = [C_6H_5NH_3^+]$; x = $[C_6H_5NH_2] = [H_3O^+]$

$K_a = \dfrac{K_w}{K_{b(C_6H_5NH_2)}} = \dfrac{1.0 \times 10^{-14}}{4.2 \times 10^{-10}} = 2.4 \times 10^{-5} = \dfrac{[C_6H_5NH_2][H_3O^+]}{[C_6H_5NH_3^+]} = \dfrac{x^2}{0.26 - x} \approx \dfrac{x^2}{0.26}$

Solving, x = 2.5×10^{-3} $\qquad$ Therefore, $[H_3O^+] = 2.5 \times 10^{-3}$ *M*; pH = **2.60**

18-88. *Refer to Section 18-10 and Appendices F and G.*

(a) Salt of a weak acid and a weak base for which $K_a = K_b$ gives a neutral solution, e.g. ammonium acetate, NH_4CH_3COO. $K_{a(CH_3COOH)} = K_{b(NH_3)} = 1.8 \times 10^{-5}$.

(b) Salts of a weak acid and weak base for which $K_a > K_b$ gives an acidic solution, e.g., pyridinium fluoride, C_5H_5NHF. $K_{a(HF)} = 7.2 \times 10^{-4}$; $K_{b(C_5H_5N)} = 1.5 \times 10^{-9}$.

(c) A salts of a weak acid and weak base for which $K_b > K_a$ gives a basic solution, e.g., methylammonium cyanide, CH_3NH_3CN. $K_{a(HCN)} = 4.0 \times 10^{-10}$; $K_{b(CH_3NH_2)} = 5.0 \times 10^{-4}$.

18-90. *Refer to Section 18-11 and Table 18-10.*

The cations that will reaction with water to form H^+ (or H_3O^+) ions are

(a) $[Be(OH_2)_4]^{2+} + H_2O \rightleftarrows [Be(OH)(OH_2)_3]^+ + H_3O^+$

(b) $[Al(OH_2)_6]^{3+} + H_2O \rightleftarrows [Al(OH)(OH_2)_5]^{2+} + H_3O^+$

(c) $[Fe(OH_2)_6]^{3+} + H_2O \rightleftarrows [Fe(OH)(OH_2)_5]^{2+} + H_3O^+$

(d) $[Cu(OH_2)_6]^{2+} + H_2O \rightleftarrows [Cu(OH)(OH_2)_5]^+ + H_3O^+$

18-92. *Refer to Section 18-11, Example 18-23 and Table 18-10.*

(a) Balanced equation: $[Al(OH_2)_6]^{3+} + H_2O \rightleftarrows [Al(OH)(OH_2)_5]^{2+} + H_3O^+ \quad K_a = 1.2 \times 10^{-5}$
Assume that the aluminum salt totally dissociated and all the Al^{3+} became $[Al(OH_2)_6]^{3+}$.

Let $x = [[Al(OH_2)_6]^{3+}]_{hydrolyzed}$ $\qquad$ Then, $0.15 - x = [[Al(OH_2)_6]^{3+}]$; $x = [[Al(OH)(OH_2)_5]^{2+}] = [H_3O^+]$

$K_a = \dfrac{[[Al(OH)(OH_2)_5]^{2+}][H_3O^+]}{[[Al(OH_2)_6]^{3+}]} = \dfrac{x^2}{0.15 - x} = 1.2 \times 10^{-5} \approx \dfrac{x^2}{0.15}$ $\qquad$ Solving, $x = 1.3 \times 10^{-3}$

Therefore, $[H_3O^+] = 1.3 \times 10^{-3}$ M; pH = **2.89**

% hydrolysis $= \dfrac{[Al(OH_2)_6]^{3+}_{hydrolyzed}}{[Al(OH_2)_6]^{3+}_{initial}} \times 100\% = \dfrac{1.3 \times 10^{-3} M}{0.15 M} \times 100\% = \textbf{0.87\%}$

(b) Balanced equation: $[Co(OH_2)_6]^{2+} + H_2O \rightleftarrows [Co(OH)(OH_2)_5]^+ + H_3O^+ \quad K_a = 5.0 \times 10^{-10}$
Assume that the cobalt(II) salt totally dissociated and all the Co^{2+} became $[Co(OH_2)_6]^{2+}$.

Let $x = [[Co(OH_2)_6]^{2+}]_{hydrolyzed}$ $\qquad$ Then, $0.075 - x = [[Co(OH_2)_6]^{2+}]$; $x = [[Co(OH)(OH_2)_5]^+] = [H_3O^+]$

$K_a = \dfrac{[[Co(OH)(OH_2)_5]^+][H_3O^+]}{[[Co(OH_2)_6]^{2+}]} = \dfrac{x^2}{0.075 - x} = 5.0 \times 10^{-10} \approx \dfrac{x^2}{0.075}$ $\qquad$ Solving, $x = 6.1 \times 10^{-6}$

Therefore, $[H_3O^+] = 6.1 \times 10^{-6}$ M; pH = **5.21**

% hydrolysis $= \dfrac{[[Co(OH_2)_6]^{2+}]_{hydrolyzed}}{[[Co(OH_2)_6]^{2+}]_{initial}} \times 100\% = \dfrac{6.1 \times 10^{-6} M}{0.075 M} \times 100\% = \textbf{8.2} \times \textbf{10}^{-3}\ \%$

(c) Balanced equation: $[Mg(OH_2)_6]^{2+} + H_2O \rightleftarrows [Mg(OH)(OH_2)_5]^+ + H_3O^+ \quad K_a = 3.0 \times 10^{-12}$
Assume that the magnesium salt totally dissociated and all the Mg^{2+} became $[Mg(OH_2)_6]^{2+}$.

Let $x = [[Mg(OH_2)_6]^{2+}]_{hydrolyzed}$ $\qquad$ Then, $0.15 - x = [[Mg(OH_2)_6]^{2+}]$; $x = [[Mg(OH)(OH_2)_5]^+] = [H_3O^+]$

$K_a = \dfrac{[[Mg(OH)(OH_2)_5]^+][H_3O^+]}{[[Mg(OH_2)_6]^{2+}]} = \dfrac{x^2}{0.15 - x} = 3.0 \times 10^{-12} \approx \dfrac{x^2}{0.15}$ $\qquad$ Solving, $x = 6.7 \times 10^{-7}$

Therefore, $[H_3O^+] = 6.7 \times 10^{-7}$ M; pH = **6.17** (ignoring the H_3O^+ produced by the ionization of water)

% hydrolysis $= \dfrac{[[Mg(OH_2)_6]^{2+}]_{hydrolyzed}}{[[Mg(OH_2)_6]^{2+}]_{initial}} \times 100\% = \dfrac{6.7 \times 10^{-7} M}{0.15 M} \times 100\% = \textbf{4.5} \times \textbf{10}^{-4}\ \%$

Note: To calculate the actual $[H_3O^+]$, let $x = [OH^-] = [H_3O^+]$ produced by the ionization of water.

Therefore $[H_3O^+]_{total}$ = $[H_3O^+]$ produced by hydrolysis + $[H_3O^+]$ produced by the ionization of water
$= 6.7 \times 10^{-7} + x$

We know that $K_w = 1.0 \times 10^{-14} = [H_3O^+][OH^-] = (6.7 \times 10^{-7} + x)(x) = (6.7 \times 10^{-7})x + x^2$

Solving the quadratic equation: $x^2 + (6.7 \times 10^{-7})x - 1.0 \times 10^{-14} = 0$, we have $x = 1.5 \times 10^{-8}$

Therefore, $[H_3O^+]_{total} = 6.7 \times 10^{-7} + x = 6.9 \times 10^{-7}$ M; pH = **6.16**

18-94. *Refer to Section 18-4 and Example 18-10.*

Balanced equation: $HA + H_2O \rightleftarrows H_3O^+ + A^-$ $\qquad\qquad$ $pK_a = 5.35$; $K_a = 4.5 \times 10^{-6}$

Let x = mol/L of HA that reacts. Then
x = mol/L of H_3O^+ produced = mol/L A^- produced.

	HA	+	H₂O	⇌	H₃O⁺	+	A⁻

	HA	H₂O	H₃O⁺	A⁻
initial	0.100 M		≈ 0 M	0 M
change	- x M		+ x M	+ x M
at equilibrium	(0.100 - x) M		x M	x M

$$K_a = \frac{[H_3O^+][A^-]}{[HA]} = \frac{x^2}{0.100 - x} = 4.5 \times 10^{-6}$$

Assume that 0.100 - x ≈ 0.100. Then $x^2/0.100 = 4.5 \times 10^{-6}$ and x = 6.7×10^{-4}. The simplifying assumption is justified since 6.7×10^{-4} is less than 5% of 0.100.

Therefore at equilibrium: [A⁻] = [H₃O⁺] = **6.7×10^{-4} M**

18-96. *Refer to Sections 18-7, 18-8, 18-9, and 18-11.*

(a) (NH₄)HSO₄   acidic (salt of weak base and a strong acid   (Section 18-9)

(b) (NH₄)₂SO₄   acidic (salt of a weak base and strong acid)   (Section 18-9)

(c) LiCl   neutral (salt of a strong base and strong acid)   (Section 18-7)

(d) LiBrO   basic (salt of a strong base and weak acid)   (Section 18-8)

(e) AlCl₃   acidic (salt of a small, highly charged cation)   (Section 18-11)

18-98. *Refer to Section 18-11.*

If a cation reacts appreciably with water, its acid strength must be greater than that of water. The pH of the solution will be less than 7.

18-100. *Refer to Sections 18-7, 18-8 and 18-9.*

Balanced equations:

(a) Na₂CO₃ → 2Na⁺ + CO₃²⁻
     CO₃²⁻ + H₂O ⇌ HCO₃⁻ + OH⁻ $K_{b1} = 2.1 \times 10^{-4}$
     HCO₃⁻ + H₂O ⇌ H₂CO₃ + OH⁻ $K_{b2} = 2.4 \times 10^{-8}$

(b) Na₂SO₄ → 2Na⁺ + SO₄²⁻
     SO₄²⁻ + H₂O ⇌ HSO₄⁻ + OH⁻ $K_{b1} = 8.3 \times 10^{-13}$
     HSO₄⁻ + H₂O ⇌ H₂SO₄ + OH⁻ K_{b2} = very small

(c) (NH₄)₂SO₄ → 2NH₄⁺ + SO₄²⁻
     NH₄⁺ + H₂O ⇌ NH₃ + H₃O⁺ $K_a = 5.6 \times 10^{-10}$
     SO₄²⁻ + H₂O ⇌ HSO₄⁻ + OH⁻ $K_{b1} = 8.3 \times 10^{-13}$
     HSO₄⁻ + H₂O ⇌ H₂SO₄ + OH⁻ K_{b2} = very small

(d) Na₃PO₄ → 3Na⁺ + PO₄³⁻
     PO₄³⁻ + H₂O ⇌ HPO₄²⁻ + OH⁻ $K_{b1} = 2.8 \times 10^{-2}$
     HPO₄²⁻ + H₂O ⇌ H₂PO₄⁻ + OH⁻ $K_{b2} = 1.6 \times 10^{-7}$
     H₂PO₄⁻ + H₂O ⇌ H₃PO₄ + OH⁻ $K_{b3} = 1.3 \times 10^{-12}$

(NH₄)₂SO₄ definitely could not be used in cleaning materials since it produces an acidic solution, not a basic solution. Also, Na₂SO₄ cannot be used either since SO₄²⁻ is an extremely weak base (has a very small K_b).

18-102. *Refer to Sections 18-9 and 18-10.*

(a) NH_4Br and NH_4NO_3 are both salts derived from monoprotic strong acids and the weak base, NH_3. Since the concentration of NH_4^+ is the same in each solution, the pH values will be identical.

(b) NH_4ClO_4 is a salt derived from the monoprotic strong acid, $HClO_4$, and the weak base, NH_3. NH_4F is the salt derived from the monoprotic weak acid, HF, and the weak base, NH_3. The concentration of NH_4F (0.010 *M*) is less than the concentration of NH_4ClO_4 (0.015 *M*). After the salts dissociate, F^- is a weak base whereas ClO_4^- is too weak a base to react with water. Therefore, the solution of NH_4ClO_4 will have a lower pH (be more acidic) than the solution of NH_4F for two reasons: higher weak acid concentration (NH_4^+), and essentially no base present.

(c) The only difference between these two solutions of NH_4Cl is their concentration. Even though the 0.010 *M* solution of NH_4^+ will hydrolyze to a greater extent, the 0.050 *M* solution of NH_4Cl will be more acidic, because it has a higher concentration of the weak acid, NH_4^+.

18-104. *Refer to Sections 10-7, 18-4 and 18-9.*

(a) NH_3 is a stronger base than PH_3 because it can accept a proton more easily. Nitrogen is more electronegative than phosphorus and so the NH_4^+ ion is more stable than the PH_4^+ ion.

(b) F^- is a stronger base than Br^- since the conjugate acid of F^-, HF, is weaker than the conjugate acid of Br^-, HBr.

(c) ClO_2^- is a stronger base than ClO_3^- since the conjugate acid of ClO_2^-, $HClO_2$, is a weaker acid than the conjugate acid of ClO_3^-, $HClO_3$.

(d) PO_4^{3-} is a stronger base than HPO_4^{2-} since the conjugate acid of PO_4^{3-}, HPO_4^{2-}, is a weaker acid than the conjugate acid of HPO_4^{2-}, $H_2PO_4^-$.

18-106. *Refer to Sections 18-4 and 4-2, and Figure 4-2.*

Dilute aqueous solutions of weak acids, such as HF and HNO_2, contain relatively few ions because they are weak electrolytes, ionize only slightly into their ions and are therefore, poor conductors of electricity. This can be demonstrated using a conductivity apparatus as shown in Chapter 4.

18-108. *Refer to Section 18-4, Example 18-14 and Table 18-6.*

(a) Recall that for a series of weak bases, as K_b increases, $[OH^-]$ increases, pOH decreases and pH increases.
   i.   highest pH - dimethylamine   ii. lowest pH - aniline
   iii. highest pOH - aniline   iv. lowest pOH - dimethylamine

(b) Consider the dissociation of a weak base: $B + H_2O \rightleftarrows BH^+ + OH^-$. As K_b increases, $[BH^+]$ increases.
   i.   highest $[BH^+]$ - dimethylamine   ii. lowest $[BH^+]$ - aniline

18-110. *Refer to Section 18-4 and Example 18-10.*

Balanced equation: $C_5H_7O_4COOH + H_2O \rightleftarrows H_3O^+ + C_5H_7O_4COO^-$ $K_a = 7.9 \times 10^{-5}$

Let x = $[C_5H_7O_4COOH]$ that ionizes. Then x = $[H_3O^+]$ produced = $[C_5H_7O_4COO^-]$ produced.

$$K_a = \frac{[H_3O^+][C_5H_7O_4COO^-]}{[C_5H_7O_4COOH]} = \frac{x^2}{0.110 - x} = 7.9 \times 10^{-5} \approx \frac{x^2}{0.110}$$ Solving, x = 2.9×10^{-3}

The simplifying assumption is justified since 2.9×10^{-3} is less than 5% of 0.110 (= 5.5×10^{-3}).

Therefore, $[H_3O^+] =$ **2.9×10^{-3} *M*; pH = 2.54**

Balanced equation: $C_2H_5OCOOH + H_2O \rightleftarrows H_3O^+ + C_2H_5OCOO^-$ $\qquad\qquad$ $K_a = 8.4 \times 10^{-4}$

Let   x = $[C_2H_5OCOOH]$ that ionizes. Then x = $[H_3O^+]$ produced = $[C_2H_5OCOO^-]$ produced.

$$K_a = \frac{[H_3O^+][C_2H_5OCOO^-]}{[C_2H_5OCOOH]} = \frac{x^2}{0.110 - x} = 8.4 \times 10^{-4} \approx \frac{x^2}{0.110} \qquad\qquad \text{Solving, x} = 9.6 \times 10^{-3}$$

However, 9.6×10^{-3} is more than 5% of 0.110. A simplifying assumption *cannot* be made; we must solve the original quadratic equation: $x^2 + (8.4 \times 10^{-4})x - 9.2 \times 10^{-5} = 0$

$$x = \frac{-(8.4 \times 10^{-4}) \pm \sqrt{(8.4 \times 10^{-4})^2 - 4(1)(-9.2 \times 10^{-5})}}{2(1)} = \frac{-8.4 \times 10^{-4} \pm 1.9 \times 10^{-2}}{2}$$

$$= 9.1 \times 10^{-3} \text{ or } -9.9 \times 10^{-3} \text{ (discard)}$$

Therefore, $[H_3O^+] = 9.1 \times 10^{-3}$ *M*; pH = **2.04**

Using the van't Hoff equation, $\quad \ln \dfrac{K_{w\ 323\ K}}{K_{w\ 273\ K}} = \dfrac{\Delta H°(T_2 - T_1)}{RT_1T_2}$

$$\ln \frac{5.47 \times 10^{-14}}{1.14 \times 10^{-15}} = \frac{\Delta H°(323\ K - 273\ K)}{(8.314\ J/K)(323\ K)(273\ K)}$$

$$3.87 = (\Delta H°)(6.8 \times 10^{-5})$$

$$\Delta H° = \textbf{+5.7} \times \textbf{10}^{\textbf{4}}\ \textbf{J/mol} \text{ or } \textbf{+57 kJ/mol}$$

Therefore, the reaction, $H_2O(\ell) + H_2O(\ell) \rightleftarrows H_3O^+(aq) + OH^-(aq)$, is endothermic with $\Delta H° = $ **+57 kJ/mol**.

19 Ionic Equilibria II: Buffers and Titration Curves

19-2. *Refer to Sections 18-2, 18-3 and 18-4, and the Key Terms for Chapter 18.*

(a) pK_a is the negative logarithm of K_a, the ionization constant, for a weak acid.

Consider hydrofluoric acid, HF, reacting with water: $HF + H_2O \rightleftarrows H_3O^+ + F^-$

$K_a = = \dfrac{[H_3O^+][F^-]}{[HF]} = 7.2 \times 10^{-4}$, so pK_a for HF is $-\log(7.2 \times 10^{-4}) = 3.14$.

(b) K_b is the ionization constant for a weak base.

Consider ammonia, NH_3, reacting with water: $NH_3 + H_2O \rightleftarrows NH_4^+ + OH^-$

$K_b = \dfrac{[NH_4^+][OH^-]}{[NH_3]} = 1.8 \times 10^{-5}$

(c) pK_w is the negative logarithm of K_w, the ion product for water.

Water ionizes to a slight extent: $H_2O + H_2O \rightleftarrows H_3O^+ + OH^-$

$K_w = [H_3O^+][OH^-] = 1.0 \times 10^{-14}$ at 25°C, so pK_w at 25°C = 14.00.

(d) pOH is the negative logarithm of the molar concentration of the hydroxide ion in solution.

If $[OH^-] = 2.0 \times 10^{-3}$ M, then $pOH = -\log(2.0 \times 10^{-3}) = 2.70$.

19-4. *Refer to Sections 18-7 and 18-10.*

(a) $NaNO_3$ $HNO_3 + NaOH \rightarrow NaNO_3 + H_2O$ neutral (Section 18-7)

(b) Na_2S $H_2S + 2NaOH \rightarrow Na_2S + 2H_2O$ basic (Section 18-8)

(c) $Al_2(SO_4)_3$ $3H_2SO_4 + 2Al(OH)_3 \rightarrow Al_2(SO_4)_3 + 6H_2O$ acidic due to Al^{3+} (Section 18-11)

(d) $Mg(CH_3COO)_2$ $2CH_3COOH + Mg(OH)_2 \rightarrow Mg(CH_3COO)_2 + 2H_2O$ basic (Section 18-8)

(e) $(NH_4)_2SO_4$ $H_2SO_4 + 2NH_3 \rightarrow (NH_4)_2SO_4$ acidic (Section 18-9)

19-6. *Refer to Sections 18-7 and 18-10.*

A neutral salt solution is produced by an acid-base reaction in two cases:

(1) a strong acid reacting with a strong base, and

(2) a weak acid reacting with a weak base, where the K_a for the acid is the same value as the K_b for the base.

19-8. *Refer to Section 19-1.*

Buffered solutions are produced by mixing together solutions of a weak acid and its soluble, ionic salt or a weak base and its soluble, ionic salt in approximately the same concentrations. The concentration of one can be no more than ten times the concentration of the other.

(a) 1.0 M HCH_3CH_2COO and 0.20 M $NaCH_3CH_2COO$ is a **buffered** solution because it is composed of a weak acid (HCH_3CH_2COO) and its salt ($NaCH_3CH_2COO$) in appropriate concentrations.

(b) 0.10 M NaOCl and 0.10 M HOCl is a **buffered** solution because it is composed of a weak acid (HOCl) and its salt (NaOCl) in appropriate concentrations.

(c) 0.10 M NH$_4$Cl and 0.90 M NH$_4$Br is **not** a buffered solution because it is composed of two salts of the weak base, NH$_3$, with no NH$_3$ present.

(d) 0.10 M NaCl and 0.20 M HF is **not** a buffered solution because it is composed of a weak acid (HF) and a salt that does not contain F$^-$ ion.

19-10. *Refer to Section 19-1, Examples 19-1 and 19-2, and Appendix F.*

(a) Balanced equations: $KF \rightarrow K^+ + F^-$ (to completion)

$HF + H_2O \rightleftarrows H_3O^+ + F^-$ (reversible) $K_a = 7.2 \times 10^{-4}$

Since KF dissociates completely, [F$^-$] from the salt = [KF]$_{initial}$ = 0.20 M

Let   x = [HF] that ionizes. Then
     x = [H$_3$O$^+$] produced from HF
     x = [F$^-$] produced from HF.

	HF	+	H$_2$O	$\rightleftarrows$	H$_3$O$^+$	+	F$^-$
initial	0.15 M				≈ 0 M		0.20 M
change	- x M				+ x M		+ x M
at equilibrium	(0.15 - x) M				x M		(0.20 + x) M

$$K_a = \frac{[H_3O^+][F^-]}{[HF]} = \frac{(x)(0.20 + x)}{(0.15 - x)} = 7.2 \times 10^{-4} \approx \frac{x(0.20)}{(0.15)}$$ Solving, x = 5.4 $\times 10^{-4}$

Therefore, [H$_3$O$^+$] = 5.4 $\times 10^{-4}$ M; pH = **3.27**

Alternatively, this problem can be solved using the Henderson-Hasselbalch equation:

$$pH = pK_a + \log \frac{[\text{conjugate base}]}{[\text{acid}]}$$

Substituting, $pH = 3.14 + \log \frac{(0.20)}{(0.15)} = \textbf{3.27}$

(b) Since Ba(CH$_3$COO)$_2$ dissociates totally, [CH$_3$COO$^-$] from the salt = 2 x [Ba(CH$_3$COO)$_2$]$_{initial}$ = 0.050 M

Let   x = [CH$_3$COOH] that ionizes. Then
     x = [H$_3$O$^+$] produced from CH$_3$COOH
     x = [CH$_3$COO$^-$] produced from CH$_3$COOH.

	CH$_3$COOH	+	H$_2$O	$\rightleftarrows$	H$_3$O$^+$	+	CH$_3$COO$^-$
initial	0.040 M				≈ 0 M		0.050 M
change	- x M				+ x M		+ x M
at equilibrium	(0.040 - x) M				x M		(0.050 + x) M

$$K_a = \frac{[H_3O^+][CH_3COO^-]}{[CH_3COOH]} = \frac{(x)(0.050 + x)}{(0.040 - x)} = 1.8 \times 10^{-5} \approx \frac{x(0.050)}{(0.040)}$$ Solving, x = 1.4 $\times 10^{-5}$

Therefore, [H$_3$O$^+$] = 1.4 $\times 10^{-5}$ M; pH = **4.84**

Alternatively, this problem can be solved using the Henderson-Hasselbalch equation:

$$pH = pK_a + \log \frac{[\text{conjugate base}]}{[\text{acid}]}$$

Substituting, $pH = 4.74 + \log \frac{(0.050)}{(0.040)} = \textbf{4.84}$

19-12. *Refer to Section 19-1, Example 19-3 and Appendix G.*

Balanced equations: $NH_4NO_3 \rightarrow NH_4^+ + NO_3^-$ (to completion)

$NH_3 + H_2O \rightleftharpoons NH_4^+ + OH^-$ (reversible) $K_b = 1.8 \times 10^{-5}$

(a) Since NH_4NO_3 is a soluble salt, $[NH_4^+]$ from the salt = $[NH_4NO_3]_{initial}$ = 0.15 M

Let $x = [NH_3]$ that ionizes. Then $x = [NH_4^+]$ produced from NH_3 and $x = [OH^-]$ produced from NH_3.

	NH_3	+	H_2O	$\rightleftharpoons$	NH_4^+	+	OH^-
initial	0.25 M				0.15 M		$\approx 0\ M$
change	- x M				+ x M		+ x M
at equilibrium	(0.25 - x) M				(0.15 + x) M		x M

$K_b = \dfrac{[NH_4^+][OH^-]}{[NH_3]} = \dfrac{(0.15 + x)(x)}{(0.25 - x)} = 1.8 \times 10^{-5} \approx \dfrac{(0.15)(x)}{(0.25)}$ Solving, x = 3.0 x 10^{-5}

Therefore, $[OH^-] = $ **3.0 x 10^{-5} M**; pOH = 4.52; pH = **9.48**

Alternatively, using the Henderson-Hasselbalch equation,

$$pOH = pK_b + \log \frac{[\text{conjugate acid}]}{[\text{base}]} = -\log(1.8 \times 10^{-5}) + \log \frac{(0.25)}{(0.15)} = 4.57;\ pH = \textbf{9.48}$$

(b) Since $(NH_4)_2SO_4$ is a soluble salt, $[NH_4^+]$ from the salt = 2 x $[(NH_4)_2SO_4]_{initial}$ = 0.20 M

Using the Henderson-Hasselbalch equation,

$$pOH = pK_b + \log \frac{[\text{conjugate acid}]}{[\text{base}]} = -\log(1.8 \times 10^{-5}) + \log \frac{(0.20)}{(0.15)} = 4.87$$

$[OH^-]$ = antilogarithm (−4.87) = **1.4 x 10^{-5} M**

pH = 14 - pOH = **9.13**

19-14. *Refer to Section 19-1.*

(a) Balanced equations: $NaHCO_3 \xrightarrow{H_2O} Na^+ + HCO_3^-$ (to completion)

$H_2CO_3 + H_2O \rightleftharpoons H_3O^+ + HCO_3^-$ (reversible)

When a small amount of base is added to the buffer: $H_2CO_3 + OH^- \rightarrow HCO_3^- + H_2O$

When a small amount of acid is added to the buffer: $HCO_3^- + H_3O^+ \rightarrow H_2CO_3 + H_2O$

(b) Balanced equations: $NaH_2PO_2 \xrightarrow{H_2O} Na^+ + H_2PO_4^-$ (to completion)

$Na_2HPO_4 \xrightarrow{H_2O} 2Na^+ + HPO_4^{2-}$ (to completion)

$H_2PO_4^- + H_2O \rightleftharpoons H_3O^+ + HPO_4^{2-}$ (reversible)

When a small amount of base is added to the buffer: $H_2PO_4^- + OH^- \rightarrow HPO_4^{2-} + H_2O$

When a small amount of acid is added to the buffer: $HPO_4^{2-} + H_3O^+ \rightarrow H_2PO_4^- + H_2O$

19-16. *Refer to Section 19-1.*

(a) Given: Solution A: 0.50 mol KCH_3COO + 0.25 mol CH_3COOH in 1.00 liter solution

Solution B: 0.25 mol KCH_3COO + 0.50 mol CH_3COOH in 1.00 liter solution

Solution B is more acidic and will have the lower pH.

(b) Buffer solutions are solutions of conjugate acid-base pairs, as defined by Brønsted-Lowry theory. In this case, the acid is CH_3COOH and the conjugate base, CH_3COO^-, is provided by the salt. Fundamentally, the more acid the buffer solution has relative to the base, the more acidic the solution is overall. Solution A has half the amount of acid as conjugate base, whereas Solution B has twice the amount of acid as conjugate base. So, Solution B is more acidic and has the lower pH.

19-18. *Refer to Sections 19-1 and 19-3, and Example 19-6.*

Balanced equations: $NaCH_3COO \rightarrow Na^+ + CH_3COO^-$ (to completion)

$\quad\quad\quad\quad\quad\quad\quad\quad CH_3COOH + H_2O \rightleftarrows H_3O^+ + CH_3COO^-$ (reversible) $\quad\quad K_a = 1.8 \times 10^{-5}$

Since pH = 4.75, $[H_3O^+] = 1.8 \times 10^{-5}$ M

Method 1: Using the equilibrium equations

Therefore, 1.8×10^{-5} $M = [CH_3COOH]_{ionized} = [H_3O^+]_{produced\ from\ CH_3COOH}$

$\quad\quad\quad\quad\quad\quad\quad\quad\quad\quad\quad\quad = [CH_3COO^-]_{produced\ from\ CH_3COOH}.$

Let x = $[NaCH_3COO] = [CH_3COO^-]_{initial}$

	CH_3COOH	+	H_2O	$\rightleftarrows$	H_3O^+	+	CH_3COO^-
initial	0.250 M				≈ 0 M		x M
change	- 1.8×10^{-5} M				+ 1.8×10^{-5} M		+ 1.8×10^{-5} M
at equilibrium	0.250 M				1.8×10^{-5} M		(x + 1.8×10^{-5}) M
	(to 3 sig. figs.)						

$K_a = \dfrac{[H_3O^+][CH_3COO^-]}{[CH_3COOH]} = \dfrac{(1.8 \times 10^{-5})(x + 1.8 \times 10^{-5})}{(0.250)} = 1.8 \times 10^{-5} \approx \dfrac{(1.8 \times 10^{-5})(x)}{(0.250)}$ $\quad$ Solving, x = 0.25

Therefore, $[NaCH_3COO] = 0.25$ M

? g $NaCH_3COO$ = 1.00 L x $\dfrac{0.25\ mol\ NaCH_3COO}{1.00\ L}$ x $\dfrac{82.0\ g\ NaCH_3COO}{1\ mol\ NaCH_3COO}$ = **20. g $NaCH_3COO$**

Method 2: Using Henderson-Hasselbalch equation

pH = pK_a + log $\dfrac{[conjugate\ base]}{[acid]}$ $\quad\quad\quad\quad$ Substituting, 4.75 = 4.74 + log $\dfrac{(x)}{(0.250)}$

$\quad\quad\quad\quad\quad\quad\quad\quad\quad\quad\quad\quad\quad\quad\quad\quad\quad\quad\quad\quad 0.01$ = log x - log 0.250

$\quad\quad\quad\quad\quad\quad\quad\quad\quad\quad\quad\quad\quad\quad\quad\quad\quad\quad\quad\quad$ log x = 0.01 + (−0.60)

$\quad\quad\quad\quad\quad\quad\quad\quad\quad\quad\quad\quad\quad\quad\quad\quad\quad\quad\quad\quad$ log x = −0.60

$\quad\quad\quad\quad\quad\quad\quad\quad\quad\quad\quad\quad\quad\quad\quad\quad\quad$ taking the inverse log,

$\quad\quad\quad\quad\quad\quad\quad\quad\quad\quad\quad\quad\quad\quad\quad\quad\quad\quad\quad\quad\quad$ x = 0.25

Therefore, $[NaCH_3COO] = 0.25$ M (to 2 significant figures)

? g $NaCH_3COO$ = 1.00 L x $\dfrac{0.25\ mol\ NaCH_3COO}{1.00\ L}$ x $\dfrac{82.0\ g\ NaCH_3COO}{1\ mol\ NaCH_3COO}$ = **20. g $NaCH_3COO$**

19-20. *Refer to Section 19-2.*

Balanced equations: $NaHCOO \rightarrow Na^+ + HCOO^-$ (to completion)

$\quad\quad\quad\quad\quad\quad\quad\quad HCOOH + H_2O \rightleftarrows H_3O^+ + HCOO^-$ (reversible)

When the soluble salt sodium formate ($NaHCOO$) is added to a formic acid solution, the salt undergoes complete dissociation in water to produce the common ion, $HCOO^-$. The original equilibrium involving the weak acid shifts to the left. As a result, the fraction of $HCOOH$ molecules that undergo ionization in aqueous solution will be **less**.

19-22. *Refer to Sections 18-3 and 19-1.*

Balanced equation: $HClO_4 + H_2O \rightarrow H_3O^+ + ClO_4^-$

$[H_3O^+] = [HClO_4] = 0.15$ M; pH = −log (0.15) = **0.82**

Perchloric acid, $HClO_4$, is a strong acid and dissociates completely into its ions, even in the presence of a supplier of common ion, $KClO_4$. A solution of 0.15 M $HClO_4$ and 0.20 M $KClO_4$ is *not* a buffer. There is no species present that could react with any added acid. A buffer must be a weak acid and its salt (its conjugate base) or a weak base and its salt (its conjugate acid).

From Section 19-1, we learned that if the concentrations of the weak acid or base and its salt are ≈ 0.05 M or greater, and the salt contains a univalent cation, then

for a weak acid buffer: $\quad [H_3O^+] = \dfrac{[acid]}{[salt]} \times K_a = \dfrac{mol\ acid}{mol\ salt} \times K_a$

for a weak base buffer: $\quad [OH^-] = \dfrac{[base]}{[salt]} \times K_b = \dfrac{mol\ base}{mol\ salt} \times K_b$

Original NH_3/NH_4^+ buffer: (K_b for $NH_3 = 1.8 \times 10^{-5}$)

$$[NH_4Cl] = \frac{12.78\ g\ NH_4Cl/L}{53.49\ g/mol} = 0.2389\ M$$

$$[OH^-] = \frac{[NH_3]}{[NH_4Cl]} \times K_b = \frac{0.400\ M}{0.2389\ M} \times (1.8 \times 10^{-5}) = 3.0 \times 10^{-5}\ M;$$

$$[H_3O^+] = \frac{K_w}{[OH^-]} = 3.3 \times 10^{-10}\ M$$

$$pH = 9.48$$

New NH_3/NH_4^+ buffer: When 0.142 mol per liter of HCl is added to the original buffer presented in (a), it reacts with the base component of the buffer, NH_3, to form more of the acid component, NH_4^+ (the conjugate acid of NH_3). Since HCl is in the gaseous phase, there is no total volume change. A new buffer solution is created with a slightly more acidic pH. In this type of problem, always perform the acid-base limiting reactant problem first, then the equilibrium calculation.

	HCl	+	NH$_3$	$\rightarrow$	NH$_4$Cl
initial	0.142 mol		0.400 mol		0.239 mol
change	- 0.142 mol		- 0.142 mol		+ 0.142 mol
after reaction	0 mol		0.258 mol		0.381 mol

$$[OH^-] = \frac{mol\ NH_3}{mol\ NH_4Cl} \times K_b = \frac{0.258\ mol}{0.381\ mol} \times (1.8 \times 10^{-5}) = 1.2 \times 10^{-5}$$

$$[H_3O^+] = \frac{K_w}{[OH^-]} = 8.3 \times 10^{-10} \ \text{and the new pH} = 9.08$$

The change in pH = final pH - initial pH = 9.08 - 9.48 = −0.40; **the pH decreases by 0.40 units**.

19-26. *Refer to Sections 19-1, 19-2 and 19-3, Example 19-4, Exercise 19-24 Solution, and Appendix G.*

(a) Original NH_3/NH_4^+ buffer: (K_b for $NH_3 = 1.8 \times 10^{-5}$)

$$[OH^-] = \frac{[NH_3]}{[NH_4Cl]} \times K_b = \frac{1.00\ M}{0.80\ M} \times (1.8 \times 10^{-5}) = 2.2 \times 10^{-5}\ M$$

$$[H_3O^+] = \frac{K_w}{[OH^-]} = 4.5 \times 10^{-10}\ M$$

$$pH = \textbf{9.34}$$

(b) New NH_3/NH_4^+ buffer: When 0.10 mol per liter of HCl is added to the original buffer presented in (a), it reacts with the base component of the buffer, NH_3, to form more of the acid component, NH_4^+ (the conjugate acid of NH_3). A new buffer solution is created with a slightly more acidic pH. In this type of problem, always perform the acid-base limiting reactant problem first, then the equilibrium calculation.

(1)		HCl	+	NH₃	→	NH₄Cl

(1)	HCl	NH₃	NH₄Cl
initial	0.10 mol	1.00 mol	0.80 mol
change	- 0.10 mol	- 0.10 mol	+ 0.10 mol
after reaction	0 mol	0.90 mol	0.90 mol

(2) $[OH^-] = \dfrac{mol\ NH_3}{mol\ NH_4Cl} \times K_b = \dfrac{0.90\ mol}{0.90\ mol} \times (1.8 \times 10^{-5}) = 1.8 \times 10^{-5}$

$[H_3O^+] = \dfrac{K_w}{[OH^-]} = 5.6 \times 10^{-10}$

pH = **9.26**

(c) This is a simple strong acid/strong base neutralization problem.

Plan: (1)   Find the concentration and the number of moles of NaOH from the pH of the solution.

(2)   Perform the limiting reactant testing for the acid-base reaction.

(3)   Determine the pH of the final solution.

(1) Since pH = 9.34, pOH = 14.00 - 9.34 = 4.66; $[OH^-] = 2.19 \times 10^{-5}\ M$

Therefore, 1.00 L would contain 2.19×10^{-5} mol NaOH

(2)		NaOH	+	HCl	→	NaCl

(2)	NaOH	HCl	NaCl
initial	2.19×10^{-5} mol	0.10 mol	0 mol
change	$- 2.19 \times 10^{-5}$ mol	$- 2.19 \times 10^{-5}$ mol	$+ 2.19 \times 10^{-5}$ mol
after reaction	0 mol	0.10 mol	2.19×10^{-5} mol

(3) The number of moles of HCl is essentially unaffected by the presence of 2.19×10^{-5} moles of NaOH.

Therefore, $[H_3O^+] = [HCl] = 0.10$ mol/L; pH = **1.00**

19-28.   *Refer to Section 19-1 and Exercise 19-24 Solution.*

Balanced equations: $NaBrCH_2COO \rightarrow Na^+ + BrCH_2COO^-$ (to completion)

$BrCH_2COOH + H_2O \rightleftarrows H_3O^+ + BrCH_2COO^-$ (reversible) $K_a = 2.0 \times 10^{-3}$

Since pH = 3.50, $[H_3O^+] = 3.2 \times 10^{-4}\ M$

Let $x = [BrCH_2COOH]$.   Then
$0.30 - x = [NaBrCH_2COO]$.

For a weak acid buffer:

$$[H_3O^+] = \frac{[BrCH_2COOH]}{[NaBrCH_2COO]} \times K_a$$

$$3.2 \times 10^{-4} = \left(\frac{x}{0.30 - x}\right)(2.0 \times 10^{-3})$$

$$9.5 \times 10^{-5} - (3.2 \times 10^{-4})x = (2.0 \times 10^{-3})x$$

$$9.5 \times 10^{-5} = (2.3 \times 10^{-3})x$$

$$x = 4.1 \times 10^{-2}$$

Therefore, $[BrCH_2COOH] = x = $ **$4.1 \times 10^{-2}\ M$**
$[NaBrCH_2COO] = 0.30 - x = $ **$0.26\ M$**

Balanced equation: $C_2H_5NH_2 + H_2O \rightleftarrows C_2H_5NH_3^+ + OH^-$ $\qquad$ $K_b = 4.7 \times 10^{-4}$

This is an example of the common ion effect; the common ion in this case is OH^-; $[OH^-]_{initial} = 0.0040\ M$

Let $\qquad$ x $= [C_2H_5NH_2]$ that ionizes. Then
$\qquad\qquad$ x $= [OH^-]$ produced from $C_2H_5NH_2$ and
$\qquad\qquad$ x $= [C_2H_5NH_3^+]$ produced from $C_2H_5NH_2$.

	$C_2H_5NH_2$	+	H_2O	$\rightleftarrows$	$C_2H_5NH_3^+$	+	OH^-
initial	0.012 M				0 M		0.0040 M
change	- x M				+ x M		+ x M
at equilibrium	(0.012 - x) M				x M		(0.0040 + x) M

$$K_b = \frac{[C_2H_5NH_3^+][OH^-]}{[C_2H_5NH_2]} = \frac{(x)(0.0040 + x)}{(0.012 - x)} = 4.7 \times 10^{-4} \approx \frac{x(0.0040)}{0.012} \qquad \text{Solving, } x = 1.4 \times 10^{-3}$$

However, x has the same order of magnitude as 0.0040, so the simplifying assumption does not hold. We must solve the original quadratic equation: $x^2 + 0.0045x - 5.6 \times 10^{-6} = 0$

$$x = \frac{-0.0045 \pm \sqrt{(0.0045)^2 - 4(1)(-5.6 \times 10^{-6})}}{2(1)} = \frac{-0.0045 \pm 0.0065}{2} = 0.0010 \text{ or } -0.0055 \text{ (discard)}$$

Therefore, $\qquad [C_2H_5NH_3^+] = x = \mathbf{0.0010\ M}$

Balanced equations: $\quad NaCH_3CH_2COO + H_2O \rightarrow Na^+ + CH_3CH_2COO^- \qquad$ (to completion)
$\qquad\qquad\qquad\qquad CH_3CH_2COOH + H_2O \rightleftarrows H_3O^+ + CH_3CH_2COO^- \qquad$ (reversible) $\qquad K_a = 1.3 \times 10^{-5}$

In this buffer system: $\qquad [H_3O^+] = 7.9 \times 10^{-6}\ M \quad$ since pH = 5.10
$\qquad\qquad\qquad\qquad\qquad [NaCH_3CH_2COO] = 0.60\ M$

$$[\text{acid}] = [CH_3CH_2COOH] = \frac{[H_3O^+][\text{salt}]}{K_a} = \frac{(7.9 \times 10^{-6})(0.60)}{(1.3 \times 10^{-5})} = \mathbf{0.36\ M} \qquad \text{since } K_a = \frac{[H_3O^+][\text{salt}]}{[\text{acid}]}$$

Plan: $\quad$ (1) Perform the acid-base neutralization limiting reactant problem.
$\qquad\quad$ (2) Determine the volumes of acetic acid and sodium hydroxide that must be mixed without adding
$\qquad\qquad\quad$ additional water by substituting into the modified K_a expression.

(1) Let $\quad V_A$ = volume (in liters) of 0.150 M CH_3COOH $\qquad\qquad V_A \times 0.150\ M$ = initial moles of CH_3COOH
$\qquad\qquad V_B$ = volume (in liters) of 0.065 M NaOH $\qquad\qquad\quad V_B \times 0.065\ M$ = initial moles of NaOH

In order to produce a buffer solution, NaOH must be consumed and is therefore the limiting reactant in the acid-base neutralization reaction.

	CH_3COOH	+	NaOH	$\rightarrow$	$NaCH_3COO$	+	H_2O
initial	0.150 $M \times V_A$ mol		0.065 $M \times V_B$ mol		0 mol		
change	- 0.065 $M \times V_B$ mol		- 0.065 $M \times V_B$ mol		+ 0.065 $M \times V_B$ mol		
after rxn	(0.150 $M \cdot V_A$ - 0.065 $M \cdot V_B$) mol		0 mol		0.065 $M \times V_B$ mol		

(2) For a weak acid buffer: pH = 4.50; $[H^+] = 3.2 \times 10^{-5}$

$$[H^+] = K_a \times \frac{\text{mol } CH_3COOH}{\text{mol } NaCH_3COO}$$

Substituting,

$$3.2 \times 10^{-5} = (1.8 \times 10^{-5}) \times \frac{0.150 \; M \cdot V_A - 0.065 \; M \cdot V_B}{0.065 \; M \times V_B}$$

$$1.8 = \frac{0.150 \; V_A}{0.065 \; V_B} - \frac{0.065 \; V_B}{0.065 \; V_B}$$

$$2.8 = \frac{0.150 \; V_A}{0.065 \; V_B}$$

$$1.2 = \frac{V_A}{V_B}$$

$$V_A = 1.2 \; V_B$$

Since $\qquad\qquad\qquad V_A + V_B = 1.00 \text{ L}$

Substituting, $\qquad\quad 1.2 \; V_B + V_B = 1.00 \text{ L}$

$$2.2 \; V_B = 1.00 \text{ L}$$

$$V_B = \text{volume of NaOH} = \textbf{0.45 L}$$

$$V_A = \text{volume of } CH_3COOH = 1.00 \text{ L} - 0.45 \text{ L} = \textbf{0.55 L}$$

19-36. *Refer to Section 19-3, Example 19-5 and Appendix F.*

Balanced equations: $Ca(CH_3COO)_2 \rightarrow Ca^{2+} + 2CH_3COO^-$ (to completion)

$\qquad\qquad\qquad\qquad CH_3COOH + H_2O \rightleftarrows H_3O^+ + CH_3COO^-$ (reversible) $\qquad K_a = 1.8 \times 10^{-5}$

Recall: for dilution, $M_1V_1 = M_2V_2$. In this instance, the acetic acid solution and the calcium acetate solution are diluting each other.

$$[CH_3COOH]_{initial} = M_2 = \frac{M_1V_1}{V_2} = \frac{1.25 \; M \times 500. \text{ mL}}{500. \text{ mL} + 500. \text{ mL}} = 0.625 \; M$$

$$[Ca(CH_3COO)_2]_{initial} = M_2 = \frac{M_1V_1}{V_2} = \frac{0.300 \; M \times 500. \text{ mL}}{500. \text{ mL} + 500. \text{ mL}} = 0.150 \; M$$

Therefore,

$$[CH_3COO^-]_{initial} = 2 \times [Ca(CH_3COO)_2]_{initial} = 2 \times 0.150 \; M = 0.300 \; M$$

Let x = $[CH_3COOH]$ that ionizes. Then
$\qquad$ x = $[H_3O^+]$ produced from CH_3COOH = $[CH_3COO^-]$ produced from CH_3COOH.

	CH_3COOH	+	H_2O	$\rightleftarrows$	H_3O^+	+	CH_3COO^-
initial	0.625 M				$\approx 0 \; M$		0.300 M
change	- x M				+ x M		+ x M
at equilibrium	(0.625 - x) M				x M		(0.300 + x) M

$$K_a = \frac{[H_3O^+][CH_3COO^-]}{[CH_3COOH]} = \frac{(x)(0.300 + x)}{(0.625 - x)} = 1.8 \times 10^{-5} \approx \frac{x(0.300)}{(0.625)} \qquad \text{Solving, x} = 3.8 \times 10^{-5}$$

The simplifying assumption is justified and we have

(a) $[CH_3COOH] = \textbf{0.625} \; \textbf{\textit{M}}$

(b) $[Ca^{2+}] = [Ca(CH_3COO)_2]_{initial} = \textbf{0.150} \; \textbf{\textit{M}}$

(c) $[CH_3COO^-] = \textbf{0.300} \; \textbf{\textit{M}}$

(d) $[H^+] = \textbf{3.8} \times \textbf{10}^{-5} \; \textbf{\textit{M}}$

(e) pH = $-\log (3.8 \times 10^{-5})$ = **4.42**

Balanced equations: $NaClCH_2COO \rightarrow Na^+ + ClCH_2COO^-$ (to completion)

$ClCH_2COOH + H_2O \rightleftharpoons H_3O^+ + ClCH_2COO^-$ (reversible) $K_a = 1.4 \times 10^{-3}$

$[ClCH_2COO^-]_{initial} = [NaClCH_2COO] = 0.015\ M$

Since pH = 2.75, $[H_3O^+] = 1.8 \times 10^{-3}\ M$

Therefore, $1.8 \times 10^{-3}\ M = [ClCH_2COOH]_{ionized} = [H_3O^+]_{produced\ from\ ClCH_2COOH}$
$= [ClCH_2COO^-]_{produced\ from\ ClCH_2COOH}$

Let x = $[ClCH_2COOH]_{initial}$

	$ClCH_2COOH$	+	H_2O	$\rightleftharpoons$	H_3O^+	+	$ClCH_2COO^-$
initial	x M				$\approx 0\ M$		$0.015\ M$
change	- $1.8 \times 10^{-3}\ M$				+ $1.8 \times 10^{-3}\ M$		+ $1.8 \times 10^{-3}\ M$
at equilibrium	(x - 1.8×10^{-3}) M				$1.8 \times 10^{-3}\ M$		$0.017\ M$

$K_a = \dfrac{[H_3O^+][ClCH_2COO^-]}{[ClCH_2COOH]} = \dfrac{(1.8 \times 10^{-3})(0.017)}{(x - 1.8 \times 10^{-3})} = 1.4 \times 10^{-3}$ Solving, x = 0.024

Therefore, $[ClCH_2COOH] = \mathbf{0.024\ M}$

(a) Acid-base indicators are organic compounds which behave as weak acids or bases and exhibit different colors in solutions with different acidities.

(b) The essential characteristic of an acid-base indicator is that the conjugate acid-base pair must exhibit different colors. Consider the weak acid indicator, HIn. In solution, HIn dissociates slightly as follows:

$$HIn + H_2O \rightleftharpoons H_3O^+ + In^-$$
acid   conjugate base

HIn dominates in more acidic solutions with one characteristic color; In⁻ dominates in more basic solutions with another color.

(c) The color of an acid-base indicator in an aqueous solution depends upon the ratio, [In⁻]/[HIn], which in turn depends upon [H⁺] and the K_a value of the indicator. A general rule of thumb: If [In⁻]/[HIn] < 0.1, then the indicator will show its true acid color. If [In⁻]/[HIn] > 10, then the indicator will show its true base color.

Balanced equation: $HIn + H_2O \rightleftharpoons H_3O^+ + In^-$, where metacresol purple is represented by HIn

At pH 8.2, [HIn] = [In⁻]; $[H_3O^+]$ = antilog (−pH) = $6 \times 10^{-9}\ M$

Substituting into the K_a expression, we have: $K_a = \dfrac{[H_3O^+][In^-]}{[HIn]} = \mathbf{6 \times 10^{-9}}$

$pK_a = \mathbf{8.2}$

Note: $[H^+] = K_a$ and pH = pK_a at the endpoint, since [In⁻] and [HIn] are equal and cancel out each other.

When a solution is colorless with phenolphthalein, pH < ~8. When a solution is blue in bromthymol blue, pH > ~7.6. When a solution is yellow in methyl orange, pH > ~4.4. Therefore, we know that this solution has a pH between ~7.6 and ~8.

Balanced equation: $CH_3COOH + NaOH \rightarrow NaCH_3COO + H_2O$

Method 1: To calculate the concentration of the salt solution: assume you have 1.00 L of 0.020 M acetic acid, so you have 1.00 L x 0.020 M = 0.020 mol CH_3COOH. Therefore, you will need to add 0.020 mol NaOH to make 0.020 mol of $NaCH_3COO$. The volume of NaOH that must be added to get to the equivalence point is

? L NaOH = 0.020 mol NaOH/0.025 M NaOH = 0.800 L   (rearrange M = mol/L to L = mol/M)

? total volume at the equivalence point = 1.00 L acetic acid + 0.80 L NaOH = 1.80 L soln

? $[NaCH_3COO]$ = 0.020 mol $NaCH_3COO$/1.80 L soln = 0.011 M $NaCH_3COO$

Method 2: The resultant solution at the equivalence point of any acid-base reaction contains only salt and water. The pH is determined from the concentration of the salt. Even when the volume of solution is given, it is not necessary to use that information. We can calculate the concentration of a salt derived from a monoprotic acid and a base with one OH group by:

$$[salt] = \frac{M_A M_B}{M_A + M_B}$$ where M_A = molarity of the acid
M_B = molarity of the base

Derivation:

Let V_A = volume (in L) of acid with molarity M_A V_B = volume (in L) of base with molarity M_B

At the equivalence point, mol acid $(M_A V_A)$ = mol base $(M_B V_B)$ = mol salt produced

$$[salt] = \frac{mol\ acid}{total\ volume} = \frac{M_A V_A}{V_A + V_B} = \frac{M_A}{\left(1 + \frac{V_B}{V_A}\right)} = \frac{M_A}{\left(1 + \frac{M_A}{M_B}\right)} = \frac{M_A M_B}{M_A + M_B}$$ since $\frac{V_B}{V_A} = \frac{M_A}{M_B}$ in this case

Plan: (1) Calculate the concentration of $NaCH_3COO$.
(2) Determine the pH.

(1) $[NaCH_3COO]$ = 0.011 M (see Method 1 above)

(2) The anion of the soluble salt hydrolyzes to form a basic solution: $CH_3COO^- + H_2O \rightleftarrows CH_3COOH + OH^-$

Let x = $[CH_3COO^-]_{hydrolyzed}$ Then, 0.011 - x = $[CH_3COO^-]$; x = $[CH_3COOH]$ = $[OH^-]$

$$K_b = \frac{K_w}{K_{a(CH_3COOH)}} = \frac{1.0 \times 10^{-14}}{1.8 \times 10^{-5}} = 5.6 \times 10^{-10} = \frac{[CH_3COOH][OH^-]}{[CH_3COO^-]} = \frac{x^2}{0.011 - x} \approx \frac{x^2}{0.011}$$

Solving, x = 2.5×10^{-6}; $[OH^-]$ = 2.5×10^{-6} M; pOH = 5.60; pH = **8.40**

(**Phenolphthalein** is an appropriate indicator)

Example of a titration of a strong acid with a strong base:

(a) When no base is added, the pH of the solution is determined by the initial strong acid concentration.

(b) At the point halfway to the equivalence point, only half of the base required to titrate all of the acid has been added. The strong base is the limiting reactant and the pH of the resulting solution is less than 7. It is calculated from the concentration of the remaining acid.

(c) At the equivalence point, only water and the salt of the strong acid and strong base are present. Since neither the cation nor the anion of the salt hydrolyzes appreciably, the pH is 7.

(d) Past the equivalence point, the strong acid is the limiting reactant and the pH of the solution is greater than 7. It is determined from the concentration of the excess strong base.

The graph compares well with Figure 19-3a.

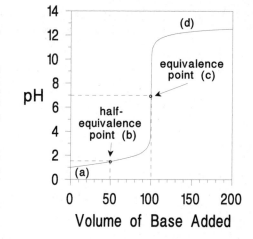

Balanced equation: $HNO_3 + NaOH \rightarrow NaNO_3 + H_2O$

A 25.00 mL sample of 0.1500 M HNO_3 is titrated with 0.100 M NaOH.

(a) Initially: $[H^+] = [HNO_3] = 0.1500$ M; pH = **0.8239**

For the rest of the exercise, the plan is straightforward: for the neutralization reaction between HNO_3 and NaOH, perform the limiting reactant problem. The pH is determined from the concentration of excess HNO_3 or NaOH. Each calculation is totally independent of the other calculations.

(b) Addition of 6.00 mL of 0.100 M NaOH:

? mmol HNO_3 = 0.1500 M x 25.00 mL = 3.750 mmol HNO_3
? mmol NaOH = 0.100 M x 6.00 mL = 0.600 mmol NaOH

Before the equivalence point, NaOH is the limiting reactant. The pH is determined from the concentration of excess HNO_3 remaining. The salt produced, $NaNO_3$, is the salt of a strong acid and a strong base. It will not affect the pH of the solution.

	HNO_3	+	NaOH	$\rightarrow$	$NaNO_3$	+	H_2O
initial	3.750 mmol		0.600 mmol		0 mmol		
change	- 0.600 mmol		- 0.600 mmol		+ 0.600 mmol		
after reaction	3.150 mmol		0 mmol		0.600 mmol		

$$[H^+] = \frac{\text{mmol excess } HNO_3}{\text{total volume (mL)}} = \frac{3.150 \text{ mmol}}{(25.00 \text{ mL} + 6.00 \text{ mL})} = 0.1016 \ M; \ \text{pH} = \mathbf{0.9931}$$

(c) Addition of 15.00 mL of 0.100 M NaOH:

? mmol HNO_3 = 0.1500 M x 25.00 mL = 3.750 mmol HNO_3
? mmol NaOH = 0.100 M x 15.00 mL = 1.50 mmol NaOH

	HNO_3	+	NaOH	$\rightarrow$	$NaNO_3$	+	H_2O
initial	3.750 mmol		1.50 mmol		0 mmol		
change	- 1.50 mmol		- 1.50 mmol		+ 1.50 mmol		
after reaction	2.25 mmol		0 mmol		1.50 mmol		

$$[H^+] = \frac{\text{mmol excess } HNO_3}{\text{total volume (mL)}} = \frac{2.25 \text{ mmol}}{(25.00 \text{ mL} + 15.00 \text{ mL})} = 0.0562 \ M; \ \text{pH} = \mathbf{1.250}$$

(d) Addition of 30.0 mL of 0.100 M NaOH:

? mmol HNO_3 = 0.1500 M x 25.00 mL = 3.750 mmol HNO_3
? mmol NaOH = 0.100 M x 30.00 mL = 3.00 mmol NaOH

	HNO_3	+	NaOH	$\rightarrow$	$NaNO_3$	+	H_2O
Initial	3.750 mmol		3.00 mmol		0 mmol		
Change	- 3.00 mmol		- 3.00 mmol		+ 3.00 mmol		
after reaction	0.75 mmol		0 mmol		3.00 mmol		

$$[H^+] = \frac{\text{mmol excess } HNO_3}{\text{total volume (mL)}} = \frac{0.75 \text{ mmol}}{(25.00 \text{ mL} + 30.00 \text{ mL})} = 0.014 \ M; \ \text{pH} = \mathbf{1.87}$$

(e) Addition of 37.44 mL of 0.100 M NaOH:

? mmol HNO_3 = 0.1500 M x 25.00 mL = 3.750 mmol HNO_3
? mmol NaOH = 0.100 M x 37.44 mL = 3.74 mmol NaOH   (use 3.744 for next calculation)

	HNO_3	+	NaOH	→	$NaNO_3$	+	H_2O
Initial	3.750 mmol		3.74 mmol		0 mmol		
change	- 3.74 mmol		- 3.74 mmol		+ 3.74 mmol		
after reaction	0.01 mmol		0 mmol		3.74 mmol		

(use 0.006 for calculation)

$[H^+] = \dfrac{\text{mmol excess } HNO_3}{\text{total volume (mL)}} = \dfrac{0.006 \text{ mmol}}{(25.00 \text{ mL} + 37.44 \text{ mL})} = 0.0001 \ M; \ pH = \textbf{4.0}$

(f) Addition of 45.00 mL of 0.100 M NaOH:

? mmol HNO_3 = 0.1500 M x 25.00 mL = 3.750 mmol HNO_3
? mmol NaOH = 0.100 M x 45.00 mL = 4.50 mmol NaOH

After the equivalence point, HNO_3 is the limiting reactant. The pH is determined from the concentration of excess NaOH.

	HNO_3	+	NaOH	→	$NaNO_3$	+	H_2O
initial	3.750 mmol		4.50 mmol		0 mmol		
change	- 3.750 mmol		- 3.750 mmol		+ 3.750 mmol		
after reaction	0 mmol		0.75 mmol		3.750 mmol		

$[OH^-] = \dfrac{\text{mmol excess NaOH}}{\text{total volume (mL)}} = \dfrac{0.75 \text{ mmol}}{(25.00 \text{ mL} + 45.00 \text{ mL})} = 0.011 \ M; \ pOH = 1.97; \ pH = \textbf{12.03}$

19-52. *Refer to Section 19-5 and Exercise 19-48 Solution.*

Balanced equation: HCl + NaOH → NaCl + H_2O

A 22.0 mL sample of 0.145 M HCl is titrated with 0.106 M NaOH.

(a) Initially: $[H^+]$ = [HCl] = 0.145 M; pH = **0.839**

For the rest of the exercise, the plan is straightforward: for the neutralization reaction between HCl and NaOH, perform the limiting reactant problem. The pH is determined from the concentration of excess HCl or NaOH.

(b) Addition of 11.10 mL of 0.106 M NaOH:

? mmol HCl = 0.145 M x 22.0 mL = 3.19 mmol HCl
? mmol NaOH = 0.106 M x 11.10 mL = 1.18 mmol NaOH

Before the equivalence point, NaOH is the limiting reactant. The pH is determined from the concentration of excess HCl remaining. The salt produced, NaCl, is the salt of a strong acid and a strong base. It will not affect the pH of the solution.

	HCl	+	NaOH	→	NaCl	+	H_2O
initial	3.19 mmol		1.18 mmol		0 mmol		
change	- 1.18 mmol		- 1.18 mmol		+ 1.18 mmol		
after reaction	2.01 mmol		0 mmol		1.18 mmol		

$[H^+] = \dfrac{\text{mmol excess HCl}}{\text{total volume (mL)}} = \dfrac{2.01 \text{ mmol}}{(22.0 \text{ mL} + 11.10 \text{ mL})} = 0.0607 \ M; \ pH = \textbf{1.217}$

(c) Addition of 24.0 mL of 0.106 M NaOH:

? mmol HCl = 0.145 M x 22.0 mL = 3.19 mmol HCl

? mmol NaOH = 0.106 M x 24.0 mL = 2.54 mmol NaOH

	HCl	+	NaOH	→	NaCl	+	H$_2$O
initial	3.19 mmol		2.54 mmol		0 mmol		
change	- 2.54 mmol		- 2.54 mmol		+ 2.54 mmol		
after reaction	0.65 mmol		0 mmol		2.54 mmol		

$$[H^+] = \frac{\text{mmol excess HCl}}{\text{total volume (mL)}} = \frac{0.65 \text{ mmol}}{(22.0 \text{ mL} + 24.0 \text{ mL})} = 0.014 \ M; \ \text{pH} = \textbf{1.85}$$

(d) Addition of 41.0 mL of 0.106 M NaOH:

? mmol HCl = 0.145 M x 22.0 mL = 3.19 mmol HCl

? mmol NaOH = 0.106 M x 41.0 mL = 4.35 mmol NaOH

After the equivalence point, HCl is the limiting reactant. The pH is determined from the concentration of excess NaOH.

	HCl	+	NaOH	→	NaCl	+	H$_2$O
initial	3.19 mmol		4.35 mmol		0 mmol		
change	- 3.19 mmol		- 3.19 mmol		+ 3.19 mmol		
after reaction	0 mmol		1.16 mmol		3.19 mmol		

$$[OH^-] = \frac{\text{mmol excess NaOH}}{\text{total volume (mL)}} = \frac{1.16 \text{ mmol}}{(22.0 \text{ mL} + 41.0 \text{ mL})} = 0.0184 \ M; \ \text{pOH} = 1.735; \ \text{pH} = \textbf{12.265}$$

(e) Addition of 54.4 mL of 0.106 M NaOH:

? mmol HCl = 0.145 M x 22.0 mL = 3.19 mmol HCl

? mmol NaOH = 0.106 M x 54.4 mL = 5.77 mmol NaOH

	HCl	+	NaOH	→	NaCl	+	H$_2$O
initial	3.19 mmol		5.77 mmol		0 mmol		
change	- 3.19 mmol		- 3.19 mmol		+ 3.19 mmol		
after reaction	0 mmol		2.58 mmol		3.19 mmol		

$$[OH^-] = \frac{\text{mmol excess NaOH}}{\text{total volume (mL)}} = \frac{2.58 \text{ mmol}}{(22.0 \text{ mL} + 54.4 \text{ mL})} = 0.0338 \ M; \ \text{pOH} = 1.471; \ \text{pH} = \textbf{12.529}$$

(f) Addition of 63.6 mL of 0.106 M NaOH:

? mmol HCl = 0.145 M x 22.0 mL = 3.19 mmol HCl

? mmol NaOH = 0.106 M x 63.6 mL = 6.74 mmol NaOH

	HNO$_3$	+	NaOH	→	NaNO$_3$	+	H$_2$O
initial	3.19 mmol		6.74 mmol		0 mmol		
change	- 3.19 mmol		- 3.19 mmol		+ 3.19 mmol		
after reaction	0 mmol		3.55 mmol		3.19 mmol		

$$[OH^-] = \frac{\text{mmol excess NaOH}}{\text{total volume (mL)}} = \frac{3.55 \text{ mmol}}{(22.0 \text{ mL} + 63.6 \text{ mL})} = 0.0415 \ M; \ \text{pOH} = 1.382; \ \text{pH} = \textbf{12.618}$$

The calculations for determining the pH at every point in the titration of 1 liter of 0.0200 M CH_3COOH with solid NaOH, assuming no volume change, can be divided into 4 types.

(1) Initially, the pH is determined by the concentration of the weak acid, CH_3COOH ($K_a = 1.8 \times 10^{-5}$). Assuming the simplifying assumption works:

$$[H^+] = \sqrt{K_a[CH_3COOH]} = \sqrt{(1.8 \times 10^{-5})(0.0200)} = 6.0 \times 10^{-4}\ M;\ pH = 3.22$$

(2) Before the equivalence point, the pH is determined by the buffer solution consisting of the unreacted CH_3COOH and $NaCH_3COO$ produced by the reaction. Each calculation is a limiting reactant problem using the original concentration of CH_3COOH. For example, at point (c) in the following table:

initial mmol CH_3COOH = 0.0200 M x 1000 mL = 20.0 mmol CH_3COOH

mmol of NaOH added = 0.00800 mol x 1000 mmol/mol = 8.00 mmol NaOH

	CH_3COOH	+	NaOH	→	$NaCH_3COO$	+	H_2O
initial	20.0 mmol		8.00 mmol		0 mmol		
change	- 8.00 mmol		- 8.00 mmol		+ 8.00 mmol		
after reaction	12.0 mmol		0 mmol		8.00 mmol		

After the reaction, we have a 1 liter buffer solution consisting of 12.0 mmol CH_3COOH and 8.00 mmol $NaCH_3COO$.

$$[H_3O^+] = K_a \times \frac{\text{mmol } CH_3COOH}{\text{mmol } NaCH_3COO} = (1.8 \times 10^{-5})\frac{12.0\ \text{mmol}}{8.00\ \text{mmol}} = 2.7 \times 10^{-5}\ M;\ pH = 4.57$$

Halfway to the equivalence point (i.e., when half of the required amount of base needed to reach the equivalence point is added), pH = pK_a. At point (d):

	CH_3COOH	+	NaOH	→	$NaCH_3COO$	+	H_2O
initial	20.0 mmol		10.00 mmol		0 mmol		
change	- 10.00 mmol		- 10.00 mmol		+ 10.00 mmol		
after reaction	10.0 mmol		0 mmol		10.00 mmol		

$$[H_3O^+] = K_a \times \frac{\text{mmol } CH_3COOH}{\text{mmol } NaCH_3COO} = (1.8 \times 10^{-5})\frac{10.0\ \text{mmol}}{10.00\ \text{mmol}} = 1.8 \times 10^{-5}\ M;\ pH = pK_a = 4.74$$

(3) At the equivalence point, there is no excess acid or base. The concentration of $NaCH_3COO$ determines the pH of the system (Refer to Exercise 19-46 Solution). At point (h):

	CH_3COOH	+	NaOH	→	$NaCH_3COO$	+	H_2O
initial	20.0 mmol		20.0 mmol		0 mmol		
change	- 20.0 mmol		- 20.0 mmol		+ 20.0 mmol		
after reaction	0 mmol		0 mmol		20.0 mmol		

$$[NaCH_3COO] = \frac{0.0200\ \text{mol}}{1.00\ \text{L}} = 0.0200\ M$$

If aqueous NaOH had been added,
$$[NaCH_3COO] = 0.0200\ \text{mol/total volume (L)}$$

Then the anion of the salt hydrolyzes to produce a basic solution: $CH_3COO^- + H_2O \rightleftarrows CH_3COOH + OH^-$

Let x = $[CH_3COO^-]_{\text{hydrolyzed}}$ Then, 0.0200 - x = $[CH_3COO^-]$; x = $[CH_3COOH]$ = $[OH^-]$

$$K_b = \frac{K_w}{K_{a(CH_3COOH)}} = 5.6 \times 10^{-10} = \frac{[CH_3COOH][OH^-]}{[CH_3COO^-]} = \frac{x^2}{0.0200 - x} \approx \frac{x^2}{0.0200} \qquad \text{Solving, } x = 3.3 \times 10^{-6}$$

$[OH^-] = x = 3.3 \times 10^{-6}\ M;\ pOH = 5.48;\ pH = 8.52$

(4) After the equivalence point, the pH is determined directly from the concentration of *excess* NaOH since CH_3COOH is now the limiting reactant. In the presence of the strong base, the effect of the weak base, CH_3COO^-, derived from the salt is negligible.

For example, at point (j):

	CH_3COOH	+	NaOH	$\rightarrow$	$NaCH_3COO$	+	H_2O
initial	20.0 mmol		24.0 mmol		0 mmol		
change	- 20.0 mmol		- 20.0 mmol		+ 20.0 mmol		
after reaction	0 mmol		4.0 mmol		20.0 mmol		

$$[OH^-] = [NaOH]_{excess} = \frac{0.0040 \text{ mol}}{1.00 \text{ L}} = 4.0 \times 10^{-3} \ M; \ pOH = 2.40; \ pH = 11.60$$

Note: If aqueous NaOH had been added, $[NaOH]_{excess}$ = 0.0040 mol/total volume (L)

Data Table:

	Mol NaOH Added	Type of Solution	$[H_3O^+]$ (M)	$[OH^-]$ (M)	pH	pOH
(a)	none	weak acid	6.0×10^{-4}	1.7×10^{-11}	3.22	10.78
(b)	0.00400	buffer	7.2×10^{-5}	1.4×10^{-10}	4.14	9.86
(c)	0.00800	buffer	2.7×10^{-5}	3.7×10^{-10}	4.57	9.43
(d)	0.01000	buffer	1.8×10^{-5}	5.6×10^{-10}	4.74 (= pK_a)	9.26
	(halfway to the equivalence point)					
(e)	0.01400	buffer	7.7×10^{-6}	1.3×10^{-9}	5.11	8.89
(f)	0.01800	buffer	2.0×10^{-6}	5.0×10^{-9}	5.70	8.30
(g)	0.01900	buffer	9.5×10^{-7}	1.1×10^{-8}	6.02	7.98
(h)	0.0200	salt	3.0×10^{-9}	3.3×10^{-6}	8.52	5.48
	(at the equivalence point)					
(i)	0.0210	strong base	1.0×10^{-11}	1.0×10^{-3}	11.00	3.0
(j)	0.0240	strong base	2.5×10^{-12}	4.0×10^{-3}	11.60	2.40
(k)	0.0300	strong base	1.0×10^{-12}	1.0×10^{-2}	12.00	2.00

Titration Curve: CH_3COOH vs. NaOH

An appropriate indicator would change color in the pH range, 7 - 10. From Table 19-4, **phenolphthalein** is the best indicator for this titration.

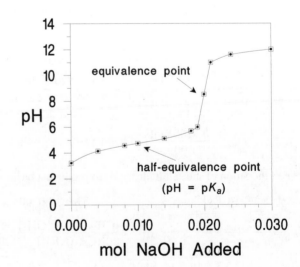

310

(a) Initially, the pH is determined by the concentration of the weak acid, CH_3COOH ($K_a = 1.8 \times 10^{-5}$). Assuming the simplifying assumption works:

$$[H^+] = \sqrt{K_a[CH_3COOH]} = \sqrt{(1.8 \times 10^{-5})(0.182)} = 1.8 \times 10^{-3} \ M; \ pH = \textbf{2.74}$$

Before the equivalence point, the pH is determined by the buffer solution consisting of the unreacted CH_3COOH and $NaCH_3COO$ produced by the reaction. Each calculation is a limiting reactant problem using the original concentration of CH_3COOH.

(b) initial mmol $CH_3COOH = 0.182 \ M \times 32.44 \ mL = 5.90 \ mmol \ CH_3COOH$
mmol of NaOH added = 0.185 $M \times 15.55 \ mL = 2.88 \ mmol \ NaOH$

	CH_3COOH	+	NaOH	→	$NaCH_3COO$	+	H_2O
initial	5.90 mmol		2.88 mmol		0 mmol		
change	- 2.88 mmol		- 2.88 mmol		+ 2.88 mmol		
after reaction	3.02 mmol		0 mmol		2.88 mmol		

After the reaction, we have a total of 47.99 mL (= 32.44 + 15.55) buffer solution containing 3.02 mmol CH_3COOH and 2.88 mmol $NaCH_3COO$.

$$[H_3O^+] = K_a \times \frac{mmol \ CH_3COOH}{mmol \ NaCH_3COO} = (1.8 \times 10^{-5}) \frac{3.02 \ mmol}{2.88 \ mmol} = 1.9 \times 10^{-5} \ M; \ pH = \textbf{4.72}$$

(c) initial mmol $CH_3COOH = 0.182 \ M \times 32.44 \ mL = 5.90 \ mmol \ CH_3COOH$
mmol of NaOH added = 0.185 $M \times 20.0 \ mL = 3.70 \ mmol \ NaOH$

	CH_3COOH	+	NaOH	→	$NaCH_3COO$	+	H_2O
initial	5.90 mmol		3.70 mmol		0 mmol		
change	- 3.70 mmol		- 3.70 mmol		+ 3.70 mmol		
after reaction	2.20 mmol		0 mmol		3.70 mmol		

After the reaction, we have a total of 52.4 mL (= 32.44 + 20.0) buffer solution containing 2.20 mmol CH_3COOH and 3.70 mmol $NaCH_3COO$.

$$[H_3O^+] = K_a \times \frac{mmol \ CH_3COOH}{mmol \ NaCH_3COO} = (1.8 \times 10^{-5}) \frac{2.20 \ mmol}{3.70 \ mmol} = 1.1 \times 10^{-5} \ M; \ pH = \textbf{4.97}$$

(d) initial mmol $CH_3COOH = 0.182 \ M \times 32.44 \ mL = 5.90 \ mmol \ CH_3COOH$
mmol of NaOH added = 0.185 $M \times 24.02 \ mL = 4.44 \ mmol \ NaOH$

	CH_3COOH	+	NaOH	→	$NaCH_3COO$	+	H_2O
initial	5.90 mmol		4.44 mmol		0 mmol		
change	- 4.44 mmol		- 4.44 mmol		+ 4.44 mmol		
after reaction	1.46 mmol		0 mmol		4.44 mmol		

After the reaction, we have a total of 56.46 mL (= 32.44 + 24.02) buffer solution containing 1.46 mmol CH_3COOH and 4.44 mmol $NaCH_3COO$.

$$[H_3O^+] = K_a \times \frac{mmol \ CH_3COOH}{mmol \ NaCH_3COO} = (1.8 \times 10^{-5}) \frac{1.46 \ mmol}{4.44 \ mmol} = 5.9 \times 10^{-6} \ M; \ pH = \textbf{5.23}$$

(e) initial mmol CH_3COOH = 0.182 M x 32.44 mL = 5.90 mmol CH_3COOH

mmol of NaOH added = 0.185 M x 27.2 mL = 5.03 mmol NaOH

	CH_3COOH	+	NaOH	→	$NaCH_3COO$	+	H_2O
initial	5.90 mmol		5.03 mmol		0 mmol		
change	- 5.03 mmol		- 5.03 mmol		+ 5.03 mmol		
after reaction	0.87 mmol		0 mmol		5.03 mmol		

After the reaction, we have a total of 59.6 mL (= 32.44 + 27.2) buffer solution containing 0.87 mmol CH_3COOH and 5.03 mmol $NaCH_3COO$.

$$[H_3O^+] = K_a \times \frac{\text{mmol } CH_3COOH}{\text{mmol } NaCH_3COO} = (1.8 \times 10^{-5}) \frac{0.87 \text{ mmol}}{5.03 \text{ mmol}} = 3.1 \times 10^{-6} \ M; \ pH = \textbf{5.51}$$

(f) initial mmol CH_3COOH = 0.182 M x 32.44 mL = 5.90 mmol CH_3COOH

mmol of NaOH added = 0.185 M x 31.91 mL = 5.90 mmol NaOH

	CH_3COOH	+	NaOH	→	$NaCH_3COO$	+	H_2O
initial	5.90 mmol		5.90 mmol		0 mmol		
change	- 5.90 mmol		- 5.90 mmol		+ 5.90 mmol		
after reaction	0 mmol		0 mmol		5.90 mmol		

After the reaction, we have a total of 64.35 mL (= 32.44 + 31.91) salt solution containing 0 mmol CH_3COOH and 5.90 mmol $NaCH_3COO$.

This is the equivalence point and there is no excess acid or base. The concentration of $NaCH_3COO$ determines the pH of the system (Refer to Exercise 19-46 Solution). At point (f):

$$[NaCH_3COO] = \frac{5.90 \text{ mmol}}{64.35 \text{ mL}} = 0.0917 \ M$$

Then the anion of the salt hydrolyzes to produce a basic solution: $CH_3COO^- + H_2O \rightleftarrows CH_3COOH + OH^-$

Let x = $[CH_3COO^-]_{hydrolyzed}$ Then, 0.0917 - x = $[CH_3COO^-]$; x = $[CH_3COOH] = [OH^-]$

$$K_b = \frac{K_w}{K_{a(CH_3COOH)}} = 5.6 \times 10^{-10} = \frac{[CH_3COOH][OH^-]}{[CH_3COO^-]} = \frac{x^2}{0.0917 - x} \approx \frac{x^2}{0.0917} \qquad \text{Solving, x} = 7.2 \times 10^{-6}$$

$[OH^-] = x = 7.2 \times 10^{-6} \ M; \ pOH = 5.14; \ pH = \textbf{8.86}$

(g) initial mmol CH_3COOH = 0.182 M x 32.44 mL = 5.90 mmol CH_3COOH

mmol of NaOH added = 0.185 M x 33.12 mL = 6.13 mmol NaOH

	CH_3COOH	+	NaOH	→	$NaCH_3COO$	+	H_2O
initial	5.90 mmol		6.13 mmol		0 mmol		
change	- 5.90 mmol		- 5.90 mmol		+ 5.90 mmol		
after reaction	0 mmol		0.23 mmol		5.90 mmol		

After the equivalence point, the pH is determined directly from the concentration of *excess* NaOH since CH_3COOH is now the limiting reactant. In the presence of the strong base, the effect of the weak base, CH_3COO^-, derived from the salt is negligible.

After the reaction, we have a total of 65.56 mL (= 32.44 + 33.12) strong base solution containing 0.23 mmol NaOH and 5.90 mmol $NaCH_3COO$.

$$[OH^-] = [NaOH]_{excess} = \frac{0.23 \text{ mmol}}{65.56 \text{ mL}} = 3.5 \times 10^{-3} \ M; \ pOH = 2.46; \ pH = \textbf{11.54}$$

Balanced equation: $CH_3COOH + KOH \rightarrow KCH_3COO + H_2O$

The resultant solution at the equivalence point of any acid-base reaction contains only salt and water. The pH is determined from the concentration of the salt.

(a) (1) initial mmol $CH_3COOH = 1.000\ M \times 100.0$ mL $= 100.0$ mmol CH_3COOH
So, at the equivalence point, mmol of KOH added = 100.0 mmol KOH, by definition

	CH_3COOH	+	KOH	$\rightarrow$	KCH_3COO	+	H_2O
initial	100.0 mmol		100.0 mmol		0 mmol		
change	- 100.0 mmol		- 100.0 mmol		+ 100.0 mmol		
after reaction	0 mmol		0 mmol		100.0 mmol		

Volume of KOH needed to reach the equivalence point $= 100.0$ mmol $\times \dfrac{1\ \text{mL}}{0.150\ \text{mmol}} = 667$ mL

Total volume at the equivalence point $= 100.0$ mL $+ 667$ mL $= 767$ mL

$[KCH_3COO] = \dfrac{100.0\ \text{mmol}}{767\ \text{mL}} = 0.130\ M$

(2) The anion of the soluble salt hydrolyzes to form a basic solution:

$$CH_3COO^- + H_2O \rightleftarrows CH_3COOH + OH^-$$

$$K_b = \frac{K_w}{K_{a(CH_3COOH)}} = \frac{1.0 \times 10^{-14}}{1.8 \times 10^{-5}} = 5.6 \times 10^{-10} = \frac{[CH_3OOH][OH^-]}{[CH_3COO^-]} = \frac{x^2}{0.130 - x} \approx \frac{x^2}{0.130}$$

Solving, $x = 8.5 \times 10^{-6}$; $[OH^-] = 8.5 \times 10^{-6}\ M$; pOH = 5.07; pH = **8.93**

(b) (1) initial mmol $CH_3COOH = 0.100\ M \times 100.0$ mL $= 10.0$ mmol CH_3COOH
So, at the equivalence point, mmol of KOH added = 10.0 mmol KOH, by definition

	CH_3COOH	+	KOH	$\rightarrow$	KCH_3COO	+	H_2O
initial	10.0 mmol		10.0 mmol		0 mmol		
change	- 10.0 mmol		- 10.0 mmol		+ 10.0 mmol		
after reaction	0 mmol		0 mmol		10.0 mmol		

Volume of KOH needed to reach the equivalence point $= 10.0$ mmol $\times \dfrac{1\ \text{mL}}{0.150\ \text{mmol}} = 66.7$ mL

Total volume at the equivalence point $= 100.0$ mL $+ 66.7$ mL $= 166.7$ mL

$[KCH_3COO] = \dfrac{10.0\ \text{mmol}}{166.7\ \text{mL}} = 0.0600\ M$

(2) $K_b = \dfrac{K_w}{K_{a(CH_3COOH)}} = \dfrac{1.0 \times 10^{-14}}{1.8 \times 10^{-5}} = 5.6 \times 10^{-10} = \dfrac{[CH_3OOH][OH^-]}{[CH_3COO^-]} = \dfrac{x^2}{0.0600 - x} \approx \dfrac{x^2}{0.0600}$

Solving, $x = 5.8 \times 10^{-6}$; $[OH^-] = 5.8 \times 10^{-6}\ M$; pOH = 5.24; pH = **8.76**

(c) (1) initial mmol $CH_3COOH = 0.0100\ M \times 100.0$ mL $= 1.00$ mmol CH_3COOH
So, at the equivalence point, mmol of KOH added = 1.00 mmol KOH, by definition

	CH_3COOH	+	KOH	$\rightarrow$	KCH_3COO	+	H_2O
initial	1.00 mmol		1.00 mmol		0 mmol		
change	- 1.00 mmol		- 1.00 mmol		+ 1.00 mmol		
after reaction	0 mmol		0 mmol		1.00 mmol		

Volume of KOH needed to reach the equivalence point $= 1.00 \text{ mmol} \times \dfrac{1 \text{ mL}}{0.150 \text{ mmol}} = 6.67 \text{ mL}$

Total volume at the equivalence point $= 100.0 \text{ mL} + 6.67 \text{ mL} = 106.7 \text{ mL}$

$[KCH_3COO] = \dfrac{1.00 \text{ mmol}}{106.7 \text{ mL}} = 0.00937 \; M$

(2) $\quad K_b = \dfrac{[CH_3OOH][OH^-]}{[CH_3COO^-]} = 5.6 \times 10^{-10} = \dfrac{x^2}{9.37 \times 10^{-3} - x} \approx \dfrac{x^2}{9.37 \times 10^{-3}}$

Solving, $x = 2.3 \times 10^{-6}$; $[OH^-] = 2.3 \times 10^{-6} \; M$; pOH $= 5.64$; pH $= \mathbf{8.36}$

19-60. *Refer to Sections 19-6 and 19-8, Figure 19-5, Table 19-4, and Exercise 19-54 Solution.*

The calculations for determining the pH in the titration of 1 liter of 0.0100 M HNO$_3$ with gaseous NH$_3$ can be divided into 4 types:

(1) Initially, the pH is determined by the concentration of the strong acid, HNO$_3$.

$[H^+] = [HNO_3] = 0.0100 \; M$; pH $= 2.00$

(2) Before the equivalence point, the pH is essentially determined by the concentration of *excess* strong acid, HNO$_3$, since NH$_3$ is the limiting reagent. In the presence of the strong acid, the effect of the weak acid, NH$_4^+$, derived from the salt is negligible. For example, consider the limiting reactant problem for point (c) in the following table:

	HNO$_3$	+	NH$_3$	→	NH$_4$NO$_3$
initial	10.0 mmol		4.00 mmol		0 mmol
change	- 4.00 mmol		- 4.00 mmol		+ 4.00 mmol
after reaction	6.0 mmol		0 mmol		4.00 mmol

$[H^+] = [HNO_3]_{excess} = \dfrac{0.0060 \text{ mol}}{1.00 \text{ L}} = 6.0 \times 10^{-3} \; M$; pH $= 2.22$

(3) At the equivalence point, there is no excess acid or base. The pH of the salt solution, NH$_4$NO$_3$, is <7 since we have a solution of the salt of a weak base and a strong acid. The concentration of NH$_4$NO$_3$ determines the pH of the system. At point (g):

	HNO$_3$	+	NH$_3$	→	NH$_4$NO$_3$
initial	10.0 mmol		10.0 mmol		0 mmol
change	- 10.0 mmol		- 10.0 mmol		+ 10.0 mmol
after reaction	0 mmol		0 mmol		10.0 mmol

$[NH_4NO_3] = \dfrac{0.0100 \text{ mol}}{1.00 \text{ L}} = 0.0100 \; M$

Then the cation of the salt hydrolyzes to produce an acidic solution:

$$NH_4^+ + H_2O \rightleftarrows NH_3 + H_3O^+$$

Let $x = [NH_4^+]_{hydrolyzed}$

Then, $0.0100 - x = [NH_4^+]$; $x = [NH_3] = [H_3O^+]$

$K_a = \dfrac{K_w}{K_{b(NH_3)}} = 5.6 \times 10^{-10} = \dfrac{[NH_3][H_3O^+]}{[NH_4^+]} = \dfrac{x^2}{0.0100 - x} \approx \dfrac{x^2}{0.0100}$

Solving, $x = 2.4 \times 10^{-6}$

$[H_3O^+] = x = 2.4 \times 10^{-6} \; M$; pH $= 5.62$

314

(4) After the equivalence point, the pH is determined by the buffer solution consisting of excess NH_3 and NH_4^+ produced by the neutralization reaction. For example, at point (i):

	HNO_3	+	NH_3	$\rightarrow$	NH_4NO_3
initial	10.0 mmol		13.0 mmol		0 mmol
change	- 10.0 mmol		- 10.0 mmol		+ 10.0 mmol
after reaction	0 mmol		3.0 mmol		10.0 mmol

After the reaction, we have a 1 liter solution of NH_3 and the soluble salt, NH_4NO_3. OH^- ion is produced by:

$$NH_3 + H_2O \rightleftarrows NH_4^+ + OH^- \qquad\qquad K_b = 1.8 \times 10^{-5}$$

$$[OH^-] = K_b \times \frac{\text{mol } NH_3}{\text{mol } NH_4NO_3} = (1.8 \times 10^{-5})\frac{3.0 \text{ mmol}}{10.0 \text{ mmol}} = 5.4 \times 10^{-6} \, M$$

Therefore, pH = 8.73

Data Table:

	Mol NH_3 Added	Type of Solution	$[H_3O^+]$ (M)	$[OH^-]$ (M)	pH	pOH
(a)	none	strong acid	1.00×10^{-2}	1.00×10^{-12}	2.00	12.00
(b)	0.00100	strong acid	9.0×10^{-3}	1.1×10^{-12}	2.05	11.95
(c)	0.00400	strong acid	6.0×10^{-3}	1.7×10^{-12}	2.22	11.78
(d)	0.00500	strong acid	5.0×10^{-3}	2.0×10^{-12}	2.30	11.70
	(halfway to the equivalence point - has no real significance in this case)					
(e)	0.00900	strong acid	1.0×10^{-3}	1.0×10^{-11}	3.00	11.00
(f)	0.00950	strong acid	5×10^{-4}	2×10^{-11}	3.3	10.7
(g)	0.0100	salt	2.4×10^{-6}	4.2×10^{-9}	5.62	8.38
	(at the equivalence point)					
(h)	0.0105	buffer	1.1×10^{-8}	9×10^{-7}	8.0	6.0
(i)	0.0130	buffer	1.9×10^{-9}	5.4×10^{-6}	8.73	5.27

Titration Curve: HNO_3 vs. NH_3

The major difference between this titration curve of the HNO_3/NH_3 reaction and the other titrations is that the system is buffered after the equivalence point, not before the equivalence point. A satisfactory indicator for this titration is methyl red.

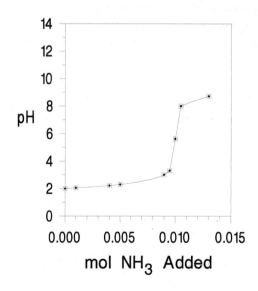

(1) NaCl is the salt of a strong acid and a strong base, so its solution has a pH = **7.00**, no matter what its concentration is.

(2) NH_4Cl is the salt of a strong acid and a weak base, so its solution has a pH < 7.

Balanced equations: $NH_4Cl \rightarrow NH_4^+ + Cl^-$ (to completion)

$NH_4^+ + H_2O \rightleftarrows NH_3 + H_3O^+$ (reversible)

Let $x = [NH_4^+]_{hydrolyzed}$ Then, $0.54 - x = [NH_4^+]$; $x = [NH_3] = [H_3O^+]$

$$K_a = \frac{K_w}{K_{b(NH_3)}} = \frac{1.0 \times 10^{-14}}{1.8 \times 10^{-5}} = 5.6 \times 10^{-10} = \frac{[NH_3][H_3O^+]}{[NH_4^+]} = \frac{x^2}{0.54 - x} \approx \frac{x^2}{0.54}$$ Solving, $x = 1.7 \times 10^{-5}$

Therefore, $[H_3O^+] = 1.7 \times 10^{-5}\ M$; pH = **4.76**

19-64. *Refer to Section 19-3.*

Buffers made of weak acids work best at the pH where the pK_a = pH. At this pH, [weak acid] = [conjugate base]. So, a buffer made of HCN and CN^- would work best at pH = $-\log(4.0 \times 10^{-10})$ = 9.40, not 8.0.

Solving for the ratio of $[CN^-]/[HCN]$ that would give a pH of 8.0, we could use the Henderson-Hasselbalch equation:

$$pH = pK_a + \log\frac{[CN^-]}{[HCN]}$$

$$8.0 = 9.4 + \log\frac{[CN^-]}{[HCN]}$$

$$\log\frac{[CN^-]}{[HCN]} = -1.4$$

$$\frac{[CN^-]}{[HCN]} = 0.040$$

Since to be an effective buffer, their concentrations must be within 0.1 - 10 of each other, one really cannot make an effective buffer using HCN at pH = 8.0.

19-66. *Refer to Section 19-4 and the Key Terms at the end of Chapter 19.*

In a titration, the equivalence point is the point at which chemically equivalent amounts of reactants have reacted, whereas the end point is the point at which an indicator changes color and a titration should be stopped. So, a chemist needs to be careful when choosing an indicator in an acid-base reaction, to be certain that the pH at which the indicator changes color is close to the pH at the equivalence point of the titration.

19-68. *Refer to Chapters 2 and 3.*

Balanced equation: $NaOH + KHSO_4 \rightarrow NaKSO_4 + H_2O$

Plan: (1) Calculate the mass of $KHSO_4$ in the mixture
         (2) Determine the % by mass of $KHSO_4$

(1) ? g $KHSO_4$ = 0.0368 L NaOH $\times \dfrac{0.115\ \text{mol NaOH}}{1\ \text{L NaOH}} \times \dfrac{1\ \text{mol } KHSO_4}{1\ \text{mol NaOH}} \times \dfrac{136.2\ \text{g } KHSO_4}{1\ \text{mol } KHSO_4} = 0.576$ g $KHSO_4$

(2) ? % $KHSO_4$ by mass = $\dfrac{0.576\ \text{g } KHSO_4}{0.738\ \text{g mixture}} \times 100 = $ **78.0%**

What we have is a solution of two weak acid, hypochlorous acid (HOCl) and hypoiodous acid (HOI):

$$HOCl + H_2O \rightleftarrows H_3O^+ + OCl^- \qquad K_a = 3.5 \times 10^{-8}$$
$$HOI + H_2O \rightleftarrows H_3O^+ + OI^- \qquad K_a = 2.3 \times 10^{-11}$$

Because the K_a for HOCl is more than 1000 times that of HOI, the pH in the solution is due only to the ionization of HOCl, following the same train of thought as for polyprotic acids.

$$K_a = \frac{[H_3O^+][OCl^-]}{[HOCl]} = \frac{x^2}{0.30 - x} = 3.5 \times 10^{-8}$$

Assume that $0.30 - x \approx 0.30$. Then $x^2/0.30 = 3.5 \times 10^{-8}$ and $x = 1.0 \times 10^{-4}$.
The simplifying assumption is justified since 1.0×10^{-4} is less than 5% of 0.30 (= 0.015).

Therefore at equilibrium: $[H_3O^+] = 1.0 \times 10^{-4}$ *M*; pH = **4.00**

Balanced equation: $HCl + NaOH \rightarrow NaCl + H_2O$

Plan: (1) Determine the concentration of the acid solution.
        (2) Calculate the necessary volume by using dilution.

(1) At the equivalence point (pH = 7 at the equivalence point of a strong acid-strong base titration):

   Therefore we can say:

$$\text{mmoles of acid} = \text{mmoles of base} \qquad \text{(when the acid has 1 acidic H and the base has 1 basic OH)}$$
$$M_{HCl} \times \text{mL of HCl} = M_{NaOH} \times \text{mL of NaOH}$$
$$M_{HCl} \times 20.00 \text{ mL} = 3.00 \ M \times 34.0 \text{ mL}$$
$$M_{HCl} = 5.10 \ M$$

(2) Solving for volume, using concepts for dilution:

$$M_1 V_1 = M_2 V_2$$
$$V_1 = \frac{M_2 V_2}{M_1}$$
$$= \frac{0.75 \ M \times 1.5 \text{ L}}{5.10 \ M}$$
$$= \textbf{0.22 L HCl}$$

20 Ionic Equilibria III: The Solubility Product Principle

20-2. *Refer to Section 20-1.*

The solubility product principle states that the solubility product expression for a slightly soluble compound is the product of the concentrations of its constituent ions, each raised to the power that corresponds to the number of ions in one formula unit of the compound. The quantity, K_{sp}, is constant at constant temperature for a saturated solution of the compound, when the system is at equilibrium. The significance of the solubility product is that it can be used to calculate the concentrations of the ions in solutions for such slightly soluble compounds.

20-4. *Refer to Section 20-1.*

The molar solubility of a compound is the number of moles of the compound that dissolve to produce one liter of saturated solution.

20-6. *Refer to Section 20-1.*

(a) $Mn_3(AsO_4)_2(s) \rightleftarrows 3Mn^{2+}(aq) + 2AsO_4^{3-}(aq)$ $K_{sp} = [Mn^{2+}]^3[AsO_4^{3-}]^2$

(b) $Hg_2I_2(s) \rightleftarrows Hg_2^{2+}(aq) + 2I^-(aq)$ $K_{sp} = [Hg_2^{2+}][I^-]^2$

(c) $AuI_3(s) \rightleftarrows Au^{3+}(aq) + 3I^-(aq)$ $K_{sp} = [Au^{3+}][I^-]^3$

(d) $SrCO_3(s) \rightleftarrows Sr^{2+}(aq) + CO_3^{2-}(aq)$ $K_{sp} = [Sr^{2+}][CO_3^{2-}]$

20-8. *Refer to Section 20-2, Examples 20-1 and 20-2, and Appendix H.*

Plan: (1) Calculate the molar solubility of the slightly soluble salt, which is the number of moles of the salt that will dissolve in 1 liter of solution.
        (2) Determine the concentrations of the ions in solution.
        (3) Substitute the ion concentrations into the K_{sp} expression to calculate K_{sp}.

(a) Balanced equation: $SrCrO_4(s) \rightleftarrows Sr^{2+}(aq) + CrO_4^{2-}(aq)$ $K_{sp} = [Sr^{2+}][CrO_4^{2-}]$

    (1) molar solubility (mol $SrCrO_4$/L) $= \dfrac{1.2 \text{ mg } SrCrO_4}{1 \text{ mL}} \times \dfrac{1000 \text{ mL}}{1 \text{ L}} \times \dfrac{1 \text{ g}}{1000 \text{ mg}} \times \dfrac{1 \text{ mol } SrCrO_4}{204 \text{ g } SrCrO_4}$

                $= 5.9 \times 10^{-3}$ mol $SrCrO_4$/L (dissolved)

    (2) $[Sr^{2+}] = [CrO_4^{2-}] =$ molar solubility $= 5.9 \times 10^{-3}$ M

    (3) $K_{sp} = [Sr^{2+}][CrO_4^{2-}] = (5.9 \times 10^{-3})^2 = $ **3.5×10^{-5}**   (3.6×10^{-5} from Appendix H)

(b) Balanced equation: $BiI_3(s) \rightleftarrows Bi^{3+}(aq) + 3I^-(aq)$ $K_{sp} = [Bi^{3+}][I^-]^3$

    (1) molar solubility $= \dfrac{7.7 \times 10^{-3} \text{ g } BiI_3}{1 \text{ L}} \times \dfrac{1 \text{ mol } BiI_3}{590 \text{ g } BiI_3} = 1.3 \times 10^{-5}$ mol BiI_3/L (dissolved)

    (2) $[Bi^{3+}] =$ molar solubility $= 1.3 \times 10^{-5}$ M
        $[I^-] = 3 \times$ molar solubility $= 3.9 \times 10^{-5}$ M

    (3) $K_{sp} = [Bi^{3+}][I^-]^3 = (1.3 \times 10^{-5})(3.9 \times 10^{-5})^3 = $ **7.7×10^{-19}**   (8.1×10^{-19} from Appendix H)

(c) Balanced equation: $Fe(OH)_2(s) \rightleftharpoons Fe^{2+}(aq) + 2OH^-(aq)$ $K_{sp} = [Fe^{2+}][OH^-]^2$

(1) molar solubility $= \dfrac{1.1 \times 10^{-3} \text{ g Fe(OH)}_2}{1 \text{ L}} \times \dfrac{1 \text{ mol Fe(OH)}_2}{89.9 \text{ g Fe(OH)}_2} = 1.2 \times 10^{-5}$ mol Fe(OH)$_2$/L (dissolved)

(2) $[Fe^{2+}]$ = molar solubility = 1.2×10^{-5} M
$[OH^-]$ = 2 x molar solubility = 2.4×10^{-5} M

(3) $K_{sp} = [Fe^{2+}][OH^-]^2 = (1.2 \times 10^{-5})(2.4 \times 10^{-5})^2 = $ **6.9×10^{-15}**   (7.9×10^{-15} from Appendix H)

(d) Balanced equation: $SnI_2(s) \rightleftharpoons Sn^{2+}(aq) + 2I^-(aq)$ $K_{sp} = [Sn^{2+}][I^-]^2$

(1) molar solubility $= \dfrac{10.9 \text{ g SnI}_2}{1 \text{ L}} \times \dfrac{1 \text{ mol SnI}_2}{372.5 \text{ g SnI}_2} = 2.93 \times 10^{-2}$ mol SnI$_2$/L (dissolved)

(2) $[Sn^{2+}]$ = molar solubility = 2.93×10^{-2} M
$[I^-]$ = 2 x molar solubility = 5.86×10^{-2} M

(3) $K_{sp} = [Sn^{2+}][I^-]^2 = (2.93 \times 10^{-2})(5.86 \times 10^{-2})^2 = $ **1.01×10^{-4}**   (1.0×10^{-4} from Appendix H)

20-10.   *Refer to Section 20-2, Table 20-1 and Exercise 20-8 Solution.*

Compound	Molar Solubility (M)	K_{sp} (calculated)
SrCrO$_4$	5.9×10^{-3}	$[Sr^{2+}][CrO_4{}^{2-}] = 3.5 \times 10^{-5}$
BiI$_3$	1.3×10^{-5}	$[Bi^{3+}][I^-]^3 = 7.7 \times 10^{-19}$
Fe(OH)$_2$	1.2×10^{-5}	$[Fe^{2+}][OH^-]^2 = 6.9 \times 10^{-15}$
SnI$_2$	2.93×10^{-2}	$[Sn^{2+}][I^-]^2 = 1.01 \times 10^{-4}$

(a) SnI$_2$ has the highest molar solubility.   (c) SnI$_2$ has the largest K_{sp}.

(b) Fe(OH)$_2$ has the lowest molar solubility.   (d) BiI$_3$ has the smallest K_{sp}.

20-12.   *Refer to Section 20-2, and Examples 20-1 and 20-2.*

Balanced equation: $CaF_2(s) \rightleftharpoons Ca^{2+}(aq) + 2F^-(aq)$ $K_{sp} = [Ca^{2+}][F^-]^2$

Plan: (1) Determine the molar solubility of CaF$_2$ at 25°C.
(2) Calculate K_{sp}.

(1) molar solubility $= \dfrac{0.0163 \text{ g CaF}_2}{1.00 \text{ L}} \times \dfrac{1 \text{ mol CaF}_2}{78.08 \text{ g CaF}_2} = 2.09 \times 10^{-4}$ mol CaF$_2$/L (dissolved)

(2) $[Ca^{2+}]$ = molar solubility = 2.09×10^{-4} M; $[F^-]$ = 2 x molar solubility = 4.18×10^{-4} M

$K_{sp} = [Ca^{2+}][F^-]^2 = (2.09 \times 10^{-4})(4.18 \times 10^{-4})^2 = $ **3.65×10^{-11}** (3.9×10^{-11} from Appendix H)

20-14.   *Refer to Section 20-3, Example 20-3 and Appendix H.*

Balanced equation: $CaCO_3(s) \rightleftharpoons Ca^{2+}(aq) + CO_3{}^{2-}(aq)$ $K_{sp} = 4.8 \times 10^{-9}$

Let x = molar solubility of CaCO$_3$. Then, $[Ca^{2+}] = [CO_3{}^{2-}] = x$

$K_{sp} = [Ca^{2+}][CO_3{}^{2-}] = x^2 = 4.8 \times 10^{-9}$ Solving, x = 6.9×10^{-5}

Therefore,   molar solubility = 6.9×10^{-5} mol CaCO$_3$/L (dissolved)

solubility (g/L) $= \dfrac{6.9 \times 10^{-5} \text{ mol CaCO}_3}{1 \text{ L}} \times \dfrac{100.09 \text{ g CaCO}_3}{1 \text{ mol CaCO}_3} = $ **6.9×10^{-3} g/L**

(a) Balanced equation: $CuCl(s) \rightleftarrows Cu^+(aq) + I^-(aq)$ $\qquad$ $K_{sp} = 1.9 \times 10^{-7}$

Let x = molar solubility of CuCl. Then, $[Cu^+] = [Cl^-] = x$

$K_{sp} = [Cu^+][Cl^-] = x^2 = 1.9 \times 10^{-7}$ $\qquad$ Solving, x = 4.4 × 10⁻⁴

Therefore, $\quad$ molar solubility = **4.4 x 10⁻⁴ mol CuCl/L** (dissolved)

$\qquad\quad [Cu^+] = $ **4.4 x 10⁻⁴ M**

$\qquad\quad [Cl^-] = $ **4.4 x 10⁻⁴ M**

$\qquad\quad$ solubility (g/L) = $\dfrac{4.4 \times 10^{-4} \text{ mol CuCl}}{1 \text{ L}} \times \dfrac{99.00 \text{ g CuCl}}{1 \text{ mol CuCl}} = $ **0.043 g/L**

(b) Balanced equation: $Ba_3(PO_4)_2(s) \rightleftarrows 3Ba^{2+}(aq) + 2PO_4^{3-}(aq)$ $\qquad$ $K_{sp} = 1.3 \times 10^{-29}$

Let x = molar solubility of $Ba_3(PO_4)_2$. Then, $[Ba^{2+}] = 3x$ and $[PO_4^{3-}] = 2x$

$K_{sp} = [Ba^{2+}]^3[PO_4^{3-}]^2 = (3x)^3(2x)^2 = 108x^5 = 1.3 \times 10^{-29}$ $\qquad$ Solving, x = 6.5 × 10⁻⁷

Therefore, $\quad$ molar solubility = **6.5 x 10⁻⁷ mol Ba₃(PO₄)₂/L** (dissolved)

$\qquad\quad [Ba^{2+}] = 3x = $ **2.0 x 10⁻⁶ M**

$\qquad\quad [PO_4^{3-}] = 2x = $ **1.3 x 10⁻⁶ M**

$\qquad\quad$ solubility (g/L) = $\dfrac{6.5 \times 10^{-7} \text{ mol Ba}_3(PO_4)_2}{1 \text{ L}} \times \dfrac{602 \text{ g Ba}_3(PO_4)_2}{1 \text{ mol Ba}_3(PO_4)_2} = $ **3.9 x 10⁻⁴ g/L**

(c) Balanced equation: $PbF_2(s) \rightleftarrows Pb^{2+}(aq) + 2F^-(aq)$ $\qquad$ $K_{sp} = 3.7 \times 10^{-8}$

Let x = molar solubility of PbF_2. Then, $[Pb^{2+}] = x$ and $[F^-] = 2x$

$K_{sp} = [Pb^{2+}][F^-]^2 = (x)(2x)^2 = 4x^3 = 3.7 \times 10^{-8}$ $\qquad$ Solving, x = 2.1 × 10⁻³

Therefore, $\quad$ molar solubility = **2.1 x 10⁻³ mol PbF₂/L** (dissolved)

$\qquad\quad [Pb^{2+}] = x = $ **2.1 x 10⁻³ M**

$\qquad\quad [F^-] = 2x = $ **4.2 x 10⁻³ M**

$\qquad\quad$ solubility (g/L) = $\dfrac{2.1 \times 10^{-3} \text{ mol PbF}_2}{1 \text{ L}} \times \dfrac{245.2 \text{ g PbF}_2}{1 \text{ mol PbF}_2} = $ **0.51 g/L**

(d) Balanced equation: $Sr_3(PO_4)_2(s) \rightleftarrows 3Sr^{2+}(aq) + 2PO_4^{3-}(aq)$ $\qquad$ $K_{sp} = 1.0 \times 10^{-31}$

Let x = molar solubility of $Sr_3(PO_4)_2$ Then, $[Sr^{2+}] = 3x$ and $[PO_4^{3-}] = 2x$

$K_{sp} = [Sr^{2+}]^3[PO_4^{3-}]^2 = (3x)^3(2x)^2 = 108x^5 = 1.0 \times 10^{-31}$ $\qquad$ Solving, x = 2.5 × 10⁻⁷ (2.47 × 10⁻⁷ for calc.)

Therefore, $\quad$ molar solubility = **2.5 x 10⁻⁷ mol Sr₃(PO₄)₂/L** (dissolved)

$\qquad\quad [Sr^{2+}] = 3x = $ **7.4 x 10⁻⁷ M** $\qquad$ (calculation done with 3 sig. fig. for x to minimize rounding errors)

$\qquad\quad [PO_4^{3-}] = 2x = $ **4.9 x 10⁻⁷ M**

$\qquad\quad$ solubility (g/L) = $\dfrac{2.5 \times 10^{-7} \text{ mol Sr}_3(PO_4)_2}{1 \text{ L}} \times \dfrac{452.8 \text{ g Sr}_3(PO_4)_2}{1 \text{ mol Sr}_3(PO_4)_2} = $ **1.1 x 10⁻⁴ g/L**

20-18. *Refer to Section 20-3, Example 20-3 and Appendix H.*

Balanced equation: $BaSO_4(s) \rightleftarrows Ba^{2+}(aq) + SO_4^{2-}(aq)$ $\qquad$ $K_{sp} = 1.1 \times 10^{-10}$

Let x = molar solubility of $BaSO_4$. Then, $[Ba^{2+}] = [SO_4^{2-}] = x$

$K_{sp} = [Ba^{2+}][SO_4^{2-}] = x^2 = 1.1 \times 10^{-10}$ $\qquad$ Solving, x = 1.0 × 10⁻⁵

Therefore,   molar solubility = 1.0×10^{-5} mol $BaSO_4$/L (dissolved)

$$\text{solubility (g/L)} = \frac{1.0 \times 10^{-5} \text{ mol } BaSO_4}{1 \text{ L}} \times \frac{233.4 \text{ g } BaSO_4}{1 \text{ mol } BaSO_4} = 2.3 \times 10^{-3} \text{ g/L}$$

mass of $BaSO_4$ that dissolves in 5.00 L = $(2.3 \times 10^{-3}$ g/L)(5.00 L) = **0.012 g**

20-20. *Refer to Section 20-3 and Example 20-4.*

This is a common ion effect problem similar to some buffer problems.

Balanced equations: $K_2SO_4(aq) \rightarrow 2K^+(aq) + SO_4^{2-}(aq)$ (to completion)

$Ag_2SO_4(s) \rightleftarrows 2Ag^+(aq) + SO_4^{2-}(aq)$ (reversible) $K_{sp} = 1.7 \times 10^{-5}$

Let   x = molar solubility of Ag_2SO_4 in 0.12 M K_2SO_4. Then,
$[Ag^+]$ = 2x (from Ag_2SO_4)
$[SO_4^{2-}]$ = x (from Ag_2SO_4) + 0.12 M (from K_2SO_4)

$K_{sp} = [Ag^+]^2[SO_4^{2-}] = (2x)^2(x + 0.12) = 1.7 \times 10^{-5} \approx (2x)^2(0.12)$ Solving, x = 6.0×10^{-3}

Therefore, molar solubility = **6.0×10^{-3} mol Ag_2SO_4/L 0.12 M K_2SO_4 soln**

20-22. *Refer to Section 20-3, Table 20-1 and Exercise 20-16 Solution.*

Compound	K_{sp}	Molar Solubility (M)	Solubility (g/L)
CuCl	$[Cu^+][Cl^-] = 1.9 \times 10^{-7}$	4.4×10^{-4}	0.043
$Ba_3(PO_4)_2$	$[Ba^{2+}]^3[PO_4^{3-}]^2 = 1.3 \times 10^{-29}$	6.5×10^{-7}	3.9×10^{-4}
PbF_2	$[Pb^{2+}][F^-]^2 = 3.7 \times 10^{-8}$	2.1×10^{-3}	0.51
$Sr_3(PO_4)_2$	$[Sr^{2+}]^3[PO_4^{3-}]^2 = 1.0 \times 10^{-31}$	2.5×10^{-7}	1.1×10^{-4}

(a) **PbF_2** has the highest molar solubility.

(b) **$Sr_3(PO_4)_2$** has the lowest molar solubility.

(c) **PbF_2** has the highest solubility (g/L).

(d) **$Sr_3(PO_4)_2$** has the lowest solubility (g/L).

20-24. *Refer to Section 20-3, Example 20-3 and Appendix H.*

(1) Balanced equation: $Ag_2CO_3(s) \rightleftarrows 2Ag^+(aq) + CO_3^{2-}(aq)$ $K_{sp} = 8.1 \times 10^{-12}$

Let x = molar solubility of Ag_2CO_3. Then, $[Ag^+]$ = 2x and $[CO_3^{2-}]$ = x

$K_{sp} = [Ag^+]^2[CO_3^{2-}] = (2x)^2(x) = 4x^3 = 8.1 \times 10^{-12}$ Solving, x = 1.3×10^{-4}

Therefore,   molar solubility = 1.3×10^{-4} mol Ag_2CO_3/L (dissolved)
(Note: use 1.265×10^{-4} to minimize rounding errors in the following calculation)

$$\text{solubility (g/L)} = \frac{1.265 \times 10^{-4} \text{ mol } Ag_2CO_3}{1 \text{ L}} \times \frac{275.8 \text{ g } Ag_2CO_3}{1 \text{ mol } Ag_2CO_3} = 0.035 \text{ g/L}$$

(2) Balanced equation: $AgCl(s) \rightleftarrows Ag^+(aq) + Cl^-(aq)$ $K_{sp} = 1.8 \times 10^{-10}$

Let x = molar solubility of AgCl. Then, $[Ag^+] = [Cl^-] = x$

$K_{sp} = [Ag^+][Cl^-] = x^2 = 1.8 \times 10^{-10}$ Solving, $x = 1.3 \times 10^{-5}$

Therefore,   molar solubility = 1.3×10^{-5} mol AgCl/L (dissolved)

$$\text{solubility (g/L)} = \frac{1.3 \times 10^{-5} \text{ mol AgCl}}{1 \text{ L}} \times \frac{143.4 \text{ g AgCl}}{1 \text{ mol AgCl}} = 1.9 \times 10^{-3} \text{ g/L}$$

(3) Balanced equation: $Pb(OH)_2(s) \rightleftarrows Pb^{2+}(aq) + 2OH^-(aq)$ $K_{sp} = 2.8 \times 10^{-16}$

Let x = molar solubility of $Pb(OH)_2$. Then, $[Pb^{2+}] = x$ and $[OH^-] = 2x$

$K_{sp} = [Pb^{2+}][OH^-]^2 = (x)(2x)^2 = 4x^3 = 2.8 \times 10^{-16}$ Solving, $x = 4.1 \times 10^{-6}$

Therefore,   molar solubility = 4.1×10^{-6} mol $Pb(OH)_2$/L (dissolved)

$$\text{solubility (g/L)} = \frac{4.1 \times 10^{-6} \text{ mol Pb(OH)}_2}{1 \text{ L}} \times \frac{241.2 \text{ g Pb(OH)}_2}{1 \text{ mol Pb(OH)}_2} = 9.9 \times 10^{-4} \text{ g/L}$$

(a) **Ag_2CO_3** has the highest molar solubility.

(c) **$Pb(OH)_2$** has the lowest molar solubility.

(c) **Ag_2CO_3** has the highest solubility expressed in grams per liter.

(d) **$Pb(OH)_2$** has the lowest solubility expressed in grams per liter.

20-26.   *Refer to Section 20-3, Example 20-4 and Appendix H.*

Plan: Determine separately the molar solubility of $BaCrO_4$ and Ag_2CrO_4 in 0.125 M K_2CrO_4. The salt with the higher molar solubility is more soluble in the K_2CrO_4 solution.

(1) Molar solubility of $BaCrO_4$ in 0.125 M K_2CrO_4:

Balanced equations: $K_2CrO_4(aq) \rightarrow 2K^+(aq) + CrO_4^{2-}(aq)$ (to completion)
                                   $BaCrO_4(s) \rightleftarrows Ba^{2+}(aq) + CrO_4^{2-}(aq)$ (reversible) $K_{sp} = 2.0 \times 10^{-10}$

Let   x = molar solubility of $BaCrO_4$ in 0.125 M K_2CrO_4. Then,
    $[Ba^{2+}]$ = x (from $BaCrO_4$)
    $[CrO_4^{2-}]$ = x (from $BaCrO_4$) + 0.125 M (from K_2CrO_4)

$K_{sp} = [Ba^{2+}][CrO_4^{2-}] = (x)(x + 0.125) = 2.0 \times 10^{-10} \approx (x)(0.125)$ Solving, $x = 1.6 \times 10^{-9}$

Therefore, molar solubility = 1.6×10^{-9} mol $BaCrO_4$/L 0.125 M K_2CrO_4

(2) Molar solubility of Ag_2CrO_4 in 0.125 M K_2CrO_4:

Balanced equations: $K_2CrO_4(aq) \rightarrow 2K^+(aq) + CrO_4^{2-}(aq)$ (to completion)
                                   $Ag_2CrO_4(s) \rightleftarrows 2Ag^+(aq) + CrO_4^{2-}(aq)$ (reversible) $K_{sp} = 9.0 \times 10^{-12}$

Let   y = molar solubility of Ag_2CrO_4 in 0.125 M K_2CrO_4. Then,
    $[Ag^+]$ = 2y (from Ag_2CrO_4)
    $[CrO_4^{2-}]$ = y (from Ag_2CrO_4) + 0.125 M (from K_2CrO_4)

$K_{sp} = [Ag^+]^2[CrO_4^{2-}] = (2y)^2(y + 0.125) = 9.0 \times 10^{-12} \approx (2y)^2(0.125)$ Solving, $y = 4.2 \times 10^{-6}$

Therefore, molar solubility = 4.2×10^{-6} mol Ag_2CrO_4/L 0.125 M K_2CrO_4

The molar solubility of **Ag_2CrO_4** in 0.125 M K_2CrO_4 is greater than that of $BaCrO_4$. We can therefore say that Ag_2CrO_4 is more soluble on a per mole basis. (Note: Ag_2CrO_4 is more soluble on a per gram basis also.)

20-28. *Refer to Section 20-3, Example 20-5 and Appendix H.*

Balanced equations: $Pb(NO_3)_2(aq) + 2NaCl(aq) \rightarrow PbCl_2(s) + 2NaNO_3(aq)$ (Will precipitation occur?)

$PbCl_2(s) \rightleftarrows Pb^{2+}(aq) + 2Cl^-(aq)$ (reversible) $K_{sp} = 1.7 \times 10^{-5}$

Plan: (1) Calculate the concentration of Pb^{2+} ions and Cl^- ions at the instant of mixing before combination occurs.

      (2) Determine the reaction quotient, Q_{sp}. If $Q_{sp} > K_{sp}$, then precipitation will occur.

(1) $[Pb^{2+}] = [Pb(NO_3)_2] = \dfrac{5.0 \text{ g } Pb(NO_3)_2/331 \text{ g/mol}}{1.00 \text{ L soln}} = 0.015 \; M \; Pb^{2+}$

     $[Cl^-] = [NaCl] = 0.010 \; M \; Cl^-$

(2) $Q_{sp} = [Pb^{2+}][Cl^-]^2 = (0.015)(0.010)^2 = 1.5 \times 10^{-6}$

     Since $Q_{sp} < K_{sp}$, precipitation will **not** occur.

20-30. *Refer to Section 20-3, Example 20-5 and Appendix H.*

Balanced equations: $CuCl_2(aq) + 2NaOH(aq) \rightarrow Cu(OH)_2(s) + 2NaCl(aq)$ (Will precipitation occur?)

$Cu(OH)_2(s) \rightleftarrows Cu^{2+}(aq) + 2OH^-(aq)$ (reversible) $K_{sp} = 1.6 \times 10^{-19}$

Plan: (1) Calculate the concentrations of Cu^{2+} ions and OH^- ions at the instant of mixing before reaction occurs using $M_1V_1 = M_2V_2$.

      (2) Determine the reaction quotient, Q_{sp}. If $Q_{sp} > K_{sp}$, a precipitate will form.

(1) $[Cu^{2+}] = [CuCl_2] = \dfrac{M_1V_1}{V_2} = \dfrac{0.010 \; M \times 1.00 \text{ L}}{(1.00 \text{ L} + 0.010 \text{ L})} = 9.9 \times 10^{-3} \; M$

     $[OH^-] = [NaOH] = \dfrac{M_1V_1}{V_2} = \dfrac{0.010 \; M \times 0.010 \text{ L}}{(1.00 \text{ L} + 0.010 \text{ L})} = 9.9 \times 10^{-5} \; M$

(2) $Q_{sp} = [Cu^{2+}][OH^-]^2 = (9.9 \times 10^{-3})(9.9 \times 10^{-5})^2 = 9.7 \times 10^{-11}$

     Since $Q_{sp} > K_{sp}$, a precipitate **will** form.

20-32. *Refer to Section 20-3, Example 20-6 and Appendix H.*

Balanced equations: $Pb^{2+}(aq) + 2NaI(aq) \rightarrow PbI_2(s) + 2Na^+(aq)$

$PbI_2(s) \rightleftarrows Pb^{2+}(aq) + 2I^-(aq)$ (reversible) $K_{sp} = 8.7 \times 10^{-9}$

Plan: (1) Do the limiting reactant problem to determine which reactant is in excess.

      (2) If Pb^{2+} is the limiting reactant, calculate its concentration in the presence of the excess amount of NaI.

      (3) Calculate $\%Pb^{2+}$ ions remaining in solution.

(1)   ? mol Pb^{2+} = 0.0100 $M \times$ 1.00 L = 0.0100 mol Pb^{2+}

     ? mol NaI = 0.103 mol NaI

	$Pb^{2+}(aq)$	+	$2NaI(aq)$	$\rightarrow$	$PbI_2(s)$	+	$2Na^+(aq)$
initial	0.0100 mol		0.103 mol		0 mol		0 mol
change	- 0.0100 mol		- 0.020 mol		+ 0.0100 mol		+ 0.0200 mol
after reaction	0 mol		0.083 mol		0.0100 mol		0.0200 mol

Pb^{2+} is the limiting reactant; NaI is in excess.

(2) In the resulting solution, $[I^-] = [NaI] = 0.083$ mol/1.00 L $= 0.083$ M

    Let   x = molar solubility of PbI_2 in 0.083 M NaI. Then,

$$[Pb^{2+}] = x \text{ (from } PbI_2)$$
$$[I^-] = 2x \text{ (from } PbI_2) + 0.083 \ M \text{ (from NaI)}$$

$K_{sp} = [Pb^{2+}][I^-]^2 = (x)(2x + 0.083)^2 = 8.7 \times 10^{-9} \approx (x)(0.083)^2$ Solving, x $= 1.3 \times 10^{-6}$

Therefore, $[Pb^{2+}] = x = 1.3 \times 10^{-6}$ M

(3) Therefore, % Pb^{2+} in solution $= \dfrac{1.3 \times 10^{-6} \ M}{0.0100 \ M} \times 100\% = \mathbf{0.013\%}$

20-34. ***Refer to Section 20-4 and the Key Terms for Chapter 20.***

Fractional precipitation refers to a separation process whereby some ions are removed from solution by precipitation, leaving other ions with similar properties in solution.

20-36. ***Refer to Section 20-4, Examples 20-8 and 20-9, and Appendix H.***

Balanced equations:
  $Cu^+(aq) + NaBr(aq) \rightarrow CuBr(s) + Na^+(aq)$
  $CuBr(s) \rightleftarrows Cu^+(aq) + Br^-(aq)$ (reversible) $K_{sp} = 5.3 \times 10^{-9}$

  $Ag^+(aq) + NaBr(aq) \rightarrow AgBr(s) + Na^+(aq)$
  $AgBr(s) \rightleftarrows Ag^+(aq) + Br^-(aq)$ (reversible) $K_{sp} = 3.3 \times 10^{-13}$

  $Au^+(aq) + NaBr(aq) \rightarrow AuBr(s) + Na^+(aq)$
  $AuBr(s) \rightleftarrows Au^+(aq) + Br^-(aq)$ (reversible) $K_{sp} = 5.0 \times 10^{-17}$

(a) Because the compounds to be precipitated are all of the same molecular type, i.e., their cation to anion ratios are the same. The one with the smallest K_{sp} is the least soluble and will precipitate first. Hence, **AuBr** ($K_{sp} = 5.0 \times 10^{-17}$) will precipitate first, then AgBr ($K_{sp} = 3.3 \times 10^{-13}$), and finally CuBr ($K_{sp} = 5.3 \times 10^{-9}$).

(b) AgBr will begin to precipitate when $Q_{sp(AgBr)} = K_{sp(AgBr)} = [Ag^+][Br^-]$. At this point,

$$[Br^-] = \frac{K_{sp(AgBr)}}{[Ag^+]} = \frac{3.3 \times 10^{-13}}{0.010} = 3.3 \times 10^{-11} \ M$$

At this concentration of $[Br^-]$, the $[Au^+]$ still in solution is governed by the K_{sp} expression for AuBr: $K_{sp(AuBr)} = [Au^+][Br^-]$.

$$[Au^+] = \frac{K_{sp(AuBr)}}{[Br^-]} = \frac{5.0 \times 10^{-17}}{3.3 \times 10^{-11}} = \mathbf{1.5 \times 10^{-6} \ M}$$

Therefore, the percentage of Au^+ that is still in solution when $[Br^-] = 3.3 \times 10^{-11}$ M is

$$\% \ Au^+ \text{ in solution} = \frac{[Au^+]}{[Au^+]_{initial}} \times 100\% = \frac{1.5 \times 10^{-6} \ M}{0.010 \ M} \times 100\% = 0.015\%$$

$$\% \ Au^+ \text{ precipitated out} = 100.000\% - 0.015\% = \mathbf{99.985\%}$$

(c) CuBr will begin to precipitate when $Q_{sp(CuBr)} = K_{sp(CuBr)} = [Cu^+][Br^-]$

$$[Br^-] = \frac{K_{sp(AuBr)}}{[Cu^+]} = \frac{5.3 \times 10^{-9}}{0.010} = 5.3 \times 10^{-7} \ M$$

At this concentration of Br^-, $[Au^+]$ and $[Ag^+]$ in solution are governed by their K_{sp} expressions:

$$[Au^+] = \frac{K_{sp(AuBr)}}{[Br^-]} = \frac{5.0 \times 10^{-17}}{5.3 \times 10^{-7}} = \textbf{9.4 x 10}^{-11} \textbf{\textit{M}}$$

$$[Ag^+] = \frac{K_{sp(AgBr)}}{[Br^-]} = \frac{3.3 \times 10^{-13}}{5.3 \times 10^{-7}} = \textbf{6.2 x 10}^{-7} \textbf{\textit{M}}$$

20-38. *Refer to Section 20-4, Example 20-8 and 20-9, and Appendix H.*

Balanced equations: $K_2SO_4(aq) + Pb(NO_3)_2(aq) \rightarrow PbSO_4(s) + 2KNO_3(aq)$

$PbSO_4(s) \rightleftarrows Pb^{2+}(aq) + SO_4^{2-}(aq)$ $\qquad\qquad K_{sp} = 1.8 \times 10^{-8}$

$K_2CrO_4(aq) + Pb(NO_3)_2(aq) \rightarrow PbCrO_4(s) + 2KNO_3(aq)$

$PbCrO_4(s) \rightleftarrows Pb^{2+}(aq) + CrO_4^{2-}(aq)$ $\qquad\qquad K_{sp} = 1.8 \times 10^{-14}$

(a) Both $PbSO_4$ and $PbCrO_4$ are of the same molecular type, i.e., their cation to anion ratio is 1:1. **$PbCrO_4$** has the smaller K_{sp}, so it is less soluble and will precipitate first.

(b) $PbCrO_4$ will begin to precipitate when $Q_{sp(PbCrO_4)} = K_{sp(PbCrO_4)} = [Pb^{2+}][CrO_4^{2-}]$. At this point,

$$[Pb^{2+}] = \frac{K_{sp(PbCrO_4)}}{[CrO_4^{2-}]} = \frac{1.8 \times 10^{-14}}{0.050} = \textbf{3.6 x 10}^{-13} \textbf{\textit{M}}$$

(c) $PbSO_4$ will begin to precipitate when $Q_{sp(PbSO_4)} = K_{sp(PbSO_4)} = [Pb^{2+}][SO_4^{2-}]$. At this point,

$$[Pb^{2+}] = \frac{K_{sp(PbSO_4)}}{[SO_4^{2-}]} = \frac{1.8 \times 10^{-8}}{0.050} = \textbf{3.6 x 10}^{-7} \textbf{\textit{M}}$$

(d) When $PbSO_4$ begins to precipitate in (c), the concentration of SO_4^{2-} in solution is still the original concentration, **0.050 *M***. The CrO_4^{2-} concentration can be calculated by substituting the concentration of Pb^{2+} obtained in (c) into the K_{sp} expression for $PbCrO_4$:

$$[CrO_4^{2-}] = \frac{K_{sp(PbCrO_4)}}{[Pb^{2+}]} = \frac{1.8 \times 10^{-14}}{3.6 \times 10^{-7}} = \textbf{5.0 x 10}^{-8} \textbf{\textit{M}}$$

20-40. *Refer to Section 20-3, Example 20-6 and Appendix H.*

Balanced equations: (i) $2KOH(aq) + Zn(NO_3)_2(aq) \rightarrow Zn(OH)_2(s) + 2KNO_3(aq)$

$Zn(OH)_2(s) \rightleftarrows Zn^{2+}(aq) + 2OH^-(aq)$ $\qquad\qquad K_{sp} = 4.5 \times 10^{-17}$

(ii) $K_2CO_3(aq) + Zn(NO_3)_2(aq) \rightarrow ZnCO_3(s) + 2KNO_3(aq)$

$ZnCO_3(s) \rightleftarrows Zn^{2+}(aq) + CO_3^{2-}(aq)$ $\qquad\qquad K_{sp} = 1.5 \times 10^{-11}$

(iii) $2KCN(aq) + Zn(NO_3)_2(aq) \rightarrow Zn(CN)_2(s) + 2KNO_3(aq)$

$Zn(CN)_2(s) \rightleftarrows Zn^{2+}(aq) + 2CN^-(aq)$ $\qquad\qquad K_{sp} = 8.0 \times 10^{-12}$

Plan: Substitute the value of K_{sp} and the given ion concentration into each K_{sp} equilibrium expression and solve for the other ion concentration. An ion concentration just greater than the calculated one will initiate precipitation.

(a) (i) $\quad K_{sp} = [Zn^{2+}][OH^-]^2$ $\qquad$ (ii) $\quad K_{sp} = [Zn^{2+}][CO_3^{2-}]$ $\qquad$ (iii) $\quad K_{sp} = [Zn^{2+}][CN^-]^2$

$4.5 \times 10^{-17} = [Zn^{2+}](0.0015)^2$ $\qquad$ $1.5 \times 10^{-11} = [Zn^{2+}](0.0015)$ $\qquad$ $8.0 \times 10^{-12} = [Zn^{2+}](0.0015)^2$

$[Zn^{2+}] = \textbf{2.0 x 10}^{-11} \textbf{\textit{M}}$ $\qquad$ $[Zn^{2+}] = \textbf{1.0 x 10}^{-8} \textbf{\textit{M}}$ $\qquad$ $[Zn^{2+}] = \textbf{3.6 x 10}^{-6} \textbf{\textit{M}}$

(b) (i) $K_{sp} = [Zn^{2+}][OH^-]^2$ (ii) $K_{sp} = [Zn^{2+}][CO_3{}^{2-}]$ (iii) $K_{sp} = [Zn^{2+}][CN^-]^2$

$\quad$ $4.5 \times 10^{-17} = (0.0015)[OH^-]^2$ $\qquad$ $1.5 \times 10^{-11} = (0.0015)[CO_3{}^{2-}]$ $\qquad$ $8.0 \times 10^{-12} = (0.0015)[CN^-]^2$

$\quad$ $[OH^-] = \textbf{1.7} \times \textbf{10}^{-7}$ $\textbf{\textit{M}}$ $\qquad\qquad$ $[CO_3{}^{2-}] = \textbf{1.0} \times \textbf{10}^{-8}$ $\textbf{\textit{M}}$ $\qquad\qquad$ $[CN^-] = \textbf{7.3} \times \textbf{10}^{-5}$ $\textbf{\textit{M}}$

This problem assumes that the ions do not appreciably hydrolyze in water.

20-42. *Refer to Section 20-5, Examples 20-10 and 20-11, and Appendices G and H.*

Plan: $\quad$ Calculate the concentrations of Mg^{2+} and OH^- ions and determine Q_{sp}. If $Q_{sp} > K_{sp}$ for $Mg(OH)_2$, then precipitation will occur.

Two equilibrium must be considered: $\quad$ $Mg(OH)_2(s) \rightleftarrows Mg^{2+}(aq) + 2OH^-(aq)$ $\qquad$ $K_{sp} = 1.5 \times 10^{-11}$

$\qquad\qquad\qquad\qquad\qquad\qquad\qquad$ $NH_3(aq) + H_2O(\ell) \rightleftarrows NH_4{}^+(aq) + OH^-(aq)$ $\qquad$ $K_b = 1.8 \times 10^{-5}$

We recognize that the given solution is a buffer. The $NH_3/NH_4{}^+$ equilibrium determines $[OH^-]$. Recall from Chapter 18:

$$[OH^-] = K_b \times \frac{[base]}{[salt]} = (1.8 \times 10^{-5}) \times \frac{0.075\ M}{3.5\ M} = 3.9 \times 10^{-7}\ M$$

Therefore, $\quad$ $[OH^-] = 3.9 \times 10^{-7}\ M$

$\qquad\qquad$ $[Mg^{2+}] = [Mg(NO_3)_2] = 0.080\ M$ $\quad$ since $Mg(NO_3)_2$ is a soluble salt

For $Mg(OH)_2$, $\quad$ $Q_{sp} = [Mg^{2+}][OH^-]^2 = (0.080)(3.9 \times 10^{-7})^2 = 1.2 \times 10^{-14}$,

$\qquad\qquad\quad$ $Q_{sp} < K_{sp}$, $Mg(OH)_2$ will **not** precipitate.

Since $[OH^-] = 3.9 \times 10^{-7}\ M$, pOH = 6.41; pH = **7.59**

20-44. *Refer to Section 20-5 and Appendices F and H.*

Balanced equations: $\quad$ $CaF_2(s) \rightleftarrows Ca^{2+}(aq) + 2F^-(aq)$ $\qquad$ $K_{sp} = 3.9 \times 10^{-11}$

$\qquad\qquad\qquad\qquad\quad$ $HF(aq) + H_2O(\ell) \rightleftarrows H_3O^+(aq) + F^-(aq)$ $\qquad$ $K_a = 7.2 \times 10^{-4}$

When equilibria are present in the same solution, all relevant equilibrium expressions must be satisfied:

$$K_{sp} = [Ca^{2+}][F^-]^2 \quad \text{and} \quad K_a = \frac{[H_3O^+][F^-]}{[HF]}$$

In this case, the solid CaF_2 is allowed to dissolve in a solution where $[H_3O^+]$ is buffered at $0.00500\ M$ and $[HF] = 0.10$ M, which is a source of common ion, F^-.

Let $\quad$ x = molar solubility of CaF_2. $\qquad$ Then, $\qquad$ x = mol/L of Ca^{2+} produced by dissolution of CaF_2

$\qquad\qquad\qquad\qquad\qquad\qquad\qquad\qquad\qquad\quad$ 2x = mol/L of F^- produced by dissolution of CaF_2

$\qquad\qquad\qquad\qquad\qquad\qquad\qquad\qquad\qquad\quad$ y = mol/L of F^- produced by dissociation of HF

Therefore, at equilibrium:

$CaF_2(s)$	$\rightleftarrows$	$Ca^{2+}(aq)$	+	$2F^-(aq)$
		x M		(2x + y) M

$HF(aq)$	+	$H_2O(\ell) \rightleftarrows$	$H_3O^+(aq)$	+	$F^-(aq)$
(0.10 - y) M			0.0050 M		(2x + y) M

Plan: $\quad$ (1) $\quad$ Use the K_a equilibrium expression to find the value of y in terms of x. Note: the H_3O^+ concentration is constant because it is buffered at 0.0050 M.

$\qquad\quad$ (2) $\quad$ Substitute for y in the K_{sp} expression to calculate x, the molar solubility.

(1) $\quad$ $K_a = \dfrac{[H_3O^+][F^-]}{[HF]} = \dfrac{(0.0050)(2x + y)}{(0.10 - y)} = 7.2 \times 10^{-4}$

Therefore,
$$(0.0050)(2x + y) = (7.2 \times 10^{-4})(0.10 - y)$$
$$0.010x + 0.0050y = 7.2 \times 10^{-5} - 7.2 \times 10^{-4}y$$
$$0.010x + 0.0057y = 7.2 \times 10^{-5}$$
$$y = \frac{7.2 \times 10^{-5} - 0.010x}{0.0057}$$
$$y = 0.0126 - 1.75x$$

(2) $K_{sp} = [Ca^{2+}][F^-]^2 = (x)(2x + y)^2 = 3.9 \times 10^{-11}$

Substituting for y,
$$3.9 \times 10^{-11} = (x)(2x + (0.0126 - 1.75x))^2$$
$$= (x)(0.0126)^2$$
$$= x(1.59 \times 10^{-4})$$
$$x = 2.5 \times 10^{-7} \quad \text{(The simplifying assumption that 2x was small was correct.)}$$

Therefore, the molar solubility of CaF_2 in this system is **2.5×10^{-7} mol CaF_2/L**

The solubility of CaF_2 (g/L) $= \dfrac{2.5 \times 10^{-7} \text{ mol } CaF_2}{1 \text{ L soln}} \times \dfrac{78.08 \text{ g } CaF_2}{1 \text{ mol } CaF_2} = \textbf{2.0} \times \textbf{10}^{-5} \textbf{ g } \textbf{CaF}_2\textbf{/L}$

20-46. *Refer to Section 20-5, Example 20-10, and Appendices G and H.*

Balanced equations: $Mn(OH)_2(s) \rightleftarrows Mn^{2+}(aq) + 2OH^-(aq)$ $K_{sp} = 4.6 \times 10^{-14}$
$\qquad\qquad\qquad\quad NH_3(aq) + H_2O(\ell) \rightleftarrows NH_4^+(aq) + OH^-(aq)$ $K_b = 1.8 \times 10^{-5}$

Plan: Calculate the concentrations of Mn^{2+} and OH^- ions and determine Q_{sp} for $Mn(OH)_2$. If $Q_{sp} > K_{sp}$, then precipitation will occur.

$[Mn^{2+}] = [Mn(NO_3)_2] = 2.0 \times 10^{-5} M$
$[OH^-]$ is determined from the ionization of NH_3.

Let $x = [NH_3]$ that ionizes.

Then, $1.0 \times 10^{-3} - x = [NH_3]$; $x = [OH^-] = [NH_4^+]$

$K_b = \dfrac{[NH_3][OH^-]}{[NH_3]} = \dfrac{x^2}{1.0 \times 10^{-3} - x} = 1.8 \times 10^{-5} \approx \dfrac{x^2}{1.0 \times 10^{-3}}$ Solving, $x = 1.3 \times 10^{-4}$

Since the value for x is greater than 5% of 1.0×10^{-3}, the simplifying assumption may not hold. When we solve the original quadratic equation, $x^2 + (1.8 \times 10^{-5})x - 1.8 \times 10^{-8} = 0$, $x = 1.3 \times 10^{-4}$. In this case, the simplifying assumption was adequate to 2 significant figures. Therefore, $[OH^-] = 1.3 \times 10^{-4} M$.

$Q_{sp} = [Mn^{2+}][OH^-]^2 = (2.0 \times 10^{-5})(1.3 \times 10^{-4})^2 = 3.4 \times 10^{-13}$

Thus, $Q_{sp} > K_{sp}$ by a factor of 7. Therefore **a precipitate will form** but will not be seen. To be seen, the Q_{sp} must be 1000 times larger than the K_{sp}.

20-48. *Refer to Section 20-3, Example 20-4 and Appendix H.*

(a) Balanced equations: $NaOH(aq) \rightarrow Na^+(aq) + OH^-(aq)$ (to completion)
$\qquad\qquad\qquad\qquad Mg(OH)_2(s) \rightleftarrows Mg^{2+}(aq) + 2OH^-(aq)$ (reversible) $K_{sp} = 1.5 \times 10^{-11}$

Let x = molar solubility of $Mg(OH)_2$ in 0.015 M NaOH. Then,
$\quad [Mg^{2+}] = x$ (from $Mg(OH)_2$)
$\quad\ \ [OH^-] = 2x$ (from $Mg(OH)_2$) + 0.015 M (from NaOH)

$K_{sp} = [Mg^{2+}][OH^-]^2 = (x)(2x + 0.015)^2 = 1.5 \times 10^{-11} \approx (x)(0.015)^2$ Solving, $x = 6.7 \times 10^{-8}$

Therefore, molar solubility = **6.7×10^{-8} mol $Mg(OH)_2$/L 0.015 M NaOH soln**

(b) Balanced equations: $MgCl_2(aq) \rightarrow Mg^{2+}(aq) + 2Cl^-(aq)$ (to completion)

$Mg(OH)_2(s) \rightleftarrows Mg^{2+}(aq) + 2OH^-(aq)$ (reversible) $K_{sp} = 1.5 \times 10^{-11}$

Let x = molar solubility of $Mg(OH)_2$ in 0.015 M $MgCl_2$. Then,

$[Mg^{2+}]$ = x (from $Mg(OH)_2$) + 0.015 M (from $MgCl_2$)
$[OH^-]$ = 2x (from $Mg(OH)_2$)

$K_{sp} = [Mg^{2+}][OH^-]^2 = (x + 0.015)(2x)^2 = 1.5 \times 10^{-11} \approx (0.015)(2x)^2$ Solving, x = 1.6 x 10^{-5}

Therefore, molar solubility = **1.6 x 10^{-5} mol $Mg(OH)_2$/L 0.015 M $MgCl_2$ soln**

20-50. *Refer to Section 20-3, Example 20-5 and Appendix H.*

Balanced equations: $Mg(NO_3)_2(aq) + 2OH^-(aq) \rightarrow Mg(OH)_2(s) + 2NO_3^-(aq)$ (Will precipitation occur?)

$MgOH_2(s) \rightleftarrows Mg^{2+}(aq) + 2OH^-(aq)$ (reversible) $K_{sp} = 1.5 \times 10^{-11}$

Plan: (1) Calculate the concentration of Mg^{2+} ions and OH^- ions at the instant of mixing before combination occurs.

(2) Determine the reaction quotient, Q_{sp}. If $Q_{sp} > K_{sp}$, then precipitation will occur.

(1) $[Mg^{2+}] = [Mg(NO_3)_2] = 0.00050$ M
Since pH = 8.70, pOH = 5.30, and $[OH^-] = 5.0 \times 10^{-6}$ M

(2) $Q_{sp} = [Mg^{2+}][OH^-]^2 = (0.00050)(5.0 \times 10^{-6})^2 = 1.3 \times 10^{-14}$

Since $Q_{sp} < K_{sp}$, precipitation will **not** occur.

20-52. *Refer to Section 20-3, Example 20-3 and Appendix H.*

Balanced equation: $Fe(OH)_2(s) \rightleftarrows Fe^{2+}(aq) + 2OH^-(aq)$ $K_{sp} = 7.9 \times 10^{-15}$

(a) Plan: Calculate the molar solubility of $Fe(OH)_2$, then determine $[OH^-]$ and pH.

Let x = molar solubility of $Fe(OH)_2$. Then
x = moles/L of Fe^{2+}
2x = moles/L of OH^-

$K_{sp} = [Fe^{2+}][OH^-]^2 = (x)(2x)^2 = 4x^3 = 7.9 \times 10^{-15}$ Solving, x = 1.3 x 10^{-5}
molar solubility = 1.3 x 10^{-5} mol $Fe(OH)_2$/L (dissolved)
$[OH^-]$ = 2x = 2.5 x 10^{-5} M; pOH = 4.60; pH = **9.40**

Note: Values for molar solubility and $[OH^-]$ were rounded to 2 significant figures after calculating them.

(b) solubility (g/100 mL) = $\dfrac{1.3 \times 10^{-5} \text{ mol } Fe(OH)_2}{1 \text{ L}}$ x $\dfrac{89.9 \text{ g}}{1 \text{ mol}}$ x $\dfrac{0.1 \text{ L}}{100 \text{ mL}}$ = **1.1 x 10^{-4} g/100 mL**

Note: Remember, do not divide by 100 mL; you want 100 mL to remain in the denominator.

20-54. *Refer to Section 20-6.*

A slightly soluble compound will dissolve when the concentration of its ions in solution are reduced to such a level that $Q_{sp} < K_{sp}$. The following hydroxides and carbonates dissolve in strong acid, such as nitric acid.

(a)
$Cu(OH)_2(s) \rightleftarrows Cu^{2+}(aq) + 2OH^-(aq)$
$\underline{2H^+(aq) + 2OH^-(aq) \rightarrow 2H_2O(\ell)}$
$Cu(OH)_2(s) + 2H^+(aq) \rightarrow Cu^{2+}(aq) + 2H_2O(\ell)$

The H⁺ from the acid reacts with OH⁻ lowering the concentration of OH⁻ by forming H_2O, a weak electrolyte in an acid/base neutralization reaction. Whenever $[OH^-]$ is low enough such that $[Cu^{2+}][OH^-]^2 < K_{sp}$, $Cu(OH)_2(s)$ will dissolve.

(b)
$$Sn(OH)_4(s) \rightleftarrows Sn^{4+}(aq) + 4OH^-(aq)$$
$$\underline{4H^+(aq) + 4OH^-(aq) \rightarrow 4H_2O(\ell)}$$
$$Sn(OH)_4(s) + 4H^+(aq) \rightarrow Sn^{4+}(aq) + 4H_2O(\ell)$$

The H⁺ from the acid reacts with OH⁻ and thus lowers the concentration of OH⁻ in an acid/base neutralization reaction. Whenever $[OH^-]$ is low enough such that $[Sn^{4+}][OH^-]^4 < K_{sp}$, $Sn(OH)_4(s)$ will dissolve.

(c)
$$ZnCO_3(s) \rightleftarrows Zn^{2+}(aq) + CO_3^{2-}(aq)$$
$$\underline{2H^+(aq) + CO_3^{2-}(aq) \rightarrow CO_2(g) + H_2O(\ell)}$$
$$ZnCO_3(s) + 2H^+(aq) \rightarrow Zn^{2+}(aq) + CO_2(g) + H_2O(\ell)$$

The H⁺ from the acid removes CO_3^{2-} from solution in a reaction which forms $CO_2(g)$ and $H_2O(\ell)$. Whenever $[CO_3^{2-}]$ is low enough such that $[Zn^{2+}][CO_3^{2-}] < K_{sp}$, $ZnCO_3(s)$ will dissolve.

(d)
$$(PbOH)_2CO_3(s) \rightleftarrows 2Pb^{2+}(aq) + 2OH^-(aq) + CO_3^{2-}(aq)$$
$$\underline{4H^+(aq) + 2OH^-(aq) + CO_3^{2-}(aq) \rightarrow 3H_2O(\ell) + CO_2(g)}$$
$$(PbOH)_2CO_3(s) + 4H^+(aq) \rightarrow 2Pb^{2+}(aq) + 3H_2O(\ell) + CO_2(g)$$

The H⁺ from the acid removes both OH⁻ and CO_3^{2-} ions from solution by forming $H_2O(\ell)$ and $CO_2(g)$. When $[OH^-]$ and $[CO_3^{2-}]$ are low enough such that $[Pb^{2+}]^2[OH^-]^2[CO_3^{2-}] < K_{sp}$, $(PbOH)_2CO_3(s)$ will dissolve.

20-56.   Refer to Section 20-6.

Nonoxidizing acids dissolve some insoluble sulfides, including MnS and CuS. The H⁺ ions react with S^{2-} ions to form gaseous H_2S, which bubbles out of the solutions. The Q_{sp} of the sulfide becomes less than the corresponding K_{sp} value and the metal sulfide dissolves.

(a)
$$MnS(s) \rightleftarrows Mn^{2+}(aq) + S^{2-}(aq)$$
$$\underline{2H^+(aq) + S^{2-}(aq) \rightarrow H_2S(g)}$$
$$MnS(s) + 2H^+(aq) \rightarrow Mn^{2+}(aq) + H_2S(g)$$

(b)
$$CuS(s) \rightleftarrows Cu^{2+}(aq) + S^{2-}(aq)$$
$$\underline{2H^+(aq) + S^{2-}(aq) \rightarrow H_2S(g)}$$
$$CuS(s) + 2H^+(aq) \rightarrow Cu^{2+}(aq) + H_2S(g)$$

20-58.   Refer to Sections 20-5 and 20-6, and Appendix F.

Balanced equation: $MnS(s) \rightleftarrows Mn^{2+}(aq) + S^{2-}(aq)$

To make MnS(s) more soluble, the above equilibrium must be shifted to the right. Applying LeChatelier's Principle, any process which will reduce either $[Mn^{2+}]$ or $[S^{2-}]$ will do this. In the presence of 0.10 M HCl (a strong acid), competing equilibria will lower $[S^{2-}]$ by producing the weak acids, HS⁻ and H_2S:

$$S^{2-} + H^+ \rightleftarrows HS^- \qquad K_{1'} = \frac{1}{K_{a2\ H_2S}} = \frac{1}{1.0 \times 10^{-19}} = 1.0 \times 10^{19}$$

$$HS^- + H^+ \rightleftarrows H_2S \qquad K_{2'} = \frac{1}{K_{a1\ H_2S}} = \frac{1}{1.0 \times 10^{-7}} = 1.0 \times 10^7$$

The equilibrium constants are very large, so the above equilibria are shifted far to the right, greatly reducing $[S^{2-}]$. In addition, $H_2S(g)$ will bubble out of solution when its solubility is exceeded, thus removing more S^{2-} from the system. On the other hand, the soluble salt, $Mn(NO_3)_2$, would **not** become more soluble in the presence of 0.10 M HCl. The H⁺ ions will not remove NO_3^- ions from solution, because HNO_3 is a strong acid and will ionize totally. The Cl⁻ ions do not remove Mn^{2+} from solution.

20-60. *Refer to Section 20-3, Example 20-5 and Appendix H.*

Balanced equations: $BaCl_2(aq) + 2NaF(aq) \rightarrow BaF_2(s) + 2NaCl(aq)$ (Will precipitation occur?)

$BaF_2(s) \rightleftarrows Ba^{2+}(aq) + 2F^-(aq)$ (reversible) $K_{sp} = 1.7 \times 10^{-6}$

Plan: (a) Calculate the concentration of Ba^{2+} ions and F^- ions at the instant of mixing before any reaction occurs, using $M_1 V_1 = M_2 V_2$.

       (b) Determine the reaction quotient, Q_{sp}. If $Q_{sp} > K_{sp}$, then precipitation will occur.

(a) $[Ba^{2+}] = [BaCl_2] = \dfrac{M_1 V_1}{V_2} = \dfrac{0.0030\ M \times 0.025\ L}{(0.025\ L + 0.050\ L)} = 1.0 \times 10^{-3}\ M$

$[F^-] = [NaF] = \dfrac{M_1 V_1}{V_2} = \dfrac{0.050\ M \times 0.050\ L}{(0.025\ L + 0.050\ L)} = 3.3 \times 10^{-2}\ M$

(b) $Q_{sp} = [Ba^{2+}][F^-]^2 = (1.0 \times 10^{-3})(3.3 \times 10^{-2})^2 = 1.1 \times 10^{-6}$

Since $Q_{sp} < K_{sp}$, precipitation will **not** occur.

20-62. *Refer to Section 20-4, Examples 20-8 and 20-9, and Exercise 20-36 and 20-38 Solutions.*

Balanced equations: $AgBr(s) \rightleftarrows Ag^+(aq) + Br^-(aq)$ (reversible) $K_{sp} = 3.3 \times 10^{-13}$

$AgI(s) \rightleftarrows Ag^+(aq) + I^-(aq)$ (reversible) $K_{sp} = 1.5 \times 10^{-16}$

(a) Because the compounds to be precipitated are all of the same molecular type, i.e., their cation to anion ratios are the same, the one with the smallest K_{sp} is the least soluble and will precipitate first. Hence, **AgI** ($K_{sp} = 1.5 \times 10^{-16}$) will precipitate first, then AgBr ($K_{sp} = 3.3 \times 10^{-13}$).

(b) AgBr will begin to precipitate when $Q_{sp(AgBr)} = K_{sp(AgBr)} = [Ag^+][Br^-]$. At this point,

$[Ag^+] = \dfrac{K_{sp(AgBr)}}{[Br^-]} = \dfrac{3.3 \times 10^{-13}}{0.015} = 2.2 \times 10^{-11}\ M$

At this concentration of $[Ag^+]$, the $[I^-]$ still in solution is governed by the K_{sp} expression for AgI: $K_{sp(AgI)} = [Ag^+][I^-]$.

$[I^-] = \dfrac{K_{sp(AgI)}}{[Ag^+]} = \dfrac{1.5 \times 10^{-16}}{2.2 \times 10^{-11}} = 6.8 \times 10^{-6}\ M$

Therefore, the percentage of I^- that is still in solution when AgBr begins to precipitate is

% I^- in solution $= \dfrac{[I^-]}{[I^-]_{initial}} \times 100\% = \dfrac{6.8 \times 10^{-6}\ M}{0.015\ M} \times 100\% = 0.045\%$

and   % I^- removed from solution $= 100.000\% - 0.045\% = \mathbf{99.955\%}$

20-64. *Refer to Section 20-3 and Appendix H.*

Balanced equation: $AuCl(s) \rightleftarrows Au^+(aq) + Cl^-(aq)$ $K_{sp} = 2.0 \times 10^{-13}$

Plan: (1) Calculate $[Au^+]$ necessary to have a saturated solution.

       (2) Determine the minimum amount of water required to give that concentration.

(1) Let x = molar solubility of AuCl. Then, $[Au^+] = [Cl^-] = x$

$K_{sp} = [Au^+][Cl^-] = x^2 = 2.0 \times 10^{-13}$ Solving, $x = 4.5 \times 10^{-7}$

Therefore, $[Au^+] = 4.5 \times 10^{-7}\ M$

(2) The volume of water (in liters) required to dissolve 1.0 g of Au^+ is:

$?\ L\ H_2O = 1.0\ g\ Au^+ \times \dfrac{1\ mol\ Au^+}{197.0\ g\ Au^+} \times \dfrac{1\ L}{4.5 \times 10^{-7}\ mol\ Au^+} = \mathbf{1.1 \times 10^4\ L\ H_2O}$

In the strict thermodynamic definition of the equilibrium constant, the activity of a component is used, not its concentration. The activity of a species in an ideal mixture is the ratio of its concentration or partial pressure to a standard concentration (1 *M*) or pressure (1 atm). The concentrations of pure solids and pure liquids are omitted from the equilibrium constant expression because their activity is taken to be 1.

20-68.   *Refer to Section 20-4, and Examples 20-8 and 20-9.*

Balanced equations: $AgCl(s)$ (white) $\rightleftarrows$ $Ag^+(aq) + Cl^-(aq)$ (reversible) $K_{sp} = 1.8 \times 10^{-10}$

$Ag_2CrO_4(s)$ (red) $\rightleftarrows$ $2Ag^+(aq) + CrO_4^{2-}(aq)$ (reversible) $K_{sp} = 9.0 \times 10^{-12}$

Because the compounds to be precipitated are not the same molecular type, i.e., their cation to anion ratios are the not the same, you cannot tell just by looking at the values of K_{sp} which salt is more soluble..

The first tint of red color appears when Ag_2CrO_4 begins to precipitate: when $Q_{sp} = K_{sp} = [Ag^+]^2[CrO_4^{2-}]$.

At this point: $[Ag^+] = \sqrt{\dfrac{K_{sp(Ag_2CrO_4)}}{[CrO_4^{2-}]}} = \sqrt{\dfrac{9.0 \times 10^{-12}}{0.010}} = 3.0 \times 10^{-5} \ M$

At this concentration of $[Ag^+]$, the $[Cl^-]$ still in solution is governed by the K_{sp} expression for AgCl:

$K_{sp(AgCl)} = [Ag^+][Cl^-]$ $[Cl^-] = \dfrac{K_{sp(AgCl)}}{[Ag^+]} = \dfrac{1.8 \times 10^{-10}}{3.0 \times 10^{-5}} = \mathbf{6.0 \times 10^{-6} \ \textit{M}}$

20-70.   *Refer to Section 20-3 and Appendix H.*

Balanced equation: $CaF_2(s) \rightleftarrows Ca^{2+}(aq) + 2F^-(aq)$ $K_{sp} = 3.9 \times 10^{-11}$

Plan: (1) Calculate $[F^-]$.
        (2) Substitute the value into the K_{sp} expression and solve for $[Ca^{2+}]$.

(1) $[F^-] = \dfrac{1 \ mg \ F^-}{1 \ L \ soln} \times \dfrac{1 \ g \ F^-}{1000 \ mg \ F^-} \times \dfrac{1 \ mol \ F^-}{19.0 \ g \ F^-} = 5.3 \times 10^{-5} \ M$ (good to 1 significant figure; will round later)

(2) $K_{sp} = [Ca^{2+}][F^-]^2$ $[Ca^{2+}] = \dfrac{[K_{sp}]}{[F^-]^2} = \dfrac{3.9 \times 10^{-11}}{(5.3 \times 10^{-5})^2} = 1.4 \times 10^{-2} \ M$

? amount of Ca^{2+} (g/L) $= \dfrac{0.014 \ mol \ Ca^{2+}}{1 \ L \ soln} \times \dfrac{40.1 \ g \ Ca^{2+}}{1 \ mol \ Ca^{2+}} = \mathbf{0.6 \ g/L}$ (1 significant figure)

20-72.   *Refer to Section 20-3, Example 20-3 and Appendix H.*

Balanced equation: $MgCO_3(s) \rightleftarrows Mg^{2+}(aq) + CO_3^{2-}(aq)$ $K_{sp} = 4.0 \times 10^{-5}$

Plan: (1) Calculate the molar solubility of $MgCO_3$.
        (2) Determine the mass of $MgCO_3$ that would dissolve in 15 L of water to produce a saturated solution.
        (3) Determine the percent loss of $MgCO_3$.

(1) Let x = molar solubility of $MgCO_3$. Then, $[Mg^{2+}] = [CO_3^{2-}] = x$
    $K_{sp} = [Mg^{2+}][CO_3^{2-}] = x^2 = 4.0 \times 10^{-5}$ Solving, x $= 6.3 \times 10^{-3}$
    Therefore,   molar solubility $= 6.3 \times 10^{-3}$ mol $MgCO_3$/L (dissolved)

(2) ? g $MgCO_3$ dissolve in 15 L water $= \dfrac{6.3 \times 10^{-3} \ mol \ MgCO_3}{1 \ L} \times \dfrac{84.3 \ g \ MgCO_3}{1 \ mol \ MgCO_3} \times 15 \ L = 8.0 \ g$

(3) % loss of $MgCO_3 = \dfrac{MgCO_3 \ lost}{initial \ MgCO_3} \times 100\% = \dfrac{8.0 \ g}{28 \ g} \times 100\% = \mathbf{29\%}$

21 Electrochemistry

21-2. *Refer to Section 4-7 and Example 4-5.*

In a redox reaction,

(a) oxidizing agents are the species that

    (1) gain or appear to gain electrons,
    (2) are reduced, and
    (3) oxidize other substances.

(b) Reducing agents are the species that

    (1) lose or appear to lose electrons,
    (2) are oxidized and
    (3) reduce other substances.

Consider the reaction:
$$\overset{+3\ -2}{Fe_2O_3}(s) + \overset{+2\ -2}{3CO}(g) \rightarrow \overset{0}{2Fe}(s) + \overset{+4\ -2}{3CO_2}(g)$$

Fe_2O_3 is the oxidizing agent because it contains Fe, which is being reduced from an oxidation state of +3 to 0.
CO is the reducing agent because it contains C, which is being oxidized from an oxidation state of +2 to +4.

21-4. *Refer to Section 11-4.*

(a) oxidation:
    reduction:
    balanced equation:

$$3(FeS + 4H_2O \rightarrow SO_4^{2-} + Fe^{2+} + 8H^+ + 8e^-)$$
$$8(3e^- + NO_3^- + 4H^+ \rightarrow NO + 2H_2O)$$
$$\overline{3FeS + 8NO_3^- + 8H^+ \rightarrow 8NO + 3SO_4^{2-} + 3Fe^{2+} + 4H_2O}$$

(b) oxidation:
    reduction:
    balanced equation:

$$6(Fe^{2+} \rightarrow Fe^{3+} + e^-)$$
$$6e^- + Cr_2O_7^{2-} + 14H^+ \rightarrow 2Cr^{3+} + 7H_2O$$
$$\overline{6Fe^{2+} + Cr_2O_7^{2-} + 14H^+ \rightarrow 6Fe^{3+} + 2Cr^{3+} + 7H_2O}$$

(c) oxidation:
    reduction:
    balanced equation:

$$S^{2-} + 8OH^- \rightarrow SO_4^{2-} + 4H_2O + 8e^-$$
$$4(2e^- + Cl_2 \rightarrow 2Cl^-)$$
$$\overline{S^{2-} + 4Cl_2 + 8OH^- \rightarrow SO_4^{2-} + 8Cl^- + 4H_2O}$$

21-6. *Refer to Sections 21-2, 21-3 and 21-8.*

The cathode is defined as the electrode at which reduction occurs, i.e., where electrons are consumed, regardless of whether the electrochemical cell is an electrolytic or voltaic cell. In both electrolytic and voltaic cells, the electrons flow through the wire from the anode, where electrons are produced, to the cathode, where electrons are consumed. In an electrolytic cell, the dc source forces the electrons to travel nonspontaneously through the wire. Thus, the electrons flow from the positive electrode (the anode) to the negative electrode (the cathode). However, in a voltaic cell, the electrons flow spontaneously, *away* from the negative electrode (the anode) and toward the positive electrode (the cathode).

(a) The statement, "The positive electrode in any electrochemical cell is the one toward which the electrons flow through the wire," is **false**. It holds for voltaic cells, but not for electrolytic cells.

(b) The statement, "The cathode in any electrochemical cell is the negative electrode," is also **false**. It holds for any electrolytic cell, but not for a voltaic cell.

21-8. *Refer to Section 11-4.*

(a) oxidation: $\qquad\qquad\qquad\qquad\qquad$ $Zn \rightarrow Zn^{2+} + 2e^-$

reduction: $\qquad\qquad\qquad\qquad$ $2e^- + Sn^{2+} \rightarrow Sn$

balanced equation: $\qquad\qquad\qquad$ $Zn + Sn^{2+} \rightarrow Zn^{2+} + Sn$

In this reaction, Zn is losing electrons while Sn^{2+} is gaining electrons.

(b) oxidation: $\qquad\qquad\qquad\qquad$ $5(Cu \rightarrow Cu^{2+} + 2e^-)$

reduction: $\qquad\qquad\quad$ $2(5e^- + MnO_4^- + 8H^+ \rightarrow Mn^{2+} + 4H_2O)$

balanced equation: $\quad$ $5Cu + 2MnO_4^- + 16H^+ \rightarrow 5Cu^{2+} + 2Mn^{2+} + 8H_2O$

In this reaction, Cu is losing electrons while MnO_4^- is gaining electrons.

(c) oxidation: $\qquad\qquad$ $2(5OH^- + Cr(OH)_3 \rightarrow CrO_4^{2-} + 4H_2O + 3e^-)$

reduction: $\qquad\qquad\qquad$ $6e^- + IO_3^- + 3H_2O \rightarrow I^- + 6OH^-)$

balanced equation: $\quad$ $2Cr(OH)_3 + IO_3^- + 4OH^- \rightarrow 2CrO_4^{2-} + I^- + 5H_2O$

In this reaction, $Cr(OH)_3$ is losing electrons while IO_3^- is gaining electrons.

21-10. *Refer to Sections 21-4, 21-14 and 21-15, and Table 21-2.*

(a) Magnesium metal is too reactive in water to be obtained by the electrolysis of $MgCl_2(aq)$. In other words, $H_2O(\ell)$ is more easily reduced to $OH^-(aq)$ and $H_2(g)$ than is $Mg^{2+}(aq)$ to $Mg(s)$. In electrochemical reactions, the species that is most easily reduced (or oxidized) will be reduced (or oxidized) first.

(b) Sodium ions do not appear in the overall cell reaction for the electrolysis of $NaCl(aq)$ because Na^+ ions are spectator ions and do not react. Since H_2O is more easily reduced than Na^+ ions, the reduction reaction involves H_2O:

reduction at cathode: $\qquad$ $2e^- + 2H_2O(\ell) \rightarrow H_2(g) + 2OH^-(aq)$

oxidation at anode: $\qquad\quad$ $2Cl^-(aq) \rightarrow Cl_2(g) + 2e^-$

overall cell reaction: $\qquad$ $2Cl^- + 2H_2O(\ell) \rightarrow H_2(g) + Cl_2(g) + 2OH^-(aq)$

21-12. *Refer to Section 21-3 and Figure 21-2.*

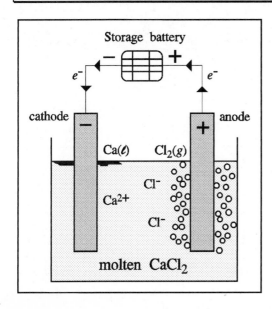

Electrolysis of molten calcium chloride:

oxidation: $\qquad\qquad\qquad$ $2Cl^-(molten) \rightarrow Cl_2(g) + 2e^-$

reduction: $\qquad\quad$ $Ca^{2+}(molten) + 2e^- \rightarrow Ca(\ell)$

overall cell reaction: $\qquad\qquad$ $CaCl_2(\ell) \rightarrow Ca(\ell) + Cl_2(g)$

333

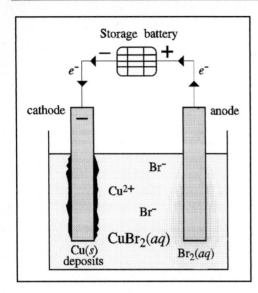

Electrolysis of aqueous copper(II) bromide:

oxidation:	$2Br^-(aq) \rightarrow Br_2(aq) + 2e^-$
reduction:	$Cu^{2+}(aq) + 2e^- \rightarrow Cu(s)$
overall cell reaction:	$Cu^{2+}(aq) + 2Br^-(aq) \rightarrow Cu(s) + Br_2(aq)$

21-16. *Refer to Section 21-6 and the Key Terms for Chapter 21.*

(a) A coulomb (C) is the amount of electrical charge that passes a given point when one ampere of current flows for one second.

(b) Electrical current is the motion of electrons or ions through a conducting medium.

(c) An ampere (A) is the practical unit of electrical current equal to the transfer of 1 coulomb per second. So, $1\ A = 1\ C/s$.

(d) A faraday of electricity corresponds to the charge on 6.022×10^{23} electrons, or 96,485 coulombs. It is the amount of electricity that reduces 1 equivalent weight of a substance at the cathode and oxidizes 1 equivalent weight of a substance at the anode.

21-18. *Refer to Section 21-6.*

(i) Recall that 1 faraday of electricity is equivalent to 1 mole of electrons passing through a system. Consider the general balanced half-reaction:

$$M^{n+} + ne^- \rightarrow M$$

In accordance with stoichiometry, the production of 1 mole of M requires n moles of electrons, hence n faradays of electricity.

Balanced Half-Reaction	No. of Faradays/1 mol Free Metal
(a) $Fe^{3+}(aq) + 3e^- \rightarrow Fe(s)$	3
(b) $Sn^{2+}(aq) + 2e^- \rightarrow Sn(s)$	2
(c) $Hg_2^{2+}(aq) + 2e^- \rightarrow 2Hg(\ell)$	1

(ii) The amount of charge required to deposit 1.00 g of each of the metals according to the reactions above:

(a) $?\ \text{coulombs} = 1.00\ \text{g Fe} \times \dfrac{1\ \text{mol Fe}}{55.85\ \text{g Fe}} \times \dfrac{3\ \text{mol}\ e^-}{1\ \text{mol Fe}} \times \dfrac{96500\ \text{C}}{1\ \text{mol}\ e^-} = \mathbf{5.18 \times 10^3\ C}$

(b) $?\ \text{coulombs} = 1.00\ \text{g Sn} \times \dfrac{1\ \text{mol Sn}}{118.7\ \text{g Sn}} \times \dfrac{2\ \text{mol}\ e^-}{1\ \text{mol Sn}} \times \dfrac{96500\ \text{C}}{1\ \text{mol}\ e^-} = \mathbf{1.63 \times 10^3\ C}$

(c) $?\ \text{coulombs} = 1.00\ \text{g Hg} \times \dfrac{1\ \text{mol Hg}}{200.6\ \text{g Hg}} \times \dfrac{2\ \text{mol}\ e^-}{2\ \text{mol Hg}} \times \dfrac{96500\ \text{C}}{1\ \text{mol}\ e^-} = \mathbf{481\ C}$

21-20. *Refer to Section 21-6 and Example 21-1.*

(a) Balanced half-reaction: $Cu^+ + e^- \rightarrow Cu$

Plan: $g\ Cu \Rightarrow mol\ Cu \Rightarrow mol\ e^- \Rightarrow coulombs \Rightarrow amperes\ (= coulombs/sec)$

$? \text{ coulombs} = 2.25 \text{ g Cu} \times \dfrac{1 \text{ mol Cu}}{63.55 \text{ g Cu}} \times \dfrac{1 \text{ mol } e^-}{1 \text{ mol Cu}} \times \dfrac{96500 \text{ C}}{1 \text{ mol } e^-} = 3420 \text{ coulombs}$

$? \text{ amperes (coulombs/s)} = \dfrac{3420 \text{ C}}{400.\ \text{min} \times (60 \text{ s}/1 \text{ min})} = \mathbf{0.142\ amperes}$

(b) Balanced half-reaction: $Cu^{2+} + 2e^- \rightarrow Cu$

$? \text{ g Cu} = 400.\ \text{min} \times \dfrac{60 \text{ s}}{1 \text{ min}} \times \dfrac{0.142 \text{ C}}{1 \text{ s}} \times \dfrac{1 \text{ mol } e^-}{96500 \text{ C}} \times \dfrac{1 \text{ mol Cu}}{2 \text{ mol } e^-} \times \dfrac{63.55 \text{ g Cu}}{1 \text{ mol Cu}} = \mathbf{1.12\ g\ Cu}$ (1/2 mass of Cu in part (a))

21-22. *Refer to Section 21-6 and Example 21-1.*

Balanced half-reaction: $Rh^{3+} + 3e^- \rightarrow Rh$

$? \text{ g Rh} = 15.0 \text{ min} \times \dfrac{60 \text{ s}}{1 \text{ min}} \times \dfrac{0.752 \text{ C}}{1 \text{ s}} \times \dfrac{1 \text{ mol } e^-}{96500 \text{ C}} \times \dfrac{1 \text{ mol Rh}}{3 \text{ mol } e^-} \times \dfrac{102.9 \text{ g Rh}}{1 \text{ mol Rh}} = \mathbf{0.241\ g\ Rh}$

21-24. *Refer to Section 21-6.*

Balanced half-reaction: $Ag^+ + e^- \rightarrow Ag$

$? \text{ coulombs} = 0.904 \text{ mg Ag} \times \dfrac{1 \text{ g Ag}}{1000 \text{ mg Ag}} \times \dfrac{1 \text{ mol Ag}}{107.9 \text{ g Ag}} \times \dfrac{1 \text{ mol } e^-}{1 \text{ mol Ag}} \times \dfrac{96500 \text{ C}}{1 \text{ mol } e^-} = \mathbf{0.808\ C}$

21-26. *Refer to Section 21-6 and Example 21-1.*

Balanced half-reaction: $Ag^+ + e^- \rightarrow Ag$

$? \text{ g Ag} = 45.0 \text{ min} \times \dfrac{60 \text{ s}}{1 \text{ min}} \times \dfrac{3.86 \text{ C}}{1 \text{ s}} \times \dfrac{1 \text{ mol } e^-}{96500 \text{ C}} \times \dfrac{1 \text{ mol Ag}}{1 \text{ mol } e^-} \times \dfrac{107.9 \text{ g Ag}}{1 \text{ mol Ag}} = \mathbf{11.7\ g\ Ag}$

21-28. *Refer to Section 21-6 and Examples 21-1 and 21-2.*

Balanced half-reactions: anode $2I^- \rightarrow I_2 + 2e^-$
cathode $2H_2O + 2e^- \rightarrow H_2 + 2OH^-$

(a) The number of faradays passing through the cell is equivalent to the number of moles of electrons passing through the cell.

$? \text{ faradays} = 37.5 \times 10^{-3} \text{ mol } I_2 \times \dfrac{2 \text{ mol } e^-}{1 \text{ mol } I_2} \times \dfrac{1 \text{ faraday}}{1 \text{ mol } e^-} = \mathbf{0.0750\ faradays}$

(b) $? \text{ coulombs} = 0.0750 \text{ faradays} \times \dfrac{96500 \text{ C}}{1 \text{ faraday}} = \mathbf{7.24 \times 10^3\ C}$

(c) $? \text{ L}_{STP}\ H_2 = 7.24 \times 10^3 \text{ C} \times \dfrac{1 \text{ mol } e^-}{96500 \text{ C}} \times \dfrac{1 \text{ mol } H_2}{2 \text{ mol } e^-} \times \dfrac{22.4 \text{ L}_{STP}\ H_2}{1 \text{ mol } H_2} = \mathbf{0.840\ L_{STP}\ H_2}$

Alternatively, $? \text{ L}_{STP}\ H_2 = 0.0750 \text{ faradays} \times \dfrac{1 \text{ mol } H_2}{2 \text{ faradays}} \times \dfrac{22.4 \text{ L}_{STP}\ H_2}{1 \text{ mol } H_2} = \mathbf{0.840\ L_{STP}\ H_2}$

(d) Plan: (1) Determine the moles of OH^- formed.
  (2) Calculate $[OH^-]$, pOH and pH.

(1) ? mol $OH^- = 7.24 \times 10^3$ C $\times \dfrac{1 \text{ mol } e^-}{96500 \text{ C}} \times \dfrac{2 \text{ mol } OH^-}{2 \text{ mol } e^-} = 0.0750$ mol OH^-

(2) $[OH^-] = \dfrac{0.0750 \text{ mol } OH^-}{0.500 \text{ L}} = 0.150$ M OH^-; pOH = 0.824; **pH = 13.176**

21-30. *Refer to Section 21-6 and Example 21-2.*

Plan: (1) Determine the half-reaction involving Cl_2.
  (2) Calculate the moles of Cl_2 produced at the experimental conditions at 83% efficiency.
  (3) Calculate the volume of Cl_2 produced using the ideal gas law, $PV = nRT$.

(1) Balanced half-reaction: $2Cl^- \rightarrow Cl_2 + 2e^-$

(2) ? mol $Cl_2 = 5.00$ hr $\times \dfrac{3600 \text{ s}}{1 \text{ hr}} \times \dfrac{1.70 \text{ C}}{1 \text{ s}} \times \dfrac{1 \text{ mol } e^-}{96500 \text{ C}} \times \dfrac{1 \text{ mol } Cl_2}{2 \text{ mol } e^-} \times \dfrac{83}{100} = 0.13$ mol Cl_2

(3) $V = \dfrac{nRT}{P} = \dfrac{(0.13 \text{ mol})(0.0821 \text{ L·atm/mol·K})(15°C + 273°)}{(752/760 \text{ atm})} = $ **3.1 L Cl_2** (to 2 significant figures)

21-32. *Refer to Section 21-6.*

Balanced half-reaction: $Cu^{2+} + 2e^- \rightarrow Cu$

(1) Plan: M,L $CuCl_2$ soln $\overset{\text{(i)}}{\Rightarrow}$ mol Cu^{2+} reacted $\overset{\text{(ii)}}{\Rightarrow}$ mol e^- reacted $\overset{\text{(iii)}}{\Rightarrow}$ time required

  (i) Original moles of $Cu^{2+} = (0.533$ $M)(0.250$ L$) = 0.133$ mol
    Final moles of $Cu^{2+} = (0.167$ $M)(0.250$ L$) = 4.18 \times 10^{-2}$ mol
    ? mol Cu^{2+} reacted $= 0.133$ mol $- 4.18 \times 10^{-2}$ mol $= 0.091$ mol Cu^{2+}

  (ii) ? mol e^- reacted $= 2 \times$ mol Cu^{2+} reacted $= 2 \times 0.091$ mol $= 0.18$ mol e^-

  (iii) ? time required (s) $= 0.18$ mol $e^- \times \dfrac{96500 \text{ C}}{1 \text{ mol } e^-} \times \dfrac{1 \text{ amp-s}}{1 \text{ C}} \times \dfrac{1}{0.750 \text{ amp}} = $ **2.3×10^4 s $\equiv$ 6.4 hr**

(2) ? mass of Cu $= 0.091$ mol Cu $\times \dfrac{63.55 \text{ g Cu}}{1 \text{ mol Cu}} = $ **5.8 g Cu**

21-34. *Refer to Section 21-6 and Example 21-1.*

Balanced half-reactions: $Cd \rightarrow Cd^{2+} + 2e^-$ $Ag^+ + e^- \rightarrow Ag$ $Fe^{2+} \rightarrow Fe^{3+} + e^-$

(a) ? faradays $= 1.20$ g Cd $\times \dfrac{1 \text{ mol Cd}}{112.4 \text{ g Cd}} \times \dfrac{2 \text{ mol } e^-}{1 \text{ mol Cd}} \times \dfrac{1 \text{ faraday}}{1 \text{ mol } e^-} = $ **0.0214 faradays**

(b) ? g Ag $= 0.0214$ faradays $\times \dfrac{1 \text{ mol } e^-}{1 \text{ faraday}} \times \dfrac{1 \text{ mol Ag}}{1 \text{ mol } e^-} \times \dfrac{107.9 \text{ g Ag}}{1 \text{ mol Ag}} = $ **2.31 g Ag**

(c) ? g $Fe(NO_3)_3 = 0.0214$ faraday $\times \dfrac{1 \text{ mol } e^-}{1 \text{ faraday}} \times \dfrac{1 \text{ mol } Fe^{3+}}{1 \text{ mol } e^-} \times \dfrac{1 \text{ mol } Fe(NO_3)_3}{1 \text{ mol } Fe^{3+}} \times \dfrac{241.9 \text{ g } Fe(NO_3)_3}{1 \text{ mol } Fe(NO_3)_3}$
  $= $ **5.18 g $Fe(NO_3)_3$**

21-36. *Refer to the Introduction to Voltaic or Galvanic Cells, Section 21-8 and Figure 21-6.*

(a) In a voltaic cell, the solutions in the two half-cells must be kept separate in order to produce usable electrical energy since electricity is only produced when electron transfer is forced to occur through the external circuit. If the two half-cells were mixed, electron transfer would happen directly in the solution and could not be exploited to give electricity.

(b) A salt bridge in a voltaic or galvanic cell has three functions: it allows electrical contact between the two solutions; it prevents mixing of the electrode solutions; and it maintains electrical neutrality in each half-cell.

21-38. *Refer to Sections 21-8, 21-9 and 21-10, and Figure 21-6.*

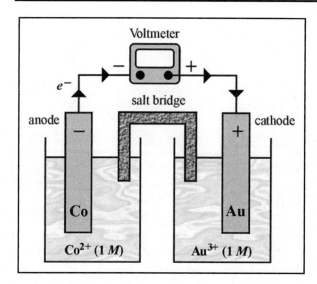

Voltaic cell:

oxidation at anode:	$3(Co \rightarrow Co^{2+} + 2e^-)$
reduction at cathode:	$2(Au^{3+} + 3e^- \rightarrow Au)$
overall cell reaction:	$3Co + 2Au^{3+} \rightarrow 3Co^{2+} + 2Au$

21-40. *Refer to Section 21-9 and Exercise 21-38.*

Shorthand notation: $Co|Co^{2+}(1\ M)\|Au^{3+}(1\ M)/Au$

21-42. *Refer to Sections 21-8, 21-9 and 21-10.*

Balanced equation: $Ni(s) + 2Ag^+(aq) \rightarrow Ni^{2+}(aq) + 2Ag(s)$

(a) reduction half-reaction: $Ag^+(aq) + e^- \rightarrow Ag(s)$

(b) oxidation half-reaction: $Ni(s) \rightarrow Ni^{2+}(aq) + 2e^-$

(c) Ni is the anode.

(d) Ag is the cathode.

(e) Refer to cell diagram at right.

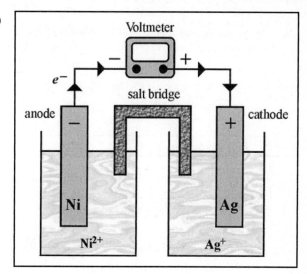

No electricity is produced when $Cu(s)$ is placed into $AgNO_3(aq)$ even though a spontaneous redox reaction occurs:

$$Cu(s) + 2AgNO_3(aq) \rightarrow Cu(NO_3)_2(aq) + 2Ag(s)$$

The electron transfer at the solution-metal interface; it is not forced to occur through an external circuit where it would produce useful electrical energy.

21-46. *Refer to Sections 21-8, 21-9 and 21-10, and Figure 21-6.*

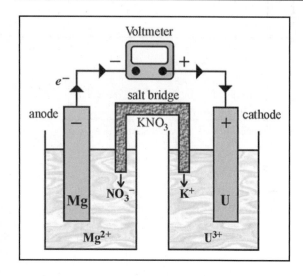

Voltaic cell:

oxidation at anode:	$3(Mg \rightarrow Mg^{2+} + 2e^-)$
reduction at cathode:	$2(U^{3+} + 3e^- \rightarrow U)$
overall cell reaction:	$3Mg + 2U^{3+} \rightarrow 3Mg^{2+} + 2U$

A note on ion flow: Because in the anodic half-cell, more positively-charged ions (Mg^{2+}) are being produced, more negatively-charged ions are required to keep the solution neutral. This is why the nitrate ions are flowing toward the anode. In the cathodic half-cell, positively-charged ions (U^{3+}) are being lost, so more positively-charged ions are needed to keep the solution neutral. This is why the potassium ions are flowing toward the cathode.

21-48. *Refer to Section 21-14 and the Introduction to Section 21-11.*

If the sign of the standard reduction potential, $E°$, of a half-reaction is positive, the half-reaction is the cathodic (reduction) reaction when connected to the standard hydrogen electrode (SHE). Half-reactions with more positive $E°$ values have greater tendencies to occur in the forward direction. Hence, the magnitude of a half-cell potential measures the spontaneity of the forward reaction. If the $E°$ of a half-reaction is negative, the half-reaction is the anodic (oxidation) reaction when connected to the SHE. Half-reactions with more negative $E°$ values have greater tendencies to occur in the reverse direction.

21-50. *Refer to Section 21-14 and Table 21-2.*

(a) The substance that is the stronger oxidizing agent is the more easily reduced and has the more positive reduction potential. Therefore, in order of increasing strength,

K^+ (−2.925 V) < Na^+ (−2.71 V) < Fe^{2+} (−0.44 V) < Cu^{2+} (0.337 V) < Cu^+ (0.521 V) < Ag^+ (0.799 V) < Cl_2 (1.36 V)

(b) Under standard state conditions, both Cl_2 and Ag^+ can oxidize Cu, since their standard reduction potentials are more positive than those for Cu^+ and Cu^{2+}. Also, Cu^+ can oxidize Cu to Cu^{2+}.

21-52. *Refer to Section 21-15.*

The activity of a metal is based on how easily it oxidizes to positively-charged ions. Therefore, a more active metal loses electrons more readily, is more easily oxidized and is a better reducing agent. The strength of a reducing agent increases as its standard reduction potential becomes more negative.

most active   Eu (−3.4 V) > Ra (−2.9 V) > Rh (0.80 V)   least active

Only Eu is more active than Li (−3.0 V). Eu and Ra are more active than H_2 (0.00 V). All are more active than Pt (1.2 V).

(a) cell diagram:

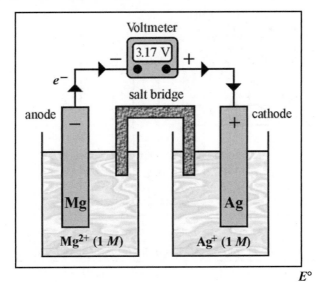

		$E°$
oxidation at anode:	$Mg \rightarrow Mg^{2+} + 2e^-$	+2.37 V
reduction at cathode:	$2(Ag^+ + e^- \rightarrow Ag)$	+0.7994 V
cell reaction:	$Mg + 2Ag^+ \rightarrow Mg^{2+} + 2Ag$ $\quad E°_{cell} =$	**+3.17 V**

(b) cell diagram:

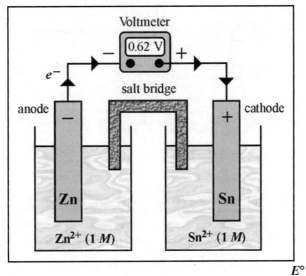

		$E°$
oxidation at anode:	$Zn \rightarrow Zn^{2+} + 2e^-$	+0.763 V
reduction at cathode:	$Sn^{2+} + 2e^- \rightarrow Sn$	−0.14 V
cell reaction:	$Zn + Sn^{2+} \rightarrow Zn^{2+} + Sn$ $\quad E°_{cell} =$	**+0.62 V**

For a standard magnesium and aluminum cell:

		$E°$
oxidation at anode:	$3(Mg \rightarrow Mg^{2+} + 2e^-)$	+2.37 V
reduction at cathode:	$2(Al^{3+} + 3e^- \rightarrow Al)$	−1.66 V
cell reaction:	$3Mg + 2Al^{3+} \rightarrow 3Mg^{2+} + 2Al$ $\quad E°_{cell} =$	**+0.71 V**

Shorthand notation: $Mg|Mg^{2+}(1\ M)||Al^{3+}(1\ M)|Al$

cell diagram:

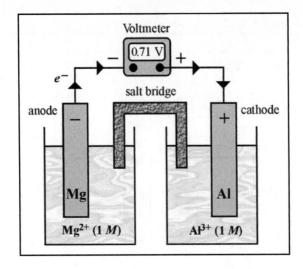

21-58. *Refer to Sections 21-15 and 21-16, Examples 21-3 and 21-4, and Appendix J.*

Plan: Calculate $E°_{cell}$ for each reaction as written. If $E°_{cell}$ is positive, the reaction is spontaneous and will go as written. If $E°_{cell}$ is negative, the reaction is nonspontaneous and will not go as written; the reverse reaction is spontaneous.

		$E°$
(a) reduction:	$2(Fe^{3+} + e^- \rightarrow Fe^{2+})$	+0.771 V
oxidation:	$Sn^{2+} \rightarrow Sn^{4+} + 2e^-$	−0.15 V
cell reaction:	$2Fe^{3+} + Sn^{2+} \rightarrow 2Fe^{2+} + Sn^{4+}$	$E°_{cell} = +0.62$ V

Yes, Fe^{3+} will oxidize Sn^{2+} to Sn^{4+} because the reaction is spontaneous ($E°_{cell} > 0$).

		$E°$
(b) reduction:	$Cr_2O_7^{2-} + 14H^+ + 6e^- \rightarrow 2Cr^{3+} + 7H_2O$	+1.33 V
oxidation:	$3(2F^- \rightarrow F_2 + 2e^-)$	−2.87 V
cell reaction:	$Cr_2O_7^{2-} + 14H^+ + 6F^- \rightarrow 2Cr^{3+} + 3F_2 + 7H_2O$	$E°_{cell} = -1.54$ V

No, $Cr_2O_7^{2-}$ ions cannot oxidize F^- ions to F_2 because the reaction is not spontaneous ($E°_{cell} < 0$).

21-60. *Refer to Sections 21-15 and 21-16, Examples 21-3 and 21-4, and Appendix J.*

Plan: Calculate $E°_{cell}$ for each reaction as written. If $E°_{cell}$ is positive, the reaction is spontaneous and will go as written. If $E°_{cell}$ is negative, the reaction is nonspontaneous and will not go as written; the reverse reaction is spontaneous.

		$E°$
(a) reduction:	$6(MnO_4^- + 8H^+ + 5e^- \rightarrow Mn^{2+} + 4H_2O)$	+1.51 V
oxidation:	$5(2Cr^{3+} + 7H_2O \rightarrow Cr_2O_7^{2-} + 14H^+ + 6e)$	−1.33 V
cell rxn:	$10 Cr^{3+} + 11H_2O + 6MnO_4^- \rightarrow 5Cr_2O_7^{2-} + 22H^+ + 6Mn^{2+}$	$E°_{cell} = +0.18$ V

Yes, MnO_4^- ions can oxidize Cr^{3+} ions to $Cr_2O_7^{2-}$ ions since the reaction is spontaneous ($E°_{cell} > 0$).

		$E°$
(b) reduction (1):	$SO_4^{2-} + 4H^+ + 2e^- \rightarrow H_2SO_3 + H_2O$	+0.17 V
reduction (2):	$SO_4^{2-} + 4H^+ + 2e^- \rightarrow SO_2 + 2H_2O$	+0.20 V
oxidation:	$H_3AsO_3 + H_2O \rightarrow H_3AsO_4 + 2H^+ + 2e^-$	−0.58 V

No matter which sulfate reduction reaction is used, $E°_{cell}$ ($= E°_{cathode} + E°_{anode}$) < 0, where $E°_{cathode}$ is the standard reduction potential and $E°_{anode}$ is the standard oxidation potential. Therefore, **no**, SO_4^{2-} ions cannot oxidize H_3AsO_3 to H_3AsO_4.

21-62. *Refer to Section 21-15, Example 21-3 and Exercise 21-38 Solution.*

Refer to the cell diagram in Exercise 21-38 Solution.

		$E°$
reduction:	$2(Au^{3+} + 3e^- \rightarrow Au)$	+1.50 V
oxidation:	$3(Co \rightarrow Co^{2+} + 2e^-)$	+0.28 V
cell reaction:	$2Au^{3+} + 3Co \rightarrow 2Au + 3Co^{2+}$	$E°_{cell} = $ **+1.78 V**

21-64. *Refer to Sections 21-9 and 21-15, and Appendix J.*

(a) Consider the voltaic cell: $Cr|Cr^{3+}||Cu^{2+}|Cu$

   (i)   cell reaction: $2Cr + 3Cu^{2+} \rightarrow 2Cr^{3+} + 3Cu$

   (ii) oxidation half-reaction at anode: $Cr \rightarrow Cr^{3+} + 3e^-$ $E° = +0.74$ V
           reduction half-reaction at cathode: $Cu^{2+} + 2e^- \rightarrow Cu$ $E° = +0.337$ V

   (iii) $E°_{cell} = E°_{anode} + E°_{cathode} = +0.74$ V $+ (+0.337$ V$) = $ **+1.08 V**
       Note: $E°_{cathode}$ is the standard reduction potential and $E°_{anode}$ is the standard oxidation potential.

   (iv) Yes, the standard reaction occurs as written since $E°_{cell} > 0$.

(b) Consider the voltaic cell: $Ag|Ag^+||Cd^{2+}|Cd$

   (i)   cell reaction: $2Ag + Cd^{2+} \rightarrow 2Ag^+ + Cd$

   (ii) oxidation half-reaction at anode: $Ag \rightarrow Ag^+ + e^-$ $E° = -0.7994$ V
           reduction half-reaction at cathode: $Cd^{2+} + 2e^- \rightarrow Cd$ $E° = -0.403$ V

   (iii) $E°_{cell} = E°_{cathode} + E°_{anode} = (-0.7994$ V$) + (-0.403$ V$) = $ **−1.202 V**
       Note: $E°_{cathode}$ is the standard reduction potential and $E°_{anode}$ is the standard oxidation potential.

   (iv) No, the standard reaction will not occur as written since $E°_{cell} < 0$; the reverse reaction will occur.

21-66. *Refer to Sections 21-15 and 21-16, and Appendix J.*

Plan: Calculate $E°_{cell}$ for each reaction as written. If $E°_{cell}$ is positive, the reaction is spontaneous.

		$E°$
(a) reduction:	$H_2 + 2e^- \rightarrow 2H^-$	−2.25 V
oxidation:	$H_2 \rightarrow 2H^+ + 2e^-$	+0.00 V
cell reaction:	$2H_2 \rightarrow 2H^- + 2H^+$	$E°_{cell} = $ −2.25 V
or	$H_2 \rightarrow H^- + H^+$	

No, the reaction is non-spontaneous; $E°_{cell} < 0$.

		$E°$
(b) reduction:	$Ag_2CrO_4 + 2e^- \rightarrow 2Ag + CrO_4^{2-}$	+0.446 V
oxidation:	$Zn + 4CN^- \rightarrow Zn(CN)_4^{2-} + 2e^-$	+1.26 V
cell reaction:	$Ag_2CrO_4 + Zn + 4CN^- \rightarrow 2Ag + CrO_4^{2-} + Zn(CN)_4^{2-}$	$E°_{cell} = $ +1.71 V

Yes, the reaction is spontaneous; $E°_{cell} > 0$.

			$E°$
(c)	reduction:	$MnO_2 + 4H^+ + 2e^- \rightarrow Mn^{2+} + 2H_2O$	$+1.23$ V
	oxidation:	$Sr \rightarrow Sr^{2+} + 2e^-$	$+2.89$ V
	cell reaction:	$MnO_2 + 4H^+ + Sr \rightarrow Mn^{2+} + 2H_2O + Sr^{2+}$	$E°_{cell} = +4.12$ V

Yes, the reaction is spontaneous; $E°_{cell} > 0$.

			$E°$
(d)	reduction:	$ZnS + 2e^- \rightarrow Zn + S^{2-}$	-1.44 V
	oxidation:	$Cl_2 + 2H_2O \rightarrow 2HOCl + 2H^+ + 2e^-$	-1.63 V
	cell reaction:	$ZnS + Cl_2 + 2H_2O \rightarrow Zn + 2HOCl + 2H^+ + S^{2-}$	$E°_{cell} = -3.07$ V
	or	$ZnS + Cl_2 + 2H_2O \rightarrow Zn + 2HOCl + H_2S$ since H_2S is a weak acid	

No, the reaction is non-spontaneous; $E°_{cell} < 0$.

21-68. *Refer to Sections 21-15 and 21-16, and Appendix J.*

The substance that is the stronger reducing agent is the more easily oxidized. The reduced form of a species is a stronger reducing agent when the half-reaction has a more negative standard reduction potential. The stronger reducing agents are given below.

(a) H_2 (0.000 V) > Ag (0.7994 V)

(b) Sn (−0.14 V) > Pb (−0.126 V)

(c) Hg (0.855 V) > Au (1.68 V or 1.50 V)

(d) Cl^- in base (0.62 V or 0.89 V) > Cl^- in acid (1.36 V)

(e) H_2S (0.14 V) > HCl, i.e., Cl^- in acid (1.36 V).

(f) Ag (0.7994 V) > Au (1.68 V or 1.50 V)

21-70. *Refer to Sections 21-21 and 15-16.*

The half-reactions can be added together so that the desired half-reaction is obtained:

reduction half-reaction:	$Yb^{3+} + 3e^- \rightarrow Yb$	(1)
oxidation half-reaction:	$Yb \rightarrow Yb^{2+} + 2e^-$	(2)
net half-reaction:	$Yb^{3+} + e^- \rightarrow Yb^{2+}$	(3)

Since ΔG is a state function, we can write:

$$\Delta G°_{rxn\,(3)} = \Delta G°_{rxn\,(1)} + \Delta G°_{rxn\,(2)}$$
$$\Delta G°_{Yb^{3+}/Yb^{2+}} = \Delta G°_{Yb^{3+}/Yb} + \Delta G°_{Yb/Yb^{2+}}$$
$$-nFE°_{Yb^{3+}/Yb^{2+}} = -nFE°_{Yb^{3+}/Yb} + (-nFE°_{Yb/Yb^{2+}})$$
$$-(1)E°_{Yb^{3+}/Yb^{2+}} = -(3)(-2.267\ V) - (2)(2.797\ V)$$
$$E°_{Yb^{3+}/Yb^{2+}} = \mathbf{-1.207\ V}$$

21-72. *Refer to Sections 21-15 and 21-16, Exercise 21-46 Solution, and Appendix J.*

Plan: Calculate $E°_{cell}$ for the reaction as written. If $E°_{cell}$ is positive, the reaction is spontaneous.

			$E°$
	reduction:	$2(U^{3+} + 3e^- \rightarrow U)$	-1.798 V
	oxidation:	$3(Mg \rightarrow Mg^{2+} + 2e^-)$	$+2.37$ V
	cell reaction:	$2U^{3+} + 3Mg \rightarrow 2U + 3Mg^{2+}$	$E°_{cell} = +0.57$ V

(a) **Yes**, the setup will work spontaneously because the reaction is spontaneous; $E°_{cell} > 0$.

(b) $E°_{cell} = \mathbf{+0.57\ V}$.

The standard reduction potentials in acidic solution for the species are listed below.

$$E°$$

$$K^+(aq) + e^- \rightarrow K(s) \qquad\qquad -2.925 \text{ V}$$

$$Ca^{2+}(aq) + 2e^- \rightarrow Ca(s) \qquad\qquad -2.87 \text{ V}$$

$$Ni^{2+}(aq) + 2e^- \rightarrow Ni(s) \qquad\qquad -0.25 \text{ V}$$

$$O_2(g) + 2H^+(aq) + 2e^- \rightarrow H_2O_2(aq) \qquad\qquad +0.682 \text{ V}$$

$$H_2O_2(aq) + 2H^+(aq) + 2e^- \rightarrow 2H_2O(\ell) \qquad\qquad +1.77 \text{ V}$$

$$F_2(g) + 2e^- \rightarrow 2F^-(aq) \qquad\qquad +2.87 \text{ V}$$

The voltaic cell with the highest voltage will be the one connecting the K^+/K half-cell with the F_2/F^- half-cell:

		$E°$
reduction:	$F_2 + 2e^- \rightarrow 2F^-$	+2.87 V
oxidation:	$2(K \rightarrow K^+ + e^-)$	+2.925 V
cell reaction:	$F_2 + 2K \rightarrow 2F^- + 2K^+$	$E°_{cell} = +5.80$ V

21-76. *Refer to Section 21-16.*

The tarnish on silver, Ag_2S, can be removed by boiling the silverware in slightly salty water (to improve the water's conductivity) in an aluminum pan. The reaction is an oxidation-reduction reaction that occurs spontaneously, similar to the redox reaction occurring in a voltaic cell. The Ag in Ag_2S is reduced back to silver, while the Al in the pan is oxidized to Al^{3+}.

reduction reaction (at surface of the silverware):	$3(Ag_2S(s) + 2e^- \rightarrow 2Ag(s) + S^{2-}(aq))$
oxidation reaction (at surface of the aluminum pan):	$2(Al(s) \rightarrow Al^{3+}(aq) + 3e^-)$
overall reaction:	$3Ag_2S(s) + 2Al \rightarrow 6Ag(s) + 3 S^{2-}(aq) + 2Al^{3+}(aq)$

21-78. *Refer to Section 21-19.*

The Nernst equation is used to calculate electrode potentials or cell potentials when the concentrations and partial pressures are other than standard state values. The Nernst equation using base 10 logarithms is given by:

$$E = E° - \frac{2.303RT}{nF} \log Q$$

or

$$E = E° - \frac{RT}{nF} \ln Q$$

where

E = potential at nonstandard conditions (V)
$E°$ = standard potential (V)
R = gas constant, 8.314 J/mol·K
T = absolute temperature (K); $T = °C + 273.15°$
F = Faraday's constant, 96485 J/V·mol e^-
n = number of moles of e^- transferred
Q = reaction quotient

Substituting at 25°C,

using base 10 logarithms:

$$E = E° - \frac{(2.303)(8.314)(298.15)}{n(96485)} \log Q$$

$$= E° - \frac{0.0592}{n} \log Q$$

using natural logarithms:

$$E = E° - \frac{(8.314)(298.15)}{n(96485)} \ln Q$$

$$= E° - \frac{0.0257}{n} \ln Q$$

21-80. *Refer to Section 21-19 and Appendix J.*

Balanced reduction half-reaction: $Zn^{2+} + 2e^- \rightarrow Zn$ $\qquad\qquad E^\circ = -0.763$ V

For the standard half-cell, $[Zn^{2+}] = 1\ M$. Substituting these data into the Nernst equation, we have

$$E = E^\circ - \frac{0.0592}{n} \log \frac{1}{[Zn^{2+}]} = -0.763\ \text{V} - \frac{0.0592}{2} \log \frac{1}{1} = -0.763\ \text{V} \qquad \text{Note: } \log 1 = 0$$

Therefore, the Nernst equation predicts that the voltage of a standard half-cell equals E°.

21-82. *Refer to Section 21-19, Example 21-7 and Appendix J.*

Balanced half-reaction: $2H^+(aq) + 2e^- \rightarrow H_2(g)$ $\qquad\qquad E^\circ = 0.000$ V

Given: $\qquad [H^+] = [HClO_4] = 2.00 \times 10^{-4}\ M \qquad\qquad P_{H_2} = 2.35$ atm

$$E = E^\circ - \frac{0.0592}{n} \log \frac{P_{H_2}}{[H^+]^2} = 0.000\ \text{V} - \frac{0.0592}{2} \log \frac{2.35}{(2.00 \times 10^{-4})^2} = 0.000\ \text{V} - 0.230\ \text{V} = \mathbf{-0.230\ V}$$

21-84. *Refer to Section 21-19, Example 21-7 and Appendix J.*

		E°
(a) oxidation half-reaction:	$Zn(s) \rightarrow Zn^{2+}(aq) + 2e^-$	$+0.763$ V
reduction half-reaction:	$Cl_2(g) + 2e^- \rightarrow 2Cl^-(aq)$	$+1.360$ V
cell reaction:	$Zn(s) + Cl_2(g) \rightarrow Zn^{2+}(aq) + 2Cl^-(aq)$	$E^\circ_{cell} = \mathbf{+2.123\ V}$

(b) $[Zn^{2+}] = [ZnCl_2] = 0.21\ M;$ $[Cl^-] = 2 \times [ZnCl_2] = 0.42\ M;$ $P_{Cl_2} = 1.0$ atm

Note: In this cell, $0.21\ M\ ZnCl_2$ is in both half-cell compartments..

$$E_{cell} = E^\circ_{cell} - \frac{0.0592}{n} \log \frac{[Zn^{2+}][Cl^-]^2}{P_{Cl_2}} = +2.123\ \text{V} - \frac{0.0592}{2} \log \frac{(0.21)(0.42)^2}{(1.0)} = \mathbf{+2.165\ V}$$

21-86. *Refer to Section 21-19, Example 21-7 and Appendix J.*

Balanced half-reaction: $\quad F_2(g) + 2e^- \rightarrow 2F^-(aq)$ $\qquad\qquad E^\circ = +2.87$ V

Applying the Nernst equation: $\qquad\qquad\qquad E = E^\circ - \dfrac{0.0592}{n} \log \dfrac{[F^-]^2}{P_{F_2}}$

Substituting, $\qquad\qquad\qquad +2.70\ \text{V} = +2.87\ \text{V} - \dfrac{0.0592}{2} \log \dfrac{(0.34)^2}{P_{F_2}}$

$$0.17\ \text{V} = \frac{0.0592}{2} \log \frac{0.12}{P_{F_2}}$$

$$5.7 = \log \frac{0.12}{P_{F_2}}$$

Taking the antilogarithm of both sides, $\qquad 5.5 \times 10^5 = \dfrac{0.12}{P_{F_2}}$

Therefore, $\qquad\qquad\qquad\qquad P_{F_2} = \mathbf{2.2 \times 10^{-7}\ atm}$

For this non-standard cell:

		$E°$
oxidation at anode:	$Sn \rightarrow Sn^{2+} + 2e^-$	+0.14 V
reduction at cathode:	$2(Ag^+ + e^- \rightarrow Ag)$	+0.7994 V
cell reaction:	$Sn + 2Ag^+ \rightarrow Sn^{2+} + 2Ag$	$E°_{cell} =$ +0.94 V

Shorthand notation: $Sn|Sn^{2+}(7.0 \times 10^{-3}\ M)||Ag^+(0.110\ M)|Ag$

$$E_{cell} = E°_{cell} - \frac{0.0592}{n} \log \frac{[Sn^{2+}]}{[Ag^+]^2} = +0.94\ V - \frac{0.0592}{2} \log \frac{(7.0 \times 10^{-3})}{(0.110)^2} = \mathbf{+0.95\ V}$$

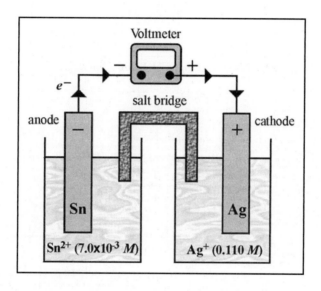

Balanced half-reactions:
(1) $2H^+(aq) + 2e^- \rightarrow H_2(g)$ $E° = 0.0000\ V$
(2) $Ag^+(aq) + e^- \rightarrow Ag(s)$ $E° = 0.7994\ V$

(a) Balanced equation: $H_2(g) + 2Ag^+(aq) \rightarrow 2H^+(aq) + 2Ag(s)$ $E° = 0.7994\ V$

$$E = E° - \frac{0.0592}{n} \log \frac{[H^+]^2}{P_{H_2}[Ag^+]^2} = 0.7994\ V - \frac{0.0592}{2} \log \frac{(1.00 \times 10^{-3})^2}{(8.00)(5.49 \times 10^{-3})^2} = \mathbf{0.870\ V}$$

(b) Balanced equation: $H_2(1.00\ atm) + 2H^+(pH = 3.47) \rightarrow 2H^+(pH = 5.97) + H_2(1.00\ atm)$

For pH = 5.97, $[H^+] = 1.07 \times 10^{-6}\ M$
For pH = 3.47, $[H^+] = 3.39 \times 10^{-4}\ M$

$$E = E° - \frac{0.0592}{n} \log \frac{[H^+]^2 P_{H_2}}{P_{H_2}[H^+]^2} = 0.0000\ V - \frac{0.0592}{2} \log \frac{(1.07 \times 10^{-6})^2(1.00)}{(1.00)(3.39 \times 10^{-4})^2} = \mathbf{0.148\ V}$$

(c) Balanced equation: $H_2(0.0361\ atm) + 2H^+(0.0175\ M) \rightarrow 2H^+(0.0175\ M) + H_2(5.98 \times 10^{-4}\ atm)$

$$E = E° - \frac{0.0592}{n} \log \frac{[H^+]^2 P_{H_2}}{P_{H_2}[H^+]^2} = 0.0000\ V - \frac{0.0592}{2} \log \frac{(0.0175)^2(5.98 \times 10^{-4})}{(0.0361)(0.0175)^2} = \mathbf{0.0527\ V}$$

21-92. *Refer to Section 21-20.*

Balanced half-reaction: $Cu^{2+} + 2e^- \rightarrow Cu$ $\qquad\qquad\qquad E° = 0.337$ V

For a concentration cell:
$$E_{cell} = E°_{cell} - \frac{0.0592}{n} \log \frac{[\text{dilute solution}]}{[\text{concentrated solution}]}$$

Substituting,
$$0.040 \text{ V} = 0 \text{ V} - \frac{0.0592}{2} \log \frac{[Cu^{2+}]_B}{0.75}$$

$$\log \frac{[Cu^{2+}]_B}{0.75} = -1.35$$

Taking the antilogarithm,
$$\frac{[Cu^{2+}]_B}{0.75} = 0.045$$

Therefore,
$$[Cu^{2+}]_B = \mathbf{0.034 \ M}$$

21-94. *Refer to Section 21-19, Exercise 21-93, Example 21-8 and Appendix J.*

(a) Plan: (1) Determine $E°_{cell}$.

(2) Use the Nernst equation to find the ratio of Zn^{2+} to Ni^{2+}.

$\qquad\qquad\qquad\qquad\qquad\qquad\qquad\qquad\qquad\qquad\qquad\qquad\qquad E°$

(1) reduction half-reaction:	$Ni^{2+} + 2e^- \rightarrow Ni$	- 0.25 V
oxidation half-reaction:	$Zn \rightarrow Zn^{2+} + 2e^-$	+0.763 V
cell reaction:	$Ni^{2+} + Zn \rightarrow Ni + Zn^{2+}$	$E°_{cell} = +0.513$ V

(2) Using the Nernst equation:
$$E_{cell} = E°_{cell} - \frac{0.0592}{n} \log \frac{[Zn^{2+}]}{[Ni^{2+}]}$$

Substituting,
$$0 = 0.513 \text{ V} - \frac{0.0592}{2} \log \frac{[Zn^{2+}]}{[Ni^{2+}]}$$

$$\log \frac{[Zn^{2+}]}{[Ni^{2+}]} = 17.33 \text{ or } 17.3 \text{ (3 significant figures)}$$

$$\frac{[Zn^{2+}]}{[Ni^{2+}]} = \mathbf{2 \times 10^{17}} \text{ (1 significant figures)}$$

(b) Since the cell starts at standard conditions, $[Ni^{2+}]_{initial} = [Zn^{2+}]_{initial} = 1.00 \ M$. Also, for every 1 mole of Zn^{2+} produced, there is 1 mole of Ni^{2+} lost. Therefore,

Let x = mol/L of Ni^{2+} that reacted. Then, x = mol/L of Zn^{2+} that were produced.

	Ni^{2+}	+	Zn	$\rightarrow$	Ni	+	Zn^{2+}
initial	1.00 *M*		-		-		1.00 *M*
change	- x *M*						+ x *M*
after reaction	(1.00 - x) *M*						(1.00 + x) *M*

Therefore, $[Zn^{2+}] + [Ni^{2+}] = (1.00 + x) \ M + (1.00 - x) \ M = 2.00 \ M$

We know from (a) that $[Zn^{2+}] = (2 \times 10^{17})[Ni^{2+}]$. Substituting for $[Zn^{2+}]$ and solving for $[Ni^{2+}]$,

$$2.00 \ M = (2 \times 10^{17})[Ni^{2+}] + [Ni^{2+}]$$

$$= [Ni^{2+}](2 \times 10^{17} + 1) \approx (2 \times 10^{17})[Ni^{2+}]$$

$$[Ni^{2+}] = \mathbf{1 \times 10^{-17} \ M}$$

$$[Zn^{2+}] = 2.00 \ M - [Ni^{2+}] = \mathbf{2.00 \ M} \text{ (to 3 significant figures)}$$

Balanced half-reaction: $2H^+(aq) + 2e^- \rightarrow H_2(g)$ $\qquad\qquad\qquad$ $E° = 0.000$ V

In a concentration cell, we have 2 half cells containing the same ions and gases, only at different concentrations and/or partial pressures. The way this question is worded, there may be 2 answers because the cell potential is a function of the square of the ratios of hydrogen ions in the two half cells.

Answer (1) - the cathodic half-cell is at pH = 1.5:

oxidation half-reaction:	$H_2(1\ atm) \rightarrow 2H^+(pH = ?) + 2e^-$
reduction half-reaction:	$2H^+(pH = 1.5) + 2e^- \rightarrow H_2(1\ atm)$
cell reaction:	$2H^+(pH = 1.5) + H_2(1\ atm) \rightarrow 2H^+(pH = ?) + H_2(1\ atm)$

For this concentration cell: $\qquad\qquad E_{cell} = E°_{cell} - \dfrac{0.0592}{n} \log \dfrac{[H^+(?)]^2 P_{H_2}(1\ atm)}{[H^+(pH = 1.5)]^2 P_{H_2}(1\ atm)}$

Substituting, $\qquad\qquad\qquad\qquad 0.275\ V = 0\ V - \dfrac{0.0592}{2} \log \dfrac{[H^+]^2(1)}{(0.032)^2(1)}$

$$\log \frac{[H^+]^2}{0.0010} = -9.29$$

Taking the antilogarithm, $\qquad\qquad \dfrac{[H^+]^2}{0.0010} = 5.1 \times 10^{-10}$

Therefore, $\qquad\qquad\qquad\qquad\qquad [H^+] = 7.2 \times 10^{-7}\ M$

$$pH = 6.15$$

Answer (2) - the anodic half-cell is at pH = 1.5:

oxidation half-reaction:	$H_2(1\ atm) \rightarrow 2H^+(pH = 1.5) + 2e^-$
reduction half-reaction:	$2H^+(pH = ?) + 2e^- \rightarrow H_2(1\ atm)$
cell reaction:	$2H^+(pH = ?) + H_2(1\ atm) \rightarrow 2H^+(pH = 1.5) + H_2(1\ atm)$

For this concentration cell: $\qquad\qquad E_{cell} = E°_{cell} - \dfrac{0.0592}{n} \log \dfrac{[H^+(pH = 1.5)]^2 P_{H_2}(1\ atm)}{[H^+(pH = ?)]^2 P_{H_2}(1\ atm)}$

Substituting, $\qquad\qquad\qquad\qquad 0.275\ V = 0\ V - \dfrac{0.0592}{2} \log \dfrac{(0.032)^2(1)}{[H^+]^2(1)}$

$$\log \frac{0.0010}{[H^+]^2} = -9.29$$

Taking the antilogarithm, $\qquad\qquad \dfrac{0.0010}{[H^+]^2} = 5.1 \times 10^{-10}$

Therefore, $\qquad\qquad\qquad\qquad\qquad [H^+] = 1.4 \times 10^3\ M$

$$pH = -3.14$$

As you can see, Answer (1) is the answer that makes sense. Answer (2) makes no sense, since it is impossible to have a solution with $[H^+] = 1.4 \times 10^3\ M$. So, the cathodic half-cell is at pH = 1.5 and the pH of the anodic half-cell must be **6.15**.

Because $\Delta G° = -nFE°_{cell}$ and $\Delta G° = -RT \ln K$, the signs and magnitudes of $E°_{cell}$, $\Delta G°$ and K are related as shown in the following table for different types of reactions under standard state conditions.

Forward Reaction	E°_{cell}	ΔG°	K
spontaneous	+	−	>1
at equilibrium	0	0	1
non-spontaneous	−	+	<1

From the above equations, it is seen that the value of K is related to the value of ΔG° and E° of the cell, but not ΔG and E of the cell. E°, ΔG° and K are indicators of the thermodynamic tendency of an oxidation-reduction reaction to occur under standard conditions.

On the other hand, E and ΔG are related to the value of Q and are indicators of the spontaneity of a reaction under any given conditions. The reaction proceeds until $Q = K$ at which point $\Delta G = 0$ and $E_{cell} = 0$. Then:

$$\log K = \frac{nFE^\circ_{cell}}{2.303RT} \qquad \text{or} \qquad \ln K = \frac{nFE^\circ_{cell}}{RT}$$

21-100. *Refer to Section 21-21 and Examples 21-9 and 21-10.*

(a)

		E°
reduction:	$MnO_4^- + 8H^+ + 5e^- \rightarrow Mn^{2+} + 4H_2O$	+1.507 V
oxidation:	$5(Fe^{2+} \rightarrow Fe^{3+} + e^-)$	−0.771 V
cell reaction:	$MnO_4^- + 8H^+ + 5Fe^{2+} \rightarrow Mn^{2+} + 4H_2O + 5Fe^{3+}$	$E^\circ_{cell} = $ **+0.736 V**

The reaction is spontaneous as written under standard conditions since $E^\circ_{cell} > 0$.

$\Delta G^\circ = -nFE^\circ_{cell} = -(5 \text{ mol } e^-)(96500 \text{ J/V·mol } e^-)(+0.736 \text{ V}) = $ **−3.55 x 10⁵ J/mol rxn or −355 kJ/mol rxn**

At 25°C, $E^\circ_{cell} = \frac{RT \ln K}{nF}$. So, $\ln K = \frac{nFE^\circ_{cell}}{RT}$

Substituting, $\ln K = \dfrac{(5 \text{ mol})(9.65 \times 10^4 \text{ J/V·mol})(+0.736 \text{ V})}{(8.314 \text{ J/mol·K})(298 \text{ K})} = 143$ 　　　 Solving, $K = e^{143}$ or **1 x 10⁶²**

(b)

		E°
reduction:	$Cu^+ + e^- \rightarrow Cu$	+0.521 V
oxidation:	$Cu^+ \rightarrow Cu^{2+} + e^-$	−0.153 V
cell reaction:	$2Cu^+ \rightarrow Cu + Cu^{2+}$	$E^\circ_{cell} = $ **+0.368 V**

The reaction is spontaneous as written under standard conditions since $E^\circ_{cell} > 0$.

$\Delta G^\circ = -nFE^\circ_{cell} = -(1 \text{ mol } e^-)(96500 \text{ J/V·mol } e^-)(0.368 \text{ V}) = $ **−3.55 x 10⁴ J/mol rxn or −35.5 kJ/mol rxn**

at 25°C, $\ln K = \dfrac{nFE^\circ_{cell}}{RT} = \dfrac{(1 \text{ mol})(9.65 \times 10^4 \text{ J/V·mol})(+0.368 \text{ V})}{(8.314 \text{ J/mol·K})(298 \text{ K})} = 14.3$

Solving, $K = e^{14.3}$ **or 1.6 x 10⁶**

(c)

		E°
reduction:	$2(MnO_4^- + 2H_2O + 3e^- \rightarrow MnO_2 + 4OH^-)$	+0.588 V
oxidation:	$3(Zn + 2OH^- \rightarrow Zn(OH)_2 + 2e^-)$	+1.245 V
cell reaction:	$2MnO_4^- + 3Zn + 4H_2O \rightarrow 2MnO_2 + 3Zn(OH)_2 + 2OH^-$	$E^\circ_{cell} = $ **+1.833 V**

The reaction is spontaneous as written under standard conditions since $E^\circ_{cell} > 0$.

$\Delta G^\circ = -nFE^\circ_{cell} = -(6 \text{ mol } e^-)(96485 \text{ J/V·mol } e^-)(1.833 \text{ V}) = $ **−1.061 x 10⁶ J/mol rxn or −1061 kJ/mol rxn**

at 25°C, $\ln K = \dfrac{nFE^\circ_{cell}}{RT} = \dfrac{(6 \text{ mol})(9.6485 \times 10^4 \text{ J/V·mol})(+1.833 \text{ V})}{(8.314 \text{ J/mol·K})(298.15 \text{ K})} = 428.1$

Solving, $K = e^{428.1}$ **or 8.6 x 10¹⁸⁵**

Note (1): Since the $E°$ value has 4 significant figures, we must use values for the constants (F, R and T) that also have at least 4 significant figures.

Note (2): It is more difficult to find K in scientific notation because most calculators cannot handle numbers this big. So, use what you know about exponents to solve for K:

$$e^{428.1} = e^{200.0} \times e^{228.1} = (7.2 \times 10^{86})(1.2 \times 10^{99}) = (7.2 \times 1.2)(10^{86} \times 10^{99}) = 8.6 \times 10^{185}$$

21-102. *Refer to Section 21-21, Example 21-10, and Appendix J.*

		$E°$
reduction half-reaction:	$Sn^{4+} + 2e^- \rightarrow Sn^{2+}$	$+0.15$ V
oxidation half-reaction:	$2(Fe^{2+} \rightarrow Fe^{3+} + e^-)$	-0.771 V
cell reaction:	$Sn^{4+} + 2Fe^{2+} \rightarrow Sn^{2+} + 2Fe^{3+}$	$E°_{cell} = -0.62$ V

at 25°C, $\ln K = \dfrac{nFE°_{cell}}{RT} = \dfrac{(2 \text{ mol})(96500 \text{ J/V·mol})(-0.62 \text{ V})}{(8.314 \text{ J/mol·K})(298 \text{ K})} = -48$

Solving, $K = e^{-48}$ or 10^{-21}

21-104. *Refer to Section 21-21 and Example 21-10.*

		$E°$
reduction half-reaction:	$PbSO_4 + 2e^- \rightarrow Pb + SO_4{}^{2-}$	-0.356 V
oxidation half-reaction:	$Pb + 2I^- \rightarrow PbI_2 + 2e^-$	$+0.365$ V
cell reaction:	$PbSO_4 + 2I^- \rightarrow PbI_2 + SO_4{}^{2-}$	$E°_{cell} = +0.009$ V

$\ln K = \dfrac{nFE°_{cell}}{RT} = \dfrac{(2 \text{ mol})(96500 \text{ J/V·mol})(+0.009 \text{ V})}{(8.314 \text{ J/mol·K})(298 \text{ K})} = 0.7$ Solving, $K = 2$ (to 1 significant figure)

21-106. *Refer to Sections 21-22, 21-23 and 21-25.*

(a) The dry cell (Leclanchè cell) is shown in Figure 21-16. The container is made of zinc, which also acts as one of the electrodes. The other electrode is a carbon rod in the center of the cell. The cell is filled with a moist mixture of NH_4Cl, MnO_2, $ZnCl_2$ and a porous inert filler. The cell is separated from the zinc container by a porous paper. Dry cells are sealed to keep moisture from evaporating. As the cell operates, the Zn electrode is the anode and is oxidized to Zn^{2+} ions. The ammonium ion is reduced to give NH_3 and H_2 at the carbon cathode. The ammonia produced combines with Zn^{2+} ion and forms a soluble compound containing the complex ion, $Zn(NH_3)_4{}^{2+}$; H_2 is removed by being oxidized by MnO_2. This type of battery cannot be recharged.

(b) The lead storage battery is shown in Figure 21-17. It consists of a group of lead plates bearing compressed spongy lead alternating with a group of lead plates bearing lead(IV) oxide, PbO_2. The electrodes are immersed in a solution of about 40% sulfuric acid. When the cell discharges, the spongy lead is oxidized to give Pb^{2+} ions which then combine with sulfate ions to form insoluble $PbSO_4$, coating the anode. Electrons produced at the anode by oxidation of spongy lead travel through the external circuit to the cathode and reduce lead(IV) to lead(II) in the presence of H^+. The cathode also becomes coated with insoluble lead sulfate. The lead storage battery can be recharged by reversal of all reactions.

(c) The hydrogen-oxygen fuel cell is shown in Figure 21-18. Hydrogen (the fuel) is supplied to the anode compartment. Oxygen is fed into the cathode compartment. Oxygen is reduced at the cathode to OH^- ions. The OH^- ions migrate through the electrolyte, an aqueous solution of a base, to the anode, where H_2 is oxidized to H_2O. The net reaction of the cell is the same as the burning of hydrogen in oxygen to form water, but combustion does not occur. Rather, most of the chemical energy, produced from the destruction of H-H and O-O bonds and the formation of O-H bonds, is converted directly into electrical energy.

(a) When attempting to recharge an Leclanchè cell (a dry cell), the electrodes are reversed; the zinc container which is the anode under normal operation becomes the cathode. The reaction expected is the reduction of Zn^{2+} to zinc metal:

$$Zn^{2+} + 2e^- \rightarrow Zn \qquad\qquad E° = -0.763 \text{ V}$$

(b) Recharging the battery means reversing the actual cell reaction to yield:

$$H_2 + 2NH_3 + Zn^{2+} \rightarrow Zn + 2NH_4^+ \qquad E_{cell} = -1.6 \text{ V}$$

This is essentially an impossible task because each of the original products have been permanently removed from the system, especially the hydrogen gas.

(1) NH_3 and Zn^{2+} have reacted together to give a very stable zinc-ammonia complex:

$$Zn^{2+} + 4NH_3 \rightarrow Zn(NH_3)_4^{2+} \qquad\qquad K = \frac{1}{K_d} = 2.9 \times 10^9 \text{ (Appendix I)}$$

The zinc complex is more difficult than the free Zn^{2+} to reduce as deduced from the more negative standard reduction potential:

$$Zn(NH_3)_4^{2+} + 2e^- \rightarrow Zn + 4NH_3. \qquad E° = -1.04 \text{ V}$$

(2) H_2 has reacted with MnO_2 to give the solid, $MnO(OH)$: $H_2 + 2MnO_2 \rightarrow 2MnO(OH)$.

Since there are essentially none of the original products, the recharging cannot occur.

A fuel cell is different from a dry cell or storage cell because:

(1) the reactant, the fuel (usually H_2) and oxygen are fed into the cell continuously and the products are constantly removed. Hence, the fuel cell creates electrical energy, but does not store it. It can operate indefinitely as long as fuel is available.

(2) The electrodes are made of an inert material such as platinum and do not react during the electrochemical process.

(3) Many fuel cells are non-polluting, e.g., the H_2/O_2 fuel cell whose only product is H_2O.

Consider: $Mg(s)|Mg^{2+}(aq)||Fe^{3+}(aq)|Fe(s)$

(a) oxidation half reaction (at anode): $\qquad\qquad 3(Mg(s) \rightarrow Mg^{2+}(aq) + 2e^-) \qquad E° = +2.37 \text{ V}$

reduction half-reaction (at cathode): $\qquad\quad 2(Fe^{3+}(aq) + 3e^- \rightarrow Fe(s)) \qquad\quad E° = -0.036 \text{ V}$

overall cell reaction: $\qquad\qquad\qquad\quad 3Mg(s) + 2Fe^{3+}(aq) \rightarrow 3Mg^{2+}(aq) + 2Fe(s)$

(b) $E°_{cell} = E°_{anode} + E°_{cathode} = (+2.37 \text{ V}) + (-0.036 \text{ V}) = \textbf{+2.33 V}$

Note: $E°_{cathode}$ is the standard reduction potential and $E°_{anode}$ is the standard oxidation potential.

(c) $E = E° - \dfrac{0.0592}{n} \log \dfrac{[Mg^{2+}]^3}{[Fe^{3+}]^2} = +2.33 \text{ V} - \dfrac{0.0592}{6} \log \dfrac{(1.00 \times 10^{-3})^3}{(10.0)^2} = +2.33 \text{ V} - (-0.109 \text{ V}) = \textbf{+2.44 V}$

(d) The minimum mass change of the magnesium electrode is the mass of Mg lost when 150 mA passes through the cell for 20.0 minutes.

$$? \text{ g Mg} = 20.0 \text{ min} \times \frac{60 \text{ s}}{1 \text{ min}} \times \frac{0.150 \text{ C}}{1 \text{ s}} \times \frac{1 \text{ mol } e^-}{96500 \text{ C}} \times \frac{1 \text{ mol Mg}}{2 \text{ mol } e^-} \times \frac{24.30 \text{ g}}{1 \text{ mol Mg}} = \textbf{0.0227 g Mg}$$

Balanced half-reaction: $Cu^{2+} + 2e^- \rightarrow Cu$

$? \text{ coulombs} = 0.0300 \text{ L soln} \times \dfrac{0.165 \text{ mol Cu}}{1 \text{ L soln}} \times \dfrac{2 \text{ mol } e^-}{1 \text{ mol Cu}} \times \dfrac{96500 \text{ C}}{1 \text{ mol } e^-} = \textbf{955 C}$

21-116. *Refer to Sections 21-19 and 21-20.*

		$E°$
reduction half-reaction:	$Mn \rightarrow Mn^{2+} + 2e^-$	$+1.18$ V
oxidation half-reaction:	$Fe^{2+} + 2e^- \rightarrow Fe$	-0.44 V
cell reaction:	$Mn + Fe^{2+} \rightarrow Mn^{2+} + Fe$	$E°_{cell} = +0.74$ V

(a) Using the Nernst equation:

$$E_{cell} = E°_{cell} - \dfrac{0.0592}{n} \log \dfrac{[Mn^{2+}]}{[Fe^{2+}]}$$

Substituting,

$$1.55 \text{ V} = 0.74 \text{ V} - \dfrac{0.0592}{2} \log \dfrac{[Mn^{2+}]}{[Fe^{2+}]}$$

$$\log \dfrac{[Mn^{2+}]}{[Fe^{2+}]} = -27$$

$$\dfrac{[Mn^{2+}]}{[Fe^{2+}]} = 10^{-27}$$

(b) The anode is manganese and the cathode is iron. Since the electrons always flow from anode to cathode, they are flowing from the manganese anode to the iron cathode.

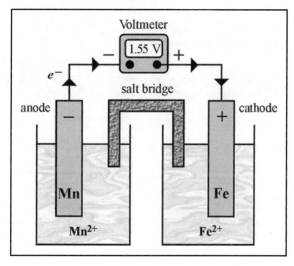

21-118. *Refer to Section 21-6.*

Balanced equations: $UO_2(s) + 4HF(g) \rightarrow UF_4(s) + 2H_2O(\ell)$
$\qquad\qquad\qquad\quad UF_4(s) + 2Mg(s) \rightarrow U(s) + 2MgF_2(s)$

(a) ox. no. U in $UO_2(s)$: $+4$

(b) ox. no. U in $UF_4(s)$: $+4$

(c) ox. no. U in $U(s)$: 0

(d) reducing agent: Mg(s)

(e) substance reduced (oxidizing agent): $UF_4(s)$

(f) U is being reduced from +4 oxidation number in $UF_4(s)$ to 0 in $U(s)$. Therefore, 4 moles of electrons are required to reduce 1 mole of $UF_4(s)$.

$$? \text{ coulombs/s} = \frac{0.500 \text{ g } UF_4}{1 \text{ min}} \times \frac{1 \text{ min}}{60 \text{ s}} \times \frac{1 \text{ mol } UF_4}{314 \text{ g } UF_4} \times \frac{4 \text{ mol } e^-}{1 \text{ mol } UF_4} \times \frac{96500 \text{ C}}{1 \text{ mol } e^-} = \textbf{10.2 C/s} \textbf{ or } \textbf{10.2 A}$$

(g) Plan: (1) Determine the number of moles of $HF(g)$.
       (2) Calculate the volume of $HF(g)$ using the ideal gas law, $PV = nRT$.

(1) $? \text{ mol HF} = 0.500 \text{ g U} \times \dfrac{1 \text{ mol U}}{238 \text{ g U}} \times \dfrac{1 \text{ mol } UF_4}{1 \text{ mol U}} \times \dfrac{4 \text{ mol HF}}{1 \text{ mol } UF_4} = 8.40 \times 10^{-3} \text{ mol HF}$

(2) $V = \dfrac{nRT}{P} = \dfrac{(8.40 \times 10^{-3} \text{ mol})(0.0821 \text{ L·atm/mol·K})(298 \text{ K})}{(10.0 \text{ atm})} = \textbf{0.0206 L HF}(g)$

(h) Plan: Determine the mass of U that can be prepared from 0.500 g Mg and compare.

$? \text{ g U} = 0.500 \text{ g Mg} \times \dfrac{1 \text{ mol Mg}}{24.30 \text{ g Mg}} \times \dfrac{1 \text{ mol U}}{2 \text{ mol Mg}} \times \dfrac{238.0 \text{ g U}}{1 \text{ mol U}} = 2.45 \text{ g U}$

Yes, 0.500 g Mg is more than enough to prepare 0.500 g U. In fact, 0.500 g Mg can ideally produce 2.45 g U.

21-120. *Refer to Section 21-7 and Figure 21-5.*

(a) Electroplating is a process that plates metal onto a cathodic surface by electrolysis.

(b) A simple silver electroplating apparatus for a jeweler consists of a dc generator (a battery) with the negative lead attached to the piece of jewelry (cathode) and the positive lead attached to a piece of silver metal (anode). The jewelry and the silver metal are both immersed in a beaker containing an aqueous solution of a silver salt such as $AgNO_3$. During electroplating, the Ag metal at the anode will be oxidized to Ag^+ ions, and the Ag^+ ions in solution will be reduced to Ag metal and plated onto the jewelry at the cathode.

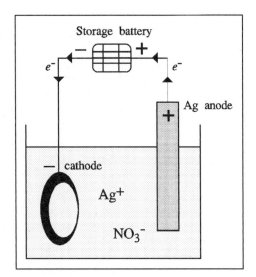

(c) Highly purified silver as the anode is not necessary in an electroplating operation. As the electrolytic cell operates, Ag and other metal impurities in a regular Ag anode oxidize to form metal cations in solution. However, only Ag^+ ions are reduced to Ag metal at the cathode because of its ease of reduction and higher concentration. Such a preference can be enhanced by controlling the operating voltage just above the threshold required to electroplate silver. So, the extra cost of a highly purified silver anode would make its purchase an unwise decision.

21-122. *Refer to Section 21-4, Table 21-2 and Appendix J.*

It is not possible to prepare F_2 by electrolysis of an aqueous NaF solution. In electrolysis, the most easily oxidized and reduced species are the ones involved. To prepare F_2, the oxidation of F^- would have to occur. However, water is more easily oxidized than is F^-, as seen by its position in the standard reduction potential chart (Appendix J and below). By inspection, H_2O is a stronger reducing agent than F^- because the reduction half-reaction has a less positive E^o. So H_2O's oxidation is preferable to F^-'s oxidation. F_2 can be prepared from molten NaF, but not aqueous NaF.

$O_2(g) + 4H^+(aq) + 4e^- \rightarrow 2H_2O(\ell)$ +1.229 V

$F_2(g) + 2e^- \rightarrow 2F^-(aq)$ +2.87 V

352

21-124. *Refer to Section 21-7, Exercise 21-120 Solution, and Figure 21-5.*

In an electroplating process, once the concentration of the element of interest, e.g. Cu^{2+}, is sufficiently low, the impurity metal cations will start plating out of solution onto the object (cathode - source of electrons). The order of plating depends both on the reduction potentials of the metals and their concentrations. If we assume that the impurity concentrations are the same in the solution, then the most easily reduced species will be the first one to plate out. This is the metal with the most positive standard reduction potential. Of the metals given in the problem, the order in which the metals will plate out is:

(1) $Au^+ + e^- \rightarrow Au$ +1.68 V or
     $Au^{3+} + 3e^- \rightarrow Au$ +1.50 V

(2) $Pt^{2+} + 2e^- \rightarrow Pt$ +1.2 V

(3) $Ag^+ + e^- \rightarrow Ag$ +0.7994 V

(4) $Fe^{3+} + 3e^- \rightarrow Fe$ −0.036 V or
     $Fe^{2+} + 2e^- \rightarrow Fe$ −0.44 V

(5) $Zn^{2+} + 2e^- \rightarrow Zn$ −0.763 V

Yes, the electrolytic process can be used to individually separate the impurity metals if their reduction potentials are sufficiently different.

21-126. *Refer to Sections 21-15 and 21-16.*

(a) Given:

			$E°$
(1) $H_2O/H_2,OH^-$	$2H_2O + 2e^- \rightarrow H_2 + 2OH^-$		−0.828 V
(2) H^+/H_2	$2H^+ + 2e^- \rightarrow H_2$		0.0000 V
(3) $O_2,H^+/H_2O$	$O_2 + 4H^+ + 4e^- \rightarrow 2H_2O$		+1.229 V
(4) $O_2,H_2O/OH^-$	$O_2 + 2H_2O + 4e^- \rightarrow 4OH^-$		+0.401 V

Combining half-reactions **(1)** and **(3)** would give the greatest voltage:

$E°_{cell} = E°_{cathode} + E°_{anode} = +1.229$ V $+ 0.828$ V $= +2.057$ V

Note: $E°_{cathode}$ is the standard reduction potential and $E°_{anode}$ is the standard oxidation potential.

(b) reduction at cathode: $O_2 + 4H^+ + 4e^- \rightarrow 2H_2O$

oxidation at anode: $2(H_2 + 2OH^- \rightarrow 2H_2O + 2e^-)$

cell reaction: $O_2 + 2H_2 + 4H^+ + 4OH^- \rightarrow 6H_2O$

     or $O_2 + 2H_2 + 4H_2O \rightarrow 6H_2O$

     or $O_2 + 2H_2 \rightarrow 2H_2O$

21-128. *Refer to Section 21-21 and Appendix J.*

(a) Plan: The K_{sp} value for $AgBr(s)$ is the equilibrium constant for: $AgBr(s) \rightleftarrows Ag^+(aq) + Br^-(aq)$. It can be estimated from data in Appendix J. Choose the appropriate oxidation and reduction half-reactions that produce the above reaction and calculate $E°_{cell}$ and K_{sp} at 25°C.

		$E°$
reduction half-reaction:	$AgBr(s) + e^- \rightarrow Ag(s) + Br^-(aq)$	+0.10 V
oxidation half-reaction:	$Ag(s) \rightarrow Ag^+(aq) + e^-$	−0.7994 V
cell reaction:	$AgBr(s) \rightarrow Ag^+(aq) + Br^-(aq)$	$E°_{cell} = -0.70$ V

$\ln K = \ln K_{sp} = \dfrac{nFE°_{cell}}{RT} = \dfrac{(1 \text{ mol})(96500 \text{ J/V·mol})(-0.70 \text{ V})}{(8.314 \text{ J/mol·K})(298 \text{ K})} = -27$ Solving, $K_{sp} = 10^{-12}$

(From Appendix H, K_{sp} for AgBr = 3.3 x 10^{-13})

(b) $\Delta G° = -nFE°_{cell} = -(1 \text{ mol})(96500 \text{ 1/V·mol})(-0.70 \text{ V}) = $ **+68,000 J/mol rxn or +68 kJ/mol rxn**

Balanced equations: (i) $\frac{1}{3}Al^{3+} + e^- \rightarrow \frac{1}{3}Al$ $\Delta G° = 160.4$ kJ/mol rxn

(ii) $Al^{3+} + 3e^- \rightarrow Al$ $\Delta G° = 481.2$ kJ/mol rxn

(i) $E° = -\dfrac{\Delta G°}{nF} = -\dfrac{(+160400 \text{ J})}{(1 \text{ mol } e^-)(96485 \text{ J/V·mol } e^-)} = \mathbf{-1.662}$ **V**

(ii) $E° = -\dfrac{\Delta G°}{nF} = -\dfrac{(+481200 \text{ J})}{(3 \text{ mol } e^-)(96485 \text{ J/V·mol } e^-)} = \mathbf{-1.662}$ **V**

21-132. *Refer to Section 21-21.*

When water is electrolyzed with copper electrodes or using other common metals, the amount of $O_2(g)$ is less than when Pt electrodes are used, but the amount of $H_2(g)$ produced is independent of electrode material. Why does this happen? In electrolysis, the most easily oxidized species is oxidized and the most easily reduced species is reduced. If we compare Cu and H_2O by looking on the standard reduction potentials chart (data given below), we see that Cu is a stronger reducing agent than H_2O, because 0.337 V is less than 0.828 V. This means that Cu is more easily oxidized than water.

$$Cu^{2+}(aq) + 2e^- \rightarrow Cu(s) \qquad E° = 0.337 \text{ V}$$
$$O_2(g) + 4H^+(aq) + 4e^- \rightarrow 2H_2O(\ell) \qquad E° = 0.828 \text{ V}$$
$$Pt^{2+}(aq) + 2e^- \rightarrow Pt(s) \qquad E° = 1.23 \text{ V}$$

When Cu is used as an electrode, Cu will oxidize more easily than water and so, Cu^{2+} (from Cu) will be formed rather than O_2 (from H_2O), thereby lessening the amount of O_2 gas produced. You can also see that H_2O ($E° = 0.828$ V) is more easily oxidized than Pt ($E° = 1.23$ V), so a platinum electrode will not oxidize during the electrolysis of water as long as there is water present.

On the other hand, the only component present in the electrolysis cell that can be reduced is water, no matter what metal the electrode material is made from. So the formation of H_2 gas by the reduction of water is unaffected when platinum, Cu or most other common metals are used as electrodes.

22 Metals I: Metallurgy

Metallurgy is the commercial extraction of metals from their ores and the preparation of metals for use. It includes

   (1) mining the ore,
   (2) pretreatment of the ore,
   (3) reduction of the ore to the free metal,
   (4) refining or purifying the metal, and
   (5) alloying, if needed.

22-4.   *Refer to Section 22-2 and Figure 22-2.*

High density sulfide ores can be separated from the less dense gangue after pulverization by several methods. One way involves a cyclone separator in which the lighter impurities are blown away.

22-6.   *Refer to Section 22-1, Table 22-1 and Figure 22-1.*

Anion Name	Formula	Example	Mineral Name
oxide	O^{2-}	Fe_2O_3	hematite
sulfide	S^{2-}	Cu_2S	chalcocite
chloride	Cl^-	$NaCl$	halite (rock salt)
carbonate	CO_3^{2-}	$CaCO_3$	limestone
sulfate	SO_4^{2-}	$BaSO_4$	barite
silicate	Si_xO_y	$Al_2(Si_2O_8)(OH)_4$	kaolinite

The silicates are the most widespread minerals. However, extraction of metals from silicates is very difficult.

22-8.   *Refer to Section 22-2.*

The flotation method of separating a crushed ore from the gangue is a physical separation method used with ores, e.g., sulfides, carbonates or silicates, which either are not "wet" by water or can be made water repellent by treatment. Their surfaces are covered by layers of oil or other flotation agents. A stream of air is blown through a swirled suspension of such an ore in a mixture of water and oil; bubbles form on the oil surfaces of the mineral particles, causing them to rise to the surface of the suspension. The bubbles are prevented from breaking and escaping by a layer of oil and emulsifying agent. A frothy ore concentrate forms on the surface. No chemical changes are involved.

22-10.   *Refer to Section 22-1.*

Metals likely to be found in the free state in nature include Cu, Ag and Au, and the less abundant metals, such as Pt, Os, Ir, Ru, Rh and Pd.

These are the less active metals with positive reduction potentials. They are transition metals and can be found in Group 8B and 1B.

22-12. _Refer to Section 22-3 and Table 22-2._

Aluminum and the metals of Groups 1A and 2A are metals which are easily oxidized to ions that are difficult to reduce. So, we predict that electrolysis would be required to obtain the free metals from the molten, anhydrous salts, KCl, Al_2O_3 and $MgSO_4$.

22-14. _Refer to Sections 22-4, 21-4 and 21-6, and Table 22-3._

For the electrolysis of a brine solution:

oxidation half-reaction:	$2Cl^- \rightarrow Cl_2 + 2e^-$
reduction half-reaction:	$2e^- + 2H_2O \rightarrow 2OH^- + H_2$
balanced net ionic equation:	$2Cl^- + 2H_2O \rightarrow 2OH^- + H_2 + Cl_2$
formula unit equation:	$2NaCl + 2H_2O \rightarrow 2NaOH + H_2 + Cl_2$ (Na^+ is the spectator ion)

If 1 mole of electrons, i.e., 1 faraday, passes through the cell at 100% efficiency, then 1 mole of NaOH, 1/2 mole of H_2 and 1/2 mole of Cl_2 are produced.

Therefore,   ? g NaOH produced = 1 mol = **39.997 g NaOH**

? g H_2 produced = 0.5 mol = **1.008 g H$_2$**

? g Cl_2 produced = 0.5 mol = **35.45 g Cl$_2$**

22-16. _Refer to Table 22-3._

(a) $2Al_2O_3$ (cryolite solution) $\xrightarrow{\text{electrolysis}}$ $4Al(\ell) + 3O_2(g)$ involves reduction of Al^{3+} to elemental Al

(b) $PbSO_4(s) + PbS(s) \rightarrow 2Pb(\ell) + 2SO_2(g)$ involves reduction of Pb^{2+} to elemental Pb

(c) $2TaCl_5(g) + 5Mg(\ell) \rightarrow 2Ta(s) + 5MgCl_2(\ell)$ involves reduction of Ta^{5+} to elemental Ta

22-18. _Refer to Section 22-7._

The basic oxygen furnace is used to purify pig iron, which is the iron obtained from the blast furnace process. It is impure and contains carbon, among other substances, but it can be converted to steel by burning out most of the carbon with oxygen in a basic oxygen furnace. The method involves blowing oxygen through the molten iron at high temperatures. The carbon is converted to carbon monoxide and finally to carbon dioxide.

22-20. _Refer to Section 22-5._

(a) Magnesium hydroxide can be precipitated from seawater by adding calcium hydroxide:

$$Mg^{2+}(aq) + Ca(OH)_2(s) \rightarrow Mg(OH)_2(s) + Ca^{2+}(aq)$$

This reaction occurs because K_{sp} for $Mg(OH)_2$, 1.5×10^{-11}, is much smaller than that for $Ca(OH)_2$, 7.9×10^{-6}. So, $Mg(OH)_2$ is much less soluble than $Ca(OH)_2$.

(b) This process cannot be used for removing sodium ions from seawater because NaOH is very soluble.

22-22.   *Refer to Sections 22-7 and 22-9.*

(a)   The procedure for obtaining Fe from Fe_2O_3 or Fe_3O_4 is as follows:

    (1)   The oxides are reduced in blast furnaces by CO.   First, coke (C), limestone ($CaCO_3$) and the crushed ore (Fe_2O_3 or Fe_3O_4 in very hard SiO_2 rock) are loaded into the top of the blast furnace.

    (2)   Most of the oxides are reduced to molten iron by CO, although some are reduced by coke directly.   Carbon dioxide, a reaction product, reacts with excess coke to provide more CO to reduce the next charge of iron ore.

    (3)   The obtained product contains C as an impurity and is called pig iron.   It can be remelted and cooled into cast iron.   Alternatively, if some C is removed and other metals, such as Mn, Cr, Ni, W, Mo and V, are added to increase the tensile strength, the mixture is known as steel.

(b)   The procedure for obtaining Au from very low grade ores by the cyanide process is as follows:

    (1)   The ore containing native Au is mixed with a solution of NaCN and converted to an aqueous slurry.

    (2)   Air is bubbled through the agitated slurry to oxidize the gold metal to a water soluble complex ion, $[Au(CN)_2]^-$.

    (3)   Free gold can then be regenerated by reduction of $[Au(CN)_2]^-$ with zinc or by electrolytic reduction.

22-24.   *Refer to Sections 22-6 and Figure 22-7.*

(a)   The Hall-Héroult process has been the standard industrial method for converting purified Al_2O_3 to pure aluminum. To avoid electrolyzing Al_2O_3(molten) at temperatures above 2045°C (its melting point), the aluminum oxide is mixed with cryolite, $Na_3[AlF_6]$.   The molten mixture is electrolyzed at the much lower temperature of 1000°C with carbon electrodes.   This method is cheaper, but there are still costs incurred.   For example, the carbon anode oxidizes to CO_2 gas and must be replaced regularly.   A new more economical commercial method is the Alcoa chlorine process   The purified Al_2O_3 is converted first to $AlCl_3$, which melts at about 190°C, then electrolyzed to form aluminum.   The chlorine is recovered and reused.

(b)   The Alcoa chlorine process uses about 30% as much electrical energy as the Hall-Héroult process. $AlCl_3$ melts at a much lower temperature than the $Al_2O_3/Na_3[AlF_6]$ mixture, so less energy is required to heat the electrolysis container.   The product, chlorine gas, is recycled in the Alcoa chlorine process, which keeps the cost down.   Also the electrodes do not have to be replaced, as they do in the Hall-Héroult process.

(c)   The Alcoa chlorine process is more dangerous to the workers because chlorine gas is toxic.

22-26.   *Refer to Section 22-7 and Table 22-1.*

Fe-containing minerals:	Fe_2O_3	hematite	oxidation number of Fe = +3
	Fe_3O_4	magnetite	average oxidation number of Fe = +8/3 (contains 2 Fe^{3+} and 1 Fe^{2+} ion)
	FeS_2	iron pyrite	oxidation number of Fe = +2
	$CuFeS_2$	chalcopyrite	oxidation number of Fe = +2

22-28.   *Refer to Section 22-7.*

(a)   Pure iron is an element consisting only of Fe atoms.   Steel is a mixture of Fe, iron carbide (Fe_3C), and other metals such as Mn, Cr, Ni, W, Mo and V.

(b)   Steel is produced from pig iron by burning out some of the carbon with O_2 in a basic oxygen furnace, converting the carbon to CO to CO_2 in an oxidizing process, then adding metals, such as Mn, Cr, Ni, W, Mo and V in an alloying process.

Impure metallic Cu obtained from the chemical reduction of Cu_2S and CuS can be refined with the following arrangement.

(1) Thin sheets of very pure Cu are made cathodes by connecting them to the negative terminal of a d.c. generator. Impure chunks of copper connected to the positive terminal function as anodes. The electrodes are immersed in a solution of $CuSO_4$ and H_2SO_4.

(2) When the cell operates, Cu from impure anodes is oxidized and goes into solution as Cu^{2+} ions; Cu^{2+} ions from the solution are reduced and plate out as metallic Cu on the pure Cu cathode.

22-32. *Refer to Section 24-11.*

Sulfur dioxide is a colorless, poisonous corrosive gas. It causes coughing and nose, throat and lung irritation, even in small quantities. A primary consequence of releasing SO_2 into the air is the production of acid rain:

$$SO_2(g) + H_2O(\ell, rain) \rightarrow H_2SO_3(aq).$$

The SO_2 gas also reacts with atmospheric O_2 to form SO_3 gas, which then reacts with H_2O to form sulfuric acid, H_2SO_4. This is another component of acid rain which attacks stone buildings and is harmful to plants.

22-34. *Refer to Section 22-7.*

The impurities in pig iron, the iron formed in a blast furnace, that make it brittle include four elements: phosphorus and silicon, two elements that came from the silicate and phosphate minerals that contaminated the original ore, and carbon and sulfur that came from the coke.

22-36. *Refer to Sections 22-7, 15-8 and 15-16.*

Balanced equation: $FeO(s) + CO(g) \rightarrow Fe(s) + CO_2(g)$

(a) $\Delta H^\circ_{800} = [\Delta H^\circ_f\, Fe(s) + \Delta H^\circ_f\, CO_2(g)] - [\Delta H^\circ_f\, FeO(s) + \Delta H^\circ_f\, CO(g)]$
    $= [(0\ kJ) + (-394\ kJ)] - [(-268\ kJ) + (-111\ kJ)]$
    $= \mathbf{-15\ kJ/mol\ rxn}$

Yes, this is a favorable enthalpy change, because the value is negative, implying an exothermic reaction.

(b) $\Delta G^\circ_{800} = [\Delta G^\circ_f\, Fe(s) + \Delta G^\circ_f\, CO_2(g)] - [\Delta G^\circ_f\, FeO(s) + \Delta G^\circ_f\, CO(g)]$
    $= [(0\ kJ) + (-396\ kJ)] - [(-219\ kJ) + (-182\ kJ)]$
    $= \mathbf{+5\ kJ/mol\ rxn}$

No, this is not a spontaneous reaction since $\Delta G > 0$.

(c) Recall the Gibbs free energy change equation equation: $\Delta G = \Delta H - T\Delta S$

   Therefore, $\Delta S^\circ_{800} = \dfrac{\Delta H^\circ_{800} - \Delta G^\circ_{800}}{T} = \dfrac{(-15\ kJ) - (+5\ kJ)}{800\ K} = \mathbf{-0.025\ kJ/(mol\ rxn)\cdot K}$ or $\mathbf{-25\ J/(mol\ rxn)\cdot K}$

22-38. *Refer to Sections 22-8 and 4-4.*

(a) Balanced equation:
$$\overset{+1\ -2}{2Cu_2S(\ell)} + \overset{0}{3O_2(g)} \rightarrow \overset{+1\ -2}{2Cu_2O(\ell)} + \overset{+4\ -2}{2SO_2(g)}$$

oxidizing agent: O_2 (O is being reduced)   reducing agent: Cu_2S (S is being oxidized)

In the balanced equation, the total increase in oxidation number equals the total decease in oxidation number.

This is true in this example:

increase in oxidation number $= |[2 \times \text{ox. no. S in } SO_2] - [2 \times \text{ox. no. S in } Cu_2S]|$

$\qquad\qquad\qquad\qquad\qquad = |[2 \times (+4)] - [2 \times (-2)]|$

$\qquad\qquad\qquad\qquad\qquad = 12$

decrease in oxidation number $= |[2 \times \text{ox no. O in } Cu_2O + 4 \times \text{ox. no. O in } SO_2] - [6 \times \text{ox. no. O in } O_2]|$

$\qquad\qquad\qquad\qquad\qquad = |[2 \times (-2) + 4 \times (-2)] - [6 \times 0]|$

$\qquad\qquad\qquad\qquad\qquad = 12$

(b) Balanced equation: $\overset{\text{+1 \ -2}}{2Cu_2O(\ell)} + \overset{\text{+1 -2}}{Cu_2S(\ell)} \rightarrow \overset{0}{6Cu(\ell)} + \overset{\text{+4 -2}}{SO_2(g)}$

oxidizing agent: Cu_2O (Cu is being reduced)
$\qquad\qquad\qquad Cu_2S$ (Cu is being reduced)

reducing agent: Cu_2S (S is being oxidized)

The total increase in oxidation number equals the total decrease in oxidation number, so the reaction is balanced.

increase in oxidation number $= |[1 \times \text{ox. no. S in } SO_2] - [1 \times \text{ox. no. S in } Cu_2S]|$

$\qquad\qquad\qquad\qquad\qquad = |[1 \times (+4)] - [1 \times (-2)]|$

$\qquad\qquad\qquad\qquad\qquad = 6$

decrease in oxidation number $= |[6 \times \text{ox. no. Cu in free Cu}]$

$\qquad\qquad\qquad\qquad\qquad\quad - [4 \times \text{ox. no. Cu in } Cu_2O + 2 \times \text{ox. no. Cu in } Cu_2S]|$

$\qquad\qquad\qquad\qquad\qquad = |[6 \times 0] - [4 \times (+1) + 2 \times (+1)]|$

$\qquad\qquad\qquad\qquad\qquad = 6$

22-40.   *Refer to Section 21-6 and Example 21-1.*

Balanced half-reaction: $Cu^{2+} + 2e^- \rightarrow Cu$

$? \text{ g Cu} = 6.50 \text{ hr} \times \dfrac{3600 \text{ s}}{1 \text{ hr}} \times \dfrac{2.50 \text{ C}}{1 \text{ s}} \times \dfrac{1 \text{ mol } e^-}{96500 \text{ C}} \times \dfrac{1 \text{ mol Cu}}{2 \text{ mol } e^-} \times \dfrac{63.55 \text{ g Cu}}{1 \text{ mol Cu}} = \textbf{19.3 g Cu}$

22-42.   *Refer to Sections 22-6 and 21-6.*

Plan: (1) Calculate the time necessary to convert all of the aluminum in 55 pounds of Al_2O_3 to aluminum metal.

$\qquad$ (2) Calculate the number of moles of diatomic oxygen in 55 pounds of Al_2O_3 and use the ideal gas law, $PV = nRT$, to determine the volume of O_2 gas produced.

(1) Balanced reduction half-reaction: $Al^{3+} + 3e^- \rightarrow Al$

$? \text{ time (s)} = 55.0 \text{ lb } Al_2O_3 \times \dfrac{454 \text{ g } Al_2O_3}{1.00 \text{ lb } Al_2O_3} \times \dfrac{1 \text{ mol } Al_2O_3}{102 \text{ g } Al_2O_3} \times \dfrac{2 \text{ mol Al}}{1 \text{ mol } Al_2O_3} \times \dfrac{3 \text{ mol } e^-}{1 \text{ mol Al}} \times \dfrac{96500 \text{ C}}{1 \text{ mol } e^-} \times \dfrac{1 \text{ s}}{0.900 \text{ C}}$

$\qquad\qquad = \textbf{1.57} \times \textbf{10}^8 \textbf{ s}$ or **4.99 years**

(2) Let us assume that all of the oxygen in Al_2O_3 is converted to $O_2(g)$.

$? \text{ mol } O_2 = 55.0 \text{ lb } Al_2O_3 \times \dfrac{454 \text{ g } Al_2O_3}{1.00 \text{ lb } Al_2O_3} \times \dfrac{1 \text{ mol } Al_2O_3}{102 \text{ g } Al_2O_3} \times \dfrac{3 \text{ mol O}}{1 \text{ mol } Al_2O_3} \times \dfrac{1 \text{ mol } O_2}{2 \text{ mol O}} = 3.67 \times 10^2 \text{ mol } O_2$

$V = \dfrac{nRT}{P} = \dfrac{(3.67 \times 10^2 \text{ mol})(0.0821 \text{ L·atm/mol·K})(125°C + 273°)}{(765/760) \text{ atm}} = \textbf{1.19} \times \textbf{10}^4 \textbf{ L } \textbf{O}_2$

Balanced equation: $HgS(s) + O_2(g) \rightarrow Hg(\ell) + SO_2(g)$

We know from the equation that for every 1 mole of HgS that reacts (232.7 g), 1 mole of SO_2 is formed (64.07 g). We can use this ratio as a unit factor. For every 232.7 tons of HgS that react, 64.07 tons of SO_2 are formed.

$$? \text{ tons } SO_2 = 355 \text{ tons of HgS} \times \frac{64.07 \text{ tons } SO_2}{232.7 \text{ tons of HgS}} = \textbf{97.7 tons } SO_2$$

22-46. *Refer to Sections 22-7 and Exercise 22-44.*

Balanced equations: $2C(coke) + O_2(g) \rightarrow 2CO(g)$
$Fe_2O_3(s) + 3CO(g) \rightarrow 2Fe(s) + 3CO_2(g)$

Plan: Rather than convert from tons to grams to moles and back again, let's find the mole-to-mole ratio between Fe_2O_3 and coke, then use a unit factor similar to that used in Exercise 22-44.

$$? \text{ mol } C = 1 \text{ mol } Fe_2O_3 \times \frac{3 \text{ mol } CO}{1 \text{ mol } Fe_2O_3} \times \frac{2 \text{ mol } C}{2 \text{ mol } CO} = 3 \text{ mol } C$$

So, 3 moles of C (3 × 12.01 g = 36.03 g) is required to react with 1 mole of Fe_2O_3 (159.7 g). Therefore, 36.03 tons of C are required to react with 159.7 tons of hematite at 100% efficiency.

$$? \text{ tons } C(coke) = 125 \text{ tons } Fe_2O_3 \times \frac{36.03 \text{ tons } C}{159.7 \text{ tons } Fe_2O_3} \times \frac{100}{98.0} = \textbf{28.8 tons } C(coke)$$

22-48. *Refer to Sections 22-3, 22-6 and 22-7, Table 22-2 and Appendix K.*

The data from Appendix K gives:

Compound	ΔH°_f		Extractive Metallurgy Method
HgS(s)	−58.2	kJ/mol	heating HgS
$Fe_2O_3(s)$	−824.2	kJ/mol	chemical reduction by CO
$Al_2O_3(s)$	−1676	kJ/mol	electrolysis of molten Al_2O_3

As the heats of formation of minerals become more exothermic, i.e., more negative, their thermodynamic stability increases. And so the difficulty by which free metals can be extracted from the minerals also increases. In other words, the more active is the metal, the easier it is to form compounds and the more difficult it is to retrieve the metal from its compounds. This relationship is obvious in the methods by which the metals are removed from their mineral matrix as shown in the third column of the above table: heating is a less severe metallurgic process, whereas electrolysis is a more severe method.

23 Metals II: Properties and Reactions

23-2. *Refer to the Introduction to Chapter 23 and Section 6-1.*

The representative elements have valence electrons in s or s and p orbitals in the outermost occupied energy level, whereas the d-transition metals must have a partially filled set of d orbitals.

23-4. *Refer to Chapter 6.*

Metals are located at the left side of the periodic table and therefore, in comparison with nonmetals, have (a) fewer outer shell electrons, (b) lower electronegativities, (c) more negative standard reduction potentials and (d) less endothermic ionization energies.

23-6. *Refer to Section 4-1 and Table 4-3.*

Malleable refers to the ability of a substance to be shaped by being beaten with a hammer or by the pressure of rollers. Ductile refers to the ability of a substance to be drawn out into wire or hammered into a thin sheet.

23-8. *Refer to Sections 23-1 and 23-4, and Tables 23-1 and 23-3.*

(a) Alkali metals are larger than alkaline earth metals in the same period due to the increased effective nuclear charge of the alkaline earth metals.

(b) Alkaline earth metals have higher densities since they are both heavier and smaller than alkali metals of the same period.

(c) Alkali metals have lower first ionization energies than alkaline earth metals of the same period due to both the increased effective nuclear charge and decreased size of the alkaline earth metals.

(d) Alkali metals have much higher second ionization energies than alkaline earth metals of the same period. This is because removal of a second electron from an alkali metal ion involves destroying the very stable noble gas electronic configuration of the ion whereas removal of a second electron from an alkaline earth metal ion involves creating a stable noble gas configuration.

23-10. *Refer to Sections 23-1 and 23-4, and Tables 23-1 and 23-3.*

(a) physical properties: Both the alkaline earth metals and the alkali metals are silvery-white, malleable and ductile, but the alkaline earths are somewhat harder than alkali metals. Both are excellent electrical and thermal conductors. The melting and boiling points of the 2A metals are higher than those for the 1A metals, which are relatively low.

chemical properties: Alkaline earth metals and alkali metals are easily oxidized and thus are strong reducing agents. The 2A metals are not as reactive as 1A metals, but both groups are too reactive to occur as free elements in nature. Alkali metals are characterized by the loss of 1 electron per metal atom and form basic metal oxides, which react with water to produce hydroxides. Alkaline earths are characterized by the loss of 2 electrons per metal atom and form basic metal oxides (except BeO) which also react with water to produce hydroxides.

(b) Alkali metals are larger than alkaline earth metals in the same period due to the increased effective nuclear charge of the alkaline earth metals. Both increase in size going down their family. Alkaline earth metals have higher densities since they are both heavier and smaller than alkali metals of the same period. Alkali metals have lower first ionization energies than alkaline earth metals of the same period due to both the increased effective nuclear charge and decreased size of the alkaline earth metals. Alkali metals have much higher second ionization energies than alkaline earth metals of the same period. This is because removal of a second electron from an alkali metal ion involves destroying the very stable noble gas electronic configuration of the ion whereas removal of a second electron from an alkaline earth metal ion involves creating a stable noble gas configuration.

23-12. *Refer to Sections 5-17 and 7-2, and Appendix B.*

(a) Ca   [Ar] $\underset{4s}{\uparrow\downarrow}$ (b) Ca^{2+} [Ar]   (c) Mg   [Ne] $\underset{3s}{\uparrow\downarrow}$ (d) Mg^{2+} [Ne]

(e) Sn   [Kr] $\underset{4d}{\uparrow\downarrow\ \uparrow\downarrow\ \uparrow\downarrow\ \uparrow\downarrow\ \uparrow\downarrow}$ $\underset{5s}{\uparrow\downarrow}$ $\underset{5p}{\uparrow\ \ \uparrow\ \ \_}$

(f) Sn^{2+} [Kr] $\underset{4d}{\uparrow\downarrow\ \uparrow\downarrow\ \uparrow\downarrow\ \uparrow\downarrow\ \uparrow\downarrow}$ $\underset{5s}{\uparrow\downarrow}$ $\_\ \_\ \_$

(g) Sn^{4+} [Kr] $\underset{4d}{\uparrow\downarrow\ \uparrow\downarrow\ \uparrow\downarrow\ \uparrow\downarrow\ \uparrow\downarrow}$ $\_$ $\_\ \_\ \_$

23-14. *Refer to Sections 23-1 and 23-4.*

The alkali metals (Group 1A) and the alkaline earth metals (Group 2A) are not found free in nature because they are so easily oxidized. Their primary sources are seawater, brines of their soluble salts and deposits of sea salt. The metals are obtained from the electrolysis of their molten salts.

23-16. *Refer to Sections 23-1, 23-2, 23-4 and 23-5, and Tables 23-1 and 23-3.*

The metals in Group 1A and Group 2A have standard reduction potentials that are more negative than that for H_2. Therefore, the alkali metals and alkaline earth metals are stronger reducing agents than H_2. They should be above H_2 in the activity series. Consequently, they all react with acids by reducing acidic H^+ ions to H_2, while they are oxidized to +1 (Group 1A) or +2 (Group 2A) ions. All but Be reduce H_2O to H_2 gas.

23-18. *Refer to Section 23-6.*

(a) Calcium metal is used (1) as a reducing agent in the metallurgy of U, Th and other metals, (2) as a scavenger to remove dissolved impurities in molten metals and residual gases in vacuum tubes, and (3) as a component in many alloys. Slaked lime, $Ca(OH)_2$, is a cheap base used in industry and is also a major component of mortar and lime plaster. Careful heating of gypsum, $CaSO_4 \cdot 2H_2O$, produces plaster of Paris, $2CaSO_4 \cdot H_2O$.

(b) Magnesium metal is used (1) in photographic flash accessories, fireworks and incendiary bombs, (2) as a component in alloys for structural purposes, and (3) as a reagent in organic syntheses. Magnesia, MgO, is an excellent heat insulator used in furnaces, ovens and crucibles. Milk of magnesia, an aqueous suspension of $Mg(OH)_2$, is a stomach antacid and laxative. Anhydrous $MgSO_4$ and $Mg(ClO_4)_2$ are used as drying agents.

23-20. *Refer to Section 23-2 and Table 23-2.*

Let M = alkali metal, X = halogen
(a) $2M + 2H_2O \rightarrow 2MOH + H_2$ (b) $12M + P_4 \rightarrow 4M_3P$ (c) $2M + X_2 \rightarrow 2MX$

23-22. *Refer to Section 23-5 and Table 23-4.*

Let M = alkaline earth metal

(a) $M + 2H_2O \rightarrow M(OH)_2 + H_2$ (b) $6M + P_4 \rightarrow 2M_3P_2$ (c) $M + Cl_2 \rightarrow MCl_2$

23-24. *Refer to Section 23-2 and the Key Terms for Chapter 23.*

Diagonal similarities refer to chemical similarities of Period 2 elements of a certain group to Period 3 elements, one group to the right. This effect is particularly evident toward the left side of the periodic table. One example is the pair, B and Si, which are both metalloids with similar properties. Another example is the pair, Li and Mg. They have similar ionic charge densities and electronegativities; their compounds are similar in many ways:

(1) Li is the only 1A metal that combines with N_2 to form a nitride, Li_3N. Mg readily forms the nitride, Mg_3N_2.
(2) Li and Mg both form carbides.
(3) The solubilities of Li compounds are similar to those of Mg compounds.
(4) Li and Mg form normal oxides, Li_2O and MgO, when oxidized in air at 1 atm pressure, while the other members of Group 1A form peroxides and superoxides.

23-26. *Refer to Sections 23-4 and 23-5, Table 23-3, and the Key Terms for Chapter 14.*

Hydration energy is the energy released when a mole of ions in the gaseous phase forms a mole of ions in the aqueous phase. The higher the charge to size ratio of a cation, the stronger is its interaction with polar water molecules and the more exothermic is its hydration energy. Therefore, hydration energies of the alkaline earth metals become less exothermic from top to bottom within a group because the size of the ions increases, whereas the charge of the ions remains 2+.

23-28. *Refer to Section 23-4 and Table 23-3.*

Standard reduction potentials of the alkaline earth metals are, in general, very negative, indicating that alkaline earth metals are easily oxidized and hence are good reducing agents. Progressing down Group 2A, the atoms are larger, the outer electrons are more easily lost, the metals become better reducing agents and standard reduction potentials become more negative.

23-30. *Refer to Section 23-2 and 23-5, and Appendix K.*

Note: The metal hydroxide product is in the solid phase because only stoichiometric amounts of water are added.

(a) Balanced equation: $Li(s) + H_2O(\ell) \rightarrow LiOH(s) + 1/2\ H_2(g)$

$$\begin{aligned}\Delta H^\circ_{rxn} &= [\Delta H^\circ_{f\ LiOH(s)} + 1/2\ \Delta H^\circ_{f\ H_2(g)}] - [\Delta H^\circ_{f\ Li(s)} + \Delta H^\circ_{f\ H_2O(\ell)}]\\ &= [(1\text{ mol})(-487.23\text{ kJ/mol}) + (1/2\text{ mol})(0\text{ kJ/mol})] - [(1\text{ mol})(0\text{ kJ/mol}) + (1\text{ mol})(-285.8\text{ kJ/mol})]\\ &= \mathbf{-201.4\ kJ/mol\ rxn}\end{aligned}$$

(b) Balanced equation: $K(s) + H_2O(\ell) \rightarrow KOH(s) + 1/2\ H_2(g)$

$$\begin{aligned}\Delta H^\circ_{rxn} &= [\Delta H^\circ_{f\ KOH(s)} + 1/2\ \Delta H^\circ_{f\ H_2(g)}] - [\Delta H^\circ_{f\ K(s)} + \Delta H^\circ_{f\ H_2O(\ell)}]\\ &= [(1\text{ mol})(-424.7\text{ kJ/mol}) + (1/2\text{ mol})(0\text{ kJ/mol})] - [(1\text{ mol})(0\text{ kJ/mol}) + (1\text{ mol})(-285.8\text{ kJ/mol})]\\ &= \mathbf{-138.9\ kJ/mol\ rxn}\end{aligned}$$

(c) Balanced equation: $Ca(s) + 2H_2O(\ell) \rightarrow Ca(OH)_2(s) + H_2(g)$

$$\Delta H^\circ_{rxn} = [\Delta H^\circ_f \, _{Ca(OH)_2(s)} + \Delta H^\circ_f \, _{H_2(g)}] - [\Delta H^\circ_f \, _{Ca(s)} + 2\Delta H^\circ_f \, _{H_2O(\ell)}]$$
$$= [(1 \text{ mol})(-986.6 \text{ kJ/mol}) + (1 \text{ mol})(0 \text{ kJ/mol})] - [(1 \text{ mol})(0 \text{ kJ/mol}) + (2 \text{ mol})(-285.8 \text{ kJ/mol})]$$
$$= \mathbf{-415.0 \text{ kJ/mol rxn}}$$

In explanation, refer to the data table:

Cation	E° (V)	ΔH°_{rxn} from above (kJ)
Li^+	−3.045	−201.4
K^+	−2.925	−138.9
Ca^{2+}	−2.87	−415.0

From the E° values, we see that the relative strengths of reducing agents are : Li > K > Ca. It is expected that for the reactions of these metals with water, ΔH°_{rxn} for Li would be more negative than ΔH°_{rxn} for K, which is in turn more negative than ΔH°_{rxn} for Ca. This trend is true only for Li and K. ΔH°_{rxn} for Ca is much more negative than predicted since it is a 2A metal (Li and K are 1A metals) and reacts with twice as much water to produce twice as much H_2 gas.

23-32.   *Refer to Section 23-8.*

The following are properties of most *d*-transition elements:

(1) All are metals.

(2) Most are harder, more brittle and have higher melting points and boiling points and higher heats of vaporization than nontransition metals.

(3) Their ions and compounds are usually colored.

(4) They form many complex ions.

(5) With few exceptions, they exhibit multiple oxidation states.

(6) Many of the metals and their compounds are paramagnetic.

(7) Many of the metals and their compounds are effective catalysts.

23-34.   *Refer to Section 23-9 and Appendix B.*

(a) Sc   [Ar] $3d^1\, 4s^2$

(b) Fe   [Ar] $3d^6\, 4s^2$

(c) Cu   [Ar] $3d^{10}\, 4s^1$

(d) Zn^{2+} [Ar] $3d^{10}$

(e) Cr^{3+} [Ar] $3d^3$

(f) Ni^{2+} [Ar] $3d^8$

(g) Ag   [Kr] $4d^{10}\, 5s^1$

(h) Ag^+ [Kr] $4d^{10}$

23-36.   *Refer to Chapters 23 and 24, and Tables 23-1, 23-3, 23-5, 23-6, 23-7, 24-4, 24-6, 24-7, and others.*

The Group A or representative elements have their last electrons filling the outer *ns* and *np* orbitals, whereas most Group B elements have their last electrons filling the $(n\text{-}1)d$ orbitals. Look to the tables to compare their properties. For example, Group 3A elements are representative elements and have the ns^2np^1 outer electron configuration, with 3 valance electrons. Boron is a metalloid and crystallizes as a covalent solid (oxidation states range from +3 to -3), while the rest are metals forming ions with +1 or +3 oxidation states. Aluminum ions only have +3 oxidation state. Group 3B elements are all transition metals, with the $ns^2(n\text{-}1)d^1$ outer electron configuration and tend to have a +3 oxidation state.

Refer to Sections 23-7, 23-9 and 23-10, and Table 23-8.

The acidity and the covalent nature of transition metal oxides generally increases with increasing oxidation state of the metals. This is shown by the oxides of chromium.

Cr Oxide	Ox. No. of Cr	Character
CrO	+2	basic
Cr_2O_3	+3	amphoteric
CrO_3	+6	weakly acidic/acidic

23-40. **Refer to Section 23-10 and Table 23-8.**

Chromium(VI) oxide, CrO_3, is the acid anhydride of chromic acid, H_2CrO_4, and dichromic acid, $H_2Cr_2O_7$. Recall that there is no change in oxidation state when an acid anhydride is converted to the corresponding acid and so the oxidation state of Cr is +6 in both acids.

$$CrO_3 + H_2O \rightarrow H_2CrO_4 \qquad\qquad 2CrO_3 + H_2O \rightarrow H_2Cr_2O_7$$

23-42. **Refer to Section 23-6 and the Internet.**

Calcium carbonate is the primary component of seashells, antacids, marble and limestone (e.g. stalactites and stalagmites in caves), blackboard chalk, scale in water pipes, and calcium supplements for people and animals. It is also used to capture SO_2 gas in fossil fuel burning boilers, thereby helping to prevent acid rain, and as a soil additive to provide pH adjustment and calcium to farmers' soil.

23-44. **Refer to Section 23-3.**

Lithium compounds, not lithium metal, are used in the treatment of some types of mental disorders. The chemical properties of lithium metal are very different from lithium compounds containing the ion, Li^+. Li metal is very reactive with water, forming the strong base, LiOH, and hydrogen gas and releasing much heat, none of which are good for the human body.

23-46. **Refer to Section 23-1 and Table 23-1.**

The two properties listed in Table 23-1 that suggest that Group 1A metals are unlikely to exist as free metals are (1) the low ionization energies, which show how easily the outermost electron can be removed and (2) very negative standard reduction potentials, which indicate that the aqueous ions are not easily reduced to metals and that the free metals are easily oxidized to 1+ cations.

23-48. **Refer to Section 3-4.**

Balanced equation: $3Co_3O_4 + 8Al \rightarrow 9Co + 4Al_2O_3$

Plan: (1) Calculate the theoretical yield of Co metal.
         (2) Calculate the mass of Co_3O_4 required to produce the theoretical yield of Co.

(1) % yield $= \dfrac{\text{actual yield}}{\text{theoretical yield}} \times 100\%$

Substituting,

$$72.5\% = \frac{190.\ g}{\text{theoretical yield}} \times 100\% \qquad\qquad \text{Solving, theoretical yield} = 262\ g\ Co$$

(2) $? \text{ g Co}_3O_4 = 262\ g\ Co \times \dfrac{1\ \text{mol Co}}{58.9\ \text{g Co}} \times \dfrac{3\ \text{mol Co}_3O_4}{9\ \text{mol Co}} \times \dfrac{241\ \text{g Co}_3O_4}{1\ \text{mol Co}_3O_4} = \mathbf{357\ g\ Co_3O_4}$

23-50. *Refer to Section 23-2, Exercise 23-49, and Appendix K.*

Balanced equation: $Rb(s) + H_2O(\ell) \rightarrow RbOH(aq) + 1/2\ H_2(g)$

$\Delta H^\circ_{rxn} = [\Delta H^\circ_{f\ RbOH(aq)} + 1/2\ \Delta H^\circ_{f\ H_2(g)}] - [\Delta H^\circ_{f\ Rb(s)} + \Delta H^\circ_{f\ H_2O(\ell)}]$

$= [(1\ \text{mol})(-481.16\ \text{kJ/mol}) + (1/2\ \text{mol})(0\ \text{kJ/mol})] - [(1\ \text{mol})(0\ \text{kJ/mol}) + (1\ \text{mol})(-285.8\ \text{kJ/mol})]$

$= \mathbf{-195.4\ kJ/mol\ Rb(s)}$

$\Delta S^\circ_{rxn} = [S^\circ_{f\ RbOH(aq)} + 1/2\ S^\circ_{f\ H_2(g)}] - [S^\circ_{f\ Rb(s)} + S^\circ_{f\ H_2O(\ell)}]$

$= [(1\ \text{mol})(110.75\ \text{J/mol·K}) + (1/2\ \text{mol})(130.60\ \text{J/mol·K})]$

$\qquad - [(1\ \text{mol})(76.78\ \text{J/mol·K}) + (1\ \text{mol})(69.91\ \text{J/mol·K})]$

$= \mathbf{29.4\ J/K\ per\ 1\ mol\ Rb(s)}$

$\Delta G^\circ_{rxn} = [\Delta G^\circ_{f\ RbOH(aq)} + 1/2\ \Delta G^\circ_{f\ H_2(g)}] - [\Delta G^\circ_{f\ Rb(s)} + \Delta G^\circ_{f\ H_2O(\ell)}]$

$= [(1\ \text{mol})(-441.24\ \text{kJ/mol}) + (1/2\ \text{mol})(0\ \text{kJ/mol})] - [(1\ \text{mol})(0\ \text{kJ/mol}) + (1\ \text{mol})(-237.2\ \text{kJ/mol})]$

$= \mathbf{-204.0\ kJ/mol\ Rb(s)}$

In Exercise 23-49, the ΔG_{rxn} was calculated for the following reaction:

$$Na(s) + H_2O(\ell) \rightarrow NaOH(aq) + 1/2\ H_2(g)$$

$\Delta G^\circ_{rxn} = [\Delta G^\circ_{f\ NaOH(aq)} + 1/2\ \Delta G^\circ_{f\ H_2(g)}] - [\Delta G^\circ_{f\ Na(s)} + \Delta G^\circ_{f\ H_2O(\ell)}]$

$= [(1\ \text{mol})(-419.2\ \text{kJ/mol}) + (1/2\ \text{mol})(0\ \text{kJ/mol})] - [(1\ \text{mol})(0\ \text{kJ/mol}) + (1\ \text{mol})(-237.2\ \text{kJ/mol})]$

$= -182.0\ \text{kJ/mol Na(s)}$

Therefore, the reaction between $Rb(s)$ and water is more favored with a greater degree of spontaneity than the reaction between $Na(s)$ and water, since the ΔG° for the reaction between $Rb(s)$ and water is more negative.

23-52. *Refer to Chapters 2, 3 and 12.*

Balanced equations: $Q_2CO_3 \rightarrow CO_2 + Q_2O$ if Q is a 1A element

$\qquad\qquad\qquad\quad QCO_3 \rightarrow CO_2 + QO$ if Q is a 2A element

Plan: (1) Determine the moles of CO_2 formed from the ideal gas law, $PV=nRT$.

(2) Determine the moles of the unknown compound and its formula weight.

(3) Determine the atomic weight of the unknown element for each case, identify the element and compound..

(1) $n = \dfrac{PV}{RT} = \dfrac{(1.00\ \text{atm})(4.48\ \text{L})}{(0.0821\ \text{L·atm/mol·K})(0°C + 273°)} = \mathbf{0.200\ mol\ CO_2}$

(2) $? \text{ mol compound} = 0.200\ \text{mol CO}_2 \times \dfrac{1\ \text{mol compound}}{1\ \text{mol CO}_2} = 0.200\ \text{mol compound}$

$FW\ (g/mol) = \dfrac{\text{g compound}}{\text{mol compound}} = \dfrac{14.78\ \text{g compound}}{0.200\ \text{mol compound}} = 74.0\ \text{g/mol}$

(3) Since the carbonate ion, CO_3^{2-}, has a formula weight of 60.0 g/mol, then the mass of the metal, Q, in one mole of compound must be

$$? \text{ g Q} = \text{g compound} - \text{g } CO_3^{2-} = 74.0 \text{ g} - 60.0 \text{ g} = 14.0 \text{ g}$$

If the formula is Q_2CO_3, then the atomic weight of Q must be 14.0/2 = 7.00 g/mol
If the formula is QCO_3, then the atomic weight of Q must be 14.0 g/mol

Since there is no metal with an atomic weight of 14.0 g/mol (nitrogen is not a metal), then the metal Q must be lithium, a IA element with an atomic weight of 6.9 g/mol), and the compound must be Li_2CO_3.

23-54. *Refer to Chapter 4, and Sections 23-2, 23-5 and 23-10.*

(a) formula unit:

$$Mg(s) + H_2O(g) \rightarrow MgO(s) + H_2(g)$$

total ionic:

$$Mg(s) + H_2O(g) \rightarrow MgO(s) + H_2(g)$$

net ionic:

$$Mg(s) + H_2O(g) \rightarrow MgO(s) + H_2(g)$$

(b) formula unit:

$$2Rb(s) + 2H_2O(\ell) \rightarrow 2RbOH(aq) + H_2(g)$$

total ionic:

$$2Rb(s) + 2H_2O(\ell) \rightarrow 2Rb^+(aq) + 2OH^-(aq) + H_2(g)$$

net ionic:

$$2Rb(s) + 2H_2O(\ell) \rightarrow 2Rb^+(aq) + 2OH^-(aq) + H_2(g)$$

(c) formula unit:

$$Cr(OH)_3(s) + NaOH(aq) \rightarrow Na[Cr(OH)_4](aq)$$

total ionic:

$$Cr(OH)_3(s) + Na^+(aq) + OH^-(aq) \rightarrow Na^+(aq) + [Cr(OH)_4]^-(aq)$$

net ionic:

$$Cr(OH)_3(s) + OH^-(aq) \rightarrow [Cr(OH)_4]^-(aq)$$

24 Some Nonmetals and Metalloids

24-2. *Refer to Section 24-1.*

The inert nature of the noble gases and their low abundances in the atmosphere were two factors causing their late discovery.

24-4. *Refer to Section 24-1.*

The noble gases in order of increasing radii are: He < Ne < Ar < Kr < Xe < Rn. The size increases as one goes down Group 8A, because the outer electrons which set their size exist in electron clouds that are farther and farther from the nucleus.

24-6. *Refer to Section 24-2.*

The accidental preparation of $O_2^+PtF_6^-$ by the reaction of O_2 with PtF_6 led Bartlett to reason that xenon should also be oxidized by PtF_6, since the first ionization energy of molecular oxygen is actually slightly larger than that of xenon. He obtained a red crystalline solid initially believed to be $Xe^+PtF_6^-$, but now known to be more complex. At present, Xe and Kr are the only noble gases known to form compounds, mostly combining with F and O. Our textbook discusses the compounds of Xe.

24-8. *Refer to Section 24-2.*

Balanced equation: $XeF_4(s) + F_2(g) \rightarrow XeF_6(s)$

$$? \text{ g } XeF_6 = 2.55 \text{ g } XeF_4 \times \frac{1 \text{ mol } XeF_4}{207 \text{ g } XeF_4} \times \frac{1 \text{ mol } XeF_6}{1 \text{ mol } XeF_4} \times \frac{245 \text{ g } XeF_6}{1 \text{ mol } XeF_6} = \mathbf{3.02 \text{ g } XeF_6}$$

24-10. *Refer to Section 24-3 and Table 24-4.*

X_2 $:\overset{..}{X}\overset{..}{X}:$ $S = N - A = [2 \times 8 \text{ (for X)}] - [2 \times 7 \text{ (for X)}] = 16 - 14 = 2$

A halogen molecule contains one nonpolar covalent single bond, as shown by its Lewis structure.

As we descend the 7A family from F_2 to I_2, the size of a halogen atom increases and so the bond length increases. The strength of the X–X bond varies; it increases from F_2 to Cl_2, then decreases from Cl_2 to Br_2 to I_2.

24-12. *Refer to Section 24-3 and Table 24-4.*

(a) In order of increasing atomic radii: F < Cl < Br < I < At

(b) In order of increasing ionic radii: F^- < Cl^- < Br^- < I^- < At^-

(c) In order of increasing electronegativity: At < I < Br < Cl < F

(d) In order of increasing melting points and boiling points: F_2 < Cl_2 < Br_2 < I_2 < At_2

In nature, the halogens exist as nonpolar diatomic molecules. London dispersion forces are the only forces of attraction acting between the molecules. These forces increase with increasing molecular size.

(e) See (d).

(f) In order of increasing standard reduction potentials: $At_2 < I_2 < Br_2 < Cl_2 < F_2$

F_2 has the most positive standard reduction potential and therefore is the strongest of all common oxidizing agents. Oxidizing strengths of the diatomic halogen molecules decrease down Group 7A.

24-14. *Refer to Section 24-4.*

Christe's preparation of F_2 is not a direct chemical oxidation, but rather it involves the formation of unstable MnF_4, which spontaneously decomposes into MnF_3 and F_2.

24-16. *Refer to Section 24-5.*

(a) $2NaI + Cl_2 \rightarrow 2NaCl + I_2$

(b) $NaCl + Br_2 \rightarrow$ no reaction

(c) $2NaI + Br_2 \rightarrow 2NaBr + I_2$

(d) $2NaBr + Cl_2 \rightarrow 2NaCl + Br_2$

(e) $NaF + I_2 \rightarrow$ no reaction

24-18. *Refer to Section 24-5.*

To the right are the electrostatic charge potential plots for F–Cl and Cl–Br. Initially you can identify which is which simply based on the relative sizes of the atoms: F < Cl < Br. However, from the greater range of colors across F–Cl, we can see that F–Cl is a more polar molecule than Cl–Br.   This is verified by the electronegativity differences. In F–Cl, ΔEN = [4.0 (for F) - 3.0 (for Cl)] = 1.0, whereas in Cl–Br, ΔEN = [3.0 (for Cl) - 2.8 (for Br)] = 0.2. Note that Cl has a $\delta+$ charge in F–Cl, but has a $\delta-$ in Cl–Br.

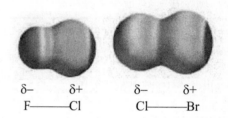

$\delta-$ $\delta+$ $\delta-$ $\delta+$

F———Cl   Cl———Br

24-20. *Refer to Section 24-6.*

Hydrogen bromide, HBr(g), is a colorless gas which dissolves in water to give hydrobromic acid, HBr(aq). The latter is a strong acid which completely dissociates in aqueous solutions giving $H_3O^+(aq)$ and $Br^-(aq)$.

24-22. *Refer to Section 24-6.*

Hydrofluoric acid is used to etch glass by reacting with the silicates in glass to produce a very volatile and thermodynamically stable compound, silicon tetrafluoride, SiF_4. For example,

$$CaSiO_3(s) + 6HF(aq) \rightarrow CaF_2(s) + SiF_4(g) + 3H_2O(\ell)$$

24-24. *Refer to Sections 24-7 and 4-6, and Table 24-5.*

(a) $KBrO_3$ potassium bromate

(b) KOBr   potassium hypobromite

(c) $NaClO_4$ sodium perchlorate

(d) $NaBrO_2$ sodium bromite

(e) HOBr   hypobromous acid

(f) $HBrO_3$ bromic acid

(g) HIO_3 iodic acid

(h) $HClO_4$ perchloric acid

(a) $X_2 + H_2O \rightarrow HX + HOX$ (X = Cl, Br, I)

(b) $X_2 + 2NaOH \rightarrow NaX + NaOX + H_2O$ (X = Cl, Br, I)

(c) $Ba(ClO_2)_2 + H_2SO_4 \rightarrow BaSO_4 + 2HClO_2$

(d) $KClO_4 + H_2SO_4 \rightarrow KHSO_4 + HClO_4$ (explosive)

The ranking of the acids from strongest to weakest is:

(a) HOCl > HOBr > HOI, since for most ternary acids containing different elements in the same oxidation state from the same group in the periodic table, acid strengths increase with increasing electronegativity of the central element.

(b) $HClO_4 > HClO_3 > HClO_2 > HOCl$, since the acid strengths of most ternary acids containing the same central element increase with increasing oxidation state of the central element and with increasing numbers of oxygen atoms.

(c) $HClO_4 > HBrO_3 > HOI$, since we know $HClO_4$ is one of our 7 strong acids and the others are not. The acid with the most electronegative central element and the most oxygen atoms, have an H-O bond that is the weakest.

oxide, O^{2-} [Ne]   or $1s^2\, 2s^2\, 2p^6$

sulfide, S^{2-} [Ar]   or $1s^2\, 2s^2\, 2p^6\, 3s^2\, 3p^6$

selenide, Se^{2-} [Kr]   or $1s^2\, 2s^2\, 2p^6\, 3s^2\, 3p^6\, 3d^{10}\, 4s^2\, 4p^6$

Every Group 6A element has six valence electrons in the highest energy level, and is therefore two electrons away from achieving an octet of electrons. This is why all of them exhibit an oxidation state of -2. In addition, Group 6A elements below oxygen, since they have empty *d* orbitals available for containing electron pairs, can share their six valence electrons to various degrees with other elements to give different positive oxidation states up to +6. An oxidation state of -3 is impossible because it would require placing an electron into the next higher energy *d* or *s* orbitals. An oxidation state of +7 is impossible because the Group 6A elements only possess six valence electrons to share or to lose.

(a) H_2S

(b) SF_6

(c) SF_4

(d) SO_2

(e) SO_3

24-36. *Refer to Section 24-9.*

(a) $E + 3F_2(\text{excess}) \rightarrow EF_6$ (E = S, Se, Te)

(b) $O_2 + 2H_2 \rightarrow 2H_2O$
$E + H_2 \rightarrow H_2E$ (E = S, Se, Te)

(c) $E + O_2 \rightarrow EO_2$ (E = S, Te; with Se - use O_2 and NO_2)

24-38. *Refer to Section 24-10.*

Group 6A hydrides dissociate in two stages. Their acid ionization constants, K_1 and K_2, are shown below:

		H_2S	H_2Se	H_2Te
$H_2E \rightleftarrows H^+ + HE^-$	K_{a1}:	1.0×10^{-7}	1.9×10^{-4}	2.3×10^{-3}
$HE^- \rightleftarrows H^+ + E^{2-}$	K_{a2}:	1.0×10^{-19}	$\approx 10^{-11}$	$\approx 1.6 \times 10^{-11}$

Acid strength increases upon descending the group: $H_2O < H_2S < H_2Se < H_2Te$. This results from the corresponding decrease in the average E-H bond energy.

24-40. *Refer to Sections 24-11 and 24-12.*

(a) $NaOH + H_2SO_4 \rightarrow NaHSO_4 + H_2O$

(b) $2NaOH + H_2SO_4 \rightarrow Na_2SO_4 + 2H_2O$

(c) $NaOH + H_2SO_3 \rightarrow NaHSO_3 + H_2O$

(d) $2NaOH + H_2SO_3 \rightarrow Na_2SO_3 + 2H_2O$

(e) $NaOH + H_2SeO_4 \rightarrow NaHSeO_4 + H_2O$

(f) $2NaOH + H_2SeO_4 \rightarrow Na_2SeO_4 + 2H_2O$

(g) $NaOH + TeO_2 \rightarrow NaHTeO_3$

(h) $2NaOH + TeO_2 \rightarrow Na_2TeO_3 + H_2O$

24-42. *Refer to Sections 24-12 and 3-5.*

Balanced equations:
(1) $4FeS_2 + 11 O_2 \rightarrow 2Fe_2O_3 + 8SO_2$
(2) $2SO_2 + O_2 \rightarrow 2SO_3$
(3) $SO_3 + H_2SO_4 \rightarrow H_2S_2O_7$
(4) $H_2S_2O_7 + H_2O \rightarrow 2H_2SO_4$

$$? \text{ ton } H_2SO_4 = 1.50 \text{ ton } FeS_2 \times \frac{2000 \text{ lb } FeS_2}{1 \text{ ton } FeS_2} \times \frac{454 \text{ g } FeS_2}{1 \text{ lb } FeS_2} \times \frac{1 \text{ mol } FeS_2}{120.0 \text{ g } FeS_2} \times \frac{8 \text{ mol } SO_2}{4 \text{ mol } FeS_2} \times \frac{2 \text{ mol } SO_3}{2 \text{ mol } SO_2}$$

$$\times \frac{1 \text{ mol } H_2S_2O_7}{1 \text{ mol } SO_3} \times \frac{2 \text{ mol } H_2SO_4}{1 \text{ mol } H_2S_2O_7} \times \frac{98.1 \text{ g } H_2SO_4}{1 \text{ mol } H_2SO_4} \times \frac{1 \text{ lb } H_2SO_4}{454 \text{ g } H_2SO_4} \times \frac{1 \text{ ton } H_2SO_4}{2000 \text{ g } H_2SO_4}$$

$$= 4.91 \text{ tons } H_2SO_4 \text{ total}$$

However, half of the H_2SO_4 was a reactant in Step (3).
Therefore, the net mass of H_2SO_4 produced = 4.09 ton/2 = **2.45 tons H_2SO_4**

24-44. *Refer to Sections 24-11, 17-2 and 17-5.*

Balanced equation: $2SO_2(g) + O_2(g) \rightleftarrows 2SO_3(g)$

Let $x = [O_2]_{\text{reacted}}$. Then,
$1.00 - 2x = [SO_2]$
$5.00 - x = [O_2]$
$2x = [SO_3]$

$$
\begin{array}{ccccc}
 & 2SO_2 & + & O_2 & \rightleftarrows & 2SO_3 \\
\text{initial} & 1.00\ M & & 5.00\ M & & 0\ M \\
\text{change} & -\ 2x\ M & & -\ x\ M & & +\ 2x\ M \\
\hline
\text{at equilibrium} & (1.00 - 2x)\ M & & (5.00 - x)\ M & & 2x\ M
\end{array}
$$

However, $[SO_3] = 2x = 77.8\%$ of $[SO_2]_{\text{initial}} = 0.778 \times 1.00\ M = 0.778\ M$

Therefore, $K_c = \dfrac{[SO_3]^2}{[SO_2]^2[O_2]} = \dfrac{(2x)^2}{(1.00 - 2x)^2(5.00 - x)} = \dfrac{(0.778)^2}{(1.00 - 0.778)^2(5.00 - 0.778/2)} = \mathbf{2.7}$

24-46. *Refer to the Introduction to Section 24-13, Table 24-7 and Appendix B.*

$N \quad 1s^2\ 2s^2\ 2p^3$

$P \quad 1s^2\ 2s^2\ 2p^6\ 3s^2\ 3p^3$

$As \quad 1s^2\ 2s^2\ 2p^6\ 3s^2\ 3p^6\ 3d^{10}\ 4s^2\ 4p^3$

$Sb \quad 1s^2\ 2s^2\ 2p^6\ 3s^2\ 3p^6\ 3d^{10}\ 4s^2\ 4p^6\ 4d^{10}\ 5s^2\ 5p^3$

$Bi \quad 1s^2\ 2s^2\ 2p^6\ 3s^2\ 3p^6\ 3d^{10}\ 4s^2\ 4p^6\ 4d^{10}\ 4f^{14}\ 5s^2\ 5p^6\ 5d^{10}\ 6s^2\ 6p^3$

$N^{3-} \quad 1s^2\ 2s^2\ 2p^6$

$P^{3-} \quad 1s^2\ 2s^2\ 2p^6\ 3s^2\ 3p^6$

24-48. *Refer to Section 24-13.*

The nitrogen cycle is the complex series of reactions by which nitrogen is slowly but continually recycled in the atmosphere, lithosphere (earth) and hydrosphere (water). Atmospheric nitrogen is made accessible to us and other life-forms in mainly two ways.

(1) A class of plants, called legumes, have bacteria which extract N_2 directly, converting it to NH_3. This nitrogen fixation process, catalyzed by an enzyme produced by the bacteria, is highly efficient at usual temperatures and pressures.

(2) N_2 and O_2 react in the atmosphere near lightning, forming NO and NO_2, which dissolve in rainwater and fall to earth. These nitrogen compounds are absorbed and incorporated into plants forming amino acids and proteins. The plants are eaten by animals or die and decay, releasing their nitrogen to the environment. The animals, in turn, excrete waste and/or die, releasing their nitrogen to the environment.

24-50. *Refer to Sections 24-14 and 17-7, and Examples 17-9 and 17-10.*

The Haber process is the economically important industrial process for making ammonia, NH_3, from atmospheric N_2, according to:

$$N_2(g) + 3H_2(g) \rightleftarrows 2NH_3(g) \qquad \Delta H^\circ = -92\ \text{kJ/mol}$$

(1) effect of temperature: The reaction is exothermic ($\Delta H < 0$), so one might expect that to increase the amount of NH_3, one would need to lower the temperature. This action would increase the *relative* amount of NH_3 present, however, the reaction rates are lowered as well. So, Haber investigated other ways to increase the yield.

(2) effect of pressure: In Chapter 17, we learned that increasing the pressure favors the reaction that produces the smaller number of moles of gas (forward in this case). This reaction is run under pressures ranging from 200 to 1000 atmospheres to increase the yield of NH_3.

(3) effect of catalyst: The addition of finely divided iron and small amounts of selected oxides speeds up both the forward and reverse reactions. This allow NH_3 to be produced not only faster but at a lower temperature, which increases the yield of NH_3 and extends the life of the equipment.

Oxidation number of N:

(a) N_2 0   (b) N_2O +1   (c) N_2O_4 +4   (d) HNO_3 +5   (e) HNO_2 +3

24-54. *Refer to Chapter 8 and the Sections as stated.*

(a) NH_2Br

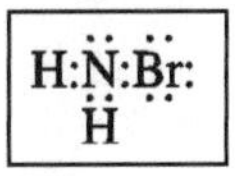

H:N:Br:
H

The Lewis dot formula predicts 4 regions of high electron density around the central N atom, a tetrahedral electronic geometry and a pyramidal molecular geometry. The N atom has sp^3 hybridization (Sections 8-8 and 24-14). The three-dimensional structure is shown below.

(b) HN_3

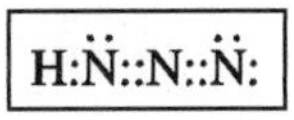

H:N::N::N:

Around the outer two N atoms, the Lewis dot formula predicts 3 regions of high electron density, a trigonal planar electronic geometry, sp^2 hybridization, and the N atom bonded to the H has a bent molecular geometry. The Lewis dot formula also predicts 2 regions of high electron density around the central N atom, a linear electronic and molecular geometry and sp hybridization for the central N atom (Sections 8-5, 8-13 and 24-14). The three-dimensional structure is shown below.

(c) N_2O_2

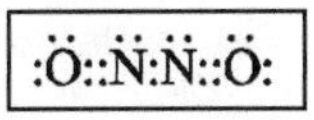

:O::N:N::O:

The Lewis dot formula predicts 3 regions of high electron density, trigonal planar electronic geometry and angular molecular geometry around each N atom. The N atoms have sp^2 hybridization (Sections 8-13 and 24-15). The three-dimensional structure is shown below.

(d) NO_2^+

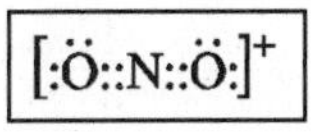

[:O::N::O:]⁺

The Lewis dot formula predicts 2 regions of high electron density, a linear electronic and ionic geometry around the N atom and sp hybridization for the N atom (Section 24-15). The three-dimensional structure is shown on the next page.

NO_3^-

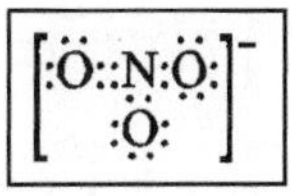

[:O::N:O:]⁻
:O:

The Lewis dot formulas for the three resonance structures (one is shown) predicts 3 regions of high electron density around the central N atom and a trigonal planar electronic and ionic geometry. The N atom has sp^2 hybridization (Section 24-16). The three-dimensional structure is shown on the next page.

(e) HNO_3

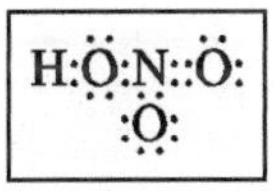

H:O:N::O:
:O:

The Lewis dot formulas for the two resonance structures (one is shown) predicts 3 regions of high electron density for the N atom, and a trigonal planar electronic and molecular geometry about the N atom. The N atom has sp^2 hybridization (Section 24-16). The three-dimensional structure is shown on the next page.

(f) NO_2^-

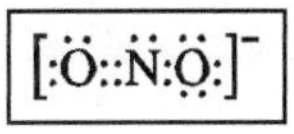

[:O::N:O:]⁻

The Lewis dot formulas for the two resonance structures (one is shown) predicts 3 regions of high electron density for the central N atom, and a trigonal planar electronic geometry and a bent ionic geometry. The N atom has sp^2 hybridization (Section 24-16). The three-dimensional structure is shown on the next page.

(a) NH_2Br

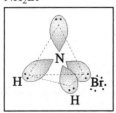

(b) HN_3

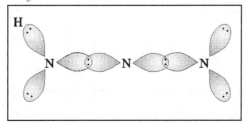

(c) N_2O_2

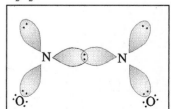

(d) NO_2^+ 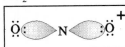 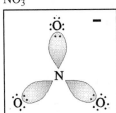 NO_3^- (e) HNO_3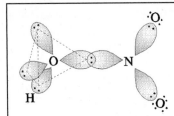

(f) NO_2^-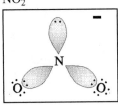

Note: Pi bonding is not shown.

24-56. *Refer to the Sections as stated.*

(a) $2KN_3(s) \xrightarrow{\text{heat}} 2K(s) + 3N_2(g)$

(b) $NH_3(g) + HCl(g) \rightarrow NH_4Cl(s)$ (Section 24-14)

(c) $NH_3(aq) + HCl(aq) \rightarrow NH_4Cl(aq)$ (Section 24-14)

(d) $2NH_4NO_3(\ell) \xrightarrow{\text{heat}} 2N_2(g) + 4H_2O(g) + O_2(g)$ (Sections 24-14 and 24-15)

(e) $4NH_3(g) + 5O_2(g) \xrightarrow[\text{Pt}]{\text{heat}} 4NO(g) + 6H_2O(g)$ (Section 24-15)

(f) $2N_2O(g) \xrightarrow{\text{heat}} 2N_2(g) + O_2(g)$ (Section 24-15)

(g) $3NO_2(g) + H_2O(\ell) \rightarrow 2HNO_3(aq) + NO(g)$ (Section 24-16)

24-58. *Refer to Section 24-14.*

$Co^{2+}(aq) + 6NH_3(aq) \rightarrow [Co(NH_3)_6]^{2+}(aq)$ $BF_3(g) + :NH_3(g) \rightarrow F_3B:NH_3(s)$

24-60. *Refer to Exercise 8-38 Solution and the Sections as stated.*

	Molecule	**Structure**	**Polarity of Molecule**	
(a)	NH_3	pyramidal	polar	(Section 8-8)
(b)	NH_2Cl	distorted pyramidal	polar	(Section 8-8)
(c)	NO_2	bent	polar	(Section 24-15)
(d)	NH_2OH	unsymmetric	polar	(Section 8-8)
(e)	HNO_3	unsymmetric	polar	(Section 24-16)

24-62. *Refer to Section 24-15.*

Nitrogen oxide, NO, is very reactive because each NO molecule contains an unpaired electron.
Note: Species with unpaired electrons, resulting from the cleavage of chemical bonds, are called radicals.

NO_2, a brown gas, is very reactive because it contains 1 unpaired electron and easily dimerizes to form the colorless gas, N_2O_4, in the following equilibrium reaction:

$$2NO_2(g) \rightleftarrows N_2O_4 (g) + heat \qquad\qquad \Delta H° = -57.2 \text{ kJ/mol rxn}$$

At room temperature, there is sufficient NO_2 present in the equilibrium mixture to give it a brown color. When the system is cooled, the equilibrium shifts to the right, brown NO_2 gas is converted to colorless N_2O_4 gas and the mixture loses color.

24-66. *Refer to Section 6-8.*

When an acid anhydride reacts with water, the corresponding acid is formed. There is no change in oxidation number of the elements.

(a) $N_2O_5(s) + H_2O(\ell) \rightarrow 2HNO_3(\ell)$

(b) $N_2O_3(g) + H_2O(\ell) \rightarrow 2HNO_2(\ell)$

(c) $P_4O_{10}(s) + 6H_2O(\ell) \rightarrow 4H_3PO_4(\ell)$

(d) $P_4O_6(s) + 6H_2O(\ell) \rightarrow 4H_3PO_3(\ell)$

24-68. *Refer to Section 24-16.*

The function of sodium nitrite, $NaNO_2$, as a food additive is two-fold:

(1) it inhibits the oxidation of blood, preventing the discoloring of red meat, and
(2) it prevents the growth of botulism bacteria.

There is now some controversy regarding this food additive because nitrites are suspected of combining with amines under the acidic conditions of the stomach to produce carcinogenic nitrosoamines.

24-70. *Refer to Sections 24-4 and 21-3.*

Elemental chlorine is produced by the electrolysis of molten NaCl in the Downs Cell. The other product is sodium metal. The products must be separated because the reaction between metallic Na and gaseous Cl_2 is very rapid, spontaneous and explosive. The product, Cl_2, is poisonous. Elemental fluorine, a corrosive and poisonous gas, is produced by the electrolysis of KHF_2, a molten mixture of KF and HF. This can be additionally dangerous because gaseous H_2 is also produced which is very explosive.

24-72. *Refer to Sections 24-8 and the Internet.*

Elemental sulfur deposits are mined beneath the earth's surface by the Frasch "hot water" process shown in Figure 24-2. Elemental sulfur is only found on the earth's surface near volcanoes and hot springs, where the water brings up H_2S which oxidizes to elemental sulfur by special anaerobic bacteria. In time, at the earth's surface, the sulfur will continue to oxidize, with the help of aerobic bacteria, to thiosulfate ($S_2O_3^{2-}$), tetrathionate ($S_4O_6^{2-}$) and sulfate (SO_4^{2-}) ions. The aerobic bacteria need oxygen to survive and therefore have far lower populations beneath the earth's surface, so the elemental sulfur is much more stable and long-lasting there.

(a) most chemically active halogen: F_2

(b) halogen most likely to be reduced from the free state: F_2

(c) best oxidizing agent of the halogens: F_2

(d) worst oxidizing agent of the halogens: At_2 (I_2 is the common worst oxidizing agent of the halogens.)

(e) halogen that is a liquid at room temperature: Br_2

(f) halogen found free in nature: none - they are too reactive

24-76. *Refer to Sections 24-2.*

Balanced equation: $XeF_4(s) + F_2(g) \rightarrow XeF_6(s)$

$\Delta H^o_{rxn} = [\Delta H^o_f \, {}_{XeF_6(s)}] - [\Delta H^o_f \, {}_{XeF_4(s)} + \Delta H^o_f \, {}_{F_2(g)}]$

$\quad\quad = [(1 \text{ mol})(-402 \text{ kJ/mol})] - [(1 \text{ mol})(-261.5 \text{ kJ/mol}) + (1 \text{ mol})(0 \text{ kJ/mol})]$

$\quad\quad = \textbf{–140 kJ/mol rxn}$

24-78. *Refer to Section 15-9, and Tables 15-2 and 15-3.*

Balanced equations: (1) $4N(g) \rightarrow 2 \text{ :N}\equiv\text{N:}(g)$ $\quad\quad\quad \Delta H_{rxn\,1}$

(2) $4N(g) \rightarrow$:N⟨N⟩N: (g) $\quad\quad\quad \Delta H_{rxn\,2}$

(3) $4P(g) \rightarrow 2 \text{ :P}\equiv\text{P:}(g)$ $\quad\quad\quad \Delta H_{rxn\,3}$

(4) $4P(g) \rightarrow$:P⟨P⟩P: (g) $\quad\quad\quad \Delta H_{rxn\,4}$

(1) $\Delta H_{rxn\,1} = \Sigma \text{ B.E.}_{reactants} - \Sigma \text{ B.E.}_{products} = 0 - (2 \text{ mol})(\text{B.E.}_{N\equiv N}) = -(2 \text{ mol})(945 \text{ kJ/mol}) = -1890 \text{ kJ}$

$\Delta H_{rxn\,2} = \Sigma \text{ B.E.}_{reactants} - \Sigma \text{ B.E.}_{products} = 0 - (6 \text{ mol})(\text{B.E.}_{N\text{-}N}) = -(6 \text{ mol})(163 \text{ kJ/mol}) = -978 \text{ kJ}$

Reaction 1 is more exothermic than Reaction 2 ($\Delta H_{rxn\,1}$ is more negative than $\Delta H_{rxn\,2}$).
Therefore, the formation of N_2 molecules by Reaction 1 is the predicted result.

(2) $\Delta H_{rxn\,3} = \Sigma \text{ B.E.}_{reactants} - \Sigma \text{ B.E.}_{products} = 0 - (2 \text{ mol})(\text{B.E.}_{P\equiv P}) = -(2 \text{ mol})(485 \text{ kJ/mol}) = -970 \text{ kJ}$

$\Delta H_{rxn\,4} = \Sigma \text{ B.E.}_{reactants} - \Sigma \text{ B.E.}_{products} = 0 - (6 \text{ mol})(\text{B.E.}_{P\text{-}P}) = -(6 \text{ mol})(201 \text{ kJ/mol}) = -1210 \text{ kJ}$

Reaction 4 is more exothermic than Reaction 3 ($\Delta H_{rxn\,4}$ is more negative than $\Delta H_{rxn\,3}$).
Therefore, the formation of P_4 molecules by Reaction 4 is the predicted result.

24-80. *Refer to Section 24-18.*

? volume of earth's crust $=$ (volume of the earth) - (volume of the earth minus the crust)

$\quad = \frac{4}{3}\pi(r_{earth})^3 - \frac{4}{3}\pi(r_{earth\;minus\;crust})^3$

$\quad = \frac{4}{3}(3.14157)\left(6400 \text{ km} \times \frac{1000 \text{ m}}{1 \text{ km}} \times \frac{100 \text{ cm}}{1 \text{ m}}\right)^3$

$\quad\quad\quad - \frac{4}{3}(3.14157)\left((6400 \text{ km} - 50 \text{ km}) \times \frac{1000 \text{ m}}{1 \text{ km}} \times \frac{100 \text{ cm}}{1 \text{ m}}\right)^3$

$\quad = 1.098 \times 10^{27} \text{ cm}^3 - 1.073 \times 10^{27} \text{ cm}^3$

$\quad = 2.5 \times 10^{25} \text{ cm}^3$

Therefore, the mass of silicon in the earth's crust is:

? mass Si $= 2.5 \times 10^{25}$ cm^3 crust $\times \dfrac{3.5 \text{ g crust}}{1 \text{ cm}^3 \text{ crust}} \times \dfrac{25.7 \text{ g Si}}{100 \text{ g crust}} = \mathbf{2.2 \times 10^{25}}$ **g Si**

24-82. *Refer to Section 13-16 and Example 13-9.*

Consider the Ar face-centered cubic structure with a unit cell edge represented as *a*, shown here. For Ar, *a* = 5.43 Å

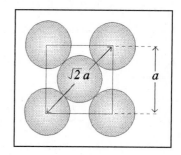

The hypotenuse, *d*, of a isosceles right angle triangle is determined by simple geometry:

$d^2 = a^2 + a^2$

$d = \sqrt{a^2 + a^2} = \sqrt{2a^2} = \sqrt{2}a = \sqrt{2} \times 5.43$ Å $= 7.68$ Å

By inspection, the hypotenuse is equal to 4 Ar radii. Therefore, the apparent radius of Ar $= \dfrac{7.68 \text{ Å}}{4} = \mathbf{1.92 \text{ Å}}$

25 Coordination Compounds

25-2. *Refer to Section 25-1 and Table 25-3.*

$NiSO_4 \cdot 6H_2O \quad \equiv \quad [Ni(OH_2)_6]SO_4$

$Cu(NO_3)_2 \cdot 4NH_3 \quad \equiv \quad [Cu(NH_3)_4](NO_3)_2$

$Co(NO_3)_2 \cdot 6NH_3 \quad \equiv \quad [Co(NH_3)_6](NO_3)_2$

25-4. *Refer to Sections 25-3 and 25-5, and Table 25-7.*

The coordination number of a metal atom or ion in a complex is the number of donor atoms to which it is coordinated. It is not necessarily equal to the number of ligands in the compound or ion.

The most common values of coordination numbers are 4 and 6.

25-6. *Refer to Sections 25-3 and 25-4.*

	Complex	Ligand(s)		Coordination Number	Ox. No. of Central Metal		
(a)	$[Co(NH_3)_2(NO_2)_4]^-$	NH_3 NO_2^-	ammine nitro	6	Co	+3
(b)	$[Cr(NH_3)_5Cl]Cl_2$	NH_3 Cl^-	ammine chloro	6	Cr	+3
(c)	$K_4[Fe(CN)_6]$	CN^-	cyano	6	Fe	+2
(d)	$[Pd(NH_3)_4]^{2+}$	NH_3	ammine	4	Pd	+2

25-8. *Refer to Section 25-3 and Table 25-5.*

Polydentate ligands, such as ethylenediamine (en) and ethylenediaminetetraacetato (edta), cause ring formation to occur in a complex. This phenomenon is called chelation, resulting from a polydentate ligand bonding to the central metal atom or ion through two or more donor atoms at the same time.

For example, consider the $[Co(en)_3]^{3+}$ ion, in which three bidentate ethylenediamine ligands bond to the Co^{3+} ion creating 3 rings in the complex.

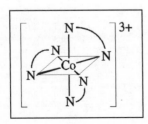

where is ethylenediamine, $H_2NCH_2CH_2NH_2$

25-10. *Refer to Section 25-1, Table 25-2 and Exercise 25-9.*

Formula	Coordination Sphere	Charge on the Complex
$[Pt(NH_3)_2Cl_4]$	$[Pt(NH_3)_2Cl_4]^0$	zero
$[Pt(NH_3)_3Cl_3]Cl$	$[Pt(NH_3)_3Cl_3]^+$	+1
$[Pt(NH_3)_4Cl_2]Cl_2$	$[Pt(NH_3)_4Cl_2]^{2+}$	+2
$[Pt(NH_3)_5Cl]Cl_3$	$[Pt(NH_3)_5Cl]^{3+}$	+3
$[Pt(NH_3)_6]Cl_4$	$[Pt(NH_3)_6]^{4+}$	+4

25-12. *Refer to Section 25-2 and Table 25-3.*

In general terms, we may represent the reaction in which a metal cation reacts in aqueous NH_3 to form an insoluble metal hydroxide by the following reaction:

$$M^{n+} + nNH_3 + nH_2O \rightarrow M(OH)_n(s) + nNH_4^+.$$

(a) $Cu^{2+} + 2NH_3 + 2H_2O \rightarrow Cu(OH)_2(s) + 2NH_4^+$

(b) $Zn^{2+} + 2NH_3 + 2H_2O \rightarrow Zn(OH)_2(s) + 2NH_4^+$

(c) $Fe^{3+} + 3NH_3 + 3H_2O \rightarrow Fe(OH)_3(s) + 3NH_4^+$

(d) $Co^{2+} + 2NH_3 + 2H_2O \rightarrow Co(OH)_2(s) + 2NH_4^+$

(e) $Ni^{2+} + 2NH_3 + 2H_2O \rightarrow Ni(OH)_2(s) + 2NH_4^+$

25-14. *Refer to Section 25-2 and Table 25-3.*

The metal hydroxides that dissolve in an excess of aqueous NH_3 to form ammine complexes are derived from the twelve metals of the cobalt, nickel, copper and zinc families. Therefore, when excess NH_3 is added:

(a) $Zn(OH)_2(s)$ will dissolve
(b) $Cr(OH)_3(s)$ will not dissolve
(c) $Fe(OH)_2(s)$ will not dissolve*
(d) $Ni(OH)_2(s)$ will dissolve
(e) $Cd(OH)_2(s)$ will dissolve

*Note: $[Fe(NH_3)_6]^{2+}$ is only stable in saturated NH_3 solution.

25-16. *Refer to Sections 25-3 and 25-4.*

(a) $K_2[Pt(NO_2)_4]$ ox. no. Pt: +2
(b) $[Co(NO_2)_3(NH_3)_3]$ ox. no. Co: +3
(c) $[Ag(NH_3)_2]I$ ox. no. Ag: +1
(d) $K_4[Ni(CN)_6]$ ox. no. Ni: +2
(e) $[Ni(NH_3)_4(H_2O)_2](NO_3)_2$ ox. no. Ni: +2
(f) $Na[Al(H_2O)_2(OH)_4]$ ox. no. Al: +3

25-18. *Refer to Sections 25-3 and 25-4.*

(a) $Ni(CO)_4$ tetracarbonylnickel(0)
(b) $Na_2[Co(OH_2)_2(OH)_4]$ sodium diaquatetrahydroxocobaltate(II)
(c) $[Ag(NH_3)_2]Br$ diamminesilver bromide (note: (I) is not needed for silver)
(d) $[Cr(en)_3](NO_3)_3$ tris(ethylenediamine)chromium(III) nitrate
(e) $[Co(NH_3)_4Cl]SO_4$ tetraamminechlorocobalt(III) sulfate

(f) $K_2[Cu(CN)_4]$ potassium tetracyanocuprate(II)

(g) $[Ni(NH_3)_4(H_2O)_2](NO_3)_2$ tetraamminediaquanickel(II) nitrate

(h) $Na[Al(H_2O)_2(OH)_4]$ sodium diaquatetrahydroxoaluminate

(i) $[Co(NH_3)_4Cl_2][Cr(C_2O_4)_2]$ tetraamminedichlorocobalt(III) bis(oxalato)chromate(III)
     or $[Co(NH_3)_4Cl_2][Cr(ox)_2]$

25-20.   *Refer to Sections 25-3 and 25-4.*

(a) $[Ag(NH_3)_2]Cl$ diamminesilver chloride   (note: (I) is not needed for silver)

(b) $Fe(en)_3]PO_4$ tris(ethylenediamine)iron(III) phosphate

(c) $[Co(NH_3)_6]SO_4$ hexaamminecobalt(II) sulfate

(d) $[Co(NH_3)_6]_2(SO_4)_3$ hexaamminecobalt(III) sulfate

(e) $[Pt(NH_3)_4]Cl_2$ tetraammineplatinum(II) chloride

(f) $(NH_4)_2[PtCl_4]$ ammonium tetrachloroplatinate(II)

(g) $[Co(NH_3)_5SO_4]NO_2$ pentaamminesulfatocobalt(III) nitrite

(h) $K_4[NiF_6]$ potassium hexafluoronickelate(II)

25-22.   *Refer to Sections 25-3 and 25-4.*

(a) diamminedichlorozinc $[Zn(NH_3)_2Cl_2]$

(b) tin(IV) hexacyanoferrate(II) $Sn[Fe(CN)_6]$

(c) tetracyanoplatinate(II) ion $[Pt(CN)_4]^{2-}$

(d) potassium hexachlorostannate(IV) $K_2[SnCl_6]$

(e) tetraammineplatinum(II) ion $[Pt(NH_3)_4]^{2+}$

(f) sodium hexachloronickelate(II) $Na_4[NiCl_6]$

(g) tetraamminecopper(II) pentacyanohydroxoferrate(III) $[Cu(NH_3)_4]_3[Fe(CN)_5(OH)]_2$

(h) diaquadicyanocopper(II) $[Cu(H_2O)_2(CN)_2]$ or $[Cu(OH_2)_2(CN)_2]$

(i) potassium hexachloropalladate(IV) $K_2[PdCl_6]$

25-24.   *Refer to Sections 25-3 and 25-4.*

(a) $[Ag(NH_3)_2]^+$ diamminesilver ion   (note: (I) is not needed for silver)
     $[Pt(NH_3)_4]^{2+}$ tetraammineplatinum(II) ion
     $[Cr(OH_2)_6]^{3+}$ hexaaquachromium(III) ion

(b) $[Ni(en)_3]^{2+}$ tris(ethylenediamine)nickel(II) ion
     $[Co(en)_3]^{3+}$ tris(ethylenediamine)cobalt(III) ion
     $[Cr(en)_3]^{3+}$ tris(ethylenediamine)chromium(III) ion

(c) $[Co(en)_2(NO_2)_2]^+$ bis(ethylenediamine)dinitrocobalt(III) ion
     $[CoBr_2(en)_2]^+$ dibromobis(ethylenediamine)cobalt(III) ion
     $[Ni(en)_2(NO_2)_2]^{2+}$ bis(ethylenediamine)dinitrosylnickel(II) ion

(d) $[FeCl(dien)(en)]^{2+}$ chloro(diethylenetriamine)(ethylenediamine)iron(III) ion
     $[Cr(OH_2)(dien)(ox)]^+$ aqua(diethylenetriamine)(oxalato)chromium(III) ion
     $[RuCl(dien)(en)]^{2+}$ chloro(diethylenetriamine)(ethylenediamine)ruthenium(III) ion

(e) $[Co(OH_2)_3(dien)]^{3+}$ triaqua(diethylenetriamine)cobalt(III) ion

   $[Cr(NH_3)_3(dien)]^{3+}$ triammine(diethylenetriamine)chromium(III) ion

   $[Fe(NH_3)_3(dien)]^{2+}$ triammine(diethylenetriamine)iron(II) ion

---

25-26.   *Refer to Section 25-7.*

(a) MA_2B_4, an octahedral complex, can exist as two geometrical isomers: *cis* and *trans*.

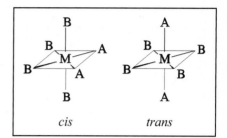

cis   *trans*

(b) The octahedral complex, MA_3B_3, can also exist in two geometrical forms. The facial isomer (*fac*) has three identical ligands at the corners of a trigonal face. The meridianal isomer (*mer*) has three identical ligands at three corners of a square plane.

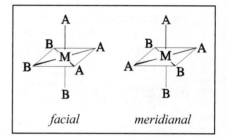

facial   *meridianal*

None of these geometric isomers are optical isomers since they all are superimposable on their mirror images, i.e., the compound and its mirror image are actually the same compound.

25-28.   *Refer to Sections 25-6 and 25-7, and Exercise 25-22 Solution.*

(a) $[Cr(NH_3)_4(OH_2)_2]^{3+}$

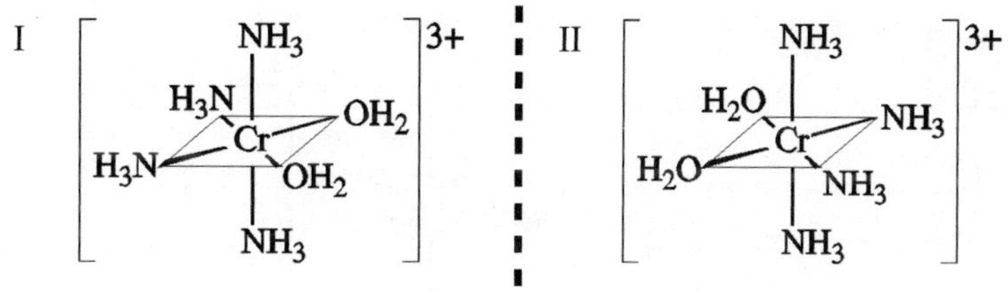

tetraammine-*cis*-diaquachromium(III) ion

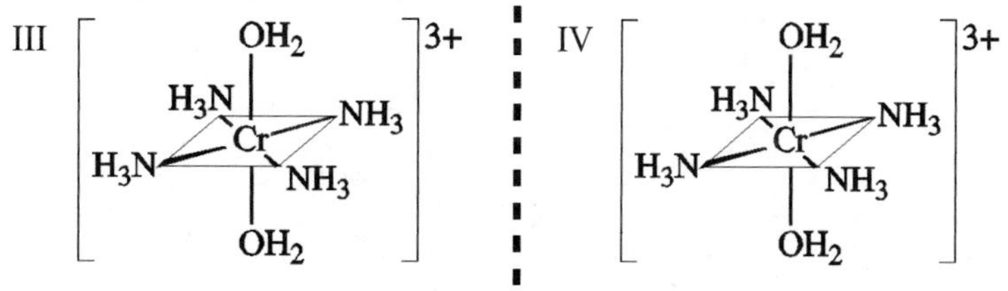

tetraammine-*trans*-diaquachromium(III) ion

Structures I and II are mirror images that are superimposable if the vertical axis is rotated 180°; therefore, they are simply different representations of the same compound. Structures III and IV are also the same compound. Hence both the *cis* and *trans* geometrical isomers of $[Cr(NH_3)_4(OH_2)_2]^{3+}$ have no optical isomers, so the total number of isomers is 2.

(b) [Cr(NH₃)₃Cl₃]

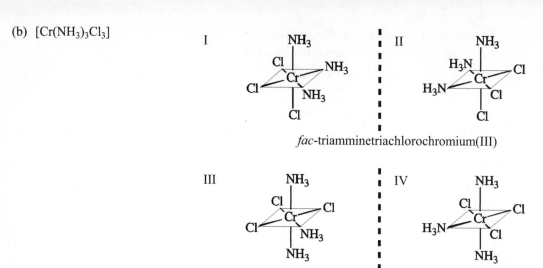

fac-triamminetriachlorochromium(III)

mer-triamminetriaquachromium(III)

Structures I and II are identical. Hence, the facial (*fac*) geometrical isomer has no optical isomer. Structures III and IV are also identical. So, the meridianal (*mer*) geometrical isomer also has no optical isomer. The total number of isomers is 2 for [Cr(NH₃)₃Cl₃].

(c) [Cr(en)₃]³⁺

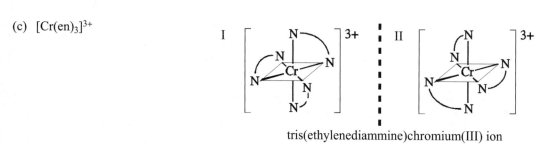

tris(ethylenediammine)chromium(III) ion

Structures I and II are mirror images of each other that are not superimposable. Therefore, [Cr(en)₃]³⁺ has 2 optical isomers.

(d) [Pt(en)₂Cl₂]Cl₂

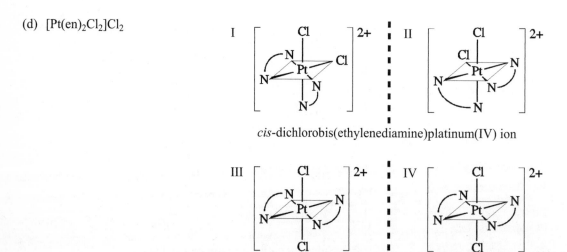

cis-dichlorobis(ethylenediamine)platinum(IV) ion

trans-dichlorobis(ethylenediamine)platinum(IV) ion

Structures I and II are mirror images that are not superimposable; hence the *cis* geometrical isomer consists of a pair of optical isomers. Structures III and IV are identical; the *trans* geometrical isomer has no optical isomer. Therefore, [Pt(en)₂Cl₂]Cl₂ has a total of 3 isomers.

(e) $[Cr(NH_3)_2Br_2Cl_2]^-$

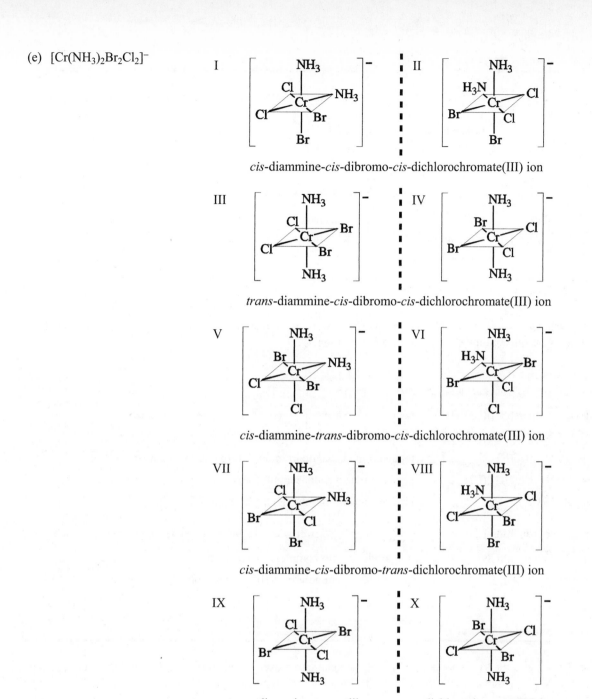

I

cis-diammine-cis-dibromo-cis-dichlorochromate(III) ion

III   IV

trans-diammine-cis-dibromo-cis-dichlorochromate(III) ion

V   VI

cis-diammine-trans-dibromo-cis-dichlorochromate(III) ion

VII   VIII

cis-diammine-cis-dibromo-trans-dichlorochromate(III) ion

IX   X

trans-diammine-trans-dibromo-trans-dichlorochromate(III) ion

Structures I and II are mirror images that are not identical and are optical isomers. The mirror image pairs of the remaining geometrical isomers are identical and therefore, they have no optical isomers. The total number of isomers for $[Cr(NH_3)_2Br_2Cl_2]^-$ is 6.

25-30. *Refer to the Introduction to Section 25-6 and the Key Terms for Chapter 25.*

Isomers are substances that have the same number and kinds of atoms, but arranged differently. Constitutional (structural) isomers, as applied to coordination compounds, are isomers whose differences involve having more than a single coordination sphere or different donor atoms on the same ligand. They contain different atom-to-atom bonding sequences. Stereoisomers, on the other hand, are isomers that differ only in the way that atoms are oriented in space, and therefore involve only one coordination sphere and the same ligands and donor atoms.

An ionization isomer results from the exchange of ions inside and outside the coordination sphere.

(a) $[Co(NH_3)_4Br_2]Br$ tetraamminedibromocobalt(III) bromide

       $[Co(NH_3)_4Br]Br_2$ tetraamminebromocobalt(III) bromide

(b) $[Ni(en)_2(NO_2)_2]Cl_2$ bis(ethylenediamine)dinitronickel(IV) chloride

       $[NiCl_2(en)_2](NO_2)_2$ dichlorobis(ethylenediamine)nickel(IV) nitrite

(c) $[Fe(NH_3)_5CN]SO_4$ pentaamminecyanoiron(III) sulfate

       $[Fe(NH_3)_5SO_4]CN$ pentaamminesulfatoiron(III) cyanide

A coordination isomer involves the exchange of ligands between a complex cation and a complex anion of the same compound, forming another complex cation and complex anion.

(a) There are 5 possible coordination isomers of $[Co(NH_3)_6][Cr(CN)_6]$, hexaamminecobalt(III) hexacyanochromate(III):

     $[Co(NH_3)_5(CN)][Cr(NH_3)(CN)_5]$ pentaamminecyanocobalt(III) amminepentacyanochromate(III)

     $[Co(NH_3)_4(CN)_2][Cr(NH_3)_2(CN)_4]$ tetraamminedicyanocobalt(III) diamminetetracyanochromate(III)

     $[Cr(NH_3)_4(CN)_2][Co(NH_3)_2(CN)_4]$ tetraamminedicyanochromium(III) diamminetetracyanocobaltate(III)

     $[Cr(NH_3)_5(CN)][Co(NH_3)(CN)_5]$ pentaamminecyanochromium(III) amminepentacyanocobaltate(III)

     $[Cr(NH_3)_6][Co(CN)_6]$ hexaamminechromium(III) hexacyanocobaltate(III)

(b) There was 1 coordination isomer of $[Ni(en)_3][Cu(CN)_4]$, tris(ethylenediamine)nickel(II) tetracyanocuprate(II)

     $[Ni(en)_2(CN)_2][Cu(en)(CN)_2]$ not possible; both complexes have no charge

     $[Cu(en)_2][Ni(en)(CN)_4]$ bis(ethylenediamine)copper(II) tetracyano(ethylenediamine)nickelate(II)

Crystal Field Theory is a theory of bonding in transition metal complexes in which the bonds between metal ions and ligands are strictly electrostatic interactions. During bonding, the repulsions between ligand electrons and metal electrons in d orbitals create an electric field, i.e., the octahedral crystal field, which splits the d orbitals into two sets, the t_{2g} set at lower energy and the e_g set at higher energy. The energy separation between the two sets in an octahedral complex is named Δ_{oct} and is proportional to the crystal field strength of the ligands, that is, how strongly the ligand electrons repel the metal electrons.

When electrons undergo transitions from a lower energy t_{2g} orbital to a higher energy e_g orbital, an amount of energy equivalent to the wavelengths of visible light are absorbed, resulting in transition metal complexes with the complementary color of the light absorbed. Δ_{oct} can be determined experimentally from the wavelength of the light absorbed:

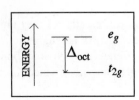

$$\Delta_{oct} = EN_A = \frac{hcN_A}{\lambda}$$

where E = energy of the absorbed photon

           N_A = Avogadro's Number

           h = Planck's constant

           c = speed of light

           λ = wavelength

It is possible to arrange the common ligands in the order of increasing crystal field strengths, by interpreting the visible spectra of many complexes. This is the spectrochemical series. From the above equation, we can deduce that transition metal complexes that are colored, e.g. red or orange, are absorbing the complementary colors, green or blue. These absorbed wavelengths are at the shorter end of the visible light range and correspond to larger Δ_{oct} values. The ligands in these complexes have larger crystal field strengths and are located at the high end of the spectrochemical series.

25-38. *Refer to Sections 25-8 and 25-9.*

A high spin complex is the Crystal Field designation when all t_{2g} and e_g orbitals are singly occupied before pairing begins. A low spin complex, on the other hand, is the Crystal Field designation, where electrons are paired in t_{2g} orbitals before e_g orbitals are occupied. However, the low spin configuration exists only for octahedral complexes having metal ions with d^1 - d^7 configurations. For d^1 - d^3 and d^8 - d^{10} ions, only one possibility exists which is designated as high spin. In the case of d^4 - d^7 configurations: (1) weak ligand field strength is associated with high spin complexes, whereas strong ligand field strength is associated with low spin complexes, and (2) the spectrochemical series ranks ligands in order of increasing ligand field strength. The following are the predictions:

	Complex Ion	Metal Ion Configuration	Complex Configuration
(a)	$[Cu(OH_2)_6]^{2+}$	d^9	high spin (by convention)
(b)	$[MnF_6]^{3-}$	d^4	high spin (weak field strength ligands)
(c)	$[Co(CN)_6]^{3-}$	d^6	low spin (strong field strength ligands)
(d)	$[Cr(NH_3)_6]^{3+}$	d^3	high spin (by convention)

25-40. *Refer to Sections 25-4, 25-8 and 25-9, and Exercise 25-39.*

Metal Ion	Ligand Field Strength	Example	
V^{2+}	weak	$[VF_6]^{4-}$	hexafluorovanadate(II) ion
		$[V(OH_2)_6]^{2+}$	hexaaquavanadium(II) ion
Mn^{2+}	strong	$[Mn(en)_3]^{2+}$	tris(ethylenediamine)manganese(II) ion
		$[Mn(NH_3)_6]^{2+}$	hexaamminemanganese(II) ion
Mn^{2+}	weak	$[MnF_6]^{4-}$	hexafluoromanganate(II) ion
		$[MnBr_6]^{4-}$	hexabromomanganate(II) ion
Ni^{2+}	weak	$[Ni(OH_2)_6]^{2+}$	hexaaquanickel(II) ion
		$[NiF_6]^{4-}$	hexafluoronickelate(II) ion
Cu^{2+}	weak	$[Cu(OH_2)_6]^{2+}$	hexaaquacopper(II) ion
		$[CuF_6]^{4-}$	hexafluorocuprate(II) ion
Fe^{3+}	strong	$[Fe(CN)_6]^{3-}$	hexacyanoferrate(III) ion
		$[Fe(NH_3)_6]^{3+}$	hexaammineiron(III) ion
Cu^+	weak	$[CuCl_6]^{5-}$	hexachlorocuprate(I) ion
		$[Cu(OH_2)_6]^+$	hexaaquacopper(I) ion
Ru^{3+}	strong	$[Ru(NH_3)_6]^{3+}$	hexaammineruthenium(III) ion
		$[Ru(en)_3]^{3+}$	tris(ethylenediamine)ruthenium(III) ion

25-42. *Refer to Sections 25-8 and 25-9, and Table 25-8.*

Electron configuration of metal ion

(a) $[Mn(NH_3)_2(OH_2)_2]^{2+}$ (low spin complex ion)　　　d^5

(b) $[Co(NO_2)_6]^{3-}$ (low spin complex ion)　　　d^6

(c) $[Cr(NH_3)_4Cl_2]^+$ (low spin complex ion)　　　d^3

(d) $[MnCl_5NO]^{3-}$ (high spin complex ion)　　　d^5

(e) paramagnetic: (a), (c), (d)

(f) diamagnetic: (b)

25-44. *Refer to Sections 25-8 and 25-9, and Table 25-8.*

A high spin complex is the Crystal Field designation when all t_{2g} and e_g orbitals are singly occupied before pairing begins. A low spin complex, on the other hand, is the Crystal Field designation, where electrons paired in t_{2g} orbitals before e_g orbitals are occupied. One can predict whether or not a complex is high or low spin by understanding that a complex is high-spin or low-spin depending on (1) the metal ion's electronic configuration and whether the ligand is a weak field or strong field ligand. The low spin configuration can exist only for octahedral complexes having metal ions with d^4 - d^7 configurations. For d^1 - d^3 and d^8 - d^{10} ions, only one possibility exists which is designated as high spin. In the case of d^4 - d^7 configurations: (1) weak ligand field strength is associated with high spin complexes, whereas strong ligand field strength is associated with low spin complexes, and (2) the spectrochemical series ranks ligands in order of increasing ligand field strength. One can determine if a complex is high or low spin by first determining whether or not the complex is diamagnetic or paramagnetic. If it is paramagnetic, one then needs to measure the extent to which it is paramagnetic to determine how many unpaired electrons it has.

25-46. *Refer to Section 25-4 and Table 25-7.*

Common metal coordination numbers found in coordination chemistry are

Coordination Number	Shape
2	linear
4	tetrahedral or square planar
5	trigonal bipyramidal or square pyramidal
6	octahedral

25-48. *Refer to Section 15-14.*

Balanced equation: $Ag^+(aq) + 3I^-(aq) \rightarrow [AgI_3]^{2-}$

A *decrease* in entropy for the above reaction is expected since the reactant side has 4 moles of ions while the product side contains only 1 mole of ions.

$\Delta S^\circ_{rxn} = [S^\circ_{[AgI_3]^{2-}(aq)}] - [S^\circ_{Ag^+(aq)} + 3S^\circ_{I^-(aq)}]$

$= [(1 \text{ mol})(253.1 \text{ J/mol·K})] - [(1 \text{ mol})(72.68 \text{ J/mol·K}) + (3 \text{ mol})(111.3 \text{ J/mol·K})]$

$= \mathbf{-153.5 \text{ J/(mol rxn)·K}}$

Since $\Delta S^\circ_{rxn} < 0$, the entropy of this reaction is indeed decreasing, as predicted.

Balanced equations: $[Cu(NH_3)_4]Cl_2 \rightarrow [Cu(NH_3)_4]^{2+} + 2Cl^-$ (to completion)

 $[Cu(NH_3)_4]^{2+} \rightleftarrows Cu^{2+} + 4NH_3$ (reversible) $K_d = 8.5 \times 10^{-13}$

 $NH_3 + H_2O \rightleftarrows NH_4^+ + OH^-$ (reversible) $K_b = 1.8 \times 10^{-5}$

Plan: (1) Calculate the concentration of NH_3 from the equilibrium expression for the dissociation of the complex ion. Note: this is possible only if we assume that the ionization of NH_3 does not appreciably alter the concentration of NH_3.

 (2) Calculate the $[OH^-]$ and pH of the solution using the equilibrium expression for the ionization of NH_3. Note: we are ignoring the effect of the hydrolysis of Cu^{2+} ion on pH.

(1) Let x = $[[Cu(NH_3)_4]^{2+}]$ that dissociates. Then

	$[Cu(NH_3)_4]^{2+}$	$\rightleftarrows$	Cu^{2+}	+	$4 NH_3$
initial	0.42 *M*		0 *M*		0 *M*
change	- x *M*		+ x *M*		+ 4x *M*
at equilibrium	(0.42 - x) *M*		x *M*		4x *M*

$K_d = \dfrac{[Cu^{2+}][NH_3]^4}{[[Cu(NH_3)_4]^{2+}]} = \dfrac{(x)(4x)^4}{(0.42 - x)} = 8.5 \times 10^{-13} \approx \dfrac{(x)(4x)^4}{0.42} = \dfrac{256x^5}{0.42}$ Solving, x = $(1.4 \times 10^{-15})^{1/5}$
$\phantom{K_d = \dfrac{[Cu^{2+}][NH_3]^4}{[[Cu(NH_3)_4]^{2+}]} = \dfrac{(x)(4x)^4}{(0.42 - x)} = 8.5 \times 10^{-13} \approx \dfrac{(x)(4x)^4}{0.42} = \dfrac{256x^5}{0.42}}$ $= 1.1 \times 10^{-3}$

Therefore, $[NH_3] = 4x = 4.3 \times 10^{-3}$ *M*

(2) Let y = $[NH_3]_{\text{ionized}}$. Then, $[NH_3] = 4.3 \times 10^{-3}$ - y; $[NH_4^+] = [OH^-] = y$

$K_b = \dfrac{[NH_4^+][OH^-]}{[NH_3]} = \dfrac{y^2}{(4.3 \times 10^{-3} - y)} = 1.8 \times 10^{-5}$

Solving the quadratic equation: $y^2 + (1.8 \times 10^{-5})y - 7.7 \times 10^{-8} = 0$,
 y = 2.9×10^{-4} or -2.7×10^{-4} (discard)

Therefore, $[OH^-] = 2.9 \times 10^{-4}$ *M*; pOH = 3.54; pH = **10.46**

(a) Balanced equation: $Zn(OH)_2(s) \rightleftarrows Zn^{2+}(aq) + 2OH^-(aq)$ $K_{sp} = 4.5 \times 10^{-17}$

Let x = molar solubility of $Zn(OH)_2$.   Then, $[Zn^{2+}] = x$; $[OH^-] = 2x$

$K_{sp} = [Zn^{2+}][OH^-]^2 = (x)(2x)^2 = 4x^3 = 4.5 \times 10^{-17}$ Solving, x = 2.2×10^{-6}

Therefore, molar solubility = **2.2×10^{-6} mol $Zn(OH)_2$/L**

(b) Balanced equations: $NaOH(aq) \rightarrow Na^+(aq) + OH^-(aq)$ (to completion)

 $Zn(OH)_2(s) \rightleftarrows Zn^{2+}(aq) + 2OH^-(aq)$ (reversible) $K_{sp} = 4.5 \times 10^{-17}$

If we proceed with part (b) as if it were an ordinary example of a common ion effect problem with $Zn(OH)_2$ dissolving into a solution with a known concentration of OH^- ion, we would get the wrong answer. We must take into account the effect of the formation of the complex ion, $Zn(OH)_4^{2-}$, on the solubility of $Zn(OH)_2$.

 $Zn(OH)_4^{2-}(aq) \rightleftarrows Zn^{2+}(aq) + 4OH^-(aq)$ (reversible) $K_d = 3.5 \times 10^{-16}$

Much more $Zn(OH)_2$ will dissolve since essentially all of the released Zn^{2+} ions are incorporated into the soluble $Zn(OH)_4^{2-}$ complex because K_d is so small. When significant complex ion formation occurs as it does in this case, the molar solubility must include the concentrations of *all* the Zn^{2+} species. Also we cannot assume that $[OH^-]$ remains at 0.30 *M* throughout this process. The net concentration of OH^- ion decreases slightly: For every one formula unit of $Zn(OH)_2$ that dissolves producing 2 OH^- ions, one formula unit of $Zn(OH)_4^{2-}$ will form, removing 4 OH^- ions. The net result is that for every 1 mol/L of $Zn(OH)_2$ that dissolves (the molar solubility), approximately 1 mol/L of $Zn(OH)_4^{2-}$ is produced, but 2 mol/L of OH^- is lost. Therefore,

$$[OH^-] = 0.30\ M - 2[Zn(OH)_4^{2-}]$$

Plan: The molar solubility of $Zn(OH)_2$ equals the sum of the concentrations of the 2 soluble Zn species, Zn^{2+} and $Zn(OH)_4^{2-}$. The concentrations can be calculated at equilibrium by solving the following 3 equations in 3 unknowns:

(1) $[OH^-] = 0.30\ M - 2[Zn(OH)_4^{2-}]$

(2) $K_{sp} = 4.5 \times 10^{-17} = [Zn^{2+}][OH^-]^2$

(3) $K_d = 3.5 \times 10^{-16} = \dfrac{[Zn^{2+}][OH^-]^4}{[Zn(OH)_4^{2-}]}$

Step 1: Using equation (1), let $x = [OH^-]$. Then, $[Zn(OH)_4^{2-}] = \dfrac{0.30 - x}{2}$

Step 2: Divide equation (3) by equation (2) to remove the $[Zn^{2+}]$ term, substitute and solve for x.

$$\frac{K_d}{K_{sp}} = \frac{3.5 \times 10^{-16}}{4.5 \times 10^{-17}} = 7.8 = \frac{\left(\frac{[Zn^{2+}][OH^-]^4}{[Zn(OH)_4^{2-}]}\right)}{[Zn^{2+}][OH^-]^2} = \frac{[OH^-]^2}{[Zn(OH)_4^{2-}]} = \frac{x^2}{\left(\frac{0.30 - x}{2}\right)} = \frac{2x^2}{0.30 - x}$$

Step 3: Solving the quadratic equation: $2x^2 + 7.8x - 2.3 = 0$, we have: $x = 0.28$ or -4.2 (discard).

Therefore, $[OH^-] = x = 0.28\ M$ (a value slightly less than $0.30\ M$, as expected)

$$[Zn^{2+}] = \frac{K_{sp}}{[OH^-]^2} = \frac{4.5 \times 10^{-17}}{(0.28)^2} = 5.7 \times 10^{-16}\ M$$

$$[Zn(OH)_4^{2-}] = \frac{[Zn^{2+}][OH^-]^4}{K_d} = \frac{(5.7 \times 10^{-16})(0.28)^4}{3.5 \times 10^{-16}} = 0.010\ M$$

molar solubility = total number of moles of $Zn(OH)_2$ that dissolved to produce 1 L of saturated solution
= $[Zn^{2+}] + [Zn(OH)_4^{2-}]$
= $(5.7 \times 10^{-16}\ M) + (0.010\ M)$
= **0.010 mol $Zn(OH)_2$/L**

(c) $[Zn(OH)_4^{2-}] = $ **0.010 M** (see part (b))

Note: If you assume that $[OH^-]$ is constant at $0.30\ M$, then an approximate value for molar solubility is calculated.

Substituting into the rearranged K_{sp} expression: $[Zn^{2+}] = \dfrac{K_{sp}}{[OH^-]^2} = \dfrac{4.5 \times 10^{-17}}{(0.30)^2} = 5.0 \times 10^{-16}\ M$

Substituting into the rearranged K_d expression: $[Zn(OH)_4^{2-}] = \dfrac{[Zn^{2+}][OH^-]^4}{K_d} = \dfrac{(5.0 \times 10^{-16})(0.30)^4}{3.5 \times 10^{-16}} = 0.012\ M$

Therefore, molar solubility = $[Zn^{2+}] + [Zn(OH)_4^{2-}] = (5.0 \times 10^{-16}\ M) + (0.012\ M) = 0.012$ mol $Zn(OH)_2$/L

26 Nuclear Chemistry

26-2. *Refer to the Introduction to Chapter 26, and Sections 26-1, 26-2 and 26-13.*

Natural radioactivity derives from spontaneous nuclear disintegrations. Induced radioactivity derives from the bombardment of nuclei with accelerated subatomic particles or other nuclei. Both cause atoms of one nuclide to be converted to another nuclide.

Using the elements mentioned in Section 26-13, induced radiation and the artificial transmutation of elements occur with both light elements, like the nonmetals ^{3}H, ^{12}C and ^{17}O as well has heavier elements, like ^{97}Tc, ^{112}Fr, ^{210}At and ^{239}U, which can be metals, metalloids or nonmetals. Transuranium elements, i.e. the elements with atomic numbers greater than 92 (uranium), must be prepared by nuclear bombardment of other elements.

26-4. *Refer to Sections 1-1 and 26-3.*

Einstein's equation relates matter and energy:

$$E = mc^2 \qquad \text{where } E = \text{amount of energy released}$$
$$m = \text{mass of matter transformed into energy}$$
$$c = \text{speed of light in a vacuum, } 3.00 \times 10^8 \text{ m/s}$$

If m is expressed in kg and c in m/s, the obtained E will be in units of J.

26-6. *Refer to Sections 26-1,5-5 and 5-7.*

Nucleons are the particles comprising the nucleus, i.e., it is a collective term for the protons and neutrons in a nucleus. The number of protons is the atomic number; the sum of the protons and the neutrons (the nucleons) is the mass number.

26-8. *Refer to Section 26-3 and Figure 26-11.*

The plot of binding energy per nucleon versus mass number for all the isotopes shows that binding energies/nucleon increase very rapidly with increasing mass number, reaching a maximum of 8.80 MeV per nucleon at mass number 56 for $^{56}_{26}$Fe, then decrease slowly.

26-10. *Refer to Sections 26-2 and 26-6.*

Potassium, with atomic number Z = 19, has three naturally occurring isotopes:

Isotope	Neutrons	Protons	Neutrons + Protons	n/p Ratio
^{39}K	20	19	39	1.05
^{40}K	21	19	40	1.11
^{41}K	22	19	41	1.16

The "magic numbers" which impart stability to a nucleus are 2, 8, 20, 28, 50, 82 or 126. The isotope, ^{39}K, has a magic number equal to its number of neutrons, so it is probably stable. The others have a larger neutron-to-proton ratio, making them neutron-rich nuclei, so ^{40}K and ^{41}K might be expected to decay by beta emission. In fact, both ^{39}K and ^{41}K are stable, and ^{40}K does decay by beta emission.

According to predictions:

^{40}K beta emission $^{40}_{19}K \rightarrow \, ^{40}_{20}Ar + \, ^{0}_{-1}e$

^{41}K beta emission $^{41}_{19}K \rightarrow \, ^{41}_{20}Ar + \, ^{0}_{-1}e$ (However, ^{41}K is not unstable and does not decay.)

26-14. *Refer to Section 26-3, Table 26-1, Examples 26-1 and 26-2, and Appendix C.*

(a) One neutral atom of $^{62}_{28}Ni$ contains 28 e^-, 28 p^+ and 34 n^0.

electrons: 28 x 0.00054858 amu = 0.015 amu
protons: 28 x 1.0073 amu = 28.204 amu
neutrons: 34 x 1.0087 amu = 34.296 amu
  sum = 62.515 amu

Δm = (sum of masses of e^-, p^+ and n^0) - (actual mass of a ^{62}Ni atom)
    = 62.515 amu - 61.9283 amu
    = 0.587 amu

Therefore, the mass deficiency for ^{62}Ni is **0.587 amu/atom** or **0.587 g/mol**.

(b) Note: 1 joule = 1 kg x $(1 \text{ m/s})^2$

The nuclear binding energy, BE = $(\Delta m)c^2$
        = $(0.587 \times 10^{-3} \text{ kg/mol})(3.00 \times 10^8 \text{ m/s})^2$
        = $5.28 \times 10^{13} \text{ kg·m}^2/\text{mol·s}^2$
        = 5.28×10^{13} J/mol or **5.28×10^{10} kJ/mol of ^{62}Ni atoms**

26-16. *Refer to Section 26-3, Table 26-1, Examples 26-1 and 26-2, and Appendix C.*

(a) A neutral atom of $^{64}_{30}Zn$ contains 30 e^-, 30 p^+ and 34 n^0.

electrons: 30 x 0.00054858 amu = 0.016 amu
protons: 30 x 1.0073 amu = 30.219 amu
neutrons: 34 x 1.0087 amu = 34.296 amu
  sum = 64.531 amu

Δm = (sum of masses of e^-, p^+ and n^0) - (actual mass of a ^{64}Zn atom)
    = 64.531 amu - 63.9291 amu
    = 0.602 amu

the mass deficiency, Δm, for ^{64}Zn is **0.602 amu/atom**

(b) This is equivalent to a mass deficiency of **0.602 g/mol**.

(c) Note: 1 J = 1 kg x $(1 \text{ m/s})^2$

The nuclear binding energy, $BE = (\Delta m)c^2$

$$= \left(\frac{0.602 \text{ g}}{1 \text{ mol}} \times \frac{1 \text{ kg}}{1000 \text{ g}} \times \frac{1 \text{ mol}}{6.02 \times 10^{23} \text{ atoms}}\right)(3.00 \times 10^8 \text{ m/s})^2$$

$$= 9.00 \times 10^{-11} \text{ kg·m}^2/\text{atom·s}^2$$

$$= \mathbf{9.00 \times 10^{-11} \text{ J/atom}}$$

(d) BE (kJ/mol)$= \dfrac{9.00 \times 10^{-11} \text{ J}}{1 \text{ atom}} \times \dfrac{6.02 \times 10^{23} \text{ atoms}}{1 \text{ mol}} \times \dfrac{1 \text{ kJ}}{1000 \text{ J}} = \mathbf{5.42 \times 10^{10} \text{ kJ/mol}}$

(e) Since there are 64 nucleons in a ^{64}Zn atom,

BE (MeV/nucleon) for ^{64}Zn$= \dfrac{9.00 \times 10^{-11} \text{ J}}{1 \text{ atom}} \times \dfrac{1 \text{ atom}}{64 \text{ nucleons}} \times \dfrac{1 \text{ MeV}}{1.60 \times 10^{-13} \text{ J}} = \mathbf{8.79 \text{ MeV/nucleon}}$

26-18. *Refer to Section 26-3 and Examples 26-1 and 26-2.*

(a) One neutral atom of $^{127}_{53}$I contains 53 e^-, 53 p^+ and 74 n^0.

electrons:	53 x 0.00054858 amu	=	0.029 amu
protons:	53 x 1.0073 amu	=	53.387 amu
neutrons:	74 x 1.0087 amu	=	74.644 amu
		sum	= 128.060 amu

Δm = (sum of masses of e^-, p^+ and n^0) - (actual mass of a ^{127}I atom)
   = 128.060 amu - 126.9044 amu
   = 1.156 amu

Therefore, the mass deficiency for ^{127}I is 1.156 amu/atom or 1.156 g/mol.

Recall: 1 joule = 1 kg x (1 m/s)^2

The nuclear binding energy, $BE = (\Delta m)c^2$
$= (1.156 \times 10^{-3} \text{ kg/mol})(3.00 \times 10^8 \text{ m/s})^2$
$= 1.04 \times 10^{14} \text{ kg·m}^2/\text{mol·s}^2$
$= 1.04 \times 10^{14} \text{ J/mol or } \mathbf{1.04 \times 10^{11} \text{ kJ/mol of } ^{127}\text{I atoms}}$

(b) One neutral atom of $^{81}_{35}$Br contains 35 e^-, 35 p^+ and 46 n^0.

electrons:	35 x 0.00054858 amu	=	0.019 amu
protons:	35 x 1.0073 amu	=	35.256 amu
neutrons:	46 x 1.0087 amu	=	46.400 amu
		sum	= 81.675 amu

Δm = 81.675 amu - 80.9163 amu = 0.759 amu/atom or 0.759 g/mol

$BE = (\Delta m)c^2 = (0.759 \times 10^{-3} \text{ kg/mol})(3.00 \times 10^8 \text{ m/s})^2 = 6.83 \times 10^{13} \text{ kg·m}^2/\text{mol·s}^2$
$= 6.83 \times 10^{13} \text{ J/mol}$
or $\mathbf{6.83 \times 10^{10} \text{ kJ/mol of } ^{81}\text{Br atoms}}$

(c) One neutral atom of $^{35}_{17}$Cl contains 17 e^-, 17 p^+ and 18 n^0.

electrons:	17 x 0.00054858 amu	=	0.0093 amu
protons:	17 x 1.0073 amu	=	17.124 amu
neutrons:	18 x 1.0087 amu	=	18.157 amu
		sum	= 35.290 amu

Δm = 35.290 amu - 34.96885 amu = 0.321 amu/atom or 0.321 g/mol

$BE = (\Delta m)c^2 = (0.321 \times 10^{-3} \text{ kg/mol})(3.00 \times 10^8 \text{ m/s})^2 = 2.89 \times 10^{13} \text{ kg·m}^2/\text{mol·s}^2$
$= 2.89 \times 10^{13} \text{ J/mol}$
or $\mathbf{2.89 \times 10^{10} \text{ kJ/mol of } ^{35}\text{Cl atoms}}$

(a) In an electric field, an alpha (α) particle (a helium nucleus with a +2 charge) will be drawn toward the negative electrode, while a beta (β^-) particle (an electron with a -1 charge) will be drawn toward the positive electrode. Gamma (γ) radiation (very high energy electromagnetic radiation) will be unaffected by the electric field.

(b) The α particle and β^- particle will be drawn in opposite directions in a magnetic field and the γ radiation will be unaffected by a magnetic field.

(c) A piece of paper will reduce the α radiation significantly, but not the β^- or γ radiation; a thick concrete slab will prevent the α particles and β^- particles from passing, and most of the γ radiation.

There are many radionuclides that have medical uses, including:

(1) cobalt-60: This is used to arrest certain types of cancer. The technique uses the gamma rays produced in the decay of ^{60}Co to destroy cancerous tissue.

(2) plutonium-238: The energy produced in its decay is converted to electrical energy which powers heart pacemakers. Its relatively long half-life allows the device to be used for ten years before replacement.

Several radioisotopes are used as radioactive tracers (rediopharmaceuticals). These are injected into the body and allow physicians to study biological processes. These include

(3) sodium-24: This is used to follow the blood flow and locate stoppages in the circulatory system.

(4) thallium-201: This isotope helps locate healthy heart tissue.

(5) technetium-99: This metastable isotope has proven to be very useful and can be used to image the bones, liver, brain, etc.

(6) iodine-131: This concentrates in the thyroid gland, liver and certain parts of the brain. It is used to monitor goiter and other thyroid problems as well as liver and brain tumors.

Positron emission tomography (PET) is another form of imaging that uses positron emitters, such as ^{11}C, ^{13}N, ^{15}O and ^{18}F. These isotopes are incorporated into chemicals that are taken up by tissue. When the isotopes decay, the emitted positron reacts with a nearby electron, giving off 2 gamma rays, which are detected and an image of the tissue is created.

(1) Photographic Detection: Radioactive substances affect photographic plates. Although the intensity of the affected spot is related to the amount of radiation, precise measurement by this method is tedious.

(2) Detection by Fluorescence: Fluorescent substances can absorb radiation and subsequently emit visible light. This is the basis for scintillation counting and can be used for quantitative detection.

(3) Cloud Chambers: A chamber containing air saturated with vapor is used. Radioactive particles ionize air molecules in the chamber. Cooling the chamber causes droplets of liquid to condense on these ions, giving observable fog-like tracks.

(4) Gas Ionization Counters: A common gas ionization counter is the Geiger-Müller counter where the electronic pulses derived from the ionization process are registered as counts.

Scientists have known that nuclides which have certain "magic numbers" of protons and neutrons are especially stable. Nuclides with a number of protons or a number of neutrons or a sum of the two equal to 2, 8, 20, 28, 50, 82 or 126 have unusual stability. Examples of this are $_2^4$He, $_8^{16}$O, $_{20}^{42}$Ca, $_{38}^{88}$Sr, and $_{82}^{208}$Pb. This suggests a shell (energy level) model for the nucleus similar to the shell model of electron configurations.

26-28. *Refer to Section 26-7 and Figure 26-1.*

A nuclide with a neutron/proton ratio which is smaller than that for a stable isotope of the element can increase its ratio by undergoing:

(1) positron emission $\qquad\qquad\qquad _1^1p \rightarrow {_0^1}n + {_{+1}^0}\beta$
(2) electron capture (K capture) $\qquad _1^1p + {_{-1}^0}e \rightarrow {_0^1}n$

The net result of both processes is the loss of one proton and the gain of one neutron, thereby increasing the n/p ratio. Also, a heavier nuclide can undergo alpha emission to increase its n/p ratio.

26-30. *Refer to Sections 26-6 and 26-7, and Figure 26-1.*

	Protons	Neutrons	n/p Ratio	Stable Mass No.	Stable n/p Ratio
(a) $_{79}^{193}$Au	79	114	1.44	197	1.49
(b) $_{75}^{189}$Re	75	114	1.52	185, 187	1.47, 1.49
(c) $_{59}^{137}$Pr	59	78	1.32	141	1.39

Both (a) ^{193}Au and (c) ^{137}Pr have n/p ratios below that which is stable, so they will either undergo positron emission or electron capture. However, (b) ^{189}Re has an n/p ratio greater than that which is stable, and so it will undergo beta emission (neutron emission is less common).

26-32. *Refer to Sections 26-6, 26-7 and 26-8, and Figure 26-1.*

(a) $_{27}^{60}$Co $\quad$ (n/p ratio too high) $\qquad$ beta emission (neutron emission is less common)

(b) $_{11}^{20}$Na $\quad$ (n/p ratio too low) $\qquad$ positron emission or electron capture (K capture)

(c) $_{86}^{222}$Rn $\qquad\qquad\qquad\qquad\qquad$ alpha emission

(d) $_{29}^{67}$Cu $\quad$ (n/p ratio too high) $\qquad$ beta emission

(e) $_{92}^{238}$U $\qquad\qquad\qquad\qquad\qquad$ alpha emission

(f) $_6^{11}$C $\quad$ (n/p ratio too low) $\qquad$ positron emission or electron capture (K capture)

26-34. *Refer to Sections 26-5, 26-6 and 26-13.*

In equations for nuclear reactions, the sums of the mass numbers and atomic numbers of the reactants must equal the sums for the products. Therefore,

(a) $_{42}^{96}$Mo $+ {_2^4}$He $\rightarrow {_{43}^{100}}$Tc $+ \boxed{_{+1}^0\beta}$ $\qquad$ (b) $_{27}^{59}$Co $+ {_0^1}n \rightarrow {_{25}^{56}}$Mn $+ \boxed{_2^4\text{He}}$ $\qquad$ (c) $_{11}^{23}$Na $+ \boxed{_1^1\text{H}} \rightarrow {_{12}^{23}}$Mg $+ {_0^1}n$

(d) $_{83}^{209}$Bi $+ \boxed{_1^2\text{H}} \rightarrow {_{84}^{210}}$Po $+ {_0^1}n$ $\qquad$ (e) $_{92}^{238}$U $+ {_8^{16}}$O $\rightarrow \boxed{_{100}^{249}\text{Fm}} + 5\,{_0^1}n$

26-36. *Refer to Sections 26-5 and 26-13.*

The equation for a nuclear reaction can be given in the following abbreviated form:
parent nucleus (bombarding particle, emitted particle) daughter nucleus.

(a) $^{60}_{28}Ni + ^{1}_{0}n \rightarrow \boxed{^{60}_{27}Co} + ^{1}_{1}H$ (b) $^{98}_{42}Mo + ^{1}_{0}n \rightarrow \boxed{^{99}_{43}Tc} + ^{0}_{-1}\beta$ (c) $^{35}_{17}Cl + ^{1}_{1}H \rightarrow \boxed{^{32}_{16}S} + ^{4}_{2}He$

26-38. *Refer to Sections 26-5 and 26-13.*

The equation for a nuclear reaction can be given in the following abbreviated form:
parent nucleus (bombarding particle, emitted particle) daughter nucleus.

(a) $^{14}_{7}N + ^{4}_{2}He \rightarrow ^{17}_{8}O + ^{1}_{1}H$ (b) $^{106}_{46}Pd + ^{1}_{0}n \rightarrow ^{106}_{45}Rh + ^{1}_{1}H$ (c) $^{23}_{11}Na + ^{1}_{0}n \rightarrow \boxed{^{24}_{12}Mg} + ^{0}_{-1}e$

26-40. *Refer to Section 26-13 and Exercise 26-38 Solution.*

(a) $^{238}_{92}U \; (n, \beta^-) \; ^{239}_{93}Np$ (b) $^{6}_{3}Li \; (n, \alpha) \; ^{3}_{1}H$ (c) $^{31}_{15}P \; (d, p) \; ^{32}_{15}P$

26-42. *Refer to Sections 26-5, 26-6 and 26-13.*

(a) $^{63}_{28}Ni \rightarrow ^{63}_{29}Cu + ^{0}_{-1}e$ (b) $2\,^{2}_{1}H \rightarrow ^{3}_{2}He + ^{1}_{0}n$ (c) $\boxed{^{10}_{5}B} + ^{1}_{0}n \rightarrow ^{7}_{3}Li + ^{4}_{2}He$

(d) $^{14}_{7}N + ^{1}_{0}n \rightarrow ^{3}_{1}H + 3\,^{4}_{2}He$

26-44. *Refer to Sections 26-5 and 26-8.*

Plan: Balance the nuclear reaction and identify the unknown "radioactinium."

$$^{235}_{92}U \rightarrow \boxed{^{227}_{90}Th} + 2\,^{4}_{2}He + 2\,^{0}_{-1}\beta$$

Therefore, "radioactinium" is the element thorium, **Th**, with atomic number **90** and mass number **227**.

26-46. *Refer to Sections 26-5, 26-8 and 26-13.*

Balanced nuclear reactions: (1) $^{249}_{98}Cf + ^{12}_{6}C \rightarrow ^{257}_{104}Rf + 4\,^{1}_{0}n$

(2) $^{257}_{104}Rf \rightarrow \boxed{^{253}_{102}No} + ^{4}_{2}He$

The element **nobelium**, No, is formed.

26-48. *Refer to Sections 26-5, 26-6 and 26-13.*

In equations for nuclear reactions, the sums of the mass numbers and atomic numbers of the reactants must equal the sums for the products. Therefore,

(a) $^{241}_{95}Am + ^{12}_{6}C \rightarrow 4\,^{1}_{0}n + \boxed{^{249}_{101}Md}$ (c) $^{235}_{92}U + ^{1}_{0}n \rightarrow ^{137}_{54}Xe + 2\,^{1}_{0}n + \boxed{^{97}_{38}Sr}$

(b) $^{14}_{7}N + \boxed{^{4}_{2}He} \rightarrow ^{17}_{8}O + ^{1}_{1}p$

The half-life of a radionuclide represents the amount of time required for half of the sample to decay. Relative stabilities of radionuclides are indicated by their half-life values. The shorter the half-life, the less stable is the radionuclide.

26-52. *Refer to Section 26-12.*

The radioisotope carbon-14 is produced continuously in the atmosphere as nitrogen atoms capture cosmic-ray neutrons:

$$^{14}_{7}N + {}^{1}_{0}n \rightarrow {}^{14}_{6}C + {}^{1}_{1}H$$

The carbon-14 atoms react with O_2 to form $^{14}CO_2$. Like ordinary $^{12}CO_2$, it is removed from the atmosphere by living plants through the process of photosynthesis. As long as the cosmic-ray intensity remains constant, the amount of $^{14}CO_2$ and therefore its ratio to $^{12}CO_2$ in the atmosphere remains constant. Consequently, a certain fraction of carbon atoms in all living substances becomes carbon-14, a beta particle emitter with a half-life of 5730 years:

$$^{14}_{6}C \rightarrow {}^{14}_{7}N + {}^{0}_{-1}e$$

A steady state ratio of $^{14}C/^{12}C$ is maintained in living plants and organisms. After death the plant no longer carries out photosynthesis, so it no longer takes up $^{14}CO_2$. The radioactive emissions from the carbon-14 in dead tissue then decrease with the passage of time. The activity per gram of carbon in the sample in comparison with that in air gives a measure of the length of time elapsed since death. This is the basis of radiocarbon dating.

This technique is useful only when dating objects that are less than 50,000 years old (roughly 10 times the half-life of carbon-14). Older objects have too little activity to be accurately dated. This technique depends on cosmic-ray intensity being constant or at least predictable in order to keep the $^{14}C/^{12}C$ known throughout the time interval. Also, the sample must not be contaminated with organic matter having a different $^{14}C/^{12}C$ ratio.

26-54. *Refer to Section 26-10.*

For first order kinetics, $\ln\left(\dfrac{A_o}{A}\right) = kt$ and $t_{1/2} = \dfrac{0.693}{k}$ where A_o = initial amount of isotope
A = amount remaining after time, t
k = rate constant (units of time^{-1})
$t_{1/2}$ = half-life of isotope

Plan: (1) Calculate the rate constant, k, from the half-life of carbon-11.
(2) Assume that the initial amount of carbon-11 is 100%. Calculate the time required for decay, t, using the first order rate equation.

(1) $k = \dfrac{0.693}{t_{1/2}} = \dfrac{0.693}{20.3 \text{ min}} = 0.0341 \text{ min}^{-1}$

(2) When 90.0% of the sample has decayed away, 10.0% of the sample remains.
Substituting into the first order rate equation,

$$\ln\left(\frac{100.0\%}{10.0\%}\right) = (0.0341 \text{ min}^{-1})t$$

$$2.303 = 0.0341t$$

$$t = \mathbf{67.5 \text{ min}}$$

When 95.0% of the sample has decayed, 5.0% of the sample remains. Substituting,

$$\ln\left(\frac{100.0\%}{5.0\%}\right) = (0.0341 \text{ min}^{-1})t$$

$$3.00 = 0.0341t$$

$$t = \mathbf{88.0 \text{ min}}$$

Balanced equation: $^{8}_{4}\text{Be} \rightarrow 2\,^{4}_{2}\text{He}$

Plan: (1) Determine the rate constant, k, from the half-life for ^{8}Be.
(2) Calculate the time required for 99.90% of ^{8}Be to decay.

(1) $k = \dfrac{0.693}{t_{1/2}} = \dfrac{0.693}{0.07 \text{ fs}} = 10 \text{ fs}^{-1}$ (Note: 1 fs $= 10^{-15}$ s)

(2) The first order rate equation: $\ln\left(\dfrac{A_o}{A}\right) = kt$

If 99.90% of ^{8}Be decayed away, then 0.10% remains. Note that the calculation does not depend on the initial amount of ^{8}Be. Substituting,

$$\ln\left(\frac{100.00\%}{0.10\%}\right) = (10 \text{ fs}^{-1})t$$

$$6.9 = 10t$$

$$t = \textbf{0.7 fs} \text{ (to 1 significant figure)}$$

Plan: (1) Calculate the first order rate constant, k, from the half-life of carbon-14.
(2) Determine the age of the wood.

(1) $k = \dfrac{0.693}{t_{1/2}} = \dfrac{0.693}{5730 \text{ yr}} = 1.21 \times 10^{-4} \text{ yr}^{-1}$

(2) The first order rate equation: $\ln\left(\dfrac{A_o}{A}\right) = kt$

Note: The ^{14}C activity is directly proportional to the amount of ^{14}C present. Therefore, we can substitute the activity values directly into the first order rate equation.

Substituting, $\ln\left(\dfrac{16.1 \text{ disintegrations per min per gram of C}}{10.8 \text{ disintegrations per min per gram of C}}\right) = (1.21 \times 10^{-4} \text{ yr}^{-1})t$

$$0.399 = (1.21 \times 10^{-4} \text{ yr}^{-1})t$$

$$t = \textbf{3.30} \times \textbf{10}^{\textbf{3}} \textbf{ yr}$$

Plan: (1) Calculate the first order rate constant, k, from the half-life of carbon-14.
(2) Determine the age of the skeleton.

(1) $k = \dfrac{0.693}{t_{1/2}} = \dfrac{0.693}{5730 \text{ yr}} = 1.21 \times 10^{-4} \text{ yr}^{-1}$

(2) The first order rate equation: $\ln\left(\dfrac{A_o}{A}\right) = kt$

Note: The ^{14}C activity is directly proportional to the amount of ^{14}C present. Therefore, we can substitute the activity values directly into the first order rate equation.

Substituting, $\ln\left(\dfrac{14.0 \text{ disintegrations per min per gram of C}}{11.5 \text{ disintegrations per min per gram of C}}\right) = (1.21 \times 10^{-4} \text{ yr}^{-1})t$

$$0.197 = (1.21 \times 10^{-4} \text{ yr}^{-1})t$$

$$t = \textbf{1630 yr}$$

26-62. *Refer to Sections 26-10 and 26-12 and Example 26-5.*

Plan: (1) Calculate the first order rate constant, k, from the half-life of carbon-14.

(2) Determine the age of the object.

(1) $k = \dfrac{0.693}{t_{1/2}} = \dfrac{0.693}{5730 \text{ yr}} = 1.21 \times 10^{-4} \text{ yr}^{-1}$

(2) The first order rate equation: $\ln\left(\dfrac{A_o}{A}\right) = kt$

Substituting, $\ln\left(\dfrac{8.50 \text{ μg}}{0.76 \text{ μg}}\right) = (1.21 \times 10^{-4} \text{ yr}^{-1})t$

$2.41 = (1.21 \times 10^{-4})t$

$t = \mathbf{2.00 \times 10^4 \text{ yr}}$

26-64. *Refer to Sections 26-14 and 26-15, and the Key Terms for Chapter 26.*

A chain reaction is a reaction that sustains itself once it has begun and may even expand. Normally, the limiting reactant is regenerated as a product to maintain the progress of the chain. Nuclear fission processes are considered chain reactions because the number of neutrons produced in the reaction equals or is greater than the number of neutrons absorbed by the fissioning nucleus.

For example:

$$^{235}_{92}\text{U} + ^1_0n \rightarrow [^{236}_{92}\text{U}] \rightarrow ^{140}_{56}\text{Ba} + ^{93}_{36}\text{Kr} + 3\,^1_0n + \text{energy}$$

The critical mass of a fissionable material is the minimum mass of a particular fissionable nuclide in a set volume that is necessary to sustain a nuclear chain reaction.

26-66. *Refer to Sections 26-16 and 26-3.*

(a) $^7_3\text{Li} + ^1_1\text{H} \rightarrow \boxed{^4_2\text{He}} + ^4_2\text{He}$ where a proton is represented by ^1_1H

(b) Plan: (1) Determine the mass difference between the products and the reactants.

The difference, Δm, is directly related to the energy involved in the reaction.

(2) Calculate the amount of energy involved.

(1) Δm = mass of products - mass of reactants

$= (2 \times \text{mass of } ^4_2\text{He}) - (\text{mass of } ^7_3\text{Li} + \text{mass of } ^1_1\text{H})$

$= (2 \times 4.00260 \text{ amu}) - (7.01600 \text{ amu} + 1.007825 \text{ amu})$

$= -0.01862 \text{ amu}$ or $-0.01862 \text{ g/mol rxn}$

(2) $\Delta E = (\Delta m)c^2$

$= (-1.862 \times 10^{-5} \text{ kg/mol rxn})(3.00 \times 10^8 \text{ m/s})^2$

$= -1.68 \times 10^{12} \text{ kg·(m/s)}^2/\text{mol rxn}$

$= -1.68 \times 10^{12} \text{ J/mol rxn}$ or $\mathbf{-1.68 \times 10^9 \text{ kJ/mol rxn}}$

Since $\Delta E < 0$, energy is being released in this fusion reaction, as expected.

26-68.   *Refer to Section 26-15.*

The primary advantage of nuclear energy is that enormous amounts of energy are liberated per unit mass of fuel. Also, the air pollution (oxides of S, N, C and particulate matter) caused by fossil fuel electric power plants is not a problem with nuclear energy plants. In European countries, where fossil fuel reserves are scarce, most of the electricity is generated by nuclear power plants for these reasons.

There are, however, some disadvantages associated with nuclear power from controlled fission reactions. The radionuclides must be properly shielded to protect the workers and the environment from radiation and contamination. Spent fuel, containing long-lived radioisotopes, must be disposed of carefully using special containers placed underground in geologically inactive areas. This is because the radiation from the fuel is biologically dangerous and must be contained until the fuel has decayed to the point when it is no longer dangerous. The problem is that the time involved could be several hundred thousand years. If there is inadequate cooling in the reactor, there is the possibility of overheating the fuel and causing a "meltdown." This cooling water can cause biological damage to aquatic life if it is returned to the natural water system while it is still too warm. Finally, it is possible that Pu-239 could be stolen and used for bomb production.

In the future, when nuclear fusion power plants are in operation, most of these disadvantages will not be a concern. Fusion reactions produce only short-lived isotopes and so there would be no long-term storage problems. An added advantage is that there is a virtually inexhaustible supply of deuterium fuel in the world's oceans.

26-70.   *Refer to Section 26-15.*

Uranium ores contain only about 0.7% ^{235}U which is fissionable. Most of the rest is nonfissionable ^{238}U. To enrich ^{235}U for use in nuclear power plants, the oxide is converted to UF_4 with HF and then oxidized to UF_6 by fluorine. The vapor of $^{235}UF_6$ and $^{238}UF_6$ is then subjected to repeated diffusion through porous barriers to concentrate $^{235}UF_6$ (Graham's Law). Gas centrifuges are now used for the concentration process which is also based upon the difference in masses of the two U isotopes.

26-72.   *Refer to Section 26-16.*

The major advantages of fusion as a potential energy source are three-fold:

(1) Fusion reactions are accompanied by much greater energy production per unit mass of reacting atoms than fission reactions.
(2) The deuterium fuel for fusion reactions is present in a virtually inexhaustible supply in the world oceans.
(3) Fusion reactions produce only short-lived radionuclides; there would be no long-term waste-disposal problem.

The only disadvantage of fusion is that extremely high temperatures are required to initiate the fusion process. A structural material that can withstand the high temperatures (4×10^7 K or more) and contain the fusion reaction, does not as yet exist.

26-74.   *Refer to Sections 26-12 and 26-3.*

(a) $^{14}_{7}N + ^{4}_{2}He \rightarrow ^{17}_{8}O + ^{1}_{1}H$

(b) Plan: (1) Determine the mass difference between the products and the reactants. The difference, Δm, is directly related to the energy involved in the reaction.
                (2) Calculate the amount of energy involved.

  (1) Δm = mass of products - mass of reactants
         = (mass of $^{17}_{8}O$ + mass of $^{1}_{1}H$) - (mass of $^{14}_{7}N$ + mass of $^{4}_{2}He$)
         = (16.99913 amu + 1.007825 amu) - (14.00307 amu + 4.00260 amu)
         = 0.00128 amu or 0.00128 g/mol rxn

(2) $\Delta E = (\Delta m)c^2$

$= (1.28 \times 10^{-6} \text{ kg/mol rxn})(3.00 \times 10^8 \text{ m/s})^2$

$= 1.15 \times 10^{11} \text{ kg}\cdot(\text{m/s})^2/\text{mol rxn}$

$= +1.15 \times 10^{11} \text{ J/mol rxn}$ or $+1.15 \times 10^8 \text{ kJ/mol rxn}$

Since $\Delta E > 0$, energy is being absorbed in this reaction, as expected.

26-76. *Refer to Sections 26-10 and 26-12.*

Plan: (1) Calculate the first order rate constant, k, from the half-life of uranium-238.
(2) Determine the age of the rock.
Assume that all the ^{206}Pb came from ^{238}U.
Because of the very long half-life of ^{238}U, 4.5 billion years,
the amounts of intermediate nuclei can be neglected.

(1) $k = \dfrac{0.693}{t_{1/2}} = \dfrac{0.693}{4.51 \times 10^9 \text{ yr}} = 1.54 \times 10^{-10} \text{ yr}^{-1}$

(2) The first order rate equation: $\ln\left(\dfrac{N_o}{N}\right) = kt$

where N = number of ^{238}U atoms remaining
N_o = number of ^{238}U atoms originally present

Therefore, N_o = number of ^{238}U atoms remaining + number of ^{238}U atoms decayed
= number of ^{238}U atoms remaining + number of ^{206}Pb atoms produced

So, $\left(\dfrac{N_o}{N}\right) = \dfrac{67.8 \ ^{238}\text{U atoms} + 32.2 \ ^{206}\text{Pb atoms}}{67.8 \ ^{238}\text{U atoms}} = \left(\dfrac{100.0}{67.8}\right)$

Substituting, $\ln\left(\dfrac{100.0}{67.8}\right) = (1.54 \times 10^{-10} \text{ yr}^{-1})t$

$0.389 = (1.54 \times 10^{-10})t$

$t = \mathbf{2.52 \times 10^9 \ yr}$

26-78. *Refer to Sections 26-14 and 26-3.*

Balanced equations: fission $^{235}_{92}\text{U} + ^{1}_{0}n \rightarrow ^{94}_{40}\text{Zr} + ^{140}_{58}\text{Ce} + 6\,^{0}_{-1}e + 2\,^{1}_{0}n$
fusion $2\,^{2}_{1}\text{H} \rightarrow ^{3}_{1}\text{H} + ^{1}_{1}\text{H}$

Plan: (1) Determine the mass difference between the products and the reactants. The difference, Δm, is directly related to the energy involved in the reaction.
(2) Calculate the amount of energy involved.

(1) fission: Δm = mass of products - mass of reactants

= [mass of $^{94}_{40}$Zr + mass of $^{140}_{58}$Ce + (6 × mass of $^{0}_{-1}e$) + (2 × mass of $^{1}_{0}n$)]
- [mass of $^{235}_{92}$U + mass of $^{1}_{0}n$]

= [93.9061 amu + 139.9053 amu + (6 × 0.000549 amu) + (2 × 1.0087 amu)]
- [235.0439 amu + 1.0087 amu)

= −0.2205 amu or −0.2205 g/mol rxn

fusion: Δm = [mass of $_1^3H$ + mass of $_1^1H$] - [2 × mass of $_1^2H$]

= [3.01605 amu + 1.007825 amu] - [2 × 2.0140 amu]

= −0.00412 amu or −0.00412 g/mol rxn

(2) fission: $\Delta E = (\Delta m)c^2 = (-2.205 \times 10^{-4}$ kg/mol rxn$)(3.00 \times 10^8$ m/s$)^2 = -1.98 \times 10^{13}$ kg·(m/s)^2/mol rxn

$= -1.98 \times 10^{13}$ J/mol rxn

$$\Delta E \text{ (J/amu }^{235}U) = \frac{-1.98 \times 10^{13} \text{ J}}{1 \text{ mol rxn}} \times \frac{1 \text{ mol rxn}}{1 \text{ mol }^{235}U} \times \frac{1 \text{ mol }^{235}U}{6.02 \times 10^{23} \text{ atoms }^{235}U} \times \frac{1 \text{ atom }^{235}U}{235.0439 \text{ amu}}$$

$= -1.40 \times 10^{-13}$ J/amu ^{235}U

fusion: $\Delta E = (\Delta m)c^2 = (-4.12 \times 10^{-6}$ kg/mol rxn$)(3.00 \times 10^8$ m/s$)^2 = -3.71 \times 10^{11}$ kg·(m/s)^2/mol rxn

$= -3.71 \times 10^{11}$ J/mol rxn

$$\Delta E \text{ (J/amu }^2H) = \frac{-3.71 \times 10^{11} \text{ J}}{1 \text{ mol rxn}} \times \frac{1 \text{ mol rxn}}{2 \text{ mol }^2H} \times \frac{1 \text{ mol }^2H}{6.02 \times 10^{23} \text{ atoms }^2H} \times \frac{1 \text{ atom }^2H}{2.0140 \text{ amu}}$$

$= -1.53 \times 10^{-13}$ J/amu ^2H

Therefore, the above **fusion** process produces about 10% more energy per amu of material than fission. However, the above fusion reaction is not a typical one because it involves the production of two particles from two particles of similar size. In general, fusion processes produce much more energy than fission processes on a per unit mass basis.

26-80. *Refer to Section 26-15.*

Both nuclear and conventional power plants produce environmentally-sensitive waste. Both use cooling water which when put into streams and rivers while still at elevated temperatures can cause significant damage to the biota.

Conventional power plants can pollute the air with particulate matter and the oxides of sulfur, nitrogen, and carbon, causing acid rain and other problems. However, with proper scrubbing and filtering at the source, this pollution has been greatly reduced.

Nuclear power produces spent fuel that contains radionuclides that will emit radiation for hundreds and thousands of years. At present, they are being stored underground indefinitely in heavy, shock-proof containers. These containers could be stolen or may corrode with time, or leak as a result of earthquakes and tremors. Transportation and reprocessing accidents could cause environmental contamination. One solution is for the United States to go to breeder reactors, as has been done in other countries, to reduce the level and amount of radioactive waste.

26-82. *Refer to Section 26-12.*

One of the limitations of radiocarbon dating artifacts is due to the half-life of the carbon-14, 5730 years. In radiochemistry, a good rule of thumb is the following: when an element decays for more than about 10 times its half-life, there is very little left to measure accurately. In the case of C-14, that time is 10 × 5730 yr or 57300 years.

What fraction of a C-14 sample remains after 10 half-lives? The answer is that only $(\frac{1}{2})^{10} = 0.00098$ or 0.098% of the C-14 in the original sample remains, and this is generally too little activity to measure accurately.

27 Organic Chemistry I: Formulas, Names and Properties

27-2. *Refer to the Introduction to Chapter 27.*

Carbon atoms bond to each other to a much greater extent than any other element. They form long chains, branched chains and rings which may also contain chains attached to them. Millions of such compounds are known which constitutes the study of organic chemistry.

27-4. *Refer to the Introduction to Chapter 27.*

(a) Most synthetic organic materials are derived from petroleum, coal and natural gas.

(b) Most geochemists believe that petroleum, natural gas and coal are derived from plant matter, buried millions of years ago. Since the source of carbon for plants is CO_2, we can say that the ultimate source of many naturally occurring organic compounds which are based on carbon, is CO_2.

27-6. *Refer to Sections 27-1, 27-3 and 27-4.*

(a) Alkanes are saturated hydrocarbons, in which each carbon atom is singly bonded to four other atoms. These atoms are either carbon atoms or hydrogen atoms. Alkenes and alkynes are unsaturated hydrocarbons in which there is a carbon-carbon double bond or a carbon-carbon triple bond, respectively.

(b) alkane: C_nH_{2n+2}
    alkene: C_nH_{2n}
    alkyne: C_nH_{2n-2}

(c) A saturated hydrocarbon is a organic compound containing only carbon and hydrogen atoms and all the carbon-carbon bonds are single bonds.

27-8. *Refer to Section 27-1 and Table 27-1.*

In alkanes such as   (a) methane, CH_4 (Figure 27-2),
                       (b) ethane, C_2H_6 (Figure 27-3),
                       (c) propane, C_3H_8 (Figure 27-4), and
                       (d) butane, C_4H_{10} (Figures 27-5),

the geometry about each C atom is tetrahedral. All the carbon atoms are connected to each other to form chains (and in the case of butane, a "straight" chain of 4 carbon atoms without branching is formed). Each of the C atoms undergoes sp^3 hybridization, and forms σ bonds with each other by using the sp^3 hybrid orbitals. The C atoms at the end of each chain are in the form of CH_3, each bonded to 3 H atoms by overlapping with their $1s$ orbitals to give σ bonds. The C atoms in the interior of each chain are in the form of CH_2, each bonded to 2 H atoms in the same fashion.

These four molecules are the first four members of the alkanes, a homologous series of saturated hydrocarbons with the general formula, C_nH_{2n+2}. The difference between them is in the number of C atoms in the compound; the formula of each alkane differs from the next by one CH_2 group.

(a) A homologous series is a series of compounds in which each member differs from the next by a specific number and kind of atoms.

(b) The alkane series contains saturated hydrocarbons such as CH_4, C_2H_6, C_3H_8 and C_4H_{10}. Each member differs from the next by CH_2 and are therefore examples of compounds that are members of a homologous series. Refer to Table 27-2 for the names and formulas of more members of this homologous series.

(c) A methylene group is a CH_2 group.

(d) The structures of homologous series members differ by a CH_2 unit from one member to the next. The properties of the members of a homologous series are closely related. For example, the boiling point of a compound in the homologous series given in (b) is higher than the compounds before it in the series, but less than the compounds after it due to increasing London dispersion forces.

(e) Homologous series that are also aliphatic hydrocarbons are the alkanes, C_nH_{2n+2}, the alkenes, C_nH_{2n}, and the alkynes, C_nH_{2n-2}; they all differ by a CH_2 unit from one member to the next.

27-12. *Refer to Sections 27-1 and the Key Terms for Chapter 27.*

Cycloalkanes are cyclic saturated hydrocarbons with the general formula C_nH_{2n}. Therefore, a substance with the formula C_3H_8 could *not* be a cycloalkane, since C_3H_8 conforms to the general formula, C_nH_{2n+2}, the molecular formula for an alkane. It is, however, too small to be a branched alkane with a methyl group attached to the longest chain. In fact, C_3H_8 is propane.

27-14. *Refer to Section 27-2 and Example 27-2.*

(a) $CH_3CH_2C(CH_3)_2CH_2CH_3$ 3,3-dimethylpentane

(b) $CH_3CH_2CH(CH_2CH_3)CH_2CH_3$ 3-ethylpentane

(c) $CH_3CH(CH_2CH_2CH_3)CH_2CH_3$ 3-methylhexane (Hint: draw it and look for the longest carbon chain.)

(d) $CH_3CH_2CH_2CH(CH_3)_2$ 2-methylpentane

(e) $CH_3CH(CH_3)CH(CH_3)CH_2CH_3$ 2,3-dimethylpentane

27-16. *Refer to Sections 27-1 and 27-2, and Example 27-5.*

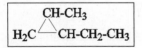

1-cyclopropylpropane   2-cyclopropylmethylethane   1,2,3-trimethylcyclopropane

1-ethyl-2-methylcyclopropane   1,1,2-trimethylcyclopropane   1-ethyl-1-methylcyclopropane

$$CH_2 \atop CH_2 \rangle CH-CH_3$$

$$CH_2 \atop CH_2 \rangle C \langle {CH_3 \atop CH_3}$$

$$CH_2 \atop CH \rangle C \langle {CH_3 \atop CH_3} \atop CH_3$$

methylcyclopropane   1,1-dimethylcyclopropane   1,1,2-trimethylcyclopropane

---

27-20. *Refer to Sections 27-1 and 27-3.*

(a) Alkenes contain C=C double bonds formed at the expense of two hydrogen atoms. Therefore, the general formula for alkenes is C_nH_{2n}, while that for alkanes is C_nH_{2n+2}.

(b) When an alkane loses two H atoms, the resulting species could undergo either ring-enclosure to give a cycloalkane, or π bond formation to give an alkene. Therefore, cycloalkanes and alkenes are isomers, both having the general formula, C_nH_{2n}.

27-22. *Refer to Section 27-3 and Table 27-1.*

(a) ethene

$$CH_2 = CH_2$$

Both carbon atoms in ethene undergo sp^2 hybridization. The C-H bonds involve overlap of sp^2 carbon orbitals with $1s$ orbitals of the H atoms. The carbon-carbon double bond involves the overlap of sp^2 orbitals from each carbon to give the σ bond and the side-on overlap of a p orbital from each carbon atom to give the π bond.

(b) methylpropene

$$\overset{4}{C}H_3 \atop CH_3 - \underset{1}{C} = \underset{3}{C}H_2 \atop _2$$

Carbon atoms (1) and (4) use sp^3 hybrid orbitals to form four sigma bonds, three by overlap with the hydrogen $1s$ orbitals and one by overlap with an sp^2 orbital from the central carbon (2). The two carbon atoms involved in the double bond undergo sp^2 hybridization. They form C-H bonds by overlapping with $1s$ orbitals of the H atoms. The C=C double bond is formed similarly to that described in (a).

(c) 2-butene

$$\underset{1}{C}H_3 - \underset{2}{C}H = \underset{3}{C}H - \underset{4}{C}H_3$$

Carbon atoms (2) and (3) undergo sp^2 hybridization, while carbon atoms (1) and (4) undergo sp^3 hybridization. The overlap of the hybrid orbitals with the $1s$ orbitals of H atoms gives the C-H bonds. The C=C double bond is formed similarly to that described in (a). The C-C single bonds involve sp^2-sp^3 overlap for the C(1)-C(2) and the C(3)-C(4) single bonds.

(d) 3-methyl-1-butene

$$\overset{5}{C}H_3 \atop \underset{1}{C}H_2 = \underset{2}{C}H - \underset{3}{C}H - \underset{4}{C}H_3$$

Carbon atoms (1) and (2) undergo sp^2 hybridization, while carbon atoms (3), (4) and (5) undergo sp^3 hybridization. The orbital overlaps are similar to those in 2-butene except carbon atoms (1) and (2) are involved in the double bond and carbon atom (2) is bonded to carbon atom (3) by sp^2-sp^3 overlap. The overlap of the hybrid orbitals with the $1s$ orbitals of H atoms gives the C-H bonds.

27-24. *Refer to Section 27-3.*

Carbon-carbon double bonds (1.34 Å) are shorter than carbon-carbon single bonds (1.54 Å). This is due to the additional side-on overlap of the p orbitals creating the π bond. Two shared electron pairs draw the atoms closer together than a single electron pair does.

27-26. *Refer to Section 27-4 and Figure 27-10.*

The C≡C triple bond consists of one σ and two π bonds. The σ bond is formed by overlapping head-on the *sp* hybridized orbitals of the corresponding carbon atoms. The π bonds are formed by overlapping side-on the two remaining sets of *p* orbitals. The triply bonded atoms and their adjacent atoms lie on a straight line.

27-28. *Refer to Sections 27-3 and 27-4, and Examples 27-8 and 27-9.*

(a) CH₃–C≡C–CH₃

      2-butyne

(b) CH₂=CH–CH=CH–CH₃

      1,3-pentadiene

(c)

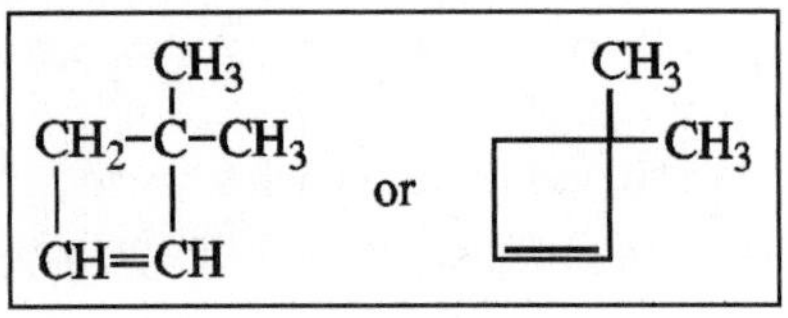

$$CH_2-\overset{\displaystyle CH_3}{\underset{\displaystyle CH=CH}{\underset{|}{\overset{|}{C}}}}-CH_3 \quad \text{or}$$

      3,3-dimethylcyclobutene

(d)

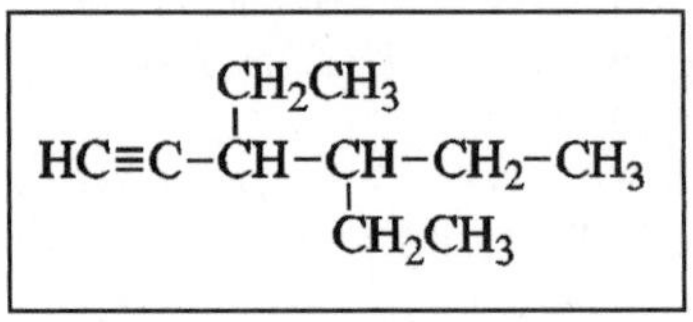

$$HC≡C-\overset{\displaystyle CH_2CH_3}{\underset{\displaystyle CH_2CH_3}{\underset{|}{\overset{|}{CH}}}}-CH-CH_2-CH_3$$

      3,4-diethyl-1-hexyne

27-30. *Refer to Sections 27-2, 27-3 and 27-4, and Example 27-9.*

(a) 1-methylcyclopentene

(b) 2-methylbutane

(c) 2-butyne

(d) 2-methyl-2-butene
     or methyl-2-butene

(e) 3-methylpentane

(f) 3-ethylhexane

27-32. *Refer to Sections 27-5 and 27-6.*

(a) The term "aromatic hydrocarbons" refer to benzene and similar condensed ring structures. They differ from other cyclic compounds because in their cyclic structures, the electrons are delocalized over the entire ring. In benzene, for example, all the carbons are *sp²* hybridized. Benzene can be represented by 2 resonance structures which describes a structure in which all the carbon-carbon bonds are the same length, the same strength and in the same plane.

(b) The principle source of aromatic hydrocarbons is petroleum refining.

27-34. *Refer to Section 27-6.*

A phenyl group, C₆H₅–, results when an H atom is removed from a benzene ring. It could take the place of any H in an organic compound. In particular, when a phenyl group replaces a hydrogen atom on a naphthalene molecule, **two** isomers of monophenylnaphthalene are possible:

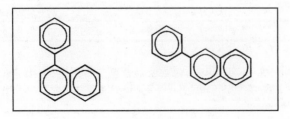

A total of 3 isomers of dibromobenzene are possible:

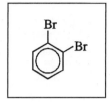

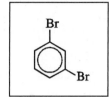

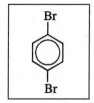

1,2-dibromobenzene	1,3-dibromobenzene	1,4-dibromobenzene
(*o*-dibromobenzene)	(*m*-dibromobenzene)	(*p*-dibromobenzene)

27-38.   *Refer to Section 27-6.*

(a)   1-ethyl-2-methylbenzene

(b)   1,2,4-trimethylbenzene

(c)   1,5-diethyl-2,4-dimethylbenzene

(d)   1,2,3,4,5-pentamethylbenzene

27-40.   *Refer to Section 27-8.*

(a)   1-chloro-2-phenylethane

(b)   1-chloro-2-methylpropane

(c)   1,2-dichloroethane

(d)   1,1,2-trichloroethene

27-42.   *Refer to Sections 27-6 and 27-8.*

(a)   1,2,3-trichlorobenzene

(b)   1-chloro-3-methylbenzene (*m*-chlorotoluene)

(c)   1,4-dibromo-2,5-diiodobenzene

(d)   2,4-dibromo-1,3,5-trichlorobenzene

27-44.   *Refer to Section 27-9.*

(a)   Alcohols and phenols are hydrocarbon derivatives which contain the hydroxyl group (–OH) as their functional group.

(b)   Alcohols are derived from aliphatic hydrocarbons by replacing at least one hydrogen atom with a hydroxyl (–OH) group. On the other hand, in phenols, the –OH group must attach directly to an aromatic ring. Phenols are weak acids, while alcohols are neutral.

(c)   Alcohols and phenols can be viewed as derivatives of hydrocarbons in which a hydrogen atom is replaced by an –OH group. On the other hand, they can also be viewed as derivatives of water in which a hydrogen atom is replaced by an organic group.

The eight saturated alcohols that contain five carbon atoms and one –OH group per molecule are:

$$CH_3-CH_2-CH_2-CH_2-CH_2-OH$$

1-pentanol
1°

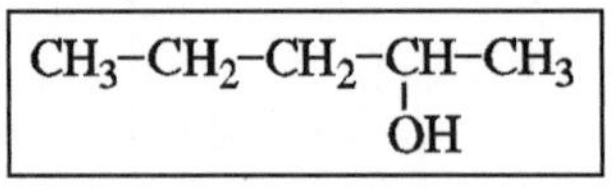

$$CH_3-CH_2-CH_2-\underset{\underset{OH}{|}}{CH}-CH_3$$

2-pentanol
2°

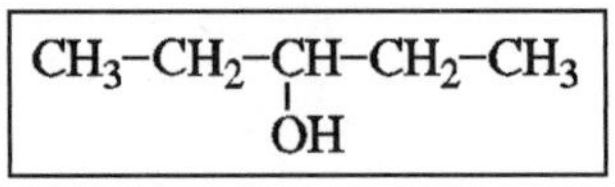

$$CH_3-CH_2-\underset{\underset{OH}{|}}{CH}-CH_2-CH_3$$

3-pentanol
2°

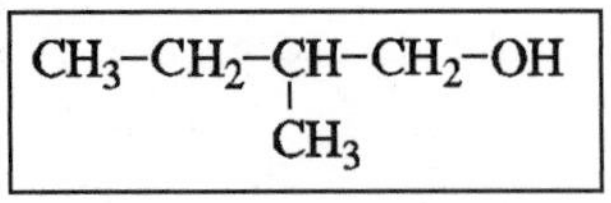

$$CH_3-CH_2-\underset{\underset{CH_3}{|}}{CH}-CH_2-OH$$

2-methyl-1-butanol
1°

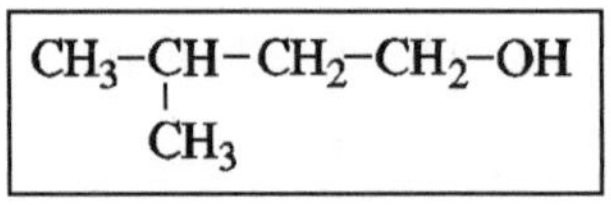

$$CH_3-\underset{\underset{CH_3}{|}}{CH}-CH_2-CH_2-OH$$

3-methyl-1-butanol
1°

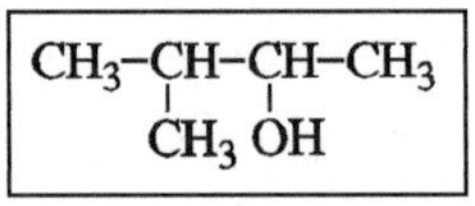

$$CH_3-\underset{\underset{CH_3}{|}}{CH}-\underset{\underset{OH}{|}}{CH}-CH_3$$

3-methyl-2-butanol
2°

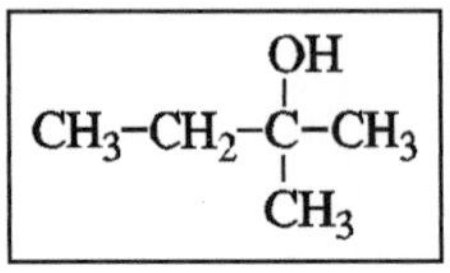

$$CH_3-CH_2-\underset{\underset{CH_3}{|}}{\overset{\overset{OH}{|}}{C}}-CH_3$$

2-methyl-2-butanol
3°

$$CH_3-\underset{\underset{CH_3}{|}}{\overset{\overset{CH_3}{|}}{C}}-CH_2-OH$$

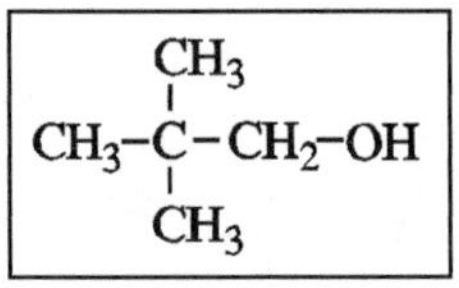

2,2-dimethyl-1-propanol
1°

Data in Table 27-8 show that the boiling points of normal primary alcohols increase and their solubilities in water decrease with increasing molecular weight. The boiling point increases because the London forces increase with the size of the molecules.

The solubility decreases because the alcohols become less polar down the list. The alcohols, ROH, have a polar hydroxyl group end and a nonpolar alkyl group end. Due to the principle, "like dissolves like," as the nonpolar end of the molecules becomes larger and larger, their solubilities in water decrease rapidly because H_2O is a very polar solvent. In fact, the C_1 - C_3 alcohols are miscible with H_2O in all proportions. Beginning with the butyl alcohols, solubility in H_2O decreases rapidly with increasing molecular weight.

Most phenols are relatively high-molecular weight compounds with large nonpolar portions and therefore, they exhibit low solubilities in water.

(a) 2-methyl-1-butanol

(b) 3,3-dimethyl-1-butanol

(c) 1,3-propanediol

(d) 2-methyl-2-propanol   or   methyl-2-propanol

(a) *p*-chlorophenol

(b) 3-nitrophenol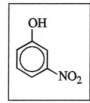

(c) *m*-nitrophenol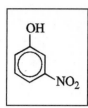

(d) The model given in the problem is represented by both (b) and (c)

27-56. *Refer to Section 27-10.*

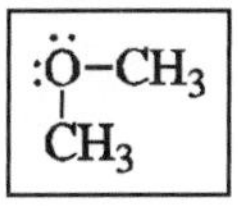

In dimethyl ether, the oxygen atom is *sp³* hybridized. Two single bonds are formed by the overlap of two of its *sp³* hybrid orbitals with the *sp³* hybrid orbitals on the adjacent carbon atoms. In each of the remaining two hybrid orbitals on the oxygen atom are contained a lone pair of electrons. The resulting molecule is polar. The intermolecular forces found operating between molecules of dimethyl ether are therefore dipole-dipole interactions and London forces.

27-58. *Refer to Section 27-10.*

(a) $H_3C-O-CH_3$

methoxymethane

(b) CH₃–CH–CH₃ / O–CH₃

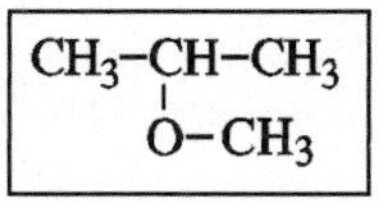

2-methoxypropane

(c) O-CH₂-CH₃ / CH₂–CH₂–CH–CH₃ / O-CH₂-CH₃

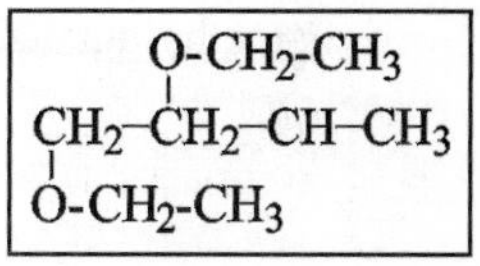

1,2-diethoxybutane

(d) ⬡–O–CH₂–CH₃

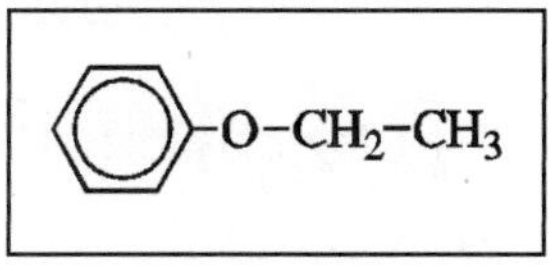

ethoxybenzene

(e)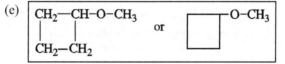

methoxycyclobutane

(f) The given model represents (a), methoxymethane.

27-60. *Refer to Section 27-12.*

(a) The amines are derivatives of ammonia, NH_3, in which one or more H atoms have been replaced by organic groups. They have the general formula: RNH_2, R_2NH or R_3N, where R is any alkyl or aryl group. Amines are basic; their basicity is derived from the lone pair of electrons on the N atoms.

(b) Amines are considered to be derivatives of ammonia. The structures of NH_3, primary, secondary and tertiary amines are shown below. From the comparison, it is obvious that amines can be treated as if one, two or three hydrogen atoms of ammonia have been replaced by organic groups.

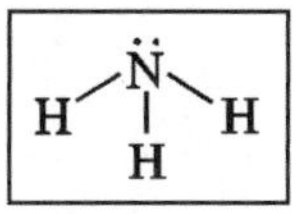

ammonia

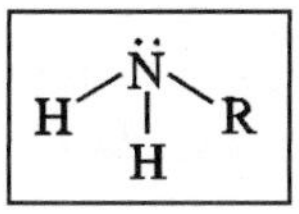

primary amine

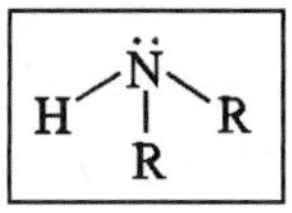

secondary amine

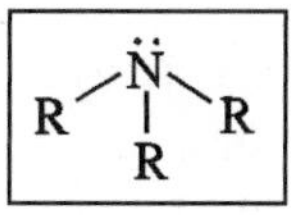

tertiary amine

27-62. *Refer to Section 27-12.*

(a) diethylamine

(b) 4-nitroaniline (*p*-nitroaniline)

(c) *N*-methylaminocyclopentane (cyclopentylmethylamine)

(d) tributylamine

(a), (b)

Classification	Examples	Sources
Aldehyde	benzaldehyde	almonds
	cinnamaldehyde	cinnamon
	vanillin	vanilla bean
Ketone	muscone	musk deer
	testosterone	male sex hormone
	camphor	camphor tree

(c) The aldehydes listed in (a) have many uses. The three aldehydes can be used to add flavor to food. Muscone is the compound that gives the scent to musk perfumes, deodorants, cologne and aftershave lotions. Testosterone is used to regulate male sexual and reproductive functions. Camphor is used in medicine as a diaphoretic, stimulant and sedative.

27-66. *Refer to Section 27-11.*

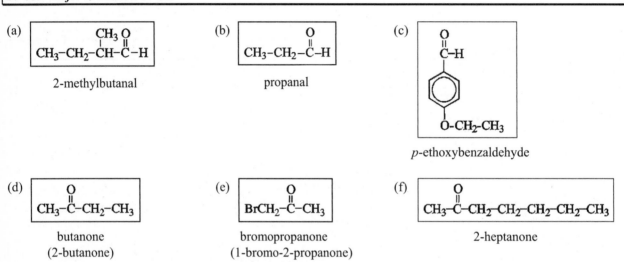

(a)

2-methylbutanal

(b)

propanal

(c)

p-ethoxybenzaldehyde

(d)

butanone
(2-butanone)

(e)

bromopropanone
(1-bromo-2-propanone)

(f)

2-heptanone

(g) The model is a representation of (e) bromopropanone.

27-68. *Refer to Section 27-13 and Table 27-11.*

(a) A carboxylic acid is an acidic organic compound containing the carboxyl group, $-\overset{\overset{\text{O}}{\|}}{\text{C}}-\text{OH}$.

(b)

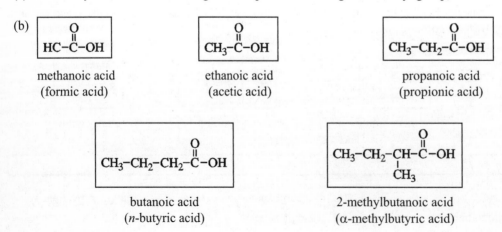

methanoic acid
(formic acid)

ethanoic acid
(acetic acid)

propanoic acid
(propionic acid)

butanoic acid
(*n*-butyric acid)

2-methylbutanoic acid
(α-methylbutyric acid)

27-70. *Refer to Section 27-14.*

(a) An acyl chloride, sometimes called an acid chloride, is a compound that is a derivative of a carboxylic acid, made by replacing the -OH with a Cl atom. It has the general formula:

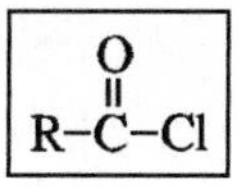

(b)

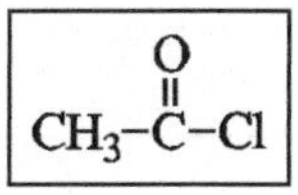

acetyl chloride

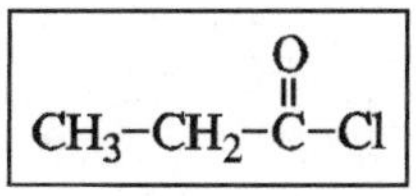

propanoyl chloride (propionyl chloride)

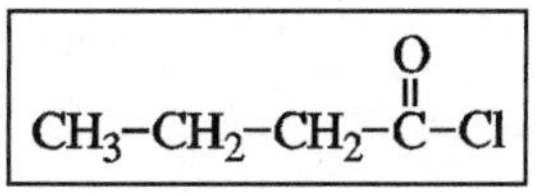

butanoyl chloride (butyryl chloride)

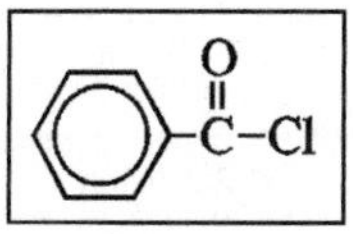

benzoyl chloride

27-72. *Refer to Section 27-14.*

(a)

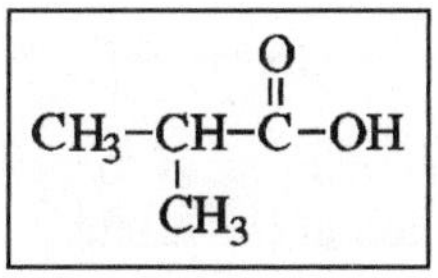

methylpropanoic acid

(b)

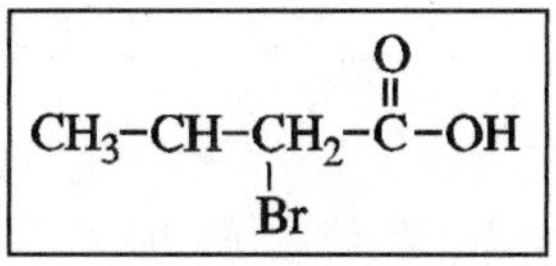

2-bromobutanoic acid

(c)

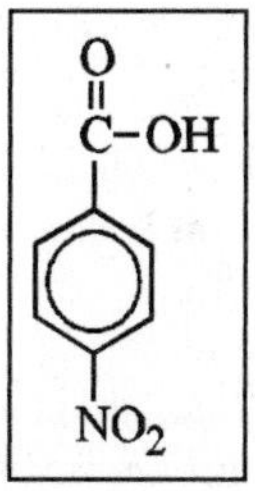

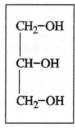

4-nitrobenzoic acid

(d)

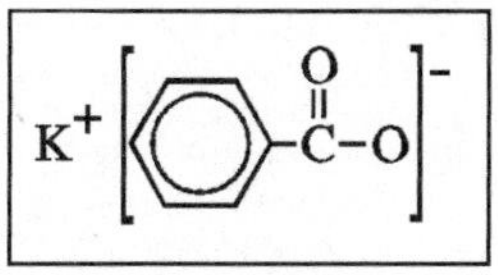

potassium benzoate

(e)

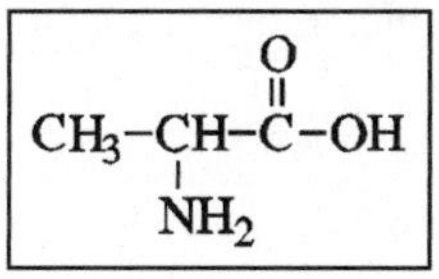

2-aminopropanoic acid

(f) The model represents (c) 4-nitrobenzoic acid.

27-74. *Refer to Section 27-14.*

(a) As shown, glycerides are the triesters of glycerol. Symmetrical glycerides are esters in which all three R groups are identical whereas mixed glycerides contain a mixture of various R groups.

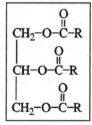

glycerol

glycerides

(b)

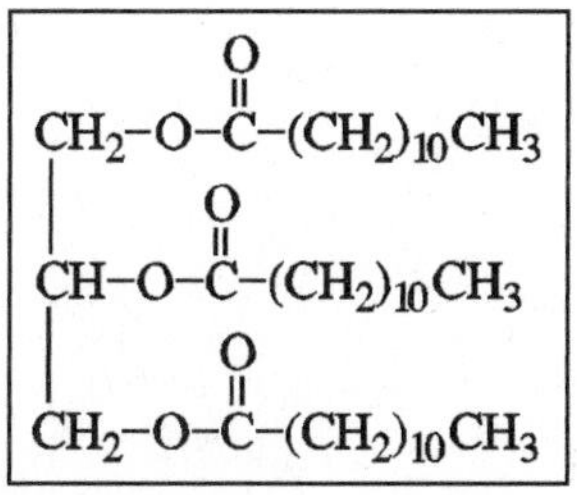

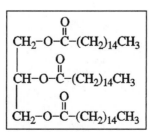

2,3-didodecanoyloxypropyl dodecanoate (glyceryl trilaurate)

2,3-dihexadecanoyloxypropyl hexadecanoate (glyceryl tripalmitate)

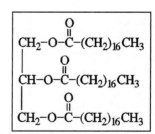

2,3-dioctadecanoyloxypropyl octadecanoate (glyceryl tristearate)

Waxes are esters of fatty acids with alcohols other than glycerol. Most of them are derived from long-chain fatty acids and long-chain monohydric alcohols, both usually having an even number of carbon atoms. Therefore, a wax usually contains an even number of carbon atoms also.

27-78. *Refer to Section 27-14.*

(a) phenyl benzoate

(b) ethyl butanoate

27-80. *Refer to Sections 27-16 and 27-17, and the Key Terms for Chapter 27.*

(a) A substitution reaction is a reaction in which an atom (or group of atoms) replaces another atom (or group of atoms) on a carbon in an organic reaction. No change occurs in the degree of saturation at the reactive carbon atom.

(b) A halogenation reaction is a substitution reaction in which one or more hydrogen atoms of an organic compound, usually a hydrocarbon, is replaced by the corresponding number of halogen atoms.

(c) An addition reaction is a reaction in which there is an increase in the number of groups attached to carbon. Two atoms or groups of atoms are added to the molecule, one on each side of a double or a triple bond. The molecule becomes more nearly saturated.

27-82. *Refer to Section 27-16.*

(a) The chlorination of ethane in ultraviolet light is a free radical chain reaction. It begins when the chlorine molecule is split into two very reactive Cl atoms, which can attack ethane, extracting one of its H atoms to form HCl and a C_2H_5 radical. This, in turn, extracts a Cl atom from Cl_2 to form a monosubstituted chloroethane. When a second hydrogen atom is replaced, a mixture of two disubstituted ethanes are produced as shown below. Subsequent substitution will eventually give C_2Cl_6 as the highest chlorinated product.

(b), (c)

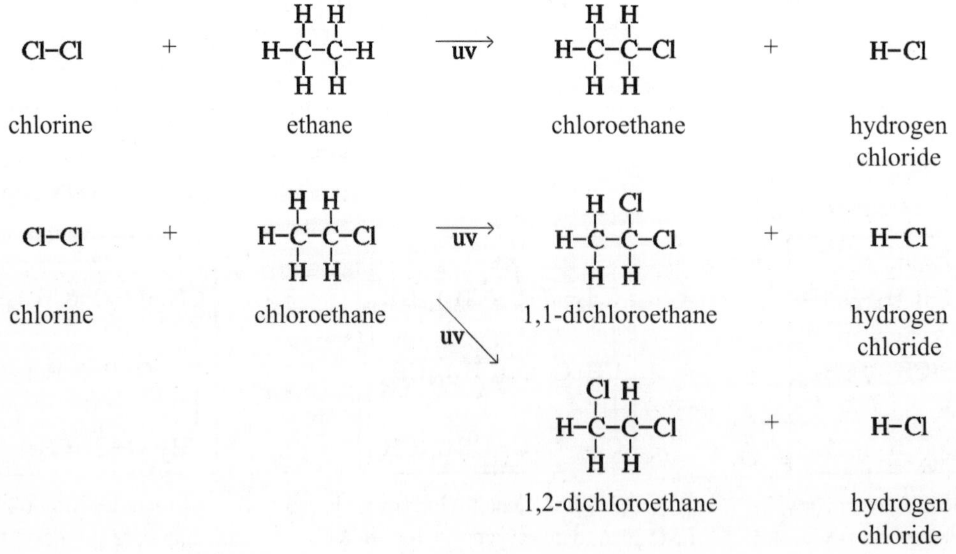

$$Cl-Cl \;+\; H_3C-CH_3 \;\xrightarrow{uv}\; H_3C-CH_2Cl \;+\; H-Cl$$

chlorine + ethane → chloroethane + hydrogen chloride

$$Cl-Cl \;+\; H_3C-CH_2Cl \;\xrightarrow{uv}\; CH_3-CHCl_2 \;+\; H-Cl$$

chlorine + chloroethane → 1,1-dichloroethane + hydrogen chloride

$$\xrightarrow{uv}\; ClCH_2-CH_2Cl \;+\; H-Cl$$

1,2-dichloroethane + hydrogen chloride

The substitution reaction will continue in the presence of excess Cl_2 to give the following chlorinated compounds:

$$
\begin{array}{c}
\text{H\ Cl} \\
\text{H–C–C–Cl} \\
\text{H\ Cl}
\end{array}
\qquad
\begin{array}{c}
\text{Cl\ Cl} \\
\text{H–C–C–Cl} \\
\text{H\ H}
\end{array}
\qquad
\begin{array}{c}
\text{Cl\ Cl} \\
\text{H–C–C–Cl} \\
\text{H\ Cl}
\end{array}
$$

  1,1,1-trichloroethane   1,1,2-trichloroethane   1,1,1,2-tetrachloroethane

$$
\begin{array}{c}
\text{Cl\ Cl} \\
\text{H–C–C–H} \\
\text{Cl\ Cl}
\end{array}
\qquad
\begin{array}{c}
\text{Cl\ Cl} \\
\text{H–C–C–Cl} \\
\text{Cl\ Cl}
\end{array}
\qquad
\begin{array}{c}
\text{Cl\ Cl} \\
\text{Cl–C–C–Cl} \\
\text{Cl\ Cl}
\end{array}
$$

  1,1,2,2-tetrachloroethane   pentachloroethane   hexachloroethane

---

27-84.   *Refer to Sections 27-16 and 27-17.*

The characteristic reaction of the relatively unreactive alkanes is the substitution reaction which involves the replacement of one σ bonded atom for another and requires heat or light.

The more reactive alkenes are characterized by addition reactions to the double bond, many of which occur easily at room temperature. The carbon-carbon double bond is a reaction site and is classified as a functional group. The π portion of the double bond can be utilized to accommodate two incoming atoms, converting the double bond into one single σ bond between the carbon atoms and the π portion into two single σ bonds between each carbon and one of the two incoming atoms.

27-86.   *Refer to Section 27-17.*

(1)   Cl–Cl   + $CH_2=CH–CH_3$ → $\begin{array}{c}\text{CH}_2\text{–CH–CH}_3\\ \text{Cl\quad Cl}\end{array}$

  chlorine   propene   1,2-dichloropropane

(2)   Br–Br   + $CH_3–CH=CH–CH_3$ → $\begin{array}{c}\text{CH}_3\text{–CH–CH–CH}_3\\ \text{Br\ \ Br}\end{array}$

  bromine   2-butene   2,3-dibromobutane

27-88.   *Refer to Section 27-17 and the Key Terms for Chapter 27.*

(a) Hydrogenation refers to the reaction in which molecular hydrogen, H_2, adds across a double or triple bond. This reaction requires elevated temperatures, high pressure and the presence of an appropriate heterogeneous catalyst (finely divided Pt, Pd or Ni).

(b) Hydrogenation is an important industrial process in many areas. For example, unsaturated hydrocarbons can be converted to saturated hydrocarbons by hydrogenation to manufacture high octane gasoline and aviation fuels. It is also employed to convert unsaturated vegetable oils to solid cooking fats.

(c),(d) $CH_2=CH_2$ + H_2 $\xrightarrow[\text{heat}]{\text{catalyst}}$ $CH_3\text{-}CH_3$
            ethene   ethane

  $CH_3\text{-}CH=CH\text{-}CH_3$ + H_2 $\xrightarrow[\text{heat}]{\text{catalyst}}$ $CH_3\text{-}CH_2\text{-}CH_2\text{-}CH_3$
            2-butene   butane

411

(a) The unsaturated π bonds are very susceptible to addition reactions because they are sources of electrons. Alkynes contain two π bonds, while alkenes contain only one π bond. Therefore, alkynes are more reactive.

(b) The most common kind of reaction that alkynes undergo is addition of atoms or groups across the triple bond.

(c), (d) (1) $CH_3-CH_2-C{\equiv}CH$ $\xrightarrow{\text{H}_2}{\text{Pt}}$ $CH_3-CH_2-CH{=}CH_2$ $\xrightarrow{\text{H}_2}{\text{Pt}}$ $CH_3-CH_2-CH_2-CH_3$
    1-butyne   1-butene   butane

  (2) $CH_3-C{\equiv}CH$ $\xrightarrow{\text{Br}_2}$ $CH_3-CBr{=}CHBr$ $\xrightarrow{\text{Br}_2}$ $CH_3-CBr_2-CHBr_2$
     propyne   1,2-dibromopropene   1,1,2,2-tetrabromopropane

  (3) $HC{\equiv}CH$ $\xrightarrow{\text{Cl}_2}$ $ClHC{=}CHCl$ $\xrightarrow{\text{Cl}_2}$ $Cl_2HC-CHCl_2$
     ethyne   1,2-dichloroethene   1,1,2,2-tetrachloroethane

(a) aliphatic substitution:

   toluene   benzyl chloride

aromatic substitution:

    toluene   *o*-chlorotoluene (60%)   *p*-chlorotoluene (40%)

(b) aliphatic substitution:

   toluene   benzyl bromide

aromatic substitution:

    toluene   *p*-bromotoluene (67%)   *o*-bromotoluene (33%)

(c) aliphatic substitution: (not found)

aromatic substitution:

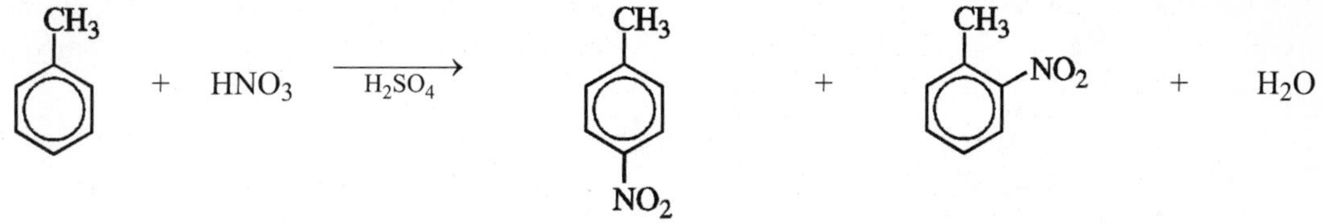

toluene   *p*-nitrotoluene (58%)   *o*-nitrotoluene (38%)

27-94. *Refer to Section 27-17.*

A simple test to distinguish between 2-pentene and cyclopentane is to add a few drops of a red Br_2 solution to the unknown liquid. The reddish color will disappear if the liquid is an alkene or alkyne, e.g., 2-pentene, due to the addition of Br_2 to the multiple bond. No such addition reaction occurs between Br_2 and cyclopentane.

27-96. *Refer to Section 27-19.*

(a) Polymerization is the combination of many monomers, usually small molecules, to form large molecules called polymers which contain repetitive units of the monomers.

(b) Examples of polymerization reactions:

(1) $nCH_2{=}CH_2$ $\xrightarrow{\text{catalyst}}$ $+CH_2{-}CH_2+_n$
      ethylene   polyethylene

(2) $nCF_2{=}CF_2$ $\xrightarrow[\text{heat}]{\text{catalyst}}$ $+CF_2{-}CF_2+_n$
    tetrafluoroethene   "Teflon"

(3) $nCH_2{=}CH{-}\overset{\displaystyle Cl}{C}{=}CH_2$ $\longrightarrow$ $+CH_2{-}CH{=}\overset{\displaystyle Cl}{C}{-}CH_2+_n$
     chloroprene   neoprene

27-98. *Refer to Section 27-19 and the Key Terms for Chapter 27.*

(a) A copolymer is a polymer formed from two different monomers.

(b) A condensation polymer is a polymer formed from a condensation reaction, in which two molecules combine by splitting out or eliminating a small molecule, such as water. For this polymer to form, the monomers must have two functional groups, one on each end.

(c) Addition polymers are polymers formed by addition reactions. Examples of this kind of polymerization are polyethylene, "Teflon", and neoprene. See Exercise 27-96 solution for structures.

27-100. *Refer to Section 27-19.*

Changes in the polymer structure that can increase its rigidity and raise the melting point include introducing (1) cross-linking between polymer chains, (2) bulky substituents or branches on the chains, and (3) groups that can interact by strong intermolecular forces such as hydrogen bonding.

27-102. *Refer to Section 27-19 and Exercise 27-100 Solution.*

(a) Natural rubber is an elastic hydrocarbon polymer obtained from the sap of the rubber tree, called latex. A molecule of rubber (MW ≈ 136,000 g/mol) is composed of approximately 2000 units of 2-methyl-1,3-butadiene, also named isoprene.

$$2n\ CH_2{=}\underset{}{\overset{CH_3}{\overset{|}{C}}}{-}CH{=}CH_2 \longrightarrow {(}CH_2{-}\overset{CH_3}{\overset{|}{C}}{=}CH{-}CH_2{-}CH_2{-}\overset{CH_3}{\overset{|}{C}}{=}CH{-}CH_2{)}_n$$

isoprene — natural rubber

(b) Vulcanization is a process in which sulfur is first added to rubber, then the system is heated to about 140°C.

(c) Vulcanization causes cross-linking between the long rubber polymer chains. The result is a stronger, more elastic rubber, which is more resistant to cold and heat.

(d) At the same time the sulfur is mixed into the rubber, fillers, such as zinc oxide, barium sulfate, titanium dioxide and antimony(V) sulfate, and a reinforcing agent, such as carbon black, are added.

(e) The purpose of fillers and reinforcing agents is to increase the durability of rubber and alter its color.

27-104. *Refer to Section 27-19.*

(a) Polyamides are polymeric amides, a class of condensation polymer. Nylon, a very important fiber product, is the best known polymeric amide.

(b) Polyamides can be formed by the condensation reactions (1) between a dicarboxylic acid with a diamine or (2) between amino acids. In each reaction, H_2O molecules are eliminated.

27-106. *Refer to Section 27-19.*

Consider the following unbalanced polymerization reaction producing a polyester:

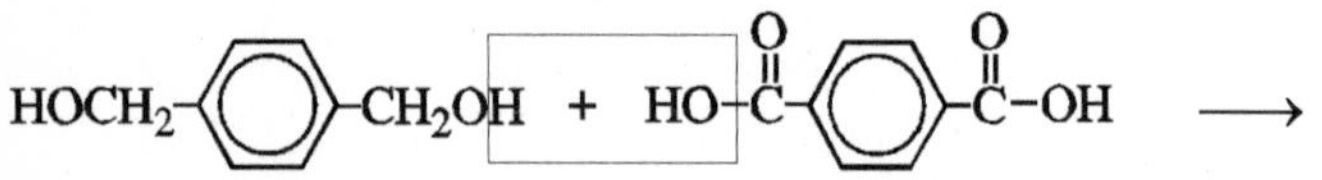

a glycol — terephthalic acid

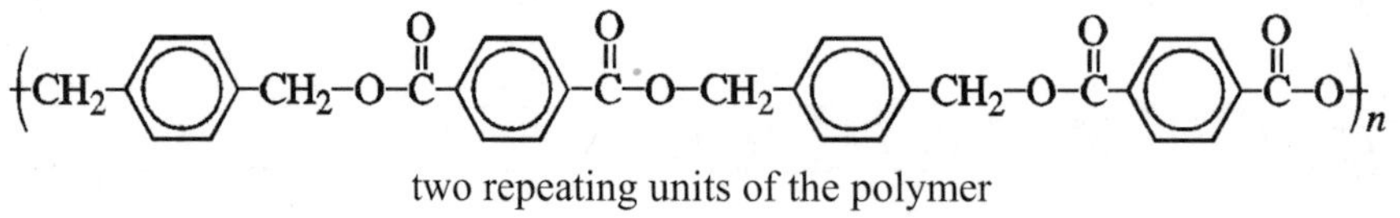

two repeating units of the polymer

27-108. *Refer to Section 27-19.*

Copolymers are formed when two different monomers are mixed and then polymerized. It is possible to produce a copolymer by addition polymerization. An example is SBR, the most important rubber produced in the United States. It is a copolymer of styrene and butadiene in a 1:3 ratio.

27-110. *Refer to Section 27-19.*

(a) Common Nylon is called Nylon 66 because the parent diamine and dicarboxylic acid of Nylon 66 each contain six carbon atoms.

(b) The parent diamine and dicarboxylic acid of Nylon xy would contain x and y carbon atoms, respectively.

Nylon 45: $-NH\!\!\left(\!\!\overset{\overset{O}{\|}}{C}-(CH_2)_3-\overset{\overset{O}{\|}}{C}-NH-(CH_2)_4-NH\!\!\right)_{\!\!n}\!\!\overset{\overset{O}{\|}}{C}-$

Nylon 64: $-NH\!\!\left(\!\!\overset{\overset{O}{\|}}{C}-(CH_2)_2-\overset{\overset{O}{\|}}{C}-NH-(CH_2)_6-NH\!\!\right)_{\!\!n}\!\!\overset{\overset{O}{\|}}{C}-$

27-112. *Refer to Section 27-15 and Figure 27-20.*

(a) carboxylic acid   (b) ester   (c) acyl chloride (acid chloride)   (d) amide

---

27-114. *Refer to Section 27-15 and Figure 27-20.*

(a) $\overset{\overset{O}{\|}}{R-C-R}$ ketone

(b) Ar-OH   phenol
R-O-R   ether
R_2-CH-OH   secondary alcohol
R_2-NH   secondary amine

(c) R-O-R   ether

(d) Ar-OH   phenol
R-O-R   ether
R_2-CH-OH   secondary alcohol
R_3-N   tertiary amine
C=C   alkene (double bond)

(e) Ar-OH   phenol
R_2-CH-OH   secondary alcohol
R_2-NH   secondary amine

27-116. *Refer to the Sections as stated.*

(a) 3-methyl-1-butanol   (Section 27-9)

(b) 2-methylcyclopentanol   (Section 27-9)

(c) 2-aminopropane (isopropylamine)   (Section 27-12)

(d) 2-chloropropene   (Section 27-3)

(e) 1,4-dibromobenzene (*p*-dibromobenzene)   (Section 27-8)

(f) triethylamine   (Section 27-12)

(g) diphenyl ether (phenoxybenzene)   (Section 27-10)

(h) 2,4,6-tribromoaniline   (Section 27-12)

27-118. *Refer to the Sections as stated.*

(a) 2-methyl-1-butanol   (Section 27-9)

(b) 1-aminobutane (*n*-butylamine)   (Section 27-12)

(c) pentanal   (Section 27-11)

(d) cyclopentanone   (Section 27-11)

(e) 2-methoxybutane   (Section 27-10)

(f) 2,2-dimethylpropanoic acid   (Section 27-13)

(a) 3-iodobenzoic acid

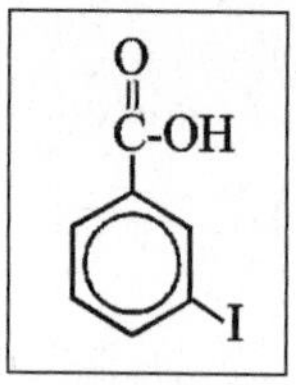

(Section 27-13)

(b) n-propylacetate

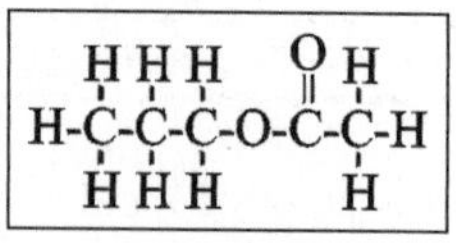

$$\begin{array}{c} \text{H H H} \quad \text{O H} \\ \text{H-C-C-C-O-C-C-H} \\ \text{H H H} \quad\quad \text{H} \end{array}$$

(Section 27-14)

(c) trans-1,2-dibromopropene

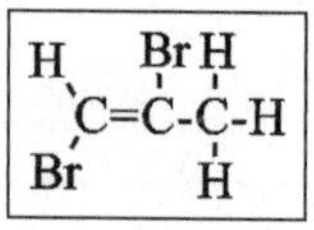

$$\begin{array}{c} \text{H} \quad \text{Br H} \\ \text{C=C-C-H} \\ \text{Br} \quad \text{H} \end{array}$$

(Section 27-3)

(d) 2-ethoxypropane

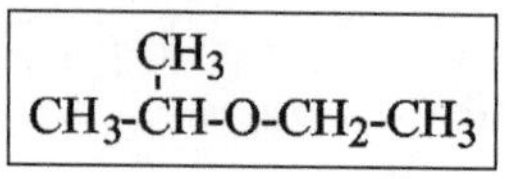

$$\begin{array}{c} \text{CH}_3 \\ \text{CH}_3\text{-CH-O-CH}_2\text{-CH}_3 \end{array}$$

(Section 27-10)

27-122. *Refer to Sections 27-1, 27-5 and 27-6.*

No, aromatic hydrocarbons cannot be saturated hydrocarbons. Saturated hydrocarbons, like hexane (C_6H_{14}), have only single bonds between the sp^3 hybridized carbon atoms.

The structures of aromatic hydrocarbons, like benzene (C_6H_6), involve sp^2 hybridized carbon atoms with pi bonds whose electrons are delocalized over the entire benzene ring.

27-124. *Refer to Section 27-1 and Figure 27-7.*

Cyclopentane has a smaller amount of strain than does cyclohexane when forced to have all its carbons in a plane (flat). Cyclopentane is nearly flat since the bond angles in a regular pentagon (108°) are near the tetrahedral angle (109.5°). The bond angles in a flat cyclohexane are 120o, which are not near the 109oC bond angle. To avoid this strain, the cyclohexane ring buckles and becomes nonplanar.

27-126. *Refer to Section 27-3.*

1,3-cyclohexadiene

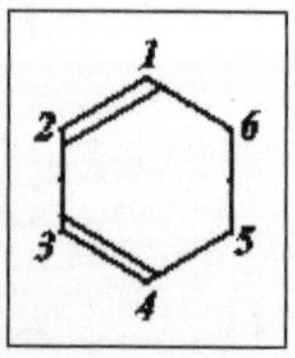

Carbon atoms (1), (2), (3) and (4) undergo sp^2 hybridization, while carbon atoms (5), and (6) undergo sp^3 hybridization. .

27-128. *Refer to Sections 27-5 and 27-6.*

The classes of compounds that must contain an aromatic group include fused ring compounds, substituted benzenes and phenols.

butanone

It is easily seen that there is a methyl group to the left of the ketone functional group and an ethyl group to the right of the ketone group.

27-132. *Refer to Section 27-12.*

Lidocaine has replaced novocain as the favored anesthetic in dentistry. Both compounds have a tertiary amine group (a diethylamino group) in common.

28 Organic Chemistry II: Shapes, Selected Reactions and Biopolymers

---

28-2. *Refer to Section 28-2 and Figures 28-1, 28-2, 28-3, 28-4, 28-5 and 28-6.*

There are two types of stereoisomerism. (1) Geometrical isomers differ only in the spatial orientation of groups about a plane or direction, i.e., they differ in orientation either (i) around a double bond (see 2-butene) or (ii) across the ring in a cyclic compound (see 1,2-dichlorocyclobutane). Both *cis* and *trans* isomers exist.

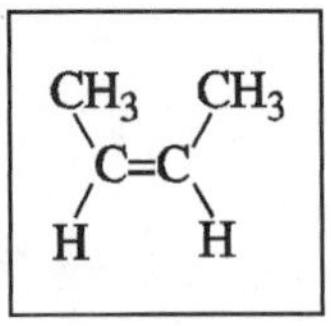

cis-2-butene

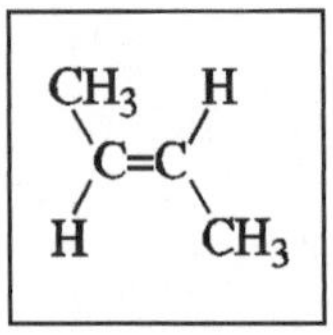

trans-2-butene

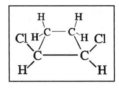

cis-1,2-dichlorocyclobutane

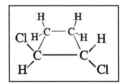

trans-1,2-dichlorocyclobutane

(2) The second type of stereoisomerism is optical isomerism, in which two molecules that are mirror images of each other are not superimposable on each other. Consider the compound 2-butanol, $CH_3CH(OH)CH_2CH_3$. It has two optical isomers, because it is not superimposable on its mirror image.

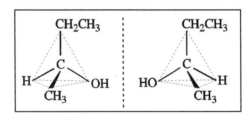

28-4. *Refer to Section 28-2, and Figures 28-1 and 28-2.*

All of the compounds, except (b) 1,2-dibromo-2-butene, can exist as *cis* and *trans* isomers:

(a)

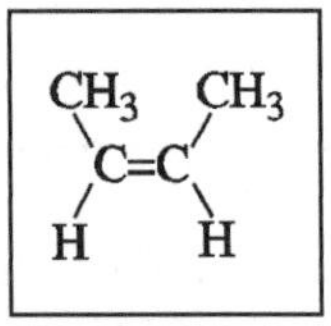

cis-2-butene

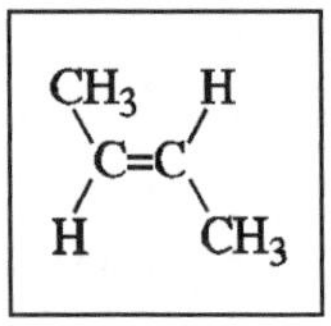

trans-2-butene

(b)

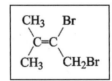

1,2-dibromo-2-butene

(c)

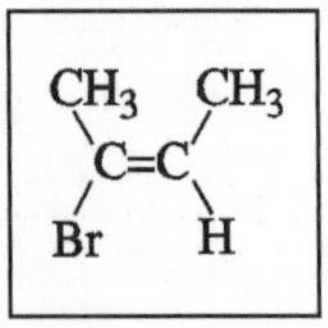

cis-2-bromo-
2-butene

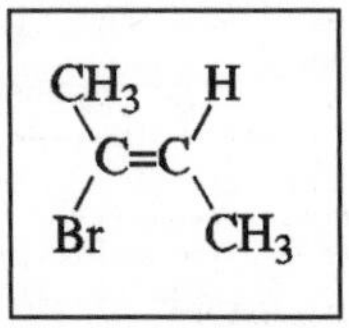

trans-2-bromo-
2-butene

(d)

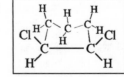

cis-1,2-
dichlorocyclopentane 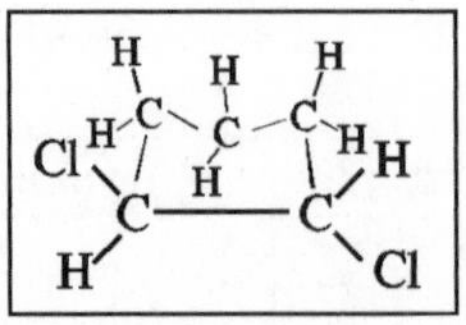   *trans*-1,2-
dichlorocyclopentane

28-6. *Refer to Section 28-2.*

(a) isomers of butanol ($C_4H_{10}O$)

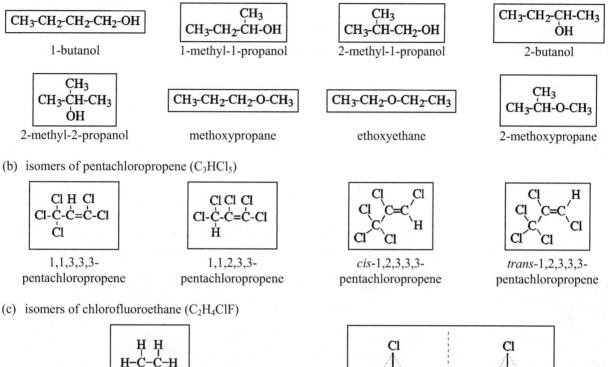

CH₃-CH₂-CH₂-CH₂-OH
1-butanol

CH₃
CH₃-CH₂-CH-OH
1-methyl-1-propanol

CH₃
CH₃-CH-CH₂-OH
2-methyl-1-propanol

CH₃-CH₂-CH-CH₃
OH
2-butanol

CH₃
CH₃-CH-CH₃
OH
2-methyl-2-propanol

CH₃-CH₂-CH₂-O-CH₃
methoxypropane

CH₃-CH₂-O-CH₂-CH₃
ethoxyethane

CH₃
CH₃-CH-O-CH₃
2-methoxypropane

(b) isomers of pentachloropropene (C_3HCl_5)

Cl H Cl
Cl-C-C=C-Cl
Cl
1,1,3,3,3-
pentachloropropene

Cl Cl Cl
Cl-C-C=C-Cl
H
1,1,2,3,3-
pentachloropropene

cis-1,2,3,3,3-
pentachloropropene

trans-1,2,3,3,3-
pentachloropropene

(c) isomers of chlorofluoroethane (C_2H_4ClF)

H H
H-C-C-H
Cl F
1-chloro-2-fluoroethane

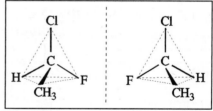

optical isomers of 1-chloro-1-fluoroethane

28-8. *Refer to Section 28-3 and Figure 28-9.*

Refer to Figure 28-9 for the chair and twist boat conformations of cyclohexane.

28-10. *Refer to Section 28-2, the Key Terms for Chapter 28 and Figures 28-3, 28-4, 28-5 and 28-6.*

Optical isomerism is exhibited by compounds that are chiral, i.e., are not superimposable on their mirror images. Such a compound and its mirror image are called optical isomers or enantiomers. They have identical physical and chemical properties except when they interact with other chiral molecules. A solution of one of the pair is capable of rotating a plane of polarized light to the right and an equimolar solution of its optical isomer will rotate the plane of polarized light by the same amount, but to the left.

28-12. *Refer to Section 28-2 and Exercise 28-11.*

The compounds exhibiting optical isomerism in Exercise 28-11 are (a) and (c):

(a)

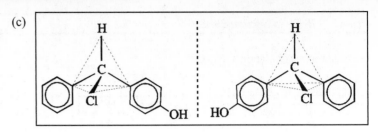

(c)

28-14. *Refer to Sections 28-4 and 18-4.*

Amines are Brønsted-Lowry bases due to the presence of a lone pair of electrons on N to accommodate incoming protons. In aqueous solution, an amine will hydrolyze in an equilibrium to produce hydroxide ions.

ammonia $\quad\quad\quad\quad NH_3(aq) + H_2O(\ell) \rightleftarrows NH_4^+(aq) + OH^-(aq)$

amine (e.g., 1°) $\quad\quad RNH_2(aq) + H_2O(\ell) \rightleftarrows RNH_3^+(aq) + OH^-(aq)$

28-16. *Refer to Sections 28-4 and 18-4, and Example 18-14.*

Balanced equation: $C_6H_5NH_2(aq) + H_2O(\ell) \rightleftarrows C_6H_5NH_2^+(aq) + OH^-(aq)$ $\quad\quad K_b = 4.2 \times 10^{-10}$

Let $x = [C_6H_5NH_2]_{ionized}$. Then, $x = [C_6H_5NH_3^+] = [OH^-]$

	$C_6H_5NH_2$	+	H_2O	$\rightleftarrows$	$C_6H_5NH_2^+$	+	OH^-
initial	0.12 M				0 M		$\approx 0\ M$
change	- x M				+ x M		+ x M
at equilibrium	(0.12 - x) M				x M		x M

$K_b = \dfrac{[C_6H_5NH_3^+][OH^-]}{[C_6H_5NH_2]} = 4.2 \times 10^{-10} = \dfrac{x^2}{0.12 - x} \approx \dfrac{x^2}{0.12}$ $\quad\quad$ Solving, $x = 7.1 \times 10^{-6}$

Therefore, $\quad [C_6H_5NH_2] = \mathbf{0.12\ M}$ $\quad\quad\quad\quad [OH^-] = \mathbf{7.1 \times 10^{-6}\ M}$

$\quad\quad\quad\quad\quad [C_6H_5NH_2^+] = \mathbf{7.1 \times 10^{-6}\ M}$ $\quad\quad [H_3O^+] = K_w/[OH^-] = \mathbf{1.4 \times 10^{-9}\ M}$

28-18. *Refer to Section 28-4.*

The compounds that are the stronger acids are:

(a) $\quad\quad$ (b) $\quad\quad$ (c) 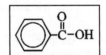 $\quad\quad$ (d)

28-20. *Refer to Section 28-4.*

(a), (b) (1) $2CH_3OH + 2Na \rightarrow 2[Na^+ + CH_3O^-] + H_2$
$\quad\quad\quad\quad$ methanol $\quad\quad\quad\quad$ sodium methoxide

$\quad\quad\quad$ (2) $2CH_3CH_2OH + 2Na \rightarrow 2[Na^+ + CH_3CH_2O^-] + H_2$
$\quad\quad\quad\quad\quad$ ethanol $\quad\quad\quad\quad\quad$ sodium ethoxide

$\quad\quad\quad$ (3) $2CH_3CH_2C_2OH + 2Na \rightarrow 2[Na^+ + CH_3CH_2CH_2O^-] + H_2$
$\quad\quad\quad\quad\quad$ 1-propanol $\quad\quad\quad\quad$ sodium propoxide

(c) The reactions of alcohols with sodium are similar to the reaction of metallic sodium with water. Both types of reactions are oxidation-reduction reactions involving the displacement of hydrogen from an O-H bond by sodium and the production of H_2 gas.

28-22. *Refer to Sections 28-4 and 18-4, and Example 18-14.*

Balanced equation: lidocaine(aq) + $H_2O(\ell)$ $\rightleftarrows$ lidocaineH$^+$(aq) + OH$^-$(aq) $K_b = 7.0 \times 10^{-6}$

$$[\text{lidocaine}] = \frac{1.2 \text{ g lidocaine}}{100 \text{ g soln}} \times \frac{1 \text{ mol lidocaine}}{234.3 \text{ g lidocaine}} \times \frac{1.00 \text{ g soln}}{1.00 \text{ mL soln}} \times \frac{1000 \text{ mL soln}}{1 \text{ L soln}} = 0.051 \ M \text{ lidocaine}$$

Let x = [lidocaine]$_{\text{ionized}}$.   Then, x = [lidocaineH$^+$] = [OH$^-$]

	lidocaine	+	H_2O	$\rightleftarrows$	lidocaineH$^+$	+	OH$^-$
initial	0.051 M				0 M		≈ 0 M
change	- x M				+ x M		+ x M
at equilibrium	(0.051 - x) M				x M		x M

$$K_b = \frac{[\text{lidocaineH}^+][\text{OH}^-]}{[\text{lidocaine}]} = 7.0 \times 10^{-6} = \frac{x^2}{0.051 - x} \approx \frac{x^2}{0.051}$$ Solving, x = 6.0×10^{-4}

Therefore,   [OH$^-$] = 6.0×10^{-4} M
        pOH = 3.22
        pH = **10.78**

28-24. *Refer to Section 28-5.*

(a) oxidation

(b) oxidation

(c) reduction

(d) reduction

28-26. *Refer to Section 28-5.*

(1)
$$\underset{\text{2-propanol}}{\overset{\displaystyle \overset{\text{OH}}{|}}{\text{CH}_3\text{-CH-CH}_3}} \quad \xrightarrow[\text{dil. H}_2\text{SO}_4]{\text{K}_2\text{Cr}_2\text{O}_7} \quad \underset{\substack{\text{propanone}\\ \text{(dimethyl ketone or acetone)}}}{\overset{\displaystyle \overset{\text{O}}{\|}}{\text{CH}_3\text{-C-CH}_3}}$$

(2)
$$\underset{\text{2-butanol}}{\overset{\displaystyle \overset{\text{OH}}{|}}{\text{CH}_3\text{-CH}_2\text{-CH-CH}_3}} \quad \xrightarrow[\text{dil. H}_2\text{SO}_4]{\text{K}_2\text{Cr}_2\text{O}_7} \quad \underset{\substack{\text{2-butanone}\\ \text{(methyl ethyl ketone)}}}{\overset{\displaystyle \overset{\text{O}}{\|}}{\text{CH}_3\text{-CH-C-CH}_3}}$$

(3)

cyclohexanol $\xrightarrow[\text{dil. H}_2\text{SO}_4]{\text{K}_2\text{Cr}_2\text{O}_7}$ cyclohexanone

(a) Elemental carbon particles (soot) are produced in burning if there is incomplete combustion. Aromatic hydrocarbons, such as benzene or toluene, are very stable due to resonance and therefore when combusted, they release less energy in the combustion process than expected. This in turn causes the carbon atoms to be less efficiently oxidized. Carbon atoms then are oxidized to an oxidation state of zero, producing soot, rather than to an oxidation state of +4, the oxidation state of carbon in CO_2.

(b) The flames would be expected to be yellow (a reducing flame), a sign of incomplete combustion, rather than blue (an oxidizing flame), a sign of complete combustion.

An inorganic ester may be thought of as a compound that contains one or more alkyl groups covalently bonded to the anion of a ternary inorganic acid.

(a), (b)   (1) $CH_3OH + HONO_2 \rightarrow CH_3ONO_2 + H_2O$
                            methyl nitrate

            (2) $CH_3CH_2OH + HONO_2 \rightarrow CH_3CH_2ONO_2 + H_2O$
                                  ethyl nitrate

            (3) $CH_3CH_2CH_2OH + HONO_2 \rightarrow CH_3CH_2CH_2ONO_2 + H_2O$
                                        propyl nitrate

(1) $CH_3COOH + CH_3CH_2OH \rightarrow CH_3COOCH_3CH_3 + H_2O$
     acetic acid   ethanol   ethyl acetate

(2) $CH_3CH_2COOH + CH_3OH \rightarrow CH_3CH_2COOCH_3 + H_2O$
     propanoic acid   methanol   methyl propanoate
     (propionic acid)   (methyl propionate)

(3) $C_6H_5COOH + CH_3CH_2OH \rightarrow C_6H_5COOCH_2CH_3 + H_2O$
     benzoic acid   ethanol   ethyl benzoate

(a) $CH_3COOCH_2CH_2CH_2CH_3 + Na^+OH^- \xrightarrow{heat} CH_3COO^-Na^+ + CH_3CH_2CH_2CH_2OH$
      butyl acetate   sodium acetate   1-butanol

(b) $HCOOCH_2CH_3 + Na^+OH^- \xrightarrow{heat} HCOO^-Na^+ + CH_3CH_2OH$
      ethyl formate   sodium formate   ethanol

(c) $CH_3COOCH_3 + Na^+OH^- \xrightarrow{heat} CH_3COO^-Na^+ + CH_3OH$
      methyl acetate   sodium acetate   methanol

(d) $CH_3COO(CH_2)_7CH_3 + Na^+OH^- \xrightarrow{heat} CH_3COO^-Na^+ + CH_3(CH_2)_7OH$
      octyl acetate   sodium acetate   1-octanol

In order to participate in polymer formation, a molecule must be able to react at both ends (difunctional) so that the polymer chain can grow in length. Three types of molecules that can polymerize are (1) alkenes (e.g., ethene molecules reacting to form polyethylene), (2) molecules with two identical functional groups (e.g., dicarboxylic acid reacting with a diamine to produce Nylon), and (3) molecules containing two different functional groups (e.g., amino acids containing an amine group and a carboxylic acid group reacting to form proteins).

28-38. *Refer to Section 28-8, and Figures 28-10 and 28-11.*

Six carbon monosaccharides:

glucose: fructose:

28-40. *Refer to Section 28-8, and Figures 28-10 and 28-11.*

fructose:

cyclic form	straight-chain form
	CH_2OH $C=O$ $HO-C-H$ $H-C-OH$ $H-C-OH$ CH_2OH

28-42. *Refer to Section 28-9.*

The link between adjacent units or monomers in a polypeptide is formed in a condensation reaction between the amine group of one amino acid and the carboxylic acid group of another with the elimination of H_2O molecules. These links are called peptide bonds.

28-44. *Refer to Section 28-9.*

A total of 9 dipeptides can be formed from the three amino acids, A, B and C:

A-A	B-A	C-A
A-B	B-B	C-B
A-C	B-C	C-C

Note: The dipeptide A-B is different from B-A. For example, consider the condensation reaction between $NH_2CHRCOOH$ and $NH_2CHR'COOH$. The two products that will form are:

The dipeptide precursor of aspartame is made from the two amino acids given here. Aspartame is the methyl ester of the dipeptide.

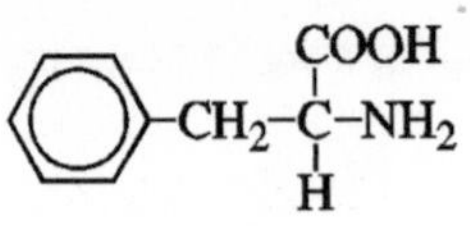

$$
\underset{\text{phenylalanine}}{\text{Ph-CH}_2\text{-}\underset{\underset{H}{|}}{\overset{\overset{\text{COOH}}{|}}{C}}\text{-NH}_2}
\qquad \text{and} \qquad
\underset{\text{aspartic acid}}{\text{HOOC-}\underset{\underset{H}{|}}{\overset{\overset{\text{CH}_2\text{COOH}}{|}}{C}}\text{-NH}_2}
$$

phenylalanine   and   aspartic acid

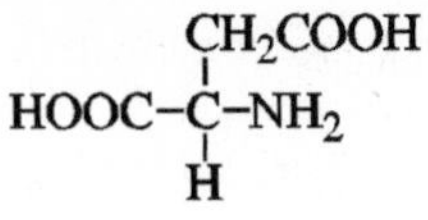

Cytosine (C)

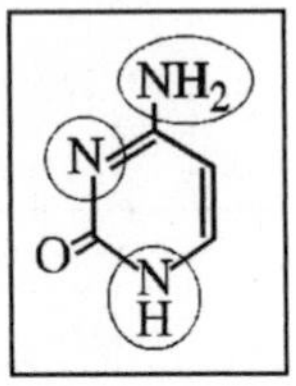

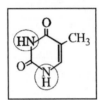
Thymine (T)

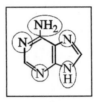
Adenine (A)

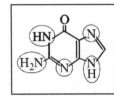
Guanine (G)

In the double strand form of DNA, the base, thymine, is found paired (hydrogen bonded) with **adenine**.

(a) 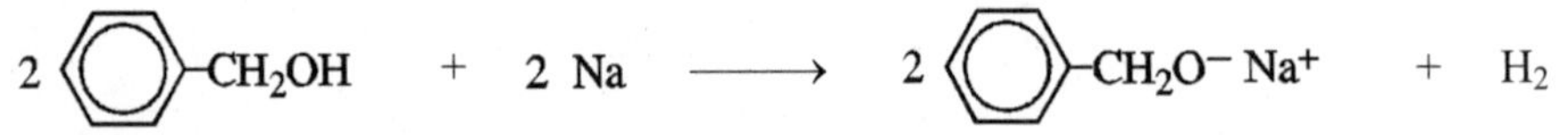 $2\ \text{Ph-CH}_2\text{OH} + 2\ \text{Na} \longrightarrow 2\ \text{Ph-CH}_2\text{O}^-\text{Na}^+ + \text{H}_2$ (Section 28-4)

an alkoxide

(b) $\text{CH}_3\text{CH}_2\overset{\overset{\text{O}}{\|}}{\text{C}}\text{OCH}_3 \xrightarrow[\text{heat}]{\text{KOH}(aq)} \text{CH}_3\text{CH}_2\overset{\overset{\text{O}}{\|}}{\text{C}}\text{O}^-\text{K}^+ + \text{CH}_3\text{OH}$ (Section 28-7)

potassium propanoate   methanol
(potassium salt of carboxylic acid)

(c) 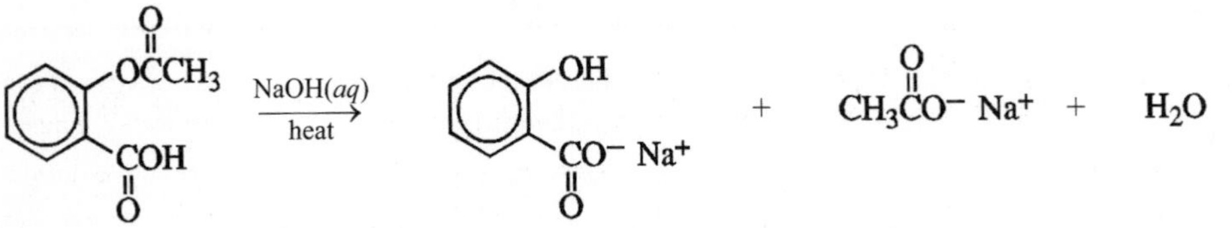 $\underset{\text{sodium 2-hydroxybenzoate}}{} \xrightarrow[\text{heat}]{\text{NaOH}(aq)} + \text{CH}_3\overset{\overset{\text{O}}{\|}}{\text{C}}\text{O}^-\text{Na}^+ + \text{H}_2\text{O}$ (Sections 28-4 and 28-7)

sodium 2-hydroxybenzoate   sodium acetate

(a) 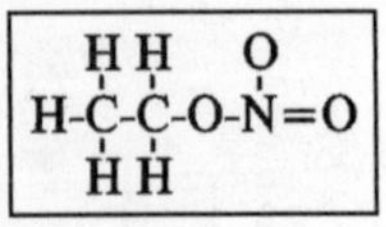
$$
\begin{array}{c}
\text{H H O}\\
\text{H-C-C-O-N=O}\\
\text{H H}
\end{array}
$$

(b)
$$
\begin{array}{c}
\text{H H H}\\
\text{H-C-C=C-H}\\
\text{H}
\end{array}
$$

(c)
$$
\begin{array}{c}
\text{H H H}\\
\text{H-C-C-C-O-H}\\
\text{H H H}
\end{array}
$$

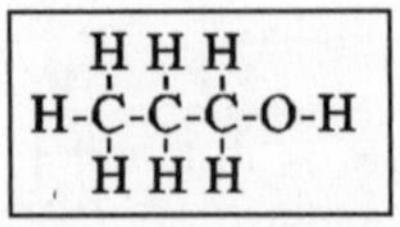

(d) 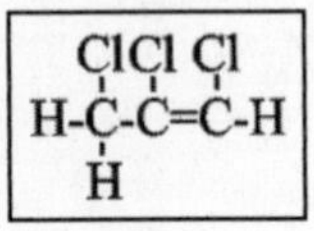
$$
\begin{array}{c}
\text{Cl Cl Cl}\\
\text{H-C-C=C-H}\\
\text{H}
\end{array}
$$

(e) 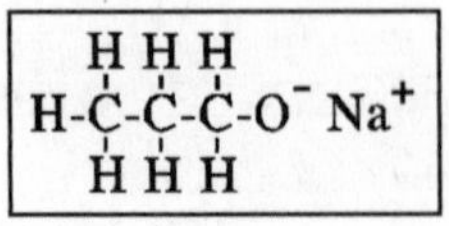
$$
\begin{array}{c}
\text{H H H}\\
\text{H-C-C-C-O}^-\ \text{Na}^+\\
\text{H H H}
\end{array}
$$

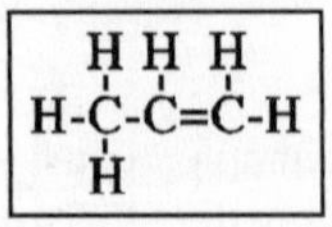

Three types of monomers found in DNA, a nucleic acid, are the phosphate group, the deoxyribose group and a base. These are illustrated by

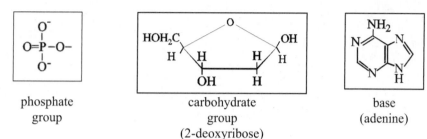

phosphate
group

carbohydrate
group
(2-deoxyribose)

base
(adenine)

28-58. *Refer to Section 28-5 and Table 28-3.*

The heat of combustion of ethanol (ethyl alcohol) is significantly lower than that of the saturated alkanes on a per gram basis. On a per mole basis, ethanol's heat of combustion is lower than those of all the saturated alkanes except methane.

28-60. *Refer to Sections 18-8 and 10-7.*

(1) Balanced equations: $NaC_6H_5COO \rightarrow Na^+ + C_6H_5COO^-$ (to completion)
$$C_6H_5COO^- + H_2O \rightleftarrows C_6H_5COOH + OH^-$$ (reversible)

Let $x = [C_6H_5COO^-]_{hydrolyzed}$.   Then, $0.12 - x = [C_6H_5COO^-]$; $x = [C_6H_5COOH] = [OH^-]$

$$K_b = \frac{K_w}{K_{a(C_6H_5COOH)}} = \frac{1.0 \times 10^{-14}}{6.3 \times 10^{-5}} = 1.6 \times 10^{-10} = \frac{[C_6H_5COOH][OH^-]}{[C_6H_5COO^-]} = \frac{x^2}{0.12 - x} \approx \frac{x^2}{0.12}$$

Solving, $x = 4.4 \times 10^{-6}$

Therefore, $[OH^-] = 4.4 \times 10^{-6}\ M$; pOH = 5.36; pH = **8.64**

(2) Acetic acid is a weaker acid than benzoic acid since the K_a for acetic acid (1.8×10^{-5}) is less than the K_a for benzoic acid (6.5×10^{-5}). Therefore, when the relative base strengths of their conjugate bases are compared, the acetate ion, CH_3COO^-, is a stronger base than the benzoate ion, $C_6H_5COO^-$. In other words, a 0.12 M solution of the benzoate ion is **more acidic** (less basic) than a 0.12 M solution of the acetate ion.

28-62. *Refer to Sections 2-9, 14-13, 27-14, 28-4, and Exercise 28-61.*

Plan: (1) For compound in Exercise 28-61, find the simplest formula, and
(2) Determine the molecular formula, by finding first the molecular weight from freezing point depression.
(3) Determine the structural formula from the observations and write the appropriate reactions.

Part (1)

Plan: (a) Use the masses of CO_2, H_2O and the sample to calculate the masses of C, H, and O respectively, in the original sample.
(b) Determine the simplest formula.

(a) $?\ g\ C = 0.352\ g\ CO_2 \times \dfrac{12.01\ g\ C}{44.01\ g\ CO_2} = 0.0961\ g\ C$

$?\ g\ H = 0.144\ g\ H_2O \times \dfrac{2.016\ g\ H}{18.02\ g\ H_2O} = 0.0161\ g\ H$

$?\ g\ O = 0.240\ g\ sample - 0.0961\ g\ C - 0.0161\ g\ H = 0.128\ g\ O$

(b) ? mol C = $\dfrac{0.0961 \text{ g C}}{12.01 \text{ g/mol}}$ = 0.00800 mol C   Ratio = $\dfrac{0.00800}{0.00800}$ = 1.00

   ? mol H = $\dfrac{0.0161 \text{ g H}}{1.0079 \text{ g/mol}}$ = 0.0160 mol H   Ratio = $\dfrac{0.0160}{0.00800}$ = 2.00

   ? mol O = $\dfrac{0.128 \text{ g O}}{16.00 \text{ g/mol}}$ = 0.00800 mol O   Ratio = $\dfrac{0.00800}{0.00800}$ = 1.00

Therefore the simplest formula is CH_2O.

Part (2)

Plan: (a) Solve for the molality and the approximate formula weight of the nonelectrolyte, using $\Delta T_f = K_f m$
    (b) Determine the molecular formula

(a) From Table 14-2, for water: $T_f = 0.00°C$ $K_f = 1.86 \text{ °C}/m$

$m = \dfrac{\Delta T_f}{K_f} = \dfrac{0.00°C - (-0.248°C)}{1.86 \text{ °C}/m} = 0.133 \ m$

Recall: $m = \dfrac{\text{mol solute}}{\text{kg solvent}} = \dfrac{\text{g solute/MW solute}}{\text{kg solvent}}$

Solving,

? MW solute = $\dfrac{\text{g solute}}{m \times \text{kg benzene}} = \dfrac{0.20 \text{ g}}{0.133 \ m \times 0.0250 \text{ kg}}$ = 60. g/mol (to 2 significant figures)

(b) $n = \dfrac{\text{molecular weight}}{\text{simplest formula weight}} = \dfrac{60. \text{ g/mol}}{30. \text{ g/mol}} = 2$

The true molecular formula for the organic compound is $(CH_2O)_2 = C_2H_4O_2$.

Part (3)

The fruity smell, characteristic of an ester, that resulted when the compound reacted with ethanol in the presence of sulfuric acid suggests that the initial compound is a carboxylic acid. As an acid, it reacts with Na metal to make hydrogen gas and a sodium salt. Based on these observations with a molecular formula of $C_2H_4O_2$, the compound can be identified as acetic acid, CH_3COOH. This also makes sense because the organic compound was soluble in water during the freezing point depression experiment.

Structural formula:

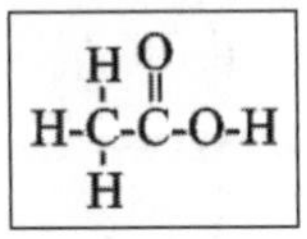

Reaction with sodium metal: $2CH_3COOH + 2Na \rightarrow 2NaCH_3COO + H_2$

Reaction with ethanol: $CH_3COOH + CH_3CH_2OH \rightarrow CH_3COOCH_3CH_3 + H_2O$